Space Mappings with
Bounded Distortion

TRANSLATIONS OF MATHEMATICAL MONOGRAPHS

VOLUME 73

Space Mappings with Bounded Distortion

YU. G. RESHETNYAK

American Mathematical Society · Providence · Rhode Island

Ю. Г. РЕШЕТНЯК

ПРОСТРАНСТВЕННЫЕ ОТОБРАЖЕНИЯ С ОГРАНИЧЕННЫМ ИСКАЖЕНИЕМ

«НАУКА», МОСКВА, 1982

Translated from the Russian by H. H. McFaden
Translation edited by Ben Silver

1980 *Mathematics Subject Classification* (1985 *Revision*). Primary 30-02, 30C60; Secondary 30C20, 30C35, 30C85, 46E35, 53B20, 31B15, 35J99, 53A30.

ABSTRACT. This monograph is an exposition of the main results obtained in recent years by Soviet and foreign mathematicians in the theory of mappings with bounded distortion. The book relates to an active direction in contemporary mathematics. The mathematical apparatus contained in it can be applied to a broad spectrum of problems that go beyond the context of the main topic of investigation. A number of questions in the theory of partial differential equations and the theory of functions with generalized derivatives are expounded for the first time in the world monograph literature. The book is intended for research workers, graduate students, and university students concerned with questions in analysis and function theory.
Bibliography: 183 titles

Library of Congress Cataloging-in-Publication Data

Reshetniak, Ĭuriĭ Grigor′evich.
  [Prostranstvennye otobrazheniĭa s ogranichennym iskazheniem. English]
  Space mappings with bounded distortion/Yu. G. Reshetnyak.
  p. cm. – (Translations of mathematical monographs; v. 73)
  Translation of: Prostranstvennye otobrazheniĭa s ogranichennym iskazheniem.
  Bibliography: p. 347
  ISBN 0-8218-4526-8 (alk. paper)
  1. Conformal mapping. I. Title. II. Series.
QA646.R3813 1989   89-72
515.9–dc 19   CIP

Copyright ©1989 by the American Mathematical Society. All rights reserved.
Translation authorized by the
All-Union Agency for Authors' Rights, Moscow
The American Mathematical Society retains all rights
except those granted to the United States Government.
Printed in the United States of America

Information on Copying and Reprinting can be found at the back of this volume.
The paper used in this book is acid-free and falls within the guidelines
established to ensure permanence and durability. ∞

CB-STK
SCIMON

Contents

Foreword to the English translation   xi

From the Author   xiii

CHAPTER I. Introduction   1
  §1. Some facts from the theory of functions of a real variable   1
    1.1. Sets in $\mathbf{R}^n$ 1
    1.2. Classes of functions in $\mathbf{R}^n$ 4
    1.3. Differentiation of measures on the space $\mathbf{R}^n$. Lebesgue points of a function and points of density of a subset of $\mathbf{R}^n$ 6
    1.4. Approximation of integrable functions by smooth functions   10
  §2. Functions with generalized derivatives   12
    2.1. Definition of a function with generalized derivatives   12
    2.2. Sobolev imbedding theorems   15
    2.3. Tests for a function to belong to the class $W^1_{p,\mathrm{loc}}(U)$ 17
    2.4. Transformations of functions with generalized derivatives   21
    2.5. Dependence of the coefficients in the imbedding theorems on the size of the domain   24
    2.6. A theorem on differentiability of $W^1_{p,\mathrm{loc}}$-functions almost everywhere   25
  §3. Möbius transformations   28
    3.1. Motions and similarity transformations of a Euclidean space   28
    3.2. Möbius transformations. Definitions   33
    3.3. Möbius transformations and cross ratios. Construction of Möbius transformations   36
    3.4. Möbius transformations and spheres   39

3.5. The hypersphere bundle and linear representations of
Möbius transformations   44

§4. Definition of a mapping with bounded distortion   53

4.1. Orthogonal invariants of linear mappings of Euclidean
spaces. A measure of nonorthogonality for a linear
mapping   53

4.2. Mappings with bounded distortion   61

4.3. Examples of mappings with bounded distortion   63

§5. Mappings with bounded distortion on Riemannian spaces   67

5.1. Riemannian metrics in domains in $\mathbf{R}^n$ 67

5.2. Mappings with bounded distortion on Riemannian
spaces   72

CHAPTER II. Main facts in the theory of mappings with bounded
distortion   79

§1. Estimates of the moduli of continuity and differentiability
almost everywhere of mappings with bounded distortion   79

1.1. Some auxiliary facts   79

1.2. An estimate of the modulus of continuity of a mapping
with bounded distortion   82

1.3. Differentiability almost everywhere of mappings with
bounded distortion   83

§2. Some facts about continuous mappings on $\mathbf{R}^n$ 85

2.1. The degree of a mapping   85

2.2. The degree of a mapping and exterior differential forms   90

2.3. Change of variables in a multiple integral   93

§3. Conformal capacity   103

3.1. The capacity of a capacitor   103

3.2. Sets of zero capacity   110

3.3. The concept of a Hausdorff measure. Cartan's lemma   114

3.4. Capacity and Hausdorff measures   118

3.5. Estimates of the capacity of certain capacitors   120

§4. The concept of the generalized differential of an exterior
form   129

4.1. General facts about exterior forms   129

4.2. The concept of generalized differential of an exterior
form   131

4.3. Properties of the generalized differential of an exterior
form   133

4.4. The homomorphism induced on the algebra of exterior
forms by a mapping of the domain   134
4.5. Weak convergence of sequences of exterior forms   138
§5. Mappings with bounded distortion and elliptic differential
equations   141
5.1. A description of a certain class of functionals of the
calculus of variations   141
5.2. Variational properties of mappings with bounded
distortion   144
5.3. The classes $W_p^1(U/A)$ and $\overset{+}{W}{}_p^1(U/A)$ 149
5.4. The Dirichlet problem, extremal functions, and
generalized solutions of the Euler equation for
functionals of the calculus of variations   158
5.5. The maximum principle for extremals of functionals of
the calculus of variations   161
5.6. Harnack's inequality and its corollaries   163
5.7. The concept of the flow of a stationary function in a
capacitor   164
5.8. The set of singular points of stationary functions for
functionals of the calculus of variations   168
5.9. Liouville's theorem on conformal mappings in space   171
5.10. The property of quasi-invariance of conformal capacity   172
§6. Topological properties of mappings with bounded distortion   173
6.1. Continuous mappings with nonnegative Jacobian   173
6.2. Satisfaction of condition N for mappings with bounded
distortion   176
6.3. Topological properties of mappings with bounded
distortion   182
6.4. A theorem on removable singularities   187
6.5. On the method of moduli   188
6.6. Bi-Lipschitz mappings   190
§7. Local structure of mappings with bounded distortion   194
7.1. Preliminary remarks   194
7.2. Some estimates of a solution of an elliptic equation
having one singular point   196
7.3. A measure of the distortion of a small sphere under a
mapping with bounded distortion   200
7.4. Behavior of a mapping with bounded distortion near an
arbitrary point of the domain   201

§8. Characterization of mappings with bounded distortion by
    the property of quasiconformality   204
    8.1. The concept of a mapping which is quasiconformal at a
         point and in a domain   204
    8.2. Differentiability almost everywhere of quasiconformal
         T-mappings   205
    8.3. The condition of absolute continuity for a real function
         of a single variable   208
    8.4. The analytic nature of quasiconformal T-mappings   210
    8.5. Main result   213
    8.6. Homeomorphic quasiconformal mappings   215
§9. Sequences of mappings with bounded distortion   216
    9.1. A theorem on local boundedness of sequences of
         mappings with bounded distortion   216
    9.2. A theorem on the limit of a sequence of mappings with
         bounded distortion   218
    9.3. A sufficient condition for relative compactness of a
         family of mappings with bounded distortion   220
§10. The set of branch points of a mapping with bounded
     distortion and locally homeomorphic mappings   221
    10.1. The measure of the set of branch points   221
    10.2. Some lemmas on local homeomorphisms   224
    10.3. The measure of the image of the set of branch points
          for mappings with bounded distortion   227
    10.4. A local homeomorphism theorem   229
    10.5. A theorem on the radius of injectivity   234
§11. Extremal properties of mappings with bounded distortion   240
    11.1. The homomorphism generated on the algebra of
          exterior forms by a mapping with bounded distortion   240
    11.2. Main theorem   242
§12. Some further results   246
    12.1. Classes of domains in $\mathbf{R}^n$ 246
    12.2. Stability in the Liouville theorem on conformal
          mappings of a space and related questions   251
    12.3. Stability of isometric and Lorentz transformations   260
    12.4. Quasiconformal and quasi-isometric deformations.
          Semigroups of quasiconformal transformations   266
    12.5. Mappings with distortion coefficient close to 1   277

12.6. The general concept of stability classes   283
12.7. A characterization of quasiconformal mappings as
    mappings preserving the space W_n^1 287

CHAPTER III. Some results in the theory of functions of a real
    variable and the theory of partial differential
    equations   289
§1. Functions with bounded mean oscillation   289
§2. Harnack's inequality for quasilinear elliptic equations   295
    2.1. Preliminary remarks   295
    2.2. Main inequalities   296
    2.3. Consequences of the integral inequalities in §2.2   300
    2.4. Boundedness of generalized solutions of equation (2.3).
        Harnack's inequality   304
§3. Theorems on semicontinuity and convergence with a
    functional for functionals of the calculus of variations 310
    3.1. Weak convergence of sequences of functions in measure
        spaces   310
    3.2. Some lemmas about convex functions   312
    3.3. Theorems about semicontinuity of functionals of the
        calculus of variations   315
    3.4. Corollaries to Theorems 3.1 and 3.2   319
    3.5. The convex envelope of a function   321
    3.6. A theorem on convergence with a functional   325
    3.7. Corollaries to the theorem on convergence with a
        functional   329
§4. Some properties of functions with generalized derivatives   330
    4.1. A theorem on differentiability almost everywhere   330
    4.2. Proof of Lemma 1.1 in Chapter II   335
    4.3. An estimate of the modulus of continuity of a
        monotone function of class W_n^1 338
§5. On the degree of a mapping   341

Bibliography   347

Foreword to the English Translation

This text differs from the Russian original in the following respects. First, Chapter I has been completely reworked. The author has tried to give a more complete exposition of the preliminary facts needed for reading the main text. In particular, complete proofs are presented for all the required properties of Möbius mappings. In Chapter II some errors in the original are corrected, and improvements in the text are introduced where it seemed possible by slight changes to strengthen individual results or to make the presentation more complete and clear. Moreover, §12 is added to Chapter II. It contains a survey of certain further investigations of questions close to the main topic of the book. Here the author has confined himself mainly to an account of work carried out in Novosibirsk and little known outside the USSR. The bibliography has been enlarged accordingly .

Chapter III remains almost without changes.

From the Author

The book before the reader is devoted to an exposition of results of investigations carried out mainly over the last 10–15 years concerning certain questions in the theory of quasiconformal mappings.

The principal objects of investigation—mappings with bounded distortion—are a kind of n-space analogue of holomorphic functions. As is known, every holomorphic function is characterized geometrically by the fact that the mapping of a planar domain it implements is conformal. In the n-space case the condition of conformality singles out a very narrow class of mappings. As Liouville showed back in 1850, already in three-dimensional Euclidean space there are no conformal mappings besides those which are compositions of finitely many inversions with respect to spheres. Such mappings are called *Möbius mappings*. They form a finite-dimensional Lie group which includes the group of motions of the space $\mathbf{R}^n$ and is only slightly broader than this group. However, if one weakens the condition of conformality, replacing it by the condition of quasiconformality, then a considerably broader class of mappings emerges.

To give the reader an idea about the subject of the book we present some explanations (the exact definitions are contained in the main text). A mapping of a domain in an n-dimensional space is called a *mapping with bounded distortion* if it satisfies definite requirements of regularity, preserves the orientation of every small domain, and (the main point) satisfies the following condition. There exists a constant q, $1 \leq q < \infty$, such that an infinitesimally small sphere is transformed by the mapping into either a point or an infinitesimally small ellipsoid for which the ratio of the largest semiaxis to the smallest does not exceed the constant q. If, moreover, the mapping is also topological (a homeomorphism), then it is said to be quasiconformal.

The regularity requirement mentioned here is that the components of the vector-valued function determining the given mapping must have first-order generalized derivatives that are locally integrable to the power n. The condition about preserving the orientation of a small domain is analytically equivalent to the Jacobian of the mapping being nonnegative. For an arbitrary mapping with bounded distortion there can exist points such that the mapping is not a homeomorphism in any neighborhood of them. These points are called *branch points* of the mapping. The dimension of the set of branch points does not exceed $n-2$. For example, the set of branch points can be a curve in three-dimensional space. For a holomorphic function of a single variable the branch points are simply the zeros of its derivative.

We note that in the two-dimensional case the study of arbitrary mappings with bounded distortion is easily reduced to the consideration of homeomorphic quasiconformal mappings and holomorphic functions of a single variable.

The theory of planar quasiconformal mappings arose at the end of the 1920's in work of Grötzsch and M. A. Lavrent'ev. This theory is now a far-advanced area of the theory of functions of a complex variable which has important applications both in function theory itself and beyond its boundaries, in particular, in applied areas.

The concept of a quasiconformal mapping in n-space introduced by Lavrent'ev in 1938 in searching for a suitable tool to construct mathematical models of certain hydrodynamics phenomena. He formulated also a number of problems whose solutions later played an essential role in the development of the theory of quasiconformal mappings in n-space. However, the beginning of intensive investigations in this area dates from 1960. Further, as in the planar case, only homeomorphic quasiconformal mappings were considered at first. The systematic study of general mappings with bounded distortion was begun in 1966.

There are two basic methods in the theory of mappings with bounded distortion. One of them goes back to the classical work of Grötzsch and is based on the use of a certain quantity characterizing a family of curves or surfaces in space and called the *modulus* of the family. This method depends on inequalities describing the behavior of the modulus of a family of curves or surfaces when the family is transformed by a given mapping, as well as on certain estimates for the moduli.

The other method consists in the use of a certain apparatus in the theory of differential equations. As is known, the real and imaginary parts of a holomorphic function of a complex variable are harmonic functions. Analogously, the components of a vector-valued function representing a mapping with bounded distortion are solutions of a certain elliptic partial differential equation. The method consists in the use of this fact (and

certain generalizations of it) and properties of elliptic equations, in particular, the maximum principle for elliptic equations, Harnack's inequality,
etc. Some estimates relating to the concept of the capacity of a capacitor
are also used.

In this monograph the second method is used to study mappings with
bounded distortion. All the needed facts about elliptic equations are given.
A significant part of the book is devoted to an exposition of material that is
auxiliary with respect to the main topic, though it is definitely of independent interest. In particular, a proof is given of the well-known theorem of
Moser and Serrin on Harnack's inequality for elliptic equations; theorems
are proved on semicontinuity and convergence with a functional for functionals of the calculus of variations; the necessary facts are given about the
concept of the degree of a mapping and the metric properties of mappings
connected with these facts; etc. In this connection the author hopes that
the book will prove to be useful not only for specialists in the theory of
mappings in n-space, but also for a broader circle of readers.

The investigation of mappings with bounded distortion is based on the
concept of the generalized differential of an exterior form [146]. In particular, a detailed study is made of the properties of the generalized differential, and this, in the author's opinion, can be of interest, for example, in
connection with certain recent investigations of the topology of Lipschitz
manifolds by analytic means [49].

Quasiconformal mappings in n-space have been used in the theory of
spaces of functions with generalized derivatives ([177], [178], [46]), as well
as in investigations of compact Riemannian spaces of constant negative
curvature [106]. The theory of mappings in n-space with bounded distortion is one of the areas in the general metric theory of space mappings
that is being intensively developed at present. Among the investigations
in this area one can cite work on the theory of quasi-isometric mappings
[60], the theory of quasi-Lorentz mappings [51], a series of investigations
in the theory of Kleinian groups in space [75], papers on the theory of
homeomorphisms of class W_n^1 ([165], [112]), and other publications.

Many interesting questions in the theory of mappings in n-space close
to the topic of the book had to be omitted for lack of space. In choosing the material the author was oriented toward results used in studying
the problem of stability in Liouville's theorem on conformal mappings in
space.

Yu. G. Reshetnyak

CHAPTER I

Introduction

§1. Some facts from the theory of functions of a real variable

1.1. Sets in $\mathbf{R}^n$. Below, $\mathbf{R}^n$ denotes the n-dimensional additive Euclidean space of points $x = (x_1, \cdots, x_n)$, $|x|$ is the length of a vector $x \in \mathbf{R}^n$, and $\langle x, y \rangle$ is the inner product of vectors x and y in $\mathbf{R}$, i.e., for $x = (x_1, \cdots, x_n)$ and $y = (y_1, \cdots, y_n)$ we have

$$\langle x, y \rangle = x_1 y_1 + \cdots + x_n y_n,$$
$$|x| = \sqrt{\langle x, y \rangle} = \sqrt{x_1^2 + \cdots + x_n^2}.$$

If a and b are two arbitrary points in $\mathbf{R}^n$, then $[a, b]$ denotes the segment joining them, i.e., $[a, b]$ is the collection of all points x of the form $x = \lambda a + \mu b$, where $\lambda + \mu = 1$ and $\lambda, \mu \geq 0$.

Denote by e_i, $i = 1, \ldots, n$, the vector in $\mathbf{R}^n$ with ith coordinate equal to 1 and the other coordinates equal to 0. The vectors $e_1, \ldots, e_n$ form a basis in $\mathbf{R}^n$, called the *canonical basis*.

Let $P(x)$ be some expression containing the variable x, and let A be a set. Then $\{x \in A | P(x)\}$ denotes the collection of all elements x in A such that the expression $P(x)$ is true. We employ the usual symbolism of mathematical logic; in particular, $\forall$ ("for all"), $\exists$ ("there exists"), $\Rightarrow$ ("if ..., then ..."), $\Leftrightarrow$ ("is equivalent to"), & ("and"), $\vee$ ("or"), and $\neg$ ("not"). The symbol $\varnothing$ denotes the empty set.

Let X be a given topological space. The closure of an arbitrary set $A \subset X$ is denoted by $\overline{A}$, CA is the complement of A, $CA = X \backslash A$, $A^\circ = C(\overline{CA})$ is the open kernel or interior of A, and $\partial A = \overline{A} \backslash A^\circ = \overline{A} \cap \overline{CA}$ is the boundary of A.

A *domain* in a topological space X is defined to be a connected open subset of X. Let U be an open set in X. We say that $A \subset X$ lies *strictly inside* U if $\overline{A}$ is compact and $U \supset \overline{A}$.

1

Let $A \subset X$. For $x \in X$ let $\chi_A(x) = 1$ if $x \in A$ and $\chi_A(x) = 0$ if $x \notin A$. The function $\chi_A \colon X \to \mathbf{R}$ so defined is called the *indicator function* of the set A.

Let X be a metric space and ρ its metric. For $x \in X$ and $A \subset X$ the distance from x to A is defined by

$$\rho(x, A) = \inf_{y \in A} \rho(x, y).$$

Obviously, $(\rho(x, A) = 0) \Leftrightarrow (x \in \overline{A})$. The function $x \mapsto \rho(x, A)$ is continuous. Further, for $x_1, x_2 \in X$

$$|\rho(x_1, A) - \rho(x_2, A)| \le \rho(x_1, x_2).$$

Let

$$d(A) = \sup_{x \in A, y \in A} \rho(x, y).$$

The quantity $d(A)$ is called the *diameter* of A. For arbitrary $A, B \subset X$ let

$$\operatorname{dist}(A, B) = \inf_{x \in A, y \in B} \rho(x, y).$$

The quantity $\operatorname{dist}(A, B)$ is called the *distance between* the sets A and B. For $A \subset X$ let

$$U_h(A) = \{x \in A \mid \rho(x, A) < h\}$$

and let $\overline{U}_h(A)$ be the closure of $U_h(A)$. The set $U_h(A)$ is called the *h-neighborhood* of the set A, and $\overline{U}_h(A)$ is called the *closed h-neighborhood* of A. Let U be an open subset of a metric space X. Then we let

$$\hat{U}_h = \{x \in X \mid \rho(x, CU) > h\}.$$

The set $\hat{U}_h$ is open. If $U \ne \varnothing$ and h is sufficiently small, then $\hat{U}_h \ne \varnothing$. The supremum of the values of h such that $\hat{U}_h$ is a nonempty set is called the *interior radius* of the nonempty set U. It is clear that $U \supset \hat{U}_h$ for every $h > 0$, $\hat{U}_{h_1} \supset \hat{U}_{h_2}$ for $h_1 < h_2$, and $\bigcup_{h>0} \hat{U}_h = U$.

For an arbitrary metric space X with metric ρ let $B_X(a, r)$ denote the open ball, $\overline{B}_X(a, r)$ the closed ball, and $S_X(a, r)$ the sphere about a with radius $r > 0$, i.e.,

$$B_X(a, r) = \{x \in X \mid \rho(x, A) < r\},$$
$$\overline{B}_X(a, r) = \{x \in X \mid \rho(x, A) \le r\},$$
$$S_X(a, r) = \{x \in X \mid \rho(x, a) = r\}.$$

The index X is omitted in this notation whenever no confusion is possible.

The area of the $(n-1)$-dimensional sphere $S(0, 1)$ in $\mathbf{R}^n$ will be denoted by ω_{n-1}, $\omega_{n-1} = 2\pi^{n/2}/\Gamma(n/2 + 1)$, and $\sigma_n = \omega_{n-1}/n$ is the volume of $B(0, 1)$ in $\mathbf{R}^n$.

Let $a = (a_1, \ldots, a_n) \in \mathbf{R}^n$. Denote by $Q(a, r)$ the n-dimensional open cube

$$\{(x_1, \ldots, x_n) \in \mathbf{R}^n | \forall i = 1, \ldots, n \; |x_i - a_i| < r/2\},$$

and by $\overline{Q}(a, r)$ the closed cube

$$\{(x_1, \ldots, x_n) \in \mathbf{R}^n | \forall i = 1, \ldots, n \; |x_i - a_i| \leq r/2\}.$$

The point a is called the *center* of each of these cubes, and the number r is the length of an edge. A *dyadic cube* in $\mathbf{R}^n$ is defined to be an n-dimensional rectangle P of the form

$$P = \left[\frac{k_1}{2^r}, \frac{k_1 + 1}{2^r}\right) \times \left[\frac{k_2}{2^r}, \frac{k_2 + 1}{2^r}\right) \times \cdots \times \left[\frac{k_n}{2^r}, \frac{k_n + 1}{2^r}\right),$$

where $k_1, \ldots, k_n$, and r are integers. The number r here is called the *rank* of the dyadic cube P. Two different dyadic cubes of the same rank do not have common points. Let P and Q be dyadic cubes of respective ranks r and s, with $r > s$; if $P \cap Q \neq \varnothing$, then $P \subset Q$. For each r and every $x \in \mathbf{R}^n$ there exists a unique dyadic cube Q of rank r such that $x \in Q$. Denote this cube by $\mathscr{D}_r(x)$.

Let U be an open subset of $\mathbf{R}^n$, and x an arbitrary point of U. Take the smallest integer r such that $r \geq 0$ and the closure of $\mathscr{D}_r(x)$ is contained in U. The cube $\mathscr{D}_r(x)$ corresponding to this value of r is denoted by $\mathscr{D}_U(x)$. The set of all cubes $\mathscr{D}_U(x)$ is countable, since the set of all dyadic cubes is countable, and the set of cubes $\mathscr{D}_U$ is obviously infinite. If the cubes $\mathscr{D}_U(x_1)$ and $\mathscr{D}_U(x_2)$ are distinct, then they do not intersect. Indeed, let r_1 be the rank of $\mathscr{D}_U(x_1)$ and r_2 the rank of $\mathscr{D}_U(x_2)$, and let $r_1 < r_2$. If $\mathscr{D}_U(x_1) \cap \mathscr{D}_U(x_2) \neq \varnothing$, then $\mathscr{D}_U(x_1) \supset \mathscr{D}_U(x_2)$, and hence $x_2 \in \mathscr{D}_U(x_1)$. We have that $\mathscr{D}_U(x_2) = \mathscr{D}_{r_2}(x_2)$, and since $x_2 \in \mathscr{D}_U(x_1)$, it follows that $\mathscr{D}_U(x_1) = \mathscr{D}_{r_1}(x_2)$. This contradicts the fact that r_2 is by definition the smallest integer $r \geq 0$ such that $\overline{\mathscr{D}_r(x_2)} \subset U$. We number all the cubes $\mathscr{D}_U(x)$ and let Q_ν be the cube with index ν. Then $U = \bigcup_1^\infty Q_\nu$, $\overline{Q}_\nu \subset U$ for each ν, and $Q_{\nu_1} \cap Q_{\nu_2} = \varnothing$ for $\nu_1 \neq \nu_2$. This partition of U will be called the *dyadic subdivision* of U. Every compact set $A \subset U$ is covered by finitely many cubes of the dyadic subdivision. Indeed, $\rho(x, CA) \geq \delta = \text{const} > 0$ for all $x \in A$; hence, for $x \in A$ the rank of the cube $\mathscr{D}_U(x)$ does not exceed some number $r_0 < \infty$, and thus the number of cubes such that $Q_\nu \cap A \neq \varnothing$ is finite.

If A is an arbitrary subset of $\mathbf{R}^n$, then $|A|$ will denote the *outer Lebesgue measure* of A; that is, $|A|$ is the infimum of the measures of the open subsets of U containing A. For every subset A of $\mathbf{R}^n$ there exists a measurable set $H \supset A$ such that $H = |A|$.

1.2. Classes of functions in $\mathbf{R}^n$. Let A be a measurable subset of $\mathbf{R}^n$, and u a measurable real-valued function on A. For $p \geq 1$ let

$$\|u\|_{p,A} = \left\{ \int_A |u(x)|^p \, dx \right\}^{1/p}. \tag{1.1}$$

Assume that U is an open subset of $\mathbf{R}^n$ and $u(x)$ is a measurable function defined almost everywhere on U. We say that $u(x)$ is *locally integrable to the power $p \geq 1$* on U (and write $u \in L_{p,\text{loc}}(U)$) if $\|u\|_{p,A} < \infty$ for every compact set $A \subset U$. By the Borel covering theorem, this is equivalent to the following: every $x \in U$ has a neighborhood V such that $\|u\|_{p,V} < \infty$. The collection of all measurable functions $u(x)$ such that $\|u\|_{p,U} < \infty$ is denoted by $L_p(U)$. In the case $U = \mathbf{R}^n$ the quantity $\|u\|_{p,U}$ is denoted by $\|u\|_p$ and L_p is written instead of $L_p(U)$.

Here we assume known all the facts in the theory of the Lebesgue integral on $\mathbf{R}^n$ and all the properties of the space $L_p(U)$, in particular, the Hölder and Minkowski inequalities, the completeness of the spaces $L_p(U)$, and so on.

Let (u_m), $m = 1, 2, \ldots$, be an arbitrary sequence of functions in the class $L_{p,\text{loc}}(U)$, where $1 \leq p \leq \infty$. Then we say that a sequence (u_m), $m = 1, 2, \ldots$ is *bounded in $L_{p,\text{loc}}(U)$*, or *locally bounded in $L_p(U)$*, if the sequence $(\|u_m\|_{p,A})$, $m = 1, 2, \ldots$, is bounded for every compact set $A \subset U$. A sequence (u_m), $m = 1, 2, \ldots$, of functions of the class $L_{p,\text{loc}}(U)$, $1 \leq p < \infty$, is said to *converge in $L_{p,\text{loc}}(U)$* to a function $u_0 \in L_{p,\text{loc}}(U)$ if $\|u_m - u\|_{p,A} \to 0$ as $m \to \infty$ for every compact set $A \subset U$.

Let X be an arbitrary topological space, and Y an arbitrary vector space. For an arbitrary function $f \colon X \to Y$ let $S^*(f)$ be the set of $x \in X$ such that $f(x) \neq 0$. The closure of $S^*(f)$ is called the *support* of f and denoted by $S(f)$. A function f is said to be *compactly supported* if its support is compact.

In the special case when X is an open subset of $\mathbf{R}^n$, endowed with the topology induced by $\mathbf{R}^n$, the support of a function $f \colon X \to Y$ is the set $\overline{S^*(f)} \cap X$. Here the closure is in the topology of $\mathbf{R}^n$.

Assume that the vector space Y is a normed space. In this case $C(X, Y)$ denotes the collection of all continuous mappings $f \colon X \to Y$. If no confusion is possible, we write $f \in C(X)$, or simply $f \in C$, instead of $f \in C(X, Y)$. The collection of all compactly supported functions $f \in C(X, Y)$ is denoted by $C_0(X, Y)$; the abbreviated notation $C_0(X)$ or C_0 will sometimes be used instead. For an arbitrary function $f \colon X \to Y$ and an arbitrary set $A \subset X$ let

$$\|f\|_{M(A)} = \sup_{x \in A} |f(x)|_Y, \tag{1.2}$$

where $|\cdot|_Y$ denotes the norm in Y. If $f \in C$ and A is compact, then clearly $\|f\|_{M(A)} < \infty$. In particular, if $f \in C_0$, then $\|f\|_{M(A)} < \infty$. Let $\|f\|_{C(A)} = \|f\|_{M(A)}$ in the case when f is a continuous mapping.

LEMMA 1.1. *Let U be an open subset of $\mathbf{R}^n$. Then for every $p \geq 1$ the set $C_0(U) = C_0(U, \mathbf{R})$ of functions is dense in $L_p(U)$.*

The proof is a simple exercise in the theory of functions of a real variable, and we omit it (see also [148]: the lemma is a direct consequence of the construction of the Lebesgue integral presented there).

Let X be a locally compact topological space. A sequence $(f_m: X \to Y)$, $m = 1, 2, \ldots$, of functions is said to *converge locally uniformly* to a function $f_0: X \to Y$ if f_m converges to f_0 uniformly on any compact set $A \subset X$ as $m \to \infty$.

Let U be an open subset of $\mathbf{R}^n$, and Y a normed vector space. We say that a function $f: U \to Y$ belongs to the class $C^k(U, Y)$, where $k \geq 1$ is an integer, if f has all partial derivatives of order at most k in U, and these derivatives are continuous on U. The collection of all functions $f: U \to Y$ of class $C^k(U, Y)$ with compact support in U is denoted by $C_0^k(U, Y)$. If $f \in C^k(U, Y)$ ($f \in C_0^k(U, Y)$) for all integers $k \geq 1$, then we say that $f \in C_0^\infty(U, Y)$ (respectively, $f \in C_0^\infty(U, Y)$). In all this notation the symbols U and Y are omitted whenever no confusion can result.

Suppose that U is an open subset of $\mathbf{R}^n$ and $f: U \to \mathbf{R}^n$ is a mapping of class $C^1(U, \mathbf{R}^n)$. (According to the conventions announced above, we can simply write $f \in C^1$.) Then at each point $x \in U$ the linear mapping $f^1(x): \mathbf{R}^n \to \mathbf{R}^n$ is defined, the derivative of f at x. For any vector $h \in \mathbf{R}^n$

$$f'(x)h = \lim_{t \to 0} \frac{f(x + th) - f(x)}{t} = \sum_{i=1}^{n} \frac{\partial f}{\partial x_i}(x)h_i.$$

The matrix of the linear mapping $f'(x)$ is the Jacobi matrix of the mapping f at x. A mapping $f: U \to \mathbf{R}^n$ is called a *diffeomorphic mapping* or a *diffeomorphism of class C^k* (where $k \geq 1$ is an integer) if f is in the class $C^k(U, \mathbf{R}^n)$, f is one-to-one, and the Jacobian $\det f'(x)$ of f is nonzero at each point $x \in U$. In this case the set $V = f(U)$ is open, and the inverse mapping f^{-1} is a diffeomorphism of the same smoothness class C^k. In particular, every diffeomorphism is a topological mapping (homeomorphism).

Let X and Y be metric spaces, with ρ_X the metric in X and ρ_Y the metric in Y. Let $f: X \to Y$ be a given function. A *modulus of continuity* of f is defined to be any function $\omega: [0, \infty) \to [0, \infty)$ satisfying the following conditions:

A) ω is nondecreasing, and $\omega(t) \to 0$ as $t \to 0$.

B) For any $x, y \in X$

$$\rho_Y[f(x), f(y)] \leq \omega[\rho_X(x, y)].$$

A function $f: X \to Y$ has a modulus of continuity if and only if it is uniformly continuous. If X is compact and $f: X \to Y$ is continuous, then f is uniformly continuous.

If a function $f: X \to Y$ has a modulus of continuity $\omega(t) = Kt^\alpha$, where $0 < K < \infty$ and $0 < \alpha \leq 1$, then f is said to *satisfy a Hölder condition with constant K and exponent* α. If f satisfies a Hölder condition with constant K and exponent $\alpha = 1$, then f is also said to *satisfy a Lipschitz condition with constant K*.

Let U be an open subset of $\mathbf{R}^n$, and Y a normed vector space. The symbol $C^{0,\alpha}(U)$, where $0 < \alpha \leq 1$, denotes the collection of all mappings $f: U \to Y$ satisfying a Hölder condition with exponent α on every compact set $A \subset U$. A mapping $f: U \to Y$ will be said to belong to the class $C^{k,\alpha}(U, Y)$ (or simply $f \in C^{k,\alpha}$), where $k \geq 1$ is an integer and $0 < \alpha \leq 1$, if $f \in C^k(U)$ and all the kth-order partial derivatives of f belong to $C^{0,\alpha}$.

1.3. Differentiation of measures on the space $\mathbf{R}^n$. Lebesgue points of a function and points of density of a subset of $\mathbf{R}^n$. Here we give a summary of classical results on differentiation of set functions on $\mathbf{R}^n$.

The symbol $\mathfrak{B}^n$ will denote the collection of all Borel subsets of $\mathbf{R}^n$. Let $A \in \mathfrak{B}^n$. The collection of all sets $E \in \mathfrak{B}^n$ contained in A will be denoted by $\mathfrak{B}(A)$. The union and intersection of any at most countable family of sets in $\mathfrak{B}(A)$ belongs to $\mathfrak{B}(A)$. The difference of any two sets in $\mathfrak{B}(A)$ belongs to $\mathfrak{B}(A)$.

A *measure* on a Borel set $A \subset \mathbf{R}^n$ is defined to be a function $\mu: \mathfrak{B}(A) \to \mathbf{R}$ such that

$$\mu(\bigcup_m E_m) = \sum_m \mu(E)$$

for any at most countable family (E_m), $m = 1, 2, \ldots$, of disjoint sets in $\mathfrak{B}(A)$. In this definition it is not assumed that the set function μ is nonnegative.

Let $\mu: \mathfrak{B}(A) \to \mathbf{R}$ be a measure on a Borel subset A of $\mathbf{R}^n$. Then a certain nonnegative measure $|\mu|$ called the *variation* of μ is defined. For an arbitrary $E \in \mathfrak{B}(A)$

$$|\mu|(E) = \sup_{E' \in \mathfrak{B}(E)} \{|\mu(E')| + |\mu(E \setminus E')|\}.$$

For every $E \in \mathfrak{B}(A)$ we have that $|\mu(E)| \leq |\mu|(E)$.

Let μ be a measure on an open set U in $\mathbf{R}^n$. We say that μ is *differentiable* at a point $x \in U$ if the limit

$$\lim_{r \to 0} \frac{\mu[\overline{B}(x,r)]}{|\overline{B}(x,r)|} = \mathscr{D}\mu(x)$$

exists and is finite. The number $\mathscr{D}\mu(x)$ is called the *density* of μ at x.

A measure μ on an open subset U of $\mathbf{R}^n$ is said to be *singular* if there exists a Borel set $A \subset U$ with Lebesgue measure zero such that $|\mu|(U \setminus A) = 0$.

THEOREM 1.1. *Every measure μ on an open subset U of $\mathbf{R}^n$ is differentiable almost everywhere on U. The function $x \mapsto \mathscr{D}\mu(x)$ is integrable, and for every $E \in \mathfrak{B}(U)$*

$$\mu(E) = \int_E \mathscr{D}\mu(x)\,dx + \sigma(E),$$

where σ is a singular measure on $\mathbf{R}^n$. (In particular, σ can be identically equal to zero).

See, for example, [149] or [48] for a proof.

Let us now define the concept of a Lebesgue point of a function. It is expedient first to introduce a certain more general concept.

Let $\mathfrak{R}$ be a Hausdorff topological space whose elements are measurable real-valued functions on the ball $\overline{B}(0,1) \subset \mathbf{R}$ such that for any $l \in \mathbf{R}$ the function identically equal to l is in $\mathfrak{R}$.

Let U be an open subset of $\mathbf{R}^n$, and let $f\colon U \to \mathbf{R}$ be a measurable function. Take an arbitrary point $a \in U$, and let $0 < h < \rho(a, \partial U)$. Then the number $F_h(X) = f(a + hX)$ is defined for almost all vectors $X \in \overline{B}(0,1)$. We thereby obtain a family of measurable functions F_h defined on the ball $\overline{B}(0,1)$. A number l is called the *limit of f as $x \to a$ in the sense of convergence in $\mathfrak{R}$* if there is an $h_0 > 0$ such that $F_h \in \mathfrak{R}$ for $0 < h < h_0$, and the functions F_h, as elements of $\mathfrak{R}$, converge to the function identically equal to the constant l as $h \to 0$. In this case we write $l = \lim_{x \to a}(\mathfrak{R})f(x)$. In the case when $f(a) = \lim_{x \to a}(\mathfrak{R})f(x)$, we say that f is continuous at a in the sense of convergence in $\mathfrak{R}$.

We give examples.

Let M be a vector space of bounded real-valued functions $F\colon \overline{B}(0,1) \to \mathbf{R}$, and let the topology on M be determined by the norm

$$\|F\| = \sup_{0 < |x| \leq 1} |F(x)|.$$

(Functions $F_1, F_2 \in M$ such that $F_1(x) = F_2(x)\ \forall x \neq 0$ are identified.) It is not hard to see that the limit in the sense of convergence in M coincides with the usual concept of limit.

Let $\mathfrak{M}$ be the collection of all measurable functions on $\overline{B}(0,1)$. For $F, G \in \mathfrak{M}$ let

$$\rho(F, G) = \int_{\overline{B}(0,1)} \frac{|F(x) - G(x)|}{1 + |F(x) - G(x)|} \, dx.$$

The function $\rho(F, G)$ is a metric on $\mathfrak{M}$. (Functions $F_1, F_2 \in \mathfrak{M}$ coinciding almost everywhere on $\overline{B}(0,1)$ are regarded as the same element of $\mathfrak{M}$.) Convergence in the sense of this metric is none other than convergence in measure. For an arbitrary measurable function $f\colon U \to \mathbf{R}$ the concept of limit in the sense of convergence in $\mathfrak{M}$ coincides with the concept of approximative limit (see the definition in [149]; the proof that these concepts are equivalent is left to the reader).

Assume that f is a function of class $L_{p,\text{loc}}(U)$, where $1 \le p < \infty$. Then for $a \in U$ and $h < \rho(a, \partial U)$ the function $F_h\colon X \mapsto f(a + hX)$ belongs to the class $L_p[\overline{B}(0,1)]$. A limit of f at a in the sense of convergence in $L_p[\overline{B}(0,1)]$ is called an L_p-*limit of f at a*. If f is continuous at a point x in the sense of convergence in $L_p[\overline{B}(0,1)]$, then one also says that a is a *Lebesgue L_p-point* of f (a *Lebesgue point* of f in the case $p = 1$). A point a is a Lebesgue L_p-point of a function f if and only if

$$\lim_{h \to 0} \int_{\overline{B}(0,1)} |f(a + hX) - f(a)|^p \, dX = 0.$$

Performing the change of variable $a + hX = x$ in the integral, we get that a is a Lebesgue L_p-point of f if and only if

$$\frac{1}{|B(a,h)|} \int_{B(a,h)} |f(x) - f(a)|^p \, dx \to 0$$

as $h \to 0$. Textbooks on the theory of functions of a real variable usually define a Lebesgue L_p-point to be one for which the last relation holds.

THEOREM 1.2 (see [183] and [48]). *Let U be an open subset of $\mathbf{R}^n$. Then for every function $f \in L_{p,\text{loc}}(U)$ ($p \ge 1$) almost all points $x \in U$ are Lebesgue L_p-points of f.*

In the theory of the Lebesgue integral all functions are regarded as defined to within values at points forming a set of measure zero. In some cases the ambiguity arising from this can lead to certain difficulties. This ambiguity can be removed by introducing the concept of a natural value of a function. Let f be a function in $L_{1,\text{loc}}(U)$. A number $L \in \mathbf{R}$ is called the *natural value* of f at a point $x \in U$ if $L = \lim_{t \to x}(L_p)f(t)$. By Theorem 1.2, if f is a function in $L_{p,\text{loc}}(U)$, then $f(x)$ is the natural value of f at x for almost all $x \in U$.

Let A be an arbitrary subset of $\mathbf{R}^n$. A point x in A is called a *point of density* of A if

$$|A \cap \overline{B}(x, r)| / |\overline{B}(x, r)| \to 1$$

as $r \to 0$. (Here it is not assumed that A is measurable.)

THEOREM 1.3. *For every set $A \subset \mathbf{R}^n$ the points $x \in A$ which are not points of density of A form a set of measure zero.*

PROOF. This is a consequence of Theorem 1.2. Assume first that A is measurable. Then the indicator function χ_A of A belongs to $L_{1,\mathrm{loc}}(\mathbf{R}^n)$. Let E_0 be the set of points $x \in \mathbf{R}^n$ which are not Lebesgue points of χ_A. By Theorem 1.2, $|E_0| = 0$. Let $x \in A \setminus E_0$ be arbitrary. Then $\chi_A(x) = 1$, and as $r \to 0$

$$\frac{1}{|B(x, r)|} \int_{B(x,r)} |\chi_A(t) - 1| \, dt \to 0,$$

which implies that

$$\frac{1}{|\overline{B}(x, r)|} \int_{\overline{B}(x,r)} \chi_A(t) \, dt = \frac{|A \cap \overline{B}(x, r)|}{|\overline{B}(x, r)|} \to 1$$

as $r \to 0$, i.e., x is a point of density of A. This proves the theorem for the case when A is measurable.

Let A be an arbitrary subset of $\mathbf{R}^n$ such that $|A| < \infty$. Then there exists a measurable set $H \supset A$ such that $|H| = |A|$. For every point $x \in A$ and any $r > 0$ we have that $|H \cap \overline{B}(x, r)| = |A \cap \overline{B}(x, r)|$. Indeed, assume on the contrary that $|H \cap \overline{B}(x, r)| > |A \cap \overline{B}(x, r)|$ for some r. We construct a measurable set $H' \supset A \cap \overline{B}(x, r)$ such that $|H'| = |A \cap \overline{B}(x, r)|$. It can obviously be assumed that $H' \subset \overline{B}(x, r)$. Then the set $H'' = H' \cup (H \setminus \overline{B}(x, r))$ is measurable, $A \subset H''$, and

$$|A| \leq |H''| = |H'| + |H \setminus \overline{B}(x, r)| < |H \cap \overline{B}(x, r)| + |H \setminus \overline{B}(x, r)| = |H| = |A|.$$

We thus get a contradiction, and hence

$$|A \cap \overline{B}(x, r)| = |H \cap \overline{B}(x, r)|.$$

Let $E_0 \subset H$ be the set of all points in H which are not points of density of H. Since H is measurable, E_0 is a set of measure zero by what has been proved. Let $x \in A \setminus E_0$. Then x is a point of density of H, i.e.,

$$\frac{|H \cap \overline{B}(x, r)|}{|\overline{B}(x, r)|} \to 0$$

as $r \to 0$. By what has been proved, $|H \cap \overline{B}(x, r)| = |A \cap \overline{B}(x, r)|$; therefore we see that x is a point of density of A, and the theorem is proved also for the case when $|A| < \infty$.

Suppose that $|A| = \infty$, and let $A_m = A \cap B(0, m)$. Then $|A_m| \leq |B(0, m)| < \infty$. Let E_m be the collection of points $x \in A_m$ which are not points of density of A_m, and let $E_0 = \bigcup_1^\infty E_m$. By what was proved, $|E_m| = 0$ for each m, so $|E_0| = 0$. Let $x \in A \setminus E_0$, and take an m such that $|x| < m$. Then $x \in A_m$, and since $x \notin E_m \subset E_0$, it follows that x is a point of density of A_m. We have that

$$1 \geq \frac{|A \cap \overline{B}(x, r)|}{|\overline{B}(x, r)|} \geq \frac{|A_m \cap \overline{B}(x, r)|}{|\overline{B}(x, r)|} \to 1$$

as $r \to 0$. Hence,

$$|A \cap \overline{B}(x, r)| / |\overline{B}(x, r)| \to 1$$

as $r \to 0$, i.e., x is a point of density of A. The theorem is proved.

1.4. Approximation of integrable functions by smooth functions. Here we describe a certain technique for smoothing functions which, in particular, is used in an essential way in studying functions with generalized derivatives.

A function $\omega \colon \mathbf{R}^n \to \mathbf{R}$ is called an *averaging kernel* if it is bounded and integrable on $\mathbf{R}^n$, has support in $B(0, 1)$, and satisfies

$$\int_{\mathbf{R}^n} \omega(x) \, dx = 1. \tag{1.3}$$

An averaging kernel ω is called a *Sobolev averaging kernel* if ω is nonnegative and belongs to $C_0^\infty(\mathbf{R}^n)$. We get an example of a Sobolev averaging kernel by first defining the function $\varphi \colon \mathbf{R} \to \mathbf{R}$ to be $\varphi(t) = e^{1/t}$ for $t < 0$ and $\varphi(t) = 0$ for $t \geq 0$, and then setting $\omega(x) = \gamma \varphi(|x|^2 - \frac{1}{4})$, where γ is chosen so that (1.3) holds.

Let U be a nonempty open subset of $\mathbf{R}^n$, and let $\hat{U}_h = \{x \in U | \rho(x, CU) > h\}$ for $h > 0$. Take an $h_0 > 0$ such that $\hat{U}_{k_0}$ is nonempty. Let ω be an arbitrary averaging kernel, and let f be a function in $L_{1,\mathrm{loc}}(U)$. For $0 < h < h_0$ we have that $\hat{U}_h \neq \varnothing$. For an arbitrary $x \in \hat{U}_h$ let

$$(\omega_h * f)(x) = \int_{|y| \leq 1} f(x + hy)\omega(y) \, dy$$

$$= \frac{1}{h^n} \int_U f(z)\omega\left(\frac{z - x}{h}\right) dz. \tag{1.4}$$

This defines a certain function $f_h = \omega_h * f$ on the set $\hat{U}_h$. It is called the *mean function* for the function f. The number h is called the *averaging parameter*. If ω is a Sobolev averaging kernel, we call f_h the *Sobolev mean function* for f.

In the case when $U = \mathbf{R}^n$ it is obvious that $\hat{U}_h = \mathbf{R}^n$, and the mean function f_h is defined everywhere in $\mathbf{R}^n$. In the general case the domain

$\hat{U}_h$ of f_h is a proper subset of U. For every compact set A there is an $h_0 = h_0(A) > 0$ such that $A \subset \hat{U}_h$ for $0 < h < h_0$.

Let x be a Lebesgue point of a function $f \in L_{1,\text{loc}}(U)$. Then as $h \to 0$ the functions $F_h(y) = f(x + hy)$ of y converge in $L_1[\overline{B}(0, 1)]$ to the function identically equal to $f(x)$. Using the first expression in (1.4) for $(\omega_h * f)(x)$, we conclude from this and the boundedness of ω that

$$(\omega_h * f)(x) \to \int_{|y| \leq 1} f(x)\omega(y) \, dy = f(x)$$

as $h \to 0$. Consequently, we get that if f is bounded and locally integrable on U, then for every averaging kernel ω

$$\lim_{h \to 0}(\omega_h * f)(x) = f(x)$$

at every Lebesgue point of f, and hence for almost all $x \in U$.

If the averaging kernel ω is a Sobolev averaging kernel, then the function $\omega_h * f$ belongs to the class $C^\infty(\hat{U}_h)$, as can be seen directly from the second expression for $(\omega_h * f)(x)$.

THEOREM 1.4. *Let ω be an averaging kernel, and let U be an open subset of $\mathbf{R}^n$. Then $\|(\omega_h * f) - f\|_{p,A} \to 0$ for every function $f \in L_{p,\text{loc}}(U)$ and for every compact set A. If $f \in C(U)$, then the functions $\omega_h * f$ are continuous, and $(\omega_h * h)(x) \to f(x)$ locally uniformly in U as $h \to 0$.*

See, for example, [161] for a proof of the theorem.

LEMMA 1.2. *Let U be a nonempty open set in $\mathbf{R}^n$. Then for any $h > 0$ it is possible to define a function $\chi_h : \mathbf{R}^n \to \mathbf{R}$ of class $C^\infty(\mathbf{R}^n)$ such that $0 \leq \chi_h \leq 1$ for all $x \in \mathbf{R}^n$, $S(\chi_h) \subset \hat{U}_h$, and $\chi_h(x) = 1$ for $x \in \hat{U}_{2h}$.*

PROOF. Let $V = \hat{U}_{(3/2)h}$, and let ω be a Sobolev averaging kernel. Let $\tau = h/4$ and

$$\chi_h(x) = (\omega_\tau * \chi_V)(x) = \frac{1}{\tau^n} \int_{\mathbf{R}^n} \chi_V(z)\omega\left(\frac{z-x}{\tau}\right) dz.$$

Since $\chi_V(z) = 1$ for $z \in V$ and $\chi_V(z) = 0$ for $z \notin V$ and since $\omega((z-x)/\tau) = 0$ if $|z - x| < \tau$, the last integral here is equal to

$$\frac{1}{\tau^n} \int_{V \cap B(x,\tau)} \omega\left(\frac{z-x}{\tau}\right) dz.$$

Assume that x is such that $\chi_h(x) \neq 0$. Then $V \cap B(x, \tau) \neq \varnothing$. Take an arbitrary $y \in V \cap B(x, \tau)$. We have that $|x - y| < \delta$ and $\rho(y, CU) > 3h/2$, from which we conclude that $\rho(x, CU) > 3h/2 - \tau = 5h/4$. This implies that if $x \in S(\chi_h)$, then $\rho(x, CU) \geq 5h/4 > h$, and hence $S(\chi_h) \subset \hat{U}_h$.

Let $x \in \hat{U}_{2h}$. For every $z \in B(x, \tau)$ we have that $\rho(z, CU) \geq \rho(x, CU) - |x - z| > 2h - h/4 > 3h/2$; thus $B(x, \tau) \subset V$, and $\chi_h(x) = 1$ for this x.

It is obvious from the definition of χ_h that $\chi_h \in C^\infty$ and $0 \leq \chi_h(x) \leq 1$ for all x. This proves the lemma.

The next statement follows from Theorem 1.4 and Lemma 1.1.

COROLLARY. *Let U be an open subset of $\mathbf{R}^n$. Then for every function $f \in L_{p,\mathrm{loc}}(U)$ there exists a family of functions $f_h: U \to \mathbf{R}$ ($h > 0$) such that $f_h \in C^\infty$ for each h, and $f_h \to f$ in $L_{p,\mathrm{loc}}(U)$ as $h \to 0$.*

Indeed, let $f_h(x) = \chi_h(x)(\omega_h * f)(x)$ for $x \in \hat{U}_h$, and let $f_h(x) = 0$ for $x \notin \hat{U}_h$. The functions f_h form the required family.

§2. Functions with generalized derivatives

2.1. Definition of a function with generalized derivatives. In the exposition to follow, essential use is made of diverse properties of functions with generalized derivatives. Here we present the needed definitions and recall some of the simplest properties of functions with generalized derivatives. The concept of a generalized derivative was introduced by Sobolev [162]. The reader can find the necessary proofs and subsequent results in the theory of functions with generalized derivatives in, for example, [1], [18], [48], [167], and elsewhere.

Let $U \subset \mathbf{R}^n$ be an arbitrary open set, and let $u: U \to \mathbf{R}$ be a function of class $L_{1,\mathrm{loc}}(U)$. Assume that a differentiation operator $\mathscr{D}^\alpha = \partial^{\alpha_1 + \cdots + \alpha_n}/\partial x_1^{\alpha_1} \cdots \partial x_n^{\alpha_n}$ is given, where $\alpha = (\alpha_1, \cdots, \alpha_n)$. In this context the vector $\alpha = (\alpha_1, \ldots, \alpha_n)$ with integer coordinates will be called a *multi-index*. The number $l = \alpha_1 + \cdots + \alpha_n$ is denoted by $|\alpha|$. A function $v \in L_{1,\mathrm{loc}}(U)$ is called the *generalized derivative $\mathscr{D}^\alpha u$* of u if there exists a sequence $u_m: U \to \mathbf{R}$ of functions of class $C^l(U)$ such that $u_m \to u$ and $\mathscr{D}^\alpha u_m \to v$ in $L_{1,\mathrm{loc}}(U)$ as $m \to \infty$.

LEMMA 2.1. *Let U be an open subset of $\mathbf{R}^n$, and $u: U \to \mathbf{R}$ a function of class $L_{1,\mathrm{loc}}(U)$. Assume that the function $v \in L_{1,\mathrm{loc}}(U)$ is the generalized derivative $\mathscr{D}^\alpha u$ of u. Let ω be an arbitrary Sobolev averaging kernel. Then for every h the equality*

$$\mathscr{D}^\alpha(\omega_h * u)(x) = (\omega_h * v)(x)$$

holds on the set $\hat{U}_h$.

PROOF. Consider a sequence $(u_m: U \to \mathbf{R})$, $m = 1, 2, \ldots$, of C^∞-functions such that $u_m \to u$ and $\mathscr{D}^\alpha u_m \to v$ in $L_{1,\mathrm{loc}}(U)$. We conclude from the equality

$$(\omega_h * u_m)(x) = \int_{|y| \leq 1} u_m(x + hy)\omega(y)\, dy$$

that for $x \in \hat{U}_h$

$$\mathscr{D}^\alpha(\omega_h * u_m)(x) = \int_{|y| \leq 1} \mathscr{D}^\alpha u_m(x + hy)\omega(y)\, dy$$

$$\rightarrow \int_{|y| \leq 1} v(x + hy)\omega(y)\, dy = (\omega_h * v)(x). \tag{2.1}$$

Using the second representation in (1.4) for the function $\omega_h * u_m$, we find that for $x \in \hat{U}_h$

$$\mathscr{D}^\alpha(\omega_h * u_m)(x) = \frac{1}{h^n} \int_U u_m(z)\mathscr{D}_x^\alpha \omega\left(\frac{z - x}{h}\right) dz$$

$$\rightarrow \frac{1}{h^n} \int_U u(z)\mathscr{D}_x^\alpha \omega\left(\frac{z - x}{h}\right) dz = \mathscr{D}^\alpha(\omega_h * u)(x) \tag{2.2}$$

as $m \rightarrow \infty$. By (2.1) and (2.2),

$$(\omega_h * v)(x) = \mathscr{D}^\alpha(\omega_h * u)(x),$$

and this proves the lemma.

COROLLARY. *The generalized derivative $\mathscr{D}^\alpha u$ of a function $u \in L_{1,\mathrm{loc}}(U)$, if it exists, is unique to within values on a set of measure zero.*

Indeed, suppose that $v_1 = \mathscr{D}^\alpha u$ and $v_2 = \mathscr{D}^\alpha u$. Then

$$(\omega_h * v_1)(x) = \mathscr{D}^\alpha(\omega_h * u)(x) \text{ and } (\omega_h * v_2)(x) = \mathscr{D}^\alpha(\omega_h * u)(x),$$

i.e., $(\omega_h * v_1)(x) = (\omega_h * v_2)(x)$ for all $x \in U$. We have that $(\omega_h * v_1)(x) \rightarrow v_1(x)$ and $(\omega_h * v_2)(x) \rightarrow v_2(x)$ as $h \rightarrow 0$ for almost all $x \in U$. This leads us to conclude that $v_1(x) = v_2(x)$ for almost all $x \in U$, which is what was to be proved.

Let $l > 0$ be an integer, and let $p \geq 1$. We say that a function of class $L_{1,\mathrm{loc}}(U)$ belongs to the class $W^l_{p,\mathrm{loc}}(U)$ if u has in U all lth-order generalized derivatives, and each is in $L_{p,\mathrm{loc}}(U)$. A function u will be said to belong to the class $W^l_{\infty,\mathrm{loc}}(U)$ if $u \in W^l_{1,\mathrm{loc}}(U)$ and for every compact set $A \subset U$ and any α with $|\alpha| = l$

$$\|\mathscr{D}^\alpha u\|_{\infty,A} = \operatorname*{ess\,sup}_{x \in A} |\mathscr{D}^\alpha u(x)| < \infty.$$

Let $(u_m: U \rightarrow \mathbf{R}), m = 1, 2, \ldots,$ be an arbitrary sequence of functions in $W^l_{p,\mathrm{loc}}(U)$, where $1 \leq p < \infty$. We say that the sequence (u_m) converges in $W^l_{p,\text{loc}}(U)$ to a function $u \in W^l_{p,\text{loc}}(U)$ if the functions u_m converge to u in $L_{1,\text{loc}}(U)$ as $m \to \infty$, and $\mathscr{D}^\alpha u_m \to \mathscr{D}^\alpha u$ in $L_{p,\text{loc}}(U)$ as $m \to \infty$ for every α with $|\alpha| = l$. A sequence (u_m), $m = 1, 2, \ldots$, is said to be bounded in $W^l_{p,\text{loc}}(U)$, where $1 \leq p \leq \infty$, if it is bounded in $L_{p,\text{loc}}(U)$ and the sequence $(\mathscr{D}^\alpha u_m)$, $m = 1, 2, \ldots$, is bounded in $L_{p,\text{loc}}(U)$ for any α with $|\alpha| = l$.

If $u(x)$ has in U all generalized derivatives of an order $l \geq 1$, then it also has in U all generalized derivatives of any order $k < l$.

LEMMA 2.2. *A function $u \in L_{1,\text{loc}}(U)$ (U an open subset of $\mathbf{R}^n$) is in $W^l_{1,\text{loc}}(U)$ if and only if for every compact set $A \subset U$ and any α with $|\alpha| = l$ the family of functions $\mathscr{D}^\alpha(\omega_h * u)$ $0 < h < \text{dist}(A, CA)$, converges in $L_1(A)$ as $h \to 0$.*

For every function $u \in W^l_{p,\text{loc}}(U)$, where $1 \leq p < \infty$, there exists a sequence (u_m), $m = 1, 2, \ldots$, of C^∞-functions converging to $u(x)$ in $W^l_{p,\text{loc}}(U)$.

PROOF. The necessity of the condition follows from Lemma 2.1 and Theorem 1.4. We prove the sufficiency. Let $u \in L_{1,\text{loc}}(U)$ be such that for every compact set $A \subset U$ the family of functions $D^\alpha(\omega_h * u)$ converges in $L_1(A)$ to some function $v_A \in L_1(A)$ as $h \to 0$. It is required to prove that $u(x)$ then has a generalized derivative $\mathscr{D}^\alpha u$ on U. Let χ_h be the function whose existence is ensured by Lemma 1.1, and define $u_h(x) = \chi_h(x)(\omega_h * u)(x)$ for $x \in \hat{U}_h$, and $u_h(x) = 0$ for $x \notin U_h$. Then $u_h \in C^\infty(U)$. Let $U = \bigcup_1^\infty Q_\nu$ be the canonical subdivision of the open set U. We choose an arbitrary cube Q_ν and let $h < \frac{1}{2}\text{dist}(\overline{Q}_\nu, CU)$. Then $\overline{Q}_\nu \subset \overline{U}_{2h}$, and hence $u_h(x) = (\omega_h * u)(x)$ for all $x \in \overline{Q}_\nu$. By assumption, for every α with $|\alpha| = l$ there exists a function $v_{\nu,\alpha} \colon \overline{Q}_\nu \to \mathbf{R}$ such that $\mathscr{D}^\alpha(\omega_h * u)(x) \to v_{\nu,\alpha}$ in $L_1(\overline{Q}_\nu)$ as $h \to 0$. Let $v_\alpha(x) = v_{\nu,\alpha}(x)$ for $x \in Q_\nu, \nu = 1, 2, \ldots$. Every compact set $A \subset U$ can be covered by finitely many cubes Q_ν, and this implies that $\mathscr{D}^\alpha u_h = \mathscr{D}^\alpha(\omega_h * u) \to v_\alpha$ in $L_1(A)$ as $h \to 0$. Let $h_m = 1/m$ and $u_m = u_{h_m}$. We have that $u_m \to u$ and $\mathscr{D}^\alpha u_m \to v_\alpha$ in $L_{1,\text{loc}}(U)$ as $m \to \infty$. This implies that $v_\alpha = \mathscr{D}^\alpha u$. Since α such that $|\alpha| = l$ was taken arbitrarily, this proves that $u \in W^l_{1,\text{loc}}(U)$.

Assume now that $u \in W^l_{p,\text{loc}}(U)$, where $1 \leq p < \infty$. Let χ_h ($h > 0$) be the function whose existence is ensured by Lemma 1.1, and define $u_h(x) = \chi_h(x)(\omega_h * u)(x)$ for $x \in \hat{U}_h$, and $u_h(x) = 0$ for $x \in \hat{U}_h$. Then $u_h \in C^\infty(U)$, and the function u_h is defined for any $h > 0$. We let A be an arbitrary compact set and find an h_0 such that $A \subset \hat{U}_{2h_0}$. If $x \in \hat{U}_{2h}$ with $h < h_0$, then $\chi_h(x) = 1$, and hence $u_h(x) = (\omega_h * u)(x)$. It follows from this and Lemma 2.1 that

$$\mathscr{D}^\alpha u_h(x) = \mathscr{D}^\alpha(\omega_h * u)(x) = \omega_h * \mathscr{D}^\alpha u(x)$$

for $x \in \hat{U}_{2h}$. On the basis of Theorem 1.4,

$$\|\omega_h * \mathscr{D}^\alpha u - \mathscr{D}^\alpha u\|_{p,A} \to 0 \quad \text{as } h \to 0.$$

The desired sequence is obtained by setting $h_m = 1/m$ and $u_m = u_{h_m}$. This proves the lemma.

If the function u belongs to $C^l(U)$, where $l \geq 1$, then $u(x)$ also belongs to $W^l_{p,\text{loc}}(U)$, as follows directly from the definition above.

Let $u(x)$ be an function in $W^1_{1,\text{loc}}(U)$, where U is an open subset of $\mathbf{R}^n$. Then the symbol $\nabla u(x)$ will denote the vector

$$\left(\frac{\partial u}{\partial x_1}(x), \dots, \frac{\partial u}{\partial x_n}(x) \right),$$

where the derivatives are understood as generalized. The vector $\nabla u(x)$ is defined almost everywhere in U.

Let U be an open subset of $\mathbf{R}^n$, and $u: U \to \mathbf{R}$ a function in the class $W^l_{p,\text{loc}}(U)$. For an arbitrary measurable set $A \subset \mathbf{R}^n$

$$\|u\|_{L^l_p(A)} = \sum_{|\alpha|=l} \|\mathscr{D}^\alpha u\|_{p,A}, \tag{2.3}$$

$$\|u\|_{l,p,A} \equiv \|u\|_{W^l_p(A)} = \|u\|_{1,A} + |A|^{1-1/p+l/n} \|u\|_{L^l_p(A)}. \tag{2.4}$$

The factor in front of $\|u\|_{L^l_p(A)}$ was chosen so that both terms on the right side of (2.4) behave the same way under similarity transformations of the space. We say that u is in the class $W^l_p(U)$ if $\|u\|_{l,p,U} < \infty$. The set $W^l_p(U)$ of functions is a Banach space with the norm $u \mapsto \|u\|_{l,p,A}$. (Formally, the elements of $W^l_p(U)$ are classes of functions differing on a set of measure zero.)

2.2. Sobolev imbedding theorems. The assertions known as the Sobolev imbedding theorems are among the main results in the theory of the spaces $W^l_p(U)$. We formulate these theorems only for the case $l = 1$, the only case used in what follows.

A domain U in $\mathbf{R}^n$ is said to be *starlike* with respect to a ball $B(a, r) \subset U$ if $[x, y] \subset U$ for any points $y \in B(a, r)$ and $x \in U$. We say that a domain U in $\mathbf{R}^n$ is a domain of class S if it is bounded and is a union of finitely many domains which are starlike with respect to a ball. Obviously, a ball and a cube in $\mathbf{R}^n$ are domains starlike with respect to a ball, and in general any convex open subset of $\mathbf{R}^n$ is starlike with respect to a ball.

A continuous linear functional $L: W^1_p(U) \to \mathbf{R}$, where $p \geq 1$, is said to be *projective* if $L(\varphi) = 1$ for the function $\varphi \equiv 1$.

We have the following theorem.

THEOREM 2.1 (First Sobolev imbedding theorem). *Let U be a bounded domain of class S in $\mathbf{R}^n$, and let $1 \leq p \leq n$. Take a number $q \geq 1$ with $q < n/(n-p)$ if $p < n$, and q arbitrary if $p = n$. Then $W_p^1(U) \subset L_q(U)$, and there exists a constant $C_1 = C(n, p, q, U) < \infty$ such that for any function $u \in W_p^1(U)$*

$$\|u\|_{q,U} \leq C_1 \|u\|_{1,p,U}. \tag{2.5}$$

For every projective continuous linear functional $L: W_p^1(U) \to \mathbf{R}$

$$\|u - Lu\|_{q,U} \leq C_2 \|u\|_{L_p^1(U)}, \tag{2.6}$$

where $C_2 = C_2(n, p, q, L, U) < \infty$ is a constant.
   Every closed bounded set in $W_p^1(U)$ is compact in $L_q(U)$.

THEOREM 2.2. *Suppose that U is a domain of class S in $\mathbf{R}^n$, and let $p > n$. Then for every function $u \in W_p^1(U)$ it is possible to construct a continuous function u^* such that $u(x) = u^*(x)$ almost everywhere in U. For all $u \in W_p^1(U)$*

$$\operatorname*{ess\,sup}_{x \in U} |u(x)| \leq C_1 \|u\|_{1,p,U}, \tag{2.7}$$

where $C_1 < \infty$ and C_1 depends only on n, p, and U. For every projective continuous linear functional $L: W_p^1(U) \to \mathbf{R}$ there exists a constant $C_2 = C_2(p, n, L, U) < \infty$ such that for $u \in W_p^1(U)$

$$\operatorname*{ess\,sup}_{x \in U} |u(x) - L(u)| \leq C_2 \|u\|_{L_p^1(U)}. \tag{2.8}$$

Let U be a domain of class S in $\mathbf{R}^n$, and let $A \subset U$ be a measurable set with $|A| \neq 0$. For $u \in L_1^1(U)$ let

$$L_A(u) = \frac{1}{|A|} \int_A u(x) \, dx.$$

Further, let the function $\varphi \in C_0^\infty(U)$ be such that $\int_U \varphi(x) \, dx = 1$. For $u \in L_1(U)$ define

$$L_\varphi(U) = \int_U u(x) \varphi(x) \, dx.$$

Obviously, L_A and L_φ are projective continuous linear functionals on $W_p^1(U)$ for any p. These are the concrete functionals L which will be used in what follows.
   Let $\overset{\circ}{W}_p^1(U)$ denote the closure of $C_0^\infty(U)$ in $W_p^1(U)$.

THEOREM 2.3 (Poincaré theorem). *Let U be a bounded domain in $\mathbf{R}^n$. Then there exists a constant $C = C(p, n, U) < \infty$ such that for every $u \in \overset{\circ}{W}_p^1(U)$*

$$\|u\|_{p,U} \leq C \|u\|_{L_p^1(U)}. \tag{2.9}$$

Proofs of Theorems 2.1, 2.2, and 2.3 can be found, for example, in [96].

We make some remarks about the theorems formulated here. In the case $p < n$ the estimate (2.5) holds also for $q = n/(n - p)$. This precise result was obtained by Il'in [56]. For this value of q a closed bounded subset of $W_p^1(U)$ can fail to be compact in $L_1(U)$.

An arbitrary function of class $W_n^1(U)$ can be discontinuous in U. It is more convenient for us to give the corresponding example a little later.

The concept of a generalized derivative can be extended in a natural way to the case of vector-valued functions.

Let U be an open subset of $\mathbf{R}^n$, and let $f: U \to \mathbf{R}^k$ be a mapping. For every $x \in U$ we can write $f(x) = (f_1(x), \ldots, f_k(x))$; hence f determines k real-valued functions $f_1, \ldots, f_k$: the components of f. The mapping f will be said to belong to the class $L_{p,\mathrm{loc}}(U)$ or to the class $W_{p,\mathrm{loc}}^l(U)$ if each of $f_1, \ldots, f_k$ belongs to $L_{p,\mathrm{loc}}(U)$ or $W_{p,\mathrm{loc}}^l(U)$, respectively.

Let $f: U \to \mathbf{R}^k$ be a mapping of class $W_{p,\mathrm{loc}}^l(U)$, and let α be an arbitrary multi-index of length n and order $|\alpha| = l$. We have that $f(x) = (f_1(x), \ldots, f_k(x))$. Now define

$$\mathscr{D}^\alpha f(x) = (\mathscr{D}^\alpha f_1(x), \mathscr{D}^\alpha f_2(x), \ldots, \mathscr{D}^\alpha f_k(x)).$$

For an arbitrary measurable set $A \subset U$ such that $0 < |A| < \infty$, let

$$\|f\|_{l,p,\alpha} = \sum_{i=1}^k \|f_i\|_{l,p,A}.$$

We consider the case $l = 1$. Let $f: U \to \mathbf{R}^n$ be a mapping of class $W_{p,\mathrm{loc}}^1(U)$. Then for almost all $x \in U$ the linear mapping $L_x: \mathbf{R}^n \to \mathbf{R}^n$ with matrix the Jacobi matrix of f at x is defined. (Here the values of the generalized derivatives at x appear instead of the ordinary derivatives.) The linear mapping L_x will be denoted henceforth by $f'(x)$. For an arbitrary vector $h \in \mathbf{R}^n$ we have that

$$f'(x)h = \sum_{i=1}^n \frac{\partial f}{\partial x_i}(x)h_i$$

at each point $x \in U$ where $f'(x)$ is defined. The linear mapping $f'(x)$ will be called the derivative of f at x.

2.3. Tests for a function to belong to the class $W_{p,\mathrm{loc}}^1(U)$.

THEOREM 2.4. *Let u be a function of class $L_{1,\mathrm{loc}}(U)$.*

Assume that there exists a sequence $(u_m: U \to \mathbf{R})$ of functions in $W_{p,\mathrm{loc}}^1(U)$, where $1 < p \le \infty$, converging to $u(x)$ in $L_{1,\mathrm{loc}}(U)$ and bounded in $W_{p,\mathrm{loc}}^1(U)$. Then $u \in W_{p,\mathrm{loc}}^1(U)$.

REMARK. The condition $p \neq 1$ is essential here.

PROOF. Take an arbitrary compact set $A \subset U$ and let η be such that $0 < \eta < \text{dist}(A, CU)$. Let $V = U_\eta(A)$. The open set V is contained strictly inside U. In view of the condition in the theorem, the sequence

$$\left(\left\| \frac{\partial u_m}{\partial x_i} \right\|_{p,V} \right), \qquad m = 1, 2, \ldots,$$

is bounded for each $i = 1, \ldots, n$. Hence, there is a subsequence (u_{m_k}), $m_1 < m_2 < \ldots$, such that as $m \to \infty$ the functions $(\partial u_{m_k}/\partial x_i)$ converge weakly in $L_p(V)$ to some function $u \in L_p(V)$ for each $i = 1, \ldots, n$. Let ω be an arbitrary Sobolev averaging kernel, and take $0 < h < \eta$. Then the functions $\omega_h * u_{m_k}$ are all defined on V. For $x \in A$

$$\frac{\partial}{\partial x_i}(\omega_h * u_{m_k})(x) = \omega_h * \frac{\partial u_{m_k}}{\partial x_i}(x) = \frac{1}{h^n} \int_V \frac{\partial u_{m_k}}{\partial x_i}(z)\omega\left(\frac{z-x}{h}\right) dz.$$

The right-hand side of this equality tends to the limit

$$\frac{1}{h^n} \int_V u_i(z)\omega\left(\frac{z-x}{h}\right) dz = (\omega_h * u_i)(x)$$

as $k \to \infty$. On the other hand, for $x \in A$

$$\frac{\partial}{\partial x_i}(\omega_h * u_{m_k}(x)) = \frac{1}{h^n} \int_V u_{m_k}(z)\frac{\partial}{\partial x_i}\omega\left(\frac{z-x}{h}\right) dz,$$

and as $k \to \infty$ the right-hand side of this equality tends to the limit

$$\frac{1}{h^n} \int_V u(z)\frac{\partial}{\partial x_i}\omega\left(\frac{z-x}{h}\right) dz = \frac{\partial}{\partial x_i}(\omega_h * u).$$

Consequently, we get that for all $x \in A$

$$\frac{\partial}{\partial x_i}(\omega_h * u)(x) = (\omega_h * u_i)(x).$$

This implies that the functions $(\partial/\partial x_i)(\omega_h * u)(x)$ converge in $L_1(A)$ to u_i as $h \to 0$. Since $i = 1, \ldots, n$ and the compact set $A \subset U$ are arbitrary, we then get that for every compact set $A \subset U$ the functions $\partial(\omega_h * u)/\partial x_i$ converge in $L_1(A)$ as $h \to 0$. By Lemma 2.2, this implies that $u \in W^1_{1,\text{loc}}(U)$.

Suppose that $A \subset U$ is a compact set, $0 < \eta < \text{dist}(A, CU)$, $V = U_\eta(A)$, and $0 < h < \eta$. By what has been proved, $\partial(\omega_h * u)/\partial x_i \to u_i$ in $L_1(A)$ as $h \to 0$, where the u_i are functions in $L_p(A)$. On the other hand,

$$\frac{\partial}{\partial x_i}(\omega_h * u) \to \frac{\partial u}{\partial x_i}$$

in $L_1(A)$ as $h \to 0$. This implies that $\partial u/\partial x_i \in L_p(A)$, i.e., we get that $u \in W^1_{p,\text{loc}}(U)$. (Here $L_p(A)$ denotes the set of all functions $v: A \to \mathbf{R}$ such that

$$\|v\|_{\infty,A} = \sup_{x \in A} |v(x)| < \infty$$

in the case $p = \infty$.) The theorem is proved.

Another important characteristic of functions in $W^1_{1,\text{loc}}(U)$ is given by Theorem 2.5 below. First we introduce some notation.

Define a mapping p_i of $\mathbf{R}^n$ into $\mathbf{R}^{n-1}$ by $p_i(x) = (x_1, \ldots, x_{i-1}, x_{i+1}, \ldots, x_n)$ for $x = (x_1, \ldots, x_n)$ (p_i amounts to crossing out the ith coordinate of a point $x \in \mathbf{R}^n$). Let

$$\sigma_i(y) = (y_1, \ldots, y_{i-1}, 0, y_i, \ldots, y_{n-1}) \in \mathbf{R}^n$$

for $y = (y_1, \ldots, y_{n-1}) \in \mathbf{R}^{n-1}$. For every $y \in \mathbf{R}^{n-1}$ we have that $p_i(\sigma_i(y)) = y$, and for any $x \in \mathbf{R}^n$ we have that

$$\sigma_i[p_i(x)] = (x_1, \ldots, x_{i-1}, 0, x_{i+1}, \ldots, x_n).$$

Let G be an open subset of $\mathbf{R}$. We say that a function $u: G \to \mathbf{R}$ is *absolutely continuous on G* if it is absolutely continuous on every closed segment $[\alpha, \beta] \subset G$.

Let U be an open subset of $\mathbf{R}^n$, and let $U_i = p_i(U)$. The set U_i is open in $\mathbf{R}^{n-1}$. For $y \in U_i$ let U^i_y denote the set of all $t \in \mathbf{R}$ such that the point $\sigma_i(y) + te_i$ belongs to U. The set U^i_y is open and hence a union of at most countably many disjoint open intervals.

A function $u: U \to \mathbf{R}$ will be said to be *absolutely continuous in the Tonelli sense*, written $U \subset \text{ACT}(U)$, if it satisfies the following conditions:

A) For each $i = 1, \ldots, n$ the partial derivative $(\partial u/\partial x_i)(x)$, understood as the limit

$$\lim_{h \to 0} \frac{u(x + he_i) - u(x)}{h},$$

exists for almost all $x \in U$, and the function $\partial u/\partial x_i$ is locally integrable on U.

B) For each $i = 1, \ldots, n$ and almost all $y \in U_i$ the function $u_i(y, t) \equiv u(\sigma_i(y) + te_i)$ of the real variable t is absolutely continuous on U^i_y.

THEOREM 2.5 [110]. *For every open set $U \subset R^n$ and any function $u \in W^1_{1,\text{loc}}(U)$ there exists a function $u^* \subset \text{ACT}(U)$ such that $u(x) = u^*(x)$ almost everywhere on U. Further, the derivatives $(\partial u^*/\partial x_i)(x)$ of u^*, which exist almost everywhere on U, are the generalized derivatives on u.*

The class of functions absolutely continuous in the Tonelli sense and with derivatives $\partial u/\partial x_i$ in $L_{p,\text{loc}}(U)$ are often denoted by $\text{ACT}_p(U)$ in the literature.

We shall also need the following theorem on removal of singularities for functions in W_1^1.

THEOREM 2.6 [173]. *Let U be a bounded open subset of $\mathbf{R}^n$, and $E \subset U$ a set closed with respect to U and with projections $E_i = p_i(E)$ of measure zero in $\mathbf{R}^{n-1}$. Let $v \colon U\backslash E \to \mathbf{R}$ be a given function. If v is in $W_p^1(U\backslash E)$, then v belongs also to $W_p^1(U)$.*

PROOF. Let $U\backslash E = V, U_i = p_i(U)$, and $V_i = p_i(V)$. Then it is clear that $U_i \supset V_i \supset U_i \setminus E_i$. The sets U_i and V_i are open in $\mathbf{R}^{n-1}$. By Theorem 2.5, it can be assumed without loss of generality that $v \in \text{ACT}(U \setminus E)$. By hypothesis the projections E_i have measure zero in $\mathbf{R}^{n-1}$ and so E is a set of measure zero. The derivative $\partial v/\partial x_i$, understood as the limit

$$\lim_{h \to 0}[v(x + he_i) - v(x)]/h,$$

is defined almost everywhere on V, and hence almost everywhere on U. This derivative is integrable to the power p on V, and thus also on U. For $y \notin E_i$ the sets U_y^i and V_y^i coincide. Since $v \in \text{ACT}(V)$, the function $t \mapsto v[\sigma_i(y) + te_i]$ is absolutely continuous on V_y^i for almost all $y \in V_i$ by Theorem 2.5. Since $U_i \setminus V_i \subset E_i$, E_i is a set of measure zero, and $U_y^i = V_y^i$ for $y \notin E_i$, we thus get that the function $v[\sigma_i(y) + te_i]$ of the variable t is absolutely continuous on U_y^i for almost all $y \in U_i$. Hence, $v(x)$ satisfies the two conditions A) and B) in the definition of a function absolutely continuous in the Tonelli sense. Further, its derivatives are integrable to the power p on U, and hence $v \in W_p^1(U)$. The theorem is proved.

REMARK. The condition of the theorem concerning E holds, in particular, if E is a set whose $(n-1)$-dimensional Hausdorff measure is zero.

THEOREM 2.7. *Every function $u \colon U \to \mathbf{R}$ of class $C^{0,1}(U)$ belongs to $W_{\infty,\text{loc}}^1(U)$.*

PROOF. The theorem can be obtained as a corollary to Theorem 2.5. However, the discussion of certain details involving measurability questions turns out to be somewhat cumbersome. We give a derivation based on Theorem 2.4.

Let $u \in C^{0,1}(U)$. This means that u is continuous and satisfies a Lipschitz condition on every compact subset of U.

Let χ_h, where $h > 0$, be the function defined in Lemma 1.1, and let $v_h(x) = \chi_h(x)(\omega_h * u)(x)$ for $x \in \hat{U}_h$, and $v_h(x) = 0$ for $x \notin \hat{U}_h$. Define $u_m(x) = v_{(1/m)}(x)$. Then $u_m \in C^\infty(U)$ for each $m = 1, 2, \ldots$, and $u_m \to u$ in $L_{1,\text{loc}}(U)$ as $m \to \infty$ (the convergence is even locally uniform in U). Let x_0 be an arbitrary point in U, and let $r > 0$ be such that $\overline{B}(x_0, r) \subset U$.

Take an m_0 with $1/m_0 < r/2$. For $m \geq m_0$ we have that $B(x_0, r/2) \subset \hat{U}_{2h}$, where $h = 1/m$, and thus for all $x \in B(x_0, r/2)$

$$u_m(x) = (\omega_{1/m} * u)(x) = \int_{|y| \leq 1} u(x + y/m)\omega(y)\,dy.$$

The ball $\overline{B}(x_0, r)$ is in U; hence there exists an $L < \infty$ such that

$$|u(x') - u(x'')| \leq L|x' - x''|$$

for any points $x', x'' \in \overline{B}(x_0, r)$. If $x \in B(x_0, r/2)$ and $|y| \leq 1$, then $x + y/m$ lies in $\overline{B}(x_0, r)$, and hence for any $x_1, x_2 \in B(x_0, r/2)$ we have

$$|u_m(x_1) - u_m(x_2)| \leq \int_{|y| \leq 1} \left| u\left(x_1 + \frac{y}{m}\right) - u\left(x_2 + \frac{y}{m}\right) \right| \omega(y)\,dy$$

$$\leq L|x_1 - x_2| \int_{|y| \leq 1} \omega(y)\,dy = L|x_1 - x_2|.$$

The function u_m belongs to the class C^∞. It thus follows from this inequality that $|\nabla u_m(x)| \leq L$ for all $x \in B(x_0, r/2)$ and all $m \geq m_0$.

Let $A \subset U$ be a compact set. By what has been proved, for every $x \in A$ there exist a ball $B(x, \delta) \subset U$, a constant $L < \infty$, and an index m_0 such that $|\nabla u_m(y)| \leq L$ for all $y \in B(x, \delta)$ when $m \geq m_0$. By the Borel theorem, A can be covered by finitely many such balls. Let $\overline{m}$ be the largest index m_0 corresponding to the balls forming such a covering, and let $\overline{L}$ be the largest of the corresponding constants L. For $m \geq \overline{m}$ it is obvious that $|\nabla u_m(x)| \leq \overline{L}$ for all $x \in A$. Since the compact set $A \subset U$ is arbitrary, this proves that the sequence (u_m), $m = 1, 2, \ldots$, is bounded in $W^1_{\infty, \text{loc}}(U)$. On the basis of Theorem 2.4, this implies that $u \in W^1_{\infty, \text{loc}}(U)$, which is what was to be proved.

REMARK. Theorem 2.7 admits a converse. Namely, if $u \in W^1_{\infty, \text{loc}}(U)$, then $u \in C^{0,1}(U)$ (more precisely, $u(x)$ is equivalent to a function in $C^{0,1}$ in the sense of Lebesgue measure in $\mathbf{R}^n$). This assertion can be obtained by the simple use of a Sobolev averaging. We leave the details of the corresponding arguments to the reader.

2.4. Transformations of functions with generalized derivatives. For all the functions of class $W^1_{1,\text{loc}}$ in Theorems 2.8 and 2.9 below, as well as for their derivatives, assume that their values are equal to the natural value at each point where it exists. The results in this subsection will be needed below in connection with the study of functions with generalized derivatives on manifolds.

THEOREM 2.8. *Let U and V be bounded open subsets of $\mathbf{R}^n$, and let $\sigma: U \to \mathbf{R}^n$ be a diffeomorphism of class C^1 such that $\sigma(U) = V$ and all* *the first-order derivatives of σ are bounded in U. Then for every function* $v \in W^1_{1,\text{loc}}(V)$ *the function* $u = v \circ \sigma$ *belongs to* $W^1_{1,\text{loc}}(U)$, *and its generalized derivatives can be expressed in terms of the derivatives of* $v(x)$ *by the formulas*

$$\frac{\partial u}{\partial x_i} = \sum_{j=1}^{n} \left(\frac{\partial v}{\partial y_j} \circ \sigma \right) \frac{\partial \sigma_j}{\partial x_i}. \tag{2.10}$$

PROOF. Equality (2.10) is true when $v \in C^1$. Let v be an arbitrary function of class $W^1_{1,\text{loc}}(V)$, and let

$$u_i = \sum_{j=1}^{n} \left(\frac{\partial v}{\partial y_j} \circ \sigma \right) \frac{\partial \sigma_j}{\partial x_i}.$$

It is required to prove that $u_i = \partial u / \partial x_i$ for every $i = 1, \ldots, n$. We construct a sequence (v_m), $m = 1, 2, \ldots$, of C^∞-functions converging to v in $W^1_{1,\text{loc}}(U)$, and let $u_m = v_m \circ \sigma$. Let $A \subset U$ be an arbitrary compact set. Since $\sigma \in C^1(U)$, the functions $\partial \sigma_j / \partial x_i$ are bounded on A. Because σ is a diffeomorphism, $\det \sigma'(x) \neq 0$ for all $x \in U$, and hence there exists a $\delta > 0$ such that $|\det \sigma'(x)| \geq \delta$ for all $x \in A$. From this we get

$$\left(\int_A \left| \frac{\partial u_m}{\partial x_i} - u_i \right|^p dx \right)^{1/p}$$

$$\leq \frac{C}{\delta^{1/p}} \sum_{i=1}^{n} \left(\int_A \left| \frac{\partial v_m}{\partial y_j} \circ \sigma - \frac{\partial v}{\partial y_j} \circ \sigma \right|^p |\det \sigma'(x)| \, dx \right)^{1/p}$$

$$= C \sum_{j=1}^{n} \left(\int_{\sigma(A)} \left| \frac{\partial v_m}{\partial y_j} - \frac{\partial v}{\partial y_j} \right|^p dy \right)^{1/p},$$

i.e.,

$$\left\| \frac{\partial u_m}{\partial x_i} - u_i \right\|_{p,A} \leq C \sum \left\| \frac{\partial v_m}{\partial y_j} - \frac{\partial v}{\partial y_j} \right\|_{p,\sigma(A)}$$

The right-hand side of the last inequality tends to zero as $m \to \infty$, and hence

$$\left\| \frac{\partial u_m}{\partial x_i} - u_i \right\|_{p,A} \to 0 \quad \text{as } m \to \infty.$$

Since the compact set $A \subset U$ is arbitrary, this proves that $\partial u_m / \partial x_i \to u_i$ in $L_{p,\text{loc}}(U)$ as $m \to \infty$. It is also obvious that $u_m \to u$ in $L_{p,\text{loc}}(U)$ as $m \to \infty$. This implies that $u \in W^1_{p,\text{loc}}(U)$ and $u_i = \partial u / \partial x_i$ for $i = 1, \ldots, n$. The theorem is proved.

THEOREM 2.9. *Let U be an open subset of $\mathbf{R}^n$, V an open subset of $\mathbf{R}^k$, and $f: U \to V$ a continuous mapping in $W^1_{p,\mathrm{loc}}(U)$. Then for any C^1-function $v: V \to \mathbf{R}$ the composition $u = v \circ f$ is a function of class $W^1_{p,\mathrm{loc}}(U)$, and its generalized derivatives are expressed in terms of the derivatives of v and f by the same formulas as in the case of smooth functions, i.e.,*

$$\frac{\partial u}{\partial x_i} = \sum_{j=1}^{k} \left(\frac{\partial v}{\partial y_j} \circ f \right) \frac{\partial f_j}{\partial x_i}$$

for each $i = 1, \ldots, n$.

PROOF. Let

$$u_i = \sum_{j=1}^{k} \left(\frac{\partial v}{\partial y_j} \circ f \right) \frac{\partial f_j}{\partial x_i},$$

$i = 1, \ldots, n$. It must be proved that $u_i = \partial u / \partial x_i$ for all $i = 1, \ldots, n$. Let ω be an arbitrary Sobolev averaging kernel, and let $f_h = \omega_h * f = (\omega_h * f_1, \ldots, \omega_h * f_k)$. Then $f_h \to f$ in $W^1_{p,\mathrm{loc}}(U)$ as $h \to 0$. Further, since f is continuous, the convergence $f_h \to f$ is also locally uniform as $h \to 0$. Let $A \subset U$ be an arbitrary compact set, and let $B = f(A)$; $B \subset V$ is compact. Take an $\eta > 0$ such that $\overline{U}_\eta(B) \subset V$, and let $H = \overline{U}_\eta(B)$; H is compact. We have that $f_h \to f$ uniformly on A as $h \to 0$; hence there exists a δ such that $|f_h(x) - f(x)| < \eta$ for all $x \in A$ if $0 < h < \delta$. Also, $f_h(x) \in H$ for $x \in A$ when $0 < h < \delta$. Since H is compact, the function $\partial v / \partial y_j$ is uniformly continuous on H. For $0 < h < \eta$ and $x \in A$

$$\left| \frac{\partial v}{\partial y_j}[f_h(x)] - \frac{\partial v}{\partial y_j}[f(x)] \right| \le \omega_j(|f_h(x) - f(x)|), \qquad (2.11)$$

where ω_j is the modulus of continuity of $\partial v / \partial y_j$ on H. This gives us that

$$\frac{\partial v}{\partial y_j} \circ f_h \to \frac{\partial f}{\partial y_j} \circ f$$

uniformly on A as $h \to 0$. We have that

$$\left| \left(\frac{\partial v}{\partial y_j} \circ f_h \right) \frac{\partial f_{j,h}}{\partial x_i} - \left(\frac{\partial v}{\partial y_j} \circ f \right) \frac{\partial f_j}{\partial x_i} \right| \le \left| \left(\frac{\partial v}{\partial y_j} \circ f_h \right) \left(\frac{\partial f_{j,h}}{\partial x_i} - \frac{\partial f_j}{\partial x_i} \right) \right|$$
$$+ \left| \frac{\partial v}{\partial y_j} \circ f_h - \frac{\partial v}{\partial y_j} \circ f \right| \left| \frac{\partial f_j}{\partial x_i} \right|.$$
$$(2.12)$$

Further,

$$\left| \frac{\partial v}{\partial y_j}[f_h(x)] \right| \le M < \infty$$

for $x \in A$ and $0 < h < \delta$, where M = const. We conclude from (2.11) and (2.12) that for $0 < h < \delta$

$$\left\| \frac{\partial}{\partial x_i}(v \circ f_h) - u_i \right\|_{p,A} \le M \sum_{j=1}^{k} \left\| \frac{\partial f_{j,h}}{\partial x_i} - \frac{\partial f_j}{\partial x_i} \right\|_{p,A}$$

$$+ \omega(\|f_h - f\|_{\infty,A}) \sum_{j=1}^{k} \left\| \frac{\partial f_j}{\partial x_i} \right\|_{p,A} .$$

The right-hand side of the inequality tends to zero as $h \to 0$, and thus the derivatives of the functions $v \circ f_h$ converge to the functions u_i in $L_{p,\mathrm{loc}}(U)$. Since the functions $v \circ f_h$ themselves converge locally uniformly to the function $u = v \circ f$, this implies that $u_i \in L_{p,\mathrm{loc}}(U)$ and $u_i = \partial u / \partial x_i$. The theorem is proved.

2.5. Dependence of the coefficients in the imbedding theorems on the size of the domain. In applications of Theorems 2.1, 2.2, and 2.3 it is useful for certain standard domains (for example, for a cube, ball, etc.) to know how the constants in (2.6), (2.8), and (2.9) depend on the sizes of the domain. This question can be answered by a simple change of variables. We confine ourselves to the case of a ball, since all the arguments go through similarly for a cube.

Let $B = B(a,r)$ be the ball about a with radius $r > 0$ in $\mathbf{R}^n$, let $B_1 = B(0,1)$, and let σ be the transformation $x \mapsto a + rx$ of $\mathbf{R}^n$. Obviously, $\sigma(B_1) = B$. The Jacobian of σ is equal to r^n. If $u \in W_p^1(B)$, then $v = u \circ \sigma \in W_p^1(B_1)$, as follows from Theorem 2.8. Further,

$$\frac{\partial v}{\partial x_i}(x) = \frac{\partial u}{\partial y_i}(a + rx)r$$

for every $i = 1, \dots, n$, and thus

$$\int_{B_1} \left| \frac{\partial v}{\partial x_i}(x) \right|^p dx = \int_{B_1} \left| \frac{\partial u}{\partial y_i}(a + rx) \right|^p r^p dx$$

$$= r^{p-n} \int_{B_1} \left| \frac{\partial u}{\partial y_i}(a + rx) \right|^p r^n dx = r^{p-n} \int_{B(a,r)} \left| \frac{\partial u}{\partial y_i} \right|^p dx.$$

From this,

$$\|v\|_{L_p^1(B_1)} = r^{1-n/p}\|u\|_{L_p^1(B)}. \tag{2.12}$$

Further,

$$\|v\|_{L_q(B_1)} = \left(\frac{1}{r^n} \int_{B_1} |u(a + rx)|^q r^n \, dx \right)^{1/q} = r^{-n/q}\|u\|_{L_q(B)}. \tag{2.13}$$

Next, let

$$\bar{u}_B = \frac{1}{|B|} \int_B u(y)\, dy.$$

Obviously,

$$\bar{v}_{B_1} = \frac{1}{|B_1|} \int_{B_1} v(x)\, dx = \bar{u}_B.$$

Let $p \le n$, and let q be such that $1 \le q$ and $(n - p)q \le n$. Choosing the L in (2.6) to be the linear functional $v \to \bar{v}_{B_1}$, we get that

$$\|v - \bar{v}_{B_1}\|_{L_q(B)} \le C_0 \|v\|_{L_p^1(B_1)}.$$

In view of (2.12) and (2.13), this implies that for the given p and q

$$\|u - \bar{u}_B\|_{L_q(B)} \le Cr^{1+n/q-n/p} \|u\|_{L_p^1(B)}, \tag{2.14}$$

where the constant C depends only on p, q, and r.

In the case $p > n$ we get the estimate

$$\operatorname*{ess\,sup}_{x \in B(a,r)} |u(x) - \bar{u}_B| \le Cr^{1-n/p} \|u\|_{L_p^1(B)} \tag{2.15}$$

from (2.6) in a similar way.

Finally, for arbitrary $p \ge 1$ inequality (2.8) allows us to conclude that for every $u \in \overset{\circ}{W}{}_p^1(B(a,r))$

$$\|u\|_{L_p(B(a,r))} \le Cr \|u\|_{L_p^1(B(a,r))}. \tag{2.16}$$

Similar estimates hold also in the case when a cube $Q(a,r)$ is taken instead of a ball in $\mathbf{R}^n$.

2.6. A theorem on differentiability of $W_{p,\mathrm{loc}}^1$-functions almost everywhere. Let U be an open subset of $\mathbf{R}^n$. A mapping $f: U \to \mathbf{R}^m$ is said to be *differentiable* at a point $a \in U$ if there exist a linear mapping $L: \mathbf{R}^n \to \mathbf{R}^m$ and a function $\alpha: U \to \mathbf{R}^n$ with $\alpha(x) \to 0$ as $x \to a$ such that

$$f(x) = f(a) + L(x - a) + \alpha(x)|x - a|$$

for all $x \in U$. In this case L is called the *derivative of f at a* and denoted by $f'(a)$.

A mapping $f: U \to \mathbf{R}^m$ is said to be *l-fold differentiable* at a point $a \in U$ if there exist functions $P: \mathbf{R}^n \to \mathbf{R}^m$ and $\alpha_l: \mathbf{R}^n \to \mathbf{R}^m$, with P a polynomial of degree at most l representable in the form $P(x) = \sum_{|\lambda| \le l} A_\lambda X^\lambda$, the A_λ being vectors in $\mathbf{R}^m$, and with $\alpha_l(x) \to 0$ as $x \to a$, such that

$$f(x) = P(x - a) + |x - a|^l \alpha_l(x)$$

for all $x \in U$.

One of the topics in the classical theory of functions of a real variable concerns theorems on differentiability almost everywhere of functions satisfying diverse conditions. A classical example is the famous theorem of Lebesgue on differentiability almost everywhere of a monotone function of one variable. Of other results in this direction we mention the theorem of Aleksandrov [11] on two-fold differentiability almost everywhere of a convex function of n variables, the theorem of Calderón and Zygmund on l-fold differentiability almost everywhere of the functions in $W^l_{p,\text{loc}}$ when $lp > n$, and the theorem of Stepanov [163] establishing necessary and sufficient conditions for differentiability almost everywhere of a function $f: U \to \mathbf{R}$, where $U \subset \mathbf{R}^n$. In the author's paper [131] it is proved that if differentiability is understood in a certain generalized sense, then every function in $W^l_{p,\text{loc}}$ is l-fold differentiable almost everywhere. As shown in [131], this result can serve as a means for establishing various theorems on differentiability almost everywhere (now in the sense of the definition given above) of functions in diverse classes. In particular, the Aleksandrov theorem and the Calderón-Zygmund theorem can be proved in this way.

Below we shall need a special case of a theorem in [131]. Let us state it and note some corollaries. The proof will be given in Chapter III.

Take an arbitrary open set U in $\mathbf{R}^n$.

Let G be a bounded measurable subset of $\mathbf{R}^n$ containing 0 as an interior point, and let $\mathfrak{R}$ be a metric space whose elements are measurable functions on G with values in $\mathbf{R}^m$. Let $f: U \to \mathbf{R}^m$ be a given function and $L: \mathbf{R}^n \to \mathbf{R}^m$ a given linear mapping. Take an arbitrary point $x \in U$. Since G is bounded, there exists an $\eta > 0$ such that if $0 < h < \eta$, then $x + hX \in U$ for any $X \in G$. The linear mapping L is called *the differential of f at x in the sense of convergence in* $\mathfrak{R}$ if for $0 < h < \eta$ the function

$$\Delta_h: X \mapsto \frac{f(x + hX) - f(x)}{h} - L(X)$$

is in $\mathfrak{R}$ and converges in the sense of the metric of $\mathfrak{R}$ to the function identically equal to zero as $h \to 0$.

If the given condition holds, then we also say that f is *differentiable at x in the sense of convergence in* $\mathfrak{R}$.

Let $\mathfrak{R}$ be the collection of all bounded measurable functions $F: G \to \mathbf{R}^m$, and let a metric be defined in $\mathfrak{R}$ by the norm

$$\|F\|_{M(G)} = \sup_{x \in G} |F(x)|.$$

Then L is the differential of a function $f: U \to \mathbf{R}^m$ in the sense of convergence in $M(G)$ if and only if

$$f(y) - f(x) - L(y - x) = o(|y - x|)$$

as $y \to x$. Indeed, let $(f(x + hx) - f(x))/h = L_h(x)$. Assume that L is the differential of f at x in the sense of convergence in $\mathfrak{R}$. Let r and R be such that $\overline{B}(0, r) \subset G \subset B(0, R)$. Define

$$\Delta(h) = \|L_h - L\|_{M(G)} = \sup_{X \in G} |L_h(X) - L(X)|.$$

Then $\Delta(h) \to 0$ as $h \to 0$. Let $h = |y - x|/r$ and $X = r(y - x)/|y - x|$. Then $X \in \overline{B}(0, r)$, and hence

$$|f(x + hX) - f(x) - hL(X)| \le h\Delta h,$$

i.e.,

$$|f(y) - f(x) - L(y - x)| \le \frac{1}{r}|y - x|\Delta\left(\frac{|y - x|}{r}\right)$$

and thus

$$|f(y) - f(x) - L(y - x)| = o(|y - x|) \quad \text{as } y \to x.$$

Conversely, let $|f(y) - f(x) - L(y - x)| = o(|y - x|)$ as $y \to x$. This means that

$$|f(y) - f(x) - L(y - x)| \le |y - x|\delta(|y - x|),$$

where $\delta(r) \to 0$ as $r \to 0$. The function $\delta(r)$ can be assumed to be nondecreasing. (This can always be made the case by replacing $\delta(r)$ by $\delta_1(r) = \sup_{0 < t \le r} \delta(t)$ if necessary; it is clear that $\delta(r) \le \delta_1(r)$ for all $r > 0$, δ_1 is nondecreasing, and $\delta_1(r) \to 0$ as $r \to 0$ if $\delta(r) \to 0$ as $r \to 0$.) Setting $y = x + hX$, where $X \in G$, we get that

$$|L_h(X) - L(X)| \le \delta(hR),$$

which implies that $\|L_h - L\|_{M(G)} \to 0$ as $h \to 0$.

THEOREM 2.10. *Let $f: U \to \mathbf{R}^m$ be a function in $W_p^1(U)$, where $U \subset R^n$ is an open set and G is a bounded open set such that $0 \in G$. Then for almost all $x \in U$ the linear mapping*

$$X \mapsto \sum_{i=1}^{n} \frac{\partial f}{\partial x_i} X_i$$

is the differential of f at x in the sense of convergence in $W_p^1(G)$.

A proof of Theorem 2.10 will be given in Chapter III, §4.1. We remark that the case of arbitrary G can be reduced trivially to the case when G is a ball.

We prove the following assertion as an example showing how Theorem 2.10 can be used in proving differentiability almost everywhere of functions in certain concrete classes.

COROLLARY 1 [22]. *Every function $f: U \to \mathbf{R}^m$ of class $W^1_{p,\text{loc}}(U)$, where $p > n$ (U an open subset of $\mathbf{R}^n$), is differentiable almost everywhere in U.*

REMARK. Here and below, every function in $W^1_{p,\text{loc}}$, where $p > n$, is assumed to be continuous. According to Theorem 2.2, this can always be made the case by changing the values of a function on a set of measure zero if necessary.

PROOF. Suppose that $x \in U$ is such that the function

$$L: X \mapsto \sum_{i=1}^{n} \frac{\partial f}{\partial x_i}(x) X_i$$

is the differential of f at x in the sense of convergence in $W^1_p(B(0,1))$. By the theorem, almost all points $x \in U$ are such points. This means that as $h \to 0$

$$\|L_h - L\|_{W^1_p[B(0,1)]},$$

where $L_h(X) = \frac{1}{h}[f(x + hX) - f(x)]$. By Theorem 2.2, this implies that

$$\|L_h - L\|_{M[B(0,1)]} \to 0 \quad \text{as } h \to 0,$$

and hence L is the differential of f at x in view of the remarks above. The corollary is proved.

COROLLARY 2. *Suppose that U is an open subset of $\mathbf{R}^n$. Every mapping $f: U \to \mathbf{R}^n$ of class $C^{0,1}$ is differentiable in U almost everywhere.*

PROOF. Let $f: U \to \mathbf{R}^m$ be a mapping of class $C^{0,1}$. Then, using Theorem 2.7, we get that $f \in W^1_{\infty,\text{loc}}(U)$, and hence $f \in W^1_{p,\text{loc}}(U)$ for every $p > n$. The assertion to be proved thus follows immediately from Corollary 1.

§3. Möbius transformations

3.1. Motions and similarity transformations of a Euclidean space. Below, E denotes a Euclidean vector space, i.e., a vector space in which each pair of vectors x, y is assigned a number $\langle x, y \rangle$ called the inner product of x and y and satisfying well-known axioms. (Namely: 1) $\langle x, y \rangle = \langle y, x \rangle$ $\forall x, y \in \mathbf{E}$; 2) $\langle \lambda_1 x_1 + \lambda_2 x_2, y \rangle = \lambda_1 \langle x_1, y \rangle + \lambda_2 \langle x_2, y \rangle$ $\forall x_1, x_2 \in \mathbf{E}, \lambda_1, \lambda_2 \in \mathbf{R}$; 3) $\langle x, x \rangle \geq 0$ $\forall x \in \mathbf{E}$, and $\langle x, x \rangle = 0 \Leftrightarrow x = 0$.) It is assumed that the dimension of E is at least 2; in other respects it can be arbitrary. In particular, E can also be infinite. In what follows we deal only with mappings of finite-dimensional spaces, but the assumption of finite-dimensionality does not lead to any essential simplifications here.

For any x, $y \in \mathbf{E}$ we have that $|\langle x, y \rangle| \leq \|x\| \|y\|$, with equality if and only if x and y are linearly dependent.

A subset P of $\mathbf{E}$ will be called a *hyperplane* if $P = \{x \in \mathbf{E} | \langle a, x \rangle - k = 0\}$ for some vector $a \neq 0$ and number k. Here the vector a and the number k are uniquely determined to within a constant factor, i.e., if $a_1 \in \mathbf{E}$ and $k_1 \in \mathbf{R}$ are such that $P = \{x \in \mathbf{E} | \langle a_1, x \rangle - k_1 = 0\}$, then there exists a $\lambda \in \mathbf{R}$, $\lambda \neq 0$, such that $a_1 = \lambda a$ and $k_1 = \lambda k$. Of course, this assertion is hardly worth stating explicitly in the case of finite-dimensional $\mathbf{E}$, but since the context here is "dimensionless", some proof must be given. Accordingly, let $\langle a, x \rangle - k = 0$ be the equation of the given plane P. Assume that the nonzero vector a_1 and the number k_1 are such that all the points of P satisfy the equation $\langle a_1, x \rangle - k_1 = 0$. Let $x_0 = ak/|a|^2$. Then $x_0 \in P$, and hence $\langle a_1, x_0 \rangle - k_1 = 0$. Let

$$y_0 = x_0 + a_1 - \frac{\langle a_1, a \rangle}{|a|^2} a.$$

Then $\langle a, y_0 \rangle - k = 0$, i.e., $y_0 \in P$, and thus $\langle a_1, y_0 \rangle - k_1 = 0$. From this we get after obvious transformations that $|a|^2 |a_1|^2 = (\langle a, a_1 \rangle)^2$, and, consequently, $a_1 = \lambda a$. Further, $k_1 = \langle a_1, x_0 \rangle = \lambda \langle a, x_0 \rangle = \lambda k$, which is what was required.

A subset S of $\mathbf{E}$ is called a *sphere* if there exist a point $a \in \mathbf{E}$ and a number $r > 0$ such that $S = \{x \in \mathbf{E} | |x - a| = r\}$. The point a and the number r are uniquely determined by the specification of the sphere S, i.e., if $a_1 \in \mathbf{E}$ and $r > 0$ are such that $S = \{x \in \mathbf{E} | |x - a_1| = r_1\}$, then $a_1 = a$ and $r_1 = r$. Indeed, assume that $a \neq a_1$. Then $h = |a - a_1| \neq 0$, and the points $x_1 = a + (r/h)(a_1 - a)$ and $x_2 = a - (r/h)(a_1 - a)$ belong to S. This implies that $|x_1 - a_1| = |x_2 - a_1| = r_1$. We have that $|x_1 - a_1| = |r - h|$ and $|x_2 - a_1| = |r + h|$, and we get a contradiction, since $|r - h| \neq |r + h|$ in view of the fact that $h \neq 0$ and $r > 0$. Thus, $a_1 = a$. For every $x \in S$ we have that $r_1 = |x - a_1| = |x - a| = r$, i.e., $r_1 = r$.

A linear transformation $P: \mathbf{E} \to \mathbf{E}$ is said to be *orthogonal* if $P(\mathbf{E}) = \mathbf{E}$ and $|Px| = |x|$ for any $x \in \mathbf{E}$. (In the case of finite-dimensional $\mathbf{E}$ the condition $P(\mathbf{E}) = \mathbf{E}$ automatically follows from the other conditions imposed on P.) If $P: \mathbf{E} \to \mathbf{E}$ is an orthogonal transformation, then $\langle Px, Py \rangle = \langle x, y \rangle$ for any $x, y \in \mathbf{E}$.

A mapping $\mathscr{D}: \mathbf{E} \to \mathbf{E}$ is called a *motion* if it is bijective and $|\mathscr{D}x - \mathscr{D}y| = |x - y|$ for any $x, y \in \mathbf{E}$.

THEOREM 3.1. *Every motion $\mathscr{D}$ of $\mathbf{E}$ admits the representation $Dx = a + Px$, where $a \in \mathbf{E}$ and P is an orthogonal transformation of $\mathbf{E}$.*

PROOF. A parallel translation, i.e., a transformation $x \mapsto a + x$, is a motion. A composition of two motions is clearly a motion. Let $\mathscr{D}: \mathbf{E} \to \mathbf{E}$ be an arbitrary motion of $\mathbf{E}$, and let $a = \mathscr{D}(0)$ and $P(x) = \mathscr{D}x - a$. The mapping P is a motion. We prove that P is an orthogonal transformation of E. For this it suffices to establish that P is linear. First of all, note that $P(0) = 0$.

Take vectors $u, v \in$ E such that $u \neq 0$, $v \neq 0$, and $|u + v| = |u| + |v|$. Then $v = \lambda u$, where $\lambda > 0$.

Let x_0, x_1, and x_2 be three different points of E lying on a single line. Define $x_i' = P(x_i)$, $i = 0, 1, 2$. We prove that the points x_0', x_1', and x_2' then also lie on a single line and are in the same order as x_0, x_1, and x_2. Assume for definiteness that x_1 is in the segment $[x_0, x_2]$. Let $u = x_1' - x_0'$ and $v = x_2' - x_1'$. Then

$$u + v = x_2' - x_0', \qquad |u| = |x_1' - x_0'| = |x_1 - x_0|,$$
$$|v| = |x_2' - x_1'| = |x_2 - x_1|$$

and hence

$$|u| + |v| = |x_1 - x_0| + |x_2 - x_1| = |x_2 - x_0| = |u + v|.$$

From this, $v = \lambda u$, where $\lambda > 0$. This allows us to conclude that x_1' belongs to $[x_0', x_2']$.

Take any $x \in$ E and $\lambda \in$ R. We prove first that $P(\lambda x) = \lambda P(x)$. This holds when $\lambda = 0$, or $\lambda = 1$, or $x = 0$. Assume that $x \neq 0$, $\lambda \neq 0$, and $\lambda \neq 1$. Let $\lambda x = y$, $x' = P(x)$, and $y' = P(y)$. We have that $|y'| = |y' - 0| = |y - 0| = |\lambda||y|$. The points x, 0, and y lie on a single line. Hence, the points x', 0, and y' also lie on a single line. If $\lambda < 0$, then 0 lies interior to the segment $[x, y]$, and thus in this case 0 also lies interior to $[x', y']$. This enables us to conclude that $y' = \mu x'$, where $\mu < 0$. Since $|y'| = |\lambda||x| = |\lambda||x'|$, it follows that $|\mu| = |\lambda|$, and hence $\mu = \lambda$. If $\lambda > 0$, then either x lies between 0 and y (when $\lambda > 1$) or y lies between 0 and x (for $\lambda < 1$). This implies that the points 0, x', and y' are arranged in the same way, which allows us to conclude that $y' = \lambda x'$ in this case.

Let x and y be two distinct points of E, z the midpoint of the segment $[x, y]$, $z = (x + y)/2$, $x' = P(x)$, $y' = P(y)$, and $z' = P(z)$. Then z' lies on $[x', y']$, and $|z' - x'| = |y' - z'|$, i.e., z' is the midpoint of $[x', y']$, $z' = (x' + y')/2$. We thus get that

$$P\left(\frac{x + y}{2}\right) = \frac{1}{2}[P(x) + P(y)].$$

By what was proved, this implies that $P(x + y) = 2P((x + y)/2) = P(x) + P(y)$. The linearity of P is established.

Since P is bijective and $|P(x)| = |x|$ for every $x \in$ E, this establishes that P is orthogonal. For every $x \in$ E we have that $\mathcal{D}(x) = a + P(x)$, and the theorem is proved.

Let E_1 and E_2 be arbitrary vector spaces. A mapping $f: E_1 \to E_2$ is said to be *affine* if there exist a vector $k \in E_2$ and a linear mapping $L: E_1 \to E_2$ such that $f(x) = k + L(x)$ for every $x \in E_1$. The mapping L is called the *linear part* of the mapping f. For every $x \in E_1$ we have that $L(x) = f(x) - f(0)$; hence L is uniquely determined by f.

According to Theorem 3.1, every motion of a Euclidean space E is an affine transformation of E.

The collection of all orthogonal transformations of E forms a group, which we denote by $O(E)$. In the case $E = R^n$ the symbol O_n is used to denote the group. The determinant of every transformation $P \in O_n$ is equal to ± 1, and the collection of $P \in O_n$ with $\det P = 1$ is denoted by O_n^+. Obviously, O_n^+ is a subgroup of O_n.

A mapping F of a Euclidean space E into itself is called a *general orthogonal transformation* if F is representable in the form $F(x) = \lambda P(x)$, where $\lambda > 0$, $\lambda \in R$, and P is an orthogonal transformation of E. A mapping $\Phi: E \to E$ is called a *similarity transformation* of E or a *similarity* if Φ is representable in the form $\Phi(x) = a + F(x)$, where F is a general orthogonal transformation. The collection of all general orthogonal transformations of E, as well as the collection of all similarities and all motions of E, form groups. We do not introduce any special notation for these groups.

A special case of similarity transformations is a *homothety*, i.e., a mapping of the form $h_\lambda: x \mapsto a + \lambda(x - a)$, where $\lambda \neq 0$. The point a is called the *center* of the homothety. A homothety for which $\lambda = -1$ is called a *symmetry with respect to the point a*. A symmetry with respect to a point is a motion. Obviously, the homotheties form a subgroup of the group of motions of E.

An arbitrary mapping $\Phi: E \to E$ is a similarity if and only if Φ is bijective and there exists a number $\lambda > 0$ such that $|\Phi(x) - \Phi(y)| = \lambda|x - y|$ for any $x, y \in E$. The fact that every similarity satisfies this condition is obvious. Conversely, let Φ be a mapping satisfying this condition. Then the transformation $\mathscr{D} = (1/\lambda)\Phi$ is obviously a motion, and hence $\mathscr{D}(x) = k + P(x)$, $\forall x \in E$, where $P \in O(E)$. From this, $\Phi(x) = \lambda k + \lambda P(x)$ $\forall x \in E$, and thus Φ is a similarity.

Let x, y, and z be three different points of E, and let

$$\langle x, y, z \rangle = |x - y|/|y - z|.$$

We call this quantity the *simple ratio* of the points x, y, and z.

THEOREM 3.2. *Let* $S: \mathbf{E} \to \mathbf{E}$ *be a bijective mapping. If for any three distinct points* $x, y, z \in \mathbf{E}$

$$\langle x, y, z \rangle = \langle S(x), S(y), S(z) \rangle,$$

then S *is a similarity.*

PROOF. Assume that $S: \mathbf{E} \to \mathbf{E}$ satisfies the condition of the theorem, and take any three distinct points $x, y, z \in \mathbf{E}$. By assumption, $\langle x, y, z \rangle = \langle S(x), S(y), S(z) \rangle$, i.e.,

$$\frac{|x - y|}{|y - z|} = \frac{|S(x) - S(y)|}{|S(y) - S(z)|},$$

which implies that

$$\frac{|S(x) - S(y)|}{|x - y|} = \frac{|S(y) - S(z)|}{|y - z|}. \tag{3.1}$$

As we now show, this equality gives us that the ratio

$$|S(x) - S(y)|/|x - y|$$

is constant. Fix any distinct points $a, b \in \mathbf{E}$ and let

$$|S(a) - S(b)|/|a - b| = \lambda.$$

Let (p, q) be an arbitrary pair of points in $\mathbf{E}$ with $p \neq q$. We show that

$$|S(p) - S(q)|/|p - q| = \lambda. \tag{3.2}$$

If $p = a$ and $q = b$, this is obvious. In the case $p = a$ and $q \neq b$ equality (3.2) follows from (3.1) if we set $x = b$, $y = a$, and $z = q$ in it. The validity of (3.2) in the case when $q = a$ is obtained analogously. Assume that $p \neq a$ and $q \neq a$. Setting $y = a, x = b$, and $z = p$ in (3.1), we get that

$$\frac{|S(p) - S(a)|}{|p - a|} = \frac{|S(b) - S(a)|}{|b - a|} = \lambda.$$

Setting $y = p, x = a$, and $z = q$ in (3.1), we find that

$$\frac{|S(p) - S(a)|}{|p - a|} = \frac{|S(p) - S(q)|}{|p - q|},$$

which implies (3.2). Consequently, we get that

$$|S(x) - S(y)|/|x - y| \equiv \text{const} > 0,$$

and hence S is a similarity. The theorem is proved.

3.2. Möbius transformations. Definitions. We extend the space **E** by adjoining to it a certain "improper" element denoted by the symbol ∞. The extended space is denoted by $\overline{\mathbf{E}}$, $\overline{\mathbf{E}} = \mathbf{E} \cup \{\infty\}$. A topology is introduced in $\overline{\mathbf{E}}$ as follows. A neighborhood of ∞ is taken to be any set U whose complement is a bounded subset of **E**. If $x \neq \infty$, then $x \in \mathbf{E}$, and in this case any set containing a ball about x is taken as a neighborhood of x.

A sequence (x_m), $m = 1, 2, \ldots$, of points in $\overline{\mathbf{E}}$ converges to ∞ if and only if $|x_m| \to \infty$ as $m \to \infty$. (If $x = \infty$, then $|x|$ is regarded as equal to ∞.) In the case when the dimension of **E** is finite and equal to n the topological space $\overline{\mathbf{E}}$ is compact.

The space $\overline{\mathbf{E}}$ constructed in the way described here will be called the Möbius space generated by **E**.

We construct a mapping of the unit sphere $S(0, 1)$ of $\mathbf{E} \times \mathbf{R}$ onto $\overline{\mathbf{E}}$. The structure of a Euclidean vector space is introduced in $\mathbf{E} \times \mathbf{R}$ in the natural way; namely, for arbitrary $u_1 = (x_1, y_1)$ and $u_2 = (x_2, y_2)$ in $\mathbf{E} \times \mathbf{R}$ (here $x_1, x_2, y_1, y_2 \in \mathbf{R}$) we let

$$\lambda u_1 + \mu u_2 = (\lambda x_1 + \mu x_2, \lambda y_1 + \mu y_2), \quad \langle u_1, u_2 \rangle = \langle x_1, x_2 \rangle + y_1 y_2.$$

The point $(x, 0) \in \mathbf{E} \times \mathbf{R}$ will henceforth be identified with the point x in **E**.

Consider the sphere $S_{\mathbf{E}} = S(0, 1) = \{u \in \mathbf{E} \times \mathbf{R} | |u| = 1\}$ in $\mathbf{E} \times \mathbf{R}$. Let $N = (0, 1)$ and $S = (0, -1)$. The respective points N and S will be called the *north* and *south poles* of the sphere $S_{\mathbf{E}}$. We construct a certain special mapping σ of $\overline{\mathbf{E}}$ into $S_{\mathbf{E}}$. Let $\sigma(\infty) = N$. For $x \in \mathbf{E}$, $x \neq \infty$, we join the point $x = (x, 0) \in \mathbf{E} \times \mathbf{R}$ to the point N in $\mathbf{E} \times \mathbf{R}$ by a straight line. Let us show that this line intersects $S_{\mathbf{E}}$ in some point different from N. The line admits the parametric representation $u(t) = tx + (1 - t)N$, where $t \in \mathbf{R}$. We have that $u(t) = (tx, 1 - t)$ and $|u(t)| = \sqrt{t^2 |x|^2 + (1 - t)^2}$. Setting $|u(t)| = 1$, we get that

$$t^2 |x|^2 + 1 - 2t + t^2 = 1.$$

This equation has two different solutions: $t_1 = 0$ and $t_2 = 2/(|x|^2 + 1)$. Let

$$\sigma(x) = u(t_2) = \left(\frac{2}{|x|^2 + 1} x, \frac{|x|^2 - 1}{|x|^2 + 1} \right). \tag{3.3}$$

The mapping σ of $\overline{\mathbf{E}}$ into $S_{\mathbf{E}}$ so defined is called the *stereographic projection*.

We mention certain properties of σ. First of all we show that σ is one-to-one. Indeed, let $x_1, x_2 \in \overline{\mathbf{E}}$, $x_1 \neq x_2$. If one of the given points x_1 and x_2 is ∞, then its image under σ is N, and, as is clear from the construction, the image of the other point is different from N; hence in this case $\sigma(x_1) \neq \sigma(x_2)$. If $x_1, x_2 \neq \infty$, then the lines $x_1 N$ and $x_2 N$

have a unique common point—the point N. From this it follows that $\sigma(x_1) \neq \sigma(x_2)$ in this case. We show that σ is a mapping of $\overline{\mathbf{E}}$ onto $S_{\mathbf{E}}$. Indeed, take an arbitrary point $u \in S_{\mathbf{E}}$. If $u = N$, then $u = \sigma(\infty)$. Assume that $u \neq N$. We have that $u = (y, z)$ and $|u|^2 = |y|^2 + z^2 = 1$, and since $u \neq N$, it follows that $z < 1$. The line passing through u and N intersects the hyperplane $\mathbf{E}$ in the point

$$x = \frac{1}{1-z} y. \tag{3.4}$$

Obviously, $u = \sigma(x)$. Each of the mappings σ and σ^{-1} is continuous. The continuity of σ at a point $x \neq \infty$ follows immediately from (3.3), and the continuity of σ^{-1} at a point $u \neq N$ follows from (3.4). We leave it to the reader to prove that σ and σ^{-1} are continuous also at the respective points ∞ and N.

Stereographic projection is thus a homeomorphism of $\overline{\mathbf{E}}$ onto $S_{\mathbf{E}}$. We endow $\overline{\mathbf{E}}$ with the structure of a differentiable manifold by agreeing that the stereographic projection is a diffeomorphism of class C^∞ from the manifold $\overline{\mathbf{E}}$ onto $S_{\mathbf{E}}$. The equalities (3.3) allow us to conclude that on $\mathbf{E}$ the mapping σ is a diffeomorphism of class C^∞ from $\mathbf{E}$ onto $S_{\mathbf{E}} \setminus \{N\}$. This shows that the structure of a differentiable manifold introduced on $\overline{\mathbf{E}}$ is consistent in the natural way with the structure on $\mathbf{E}$.

Let $S(a, r)$ be the sphere about a with radius r in $\mathbf{E}$. The inversion with respect to $S(a, r)$ is defined to be the following transformation of $\overline{\mathbf{E}}$. First of all, let $j(a) = \infty$ and $j(\infty) = a$. But if $x \neq a$ and $x \neq \infty$, then $y = j(x)$ is determined as follows: y lies on the ray from the point a passing through the point x, and is such that $|x-a| \, |y-a| = r^2$. These conditions determine y uniquely. The first condition implies that $y - a = \lambda(x - a)$, where $\lambda > 0$, and the second implies that $\lambda = r^2/|x - a|^2$; hence

$$y = j(x) = a + \frac{r^2(x - a)}{|x - a|^2}. \tag{3.5}$$

It is obvious from the definition of j that if $y = j(x)$, then $j(y) = x$; therefore, $j^{-1} = j$, i.e., j is a so-called involution. If $x \in S(a, r)$, then $j(y) = x$.

The inversion j with respect to $S(a, r)$ is a homeomorphism of $\mathbf{E}$ onto itself. Indeed, it follows from the definition that j is bijective. Equality (3.5) enables us to establish that j is continuous at a point $x \neq a$, $x \neq \infty$. The ball $B(a, \varepsilon)$ is mapped under j onto the complement in $\overline{\mathbf{E}}$ of the ball $B(a, r^2/\varepsilon)$. This clearly implies that j is continuous also at each of the points a and ∞.

The space $\overline{E}$ has been equipped with the structure of a C^∞-differentiable manifold. It is easy to establish that inversion with respect to $S(a, r)$ is a C^∞-transformation. This is a simple consequence of the definitions above. (It suffices to note that $\sigma \circ j \circ \sigma^{-1}$ is a C^∞-mapping of the sphere $S(0, 1)$ in $E \times R$.)

Let Φ be an arbitrary similarity of E. We extend Φ to $\overline{E}$ by setting $\Phi(\infty) = \infty$.

A mapping $\varphi: \overline{E} \to \overline{E}$ is called a *Möbius transformation* if it can be represented as a composition of finitely many inversions and similarities. If φ and ψ are Möbius mappings, then $\varphi \circ \psi^{-1}$ is obviously also a Möbius transformation, and hence the collection of all Möbius transformations of $\overline{E}$ forms a group, which we denote by the symbol $M(\overline{E})$. This group will be denoted by M_n in the case $E = R^n$.

Every Möbius transformation is a C^∞-diffeomorphism of the manifold $\overline{E}$.

We mention some properties of Möbius transformations.

Let j be inversion with respect to the sphere $S(a, r)$, $j(x) = a + r^2(x - a)/|x - a|^2$. Denote by h the homothety $x \mapsto a + rx$. Obviously, $S(a, r) = h[S(0, 1)]$. Denote by j_0 the inversion with respect to $S(0, 1)$. We have that

$$j_0(x) = \frac{x}{|x|^2} \quad \forall x \in E \setminus \{0\}.$$

It is not hard to see that $j = h \circ j_0 \circ h^{-1}$. Each of the transformations h and h^{-1} is a similarity. In view of the remark made above, every Möbius transformation admits a representation $\varphi = \varphi_1 \circ \cdots \circ \varphi_m$, where each of the transformations φ_i is either a similarity or the transformation j_0.

Let U be an open subset of E, and $f: U \to E$ a C^1-mapping. The mapping f is said to be *conformal at a point* $x_0 \in U$ if the derivative $f'(x_0)$ of f at x_0 is a general orthogonal transformation. A mapping $f: U \to E$ is said to be *conformal* if it is conformal at each point $x \in E$.

We show that every mapping $\varphi \in M(\overline{E})$ is conformal at each point $x \in E \setminus \varphi^{-1}(\infty)$. This is clearly true if φ is a similarity. Since a composition of general orthogonal transformations is a transformation of the same type, and every Möbius transformation is a composition of finitely many inversions and similarity transformations, it suffices to see that an inversion is a conformal mapping. Accordingly, let

$$f(x) = a + \frac{r^2(x - a)}{|x - a|^2}.$$

Simple computations show that at every point $x \neq a$ and for any vector h

$$f'(x)h = \frac{r^2}{|x-a|^2}(h - 2\langle e, h\rangle e),$$

where $e = (x-a)/|x-a|$. The mapping $P: h \mapsto h - 2\langle e, h\rangle e$ is a symmetry with respect to the plane $\langle e, x\rangle = 0$, and hence is an orthogonal transformation. (The fact that $|P(h)| = |h|$ is also easy to establish directly.) This implies that $f'(x)$ is a general orthogonal transformation, as was required to prove.

3.3. Möbius transformations and cross ratios. Construction of Möbius transformations.

Let x_1, x_2, x_3, x_4 be an arbitrary quadruple of points in E such that $x_1 \neq x_2$, $x_2 \neq x_3$, $x_3 \neq x_4$, and $x_4 \neq x_1$. The quantity

$$\langle x_1, x_2, x_3, x_4\rangle = \frac{|x_1 - x_2|\,|x_3 - x_4|}{|x_1 - x_4|\,|x_2 - x_3|} \tag{3.6}$$

is called the *cross ratio* of the points x_1, x_2, x_3, x_4. We join x_1 and x_2, x_2 and x_3, x_3 and x_4, and x_4 and x_1 by successive segments, obtaining a quadrangle. The numerator of the quotient defining $\langle x_1, x_2, x_3, x_4\rangle$ contains the product of the lengths of one pair of its opposite sides, and the denominator contains the product of the lengths of the other pair of opposite sides. Obviously,

$$\langle x_1, x_2, x_3, x_4\rangle = 1/\langle x_2, x_3, x_4, x_1\rangle = \langle x_3, x_4, x_1, x_2\rangle$$
$$= 1/\langle x_4, x_1, x_2, x_3\rangle = \langle x_2, x_1, x_4, x_3\rangle. \tag{3.7}$$

Let x_1, x_2, x_3, x_4 be an arbitrary quadruple of points in $\overline{E}$ satisfying the same conditions $x_1 \neq x_2, x_2 \neq x_3, x_3 \neq x_4$, and $x_4 \neq x_1$, with one of the points x_i equal to ∞. In this case the quantity $\langle x_1, x_2, x_3, x_4\rangle$ is defined as the limit $\lim\langle x_1', x_2', x_3', x_4'\rangle$, where $x_i' = x_i$ if x_i is finite, and $x_i' \to \infty$ if $x_i = \infty$, $i = 1, 2, 3, 4$. It is not hard to show that this limit always exists. In particular, we have that

$$\langle x_1, x_2, x_3, \infty\rangle = \lim_{x \to \infty} \frac{|x_1 - x_2|\,|x_3 - x|}{|x_2 - x_3|\,|x_1 - x|}$$
$$= \frac{|x_1 - x_2|}{|x_2 - x_3|} = \langle x_1, x_2, x_3\rangle. \tag{3.8}$$

It is frequently useful to give the quantity $\langle x_1, x_2, x_3, x_4\rangle$ a definite meaning also when the conditions $x_1 \neq x_2$, $x_2 \neq x_3$, $x_3 \neq x_4$, and $x_4 \neq x_1$ are not satisfied. In this case we set

$$\langle x_1, x_2, x_3, x_4\rangle = \lim\langle x_1', x_2', x_3', x_4'\rangle,$$

where $x_1' \neq x_2'$, $x_2' \neq x_3'$, $x_3' \neq x_4'$, $x_4' \neq x_1'$, and the limit is taken under the condition that $x_i' \to x_i$ for each $i = 1, \ldots, 4$ under the assumption that this limit exists. Here the values 0 and ∞ are allowed for $\langle x_1, x_2, x_3, x_4 \rangle$. If x_1, x_2, x_3, and x_4 are all different from ∞, then $\langle x_1, x_2, x_3, x_4 \rangle$ is defined if and only if the quotient on the right-hand side of (3.6) is not an expression of the form $0/0$, and (3.6) remains valid if we set $\langle x_1, x_2, x_3, x_4 \rangle = \infty$ when the denominator of (3.6) is zero.

THEOREM 3.3. *Let x_1, x_2, x_3, x_4 be an arbitrary quadruple of points in $\overline{\mathbf{E}}$ for which $\langle x_1, x_2, x_3, x_4 \rangle$ is defined. Then for every $\varphi \in \mathbf{M}(\overline{\mathbf{E}})$*

$$\langle \varphi(x_1), \varphi(x_2), \varphi(x_3), \varphi(x_4) \rangle = \langle x_1, x_2, x_3, x_4 \rangle. \tag{3.9}$$

PROOF. This equality obviously holds when f is a similarity transformation. Since an arbitrary Möbius transformation is a composition of finitely many inversions and similarity transformations, it suffices to consider the case when φ is an inversion. Further, we can confine ourselves to the case when φ is inversion with respect to the unit sphere $S(0, 1)$, $\varphi = j_0$.

Assume first that $x_1 \neq x_2$, $x_2 \neq x_3$, $x_3 \neq x_4$, and $x_4 \neq x_1$, and all these points are different from 0 and ∞. Let $e_i = x_i/|x_i|$, $i = 1, 2, 3, 4$. Then $j_0(x_i) = e_i/|x_i|$ and

$\langle j_0(x_1), j_0(x_2), j_0(x_3), j_0(x_4) \rangle$

$$= \left| \frac{e_1}{|x_1|} - \frac{e_2}{|x_2|} \right| \left| \frac{e_3}{|x_3|} - \frac{e_4}{|x_4|} \right| : \left| \frac{e_1}{|x_1|} - \frac{e_4}{|x_4|} \right| \left| \frac{e_2}{|x_2|} - \frac{e_3}{|x_3|} \right|$$

$$= \||x_2|e_1 - |x_2|e_2| \, ||x_4|e_3 - |x_3|e_4|/||x_4|e_1 - |x_1|e_4| \, ||x_3|e_2 - |x_2|e_3|.$$

We have that

$$\||x_i|e_j - |x_j|e_i|^2 = |x_i|^2 - 2|x_i||x_j|\langle e_i, e_j \rangle + |x_j|^2 = |x_i - x_j|^2,$$

which clearly implies (3.9). The required result can be derived from the proven passage to the limit for the case when the conditions $x_1 \neq x_2$, $x_2 \neq x_3$, $x_3 \neq x_4$, and $x_4 \neq x_1$ are not satisfied, as well as for the case when some of the given points coincide with 0 or with ∞. The theorem is proved.

COROLLARY 1. *Every transformation $\varphi \in \mathbf{M}(\overline{\mathbf{E}})$ such that $\varphi(\infty) = \infty$ is a similarity.*

PROOF. Let $\varphi \in \mathbf{M}(\overline{\mathbf{E}})$ be such that $\varphi(\infty) = \infty$. Take three arbitrary distinct points $x_1, x_2, x_3 \in \mathbf{E}$. We have that

$$\langle x_1, x_2, x_3 \rangle = \langle x_1, x_2, x_3, \infty \rangle = \langle \varphi(x_1), \varphi(x_2), \varphi(x_3), \varphi(\infty) \rangle$$
$$= \langle \varphi(x_1), \varphi(x_2), \varphi(x_3) \rangle.$$

The required result follows immediately from this, in view of Theorem 3.2.

COROLLARY 2. *Every transformation $\varphi \in M(\overline{E})$ either is a similarity or admits a representation of the form $\varphi = \alpha \circ \beta$, where α is an inversion and β a motion of E. Such a representation of φ is unique.*

PROOF. Let $\varphi \in M(\overline{E})$. If $\varphi(\infty) = \infty$, then φ is a similarity transformation. Assume that $\varphi(\infty) = a \neq \infty$. Let α_1 be an inversion with respect to the sphere $S(a, 1)$, and let $\beta_1 = \alpha_1 \circ \varphi$. We have that $\beta_1^{(\infty)} = \alpha_1[\varphi(\infty)] = \alpha_1(a) = \infty$, and hence β_1 is a similarity. Let h be the dilation coefficient of the mapping β_1,

$$\beta_1, \frac{|\beta_1(x) - \beta_1(y)|}{|x - y|} = h,$$

for any $x, y \in E$. Let $r = 1/\sqrt{h}$, and let α be inversion with respect to the sphere $S(a, r)$. A simple computation shows that the mapping $\gamma = \alpha \circ \alpha_1$ is a homothety with respect to a with dilation coefficient $r^2 = 1/h$. We have that $\alpha_1 = \alpha^{-1} \circ \gamma = \alpha \circ \gamma$ and $\varphi = \alpha_1^{-1} \circ \beta_1 = \alpha_1 \circ \beta_1 = \alpha \circ (\gamma \circ \beta_1)$. (Here we have used the fact that $\alpha^{-1} = \alpha$ and $\alpha_1^{-1} = \alpha_1$.) Let $\beta = \gamma \circ \beta_1$. The mapping β is a similarity. Its dilation coefficient is obviously equal to $(1/h) \cdot h = 1$, i.e., β is a motion. The existence of the required representation is proved. We prove its uniqueness. Let $\varphi = \alpha_1 \circ \beta_1 = \alpha_2 \circ \beta_2$, where α_1 and α_2 are inversions and β_1 and β_2 are motions. From this, $\alpha_1 \circ \alpha_2 = \beta_2 \circ \beta_1^{-1}$. The mapping on the right-hand side is a motion and thus maps ∞ into itself. Let a be the center of the inversion α_2. Then $\alpha_2(\infty) = a$, and hence $\alpha_1(a) = \infty$, i.e., a is also the center of the inversion α_1. Thus, α_1 and α_2 are inversions with a common center. Suppose that α_1 is inversion with respect to the sphere of radius r_1, and α_2 inversion with respect to the sphere of radius r_2. Then $\alpha_1 \circ \alpha_2$ is a homothety with center a and coefficient $h = r_2^2/r_1^2$. Since, on the other hand, $\alpha_1 \circ \alpha_2$ is a motion, it follows that $h = 1$, i.e., $r_1 = r_2$, and hence $\alpha_1 \circ \alpha_2 = I_E$, which implies that $\alpha_1 = \alpha_2$ and $\beta_1 = \beta_2$, as was to be proved.

COROLLARY 3. *Let $\varphi \in M(\overline{E})$. Then φ either is a similarity or admits a representation $\varphi = \beta \circ \alpha$, where α is an inversion and β a motion of E. Such a representation of φ is unique.*

PROOF. Suppose that $\varphi \in M(\overline{E})$, and that φ is not a similarity. Then φ^{-1} is also not a similarity, and hence φ^{-1} admits a representation $\varphi^{-1} = \alpha_1 \circ \beta_1$ by Corollary 2, where α_1 is an inversion and β_1 a motion. From this, $\varphi = \beta_1^{-1} \circ \alpha_1^{-1} = \beta \circ \alpha$, where $\beta = \beta_1^{-1}$ is a motion and $\alpha = \alpha_1^{-1} = \alpha_1$ an inversion. Conversely, if $\varphi = \beta \circ \alpha$, where β is a motion and α an inversion, then $\varphi^{-1} = \alpha \circ \beta^{-1}$. Since such a representation of the transformation φ^{-1} is unique by Corollary 2, this implies the uniqueness of the desired representation of φ.

COROLLARY 4. *Every Möbius transformation φ of $\overline{E}$ either is a similarity or is representable in the form $\varphi = \varphi_1 \circ j_0 \circ \varphi_2$, where j_0 is inversion with respect to the sphere $S(0, 1)$, and φ_1 and φ_2 are similarities.*

PROOF. Let $\varphi = \alpha \circ \beta$, where α is inversion with respect to a sphere $S(a, r)$ and β is a motion. We have that $\alpha = h \circ j_0 \circ h^{-1}$, where h is the homothety $x \mapsto a + rx$. From this it follows that $\varphi = \varphi_1 \circ j_0 \circ \varphi_2$, where $\varphi_1 = h$ and $\varphi_2 = h^{-1} \circ \beta$. The corollary is proved.

COROLLARY 5. *Suppose that $\varphi: \overline{E} \to \overline{E}$ is a bijective mapping such that (3.9) holds for any quadruple x_1, x_2, x_3, x_4 of points in E with $x_1 \neq x_2$, $x_2 \neq x_3$, $x_3 \neq x_4$, and $x_4 \neq x_1$. Then φ is a Möbius transformation.*

PROOF. Let $\varphi: \overline{E} \to \overline{E}$ satisfy the condition of the corollary. Assume first that $\varphi(\infty) = \infty$. Then for every triple $x_1, x_2, x_3 \in E$ we have

$$\langle \varphi(x_1), \varphi(x_2), \varphi(x_3) \rangle = \langle \varphi(x_1), \varphi(x_2), \varphi(x_3), \infty \rangle$$
$$= \langle \varphi(x_1), \varphi(x_2), \varphi(x_3), \varphi(\infty) \rangle = \langle x_1, x_2, x_3, \infty \rangle = \langle x_1, x_2, x_3 \rangle.$$

On the basis of Theorem 3.2 this implies that the restriction of φ to E is a similarity. Suppose that $\varphi(\infty) = a \neq \infty$. Let α be an inversion with center a, and let $\psi = \alpha \circ \varphi$. The mapping ψ is bijective, and clearly

$$\langle \psi(x_1), \psi(x_2), -\psi(x_3), \psi(x_4) \rangle = \langle x_1, x_2, x_3, x_4 \rangle$$

for any x_1, x_2, x_3, x_4. Further, $\psi(\infty) = \infty$, and hence ψ is a similarity by what was proved. We have that $\varphi = \alpha \circ \psi$, and thus φ is a Möbius transformation, which is what was to be proved.

3.4. Möbius transformations and spheres. A subset H of E is called a *sphere* in $\overline{E}$ if H is the closure in the topology of $\overline{E}$ of a set

$$H' = \{x \in E | \lambda |x|^2 - 2\langle \xi, x \rangle + \mu = 0\}, \tag{3.10}$$

where the vector ξ in E and the numbers $\lambda, \mu \in \mathbf{R}$ are such that

$$|\xi|^2 - \lambda\mu > 0. \tag{3.11}$$

For brevity we say in this case that the sphere H in $\overline{E}$ is *determined by the equation* $\lambda |x|^2 - 2\langle \xi, x \rangle + \mu = 0$.

Some term such as "generalized" sphere, etc., would probably be more appropriate. We say "sphere in $\overline{E}$". Here the term "sphere in E" is understood in the same sense as before. We explain what such a sphere in $\overline{E}$ is geometrically, and at the same time we establish also the meaning of condition (3.11). If $\lambda = 0$, then $\xi \neq 0$ in view of (3.11), and the set

$$H' = \{x \in E | \langle \xi, x \rangle - \mu/2 = 0\}$$

is a hyperplane in $\mathbf{E}$. The only limit point of H' in $\overline{\mathbf{E}}$ which is not in H' is the point ∞, and hence $H = H' \cup \{\infty\}$ in this case. Let $\lambda \neq 0$. We have that

$$\lambda |x|^2 - 2\langle \xi, x \rangle + \mu = \lambda(|x|^2 - 2\langle \xi/\lambda, x \rangle + \mu/\lambda)$$
$$= \lambda(|x - \xi/\lambda|^2 - (|\xi|^2 - \lambda\mu)/\lambda^2).$$

In this case the set H' is the sphere in $\mathbf{E}$ with radius $r = (1/\lambda)\sqrt{|\xi|^2 - \lambda\mu}$ about the point ξ/λ. Obviously, the closure of H' in this case coincides with H'. We thus get that if H is a sphere in $\overline{\mathbf{E}}$, then either $H = H' \cup \{\infty\}$, where H' is a hyperplane in $\mathbf{E}$, or H is an ordinary sphere in $\mathbf{E}$. It is clear from the computations that the condition $|\xi|^2 - \lambda\mu > 0$ says that H' is nonempty and does not degenerate into a single point.

The remarks in §3.1 about the arbitrariness with which the equation of a hyperplane, as well as the center and radius of a sphere, are determined imply that if

$$H' = \{x \in \mathbf{E} | \lambda |x|^2 - 2\langle \xi, x \rangle + \mu\}$$
$$= \{x \in \mathbf{E} | \lambda_1 |x|^2 - 2\langle \xi_1, x \rangle + \mu_1 = 0\},$$

and $|\xi|^2 - \lambda\mu > 0$, then there exists a $k \neq 0$ such that $\lambda_1 = k\lambda$, $\xi_1 = k\xi$, and $\mu_1 = k\mu$.

THEOREM 3.4. *The image of a sphere in $\overline{\mathbf{E}}$ under any Möbius transformation is a sphere in $\overline{\mathbf{E}}$.*

PROOF. Suppose that $\varphi \subset \mathbf{M}(\overline{\mathbf{E}})$. If φ is a similarity, then it transforms every hyperplane in $\mathbf{E}$ into a hyperplane and every sphere in $\mathbf{E}$ into a sphere, and this implies the theorem for the given case. Assume that φ is not a similarity. Then $\varphi = \varphi_1 \circ j_0 \circ \varphi_2$, where φ_1 and φ_2 are similarities, and j_0 is inversion with respect to $S(0, 1)$. Therefore, it suffices to establish that the image of every sphere in $\overline{\mathbf{E}}$ under j_0 is a sphere. Accordingly, let H be a sphere in $\overline{\mathbf{E}}$. Assume that H is given by the equation $\lambda |x|^2 - 2\langle \xi, x \rangle + \mu = 0$, where $|\xi|^2 - \lambda\mu > 0$. Denote by H' the set of points $x \in H$ such that $0 < |x| < \infty$. The sphere H is the closure in $\overline{\mathbf{E}}$ of the set H', and since j_0 is a homeomorphism of $\overline{\mathbf{E}}$ onto $\overline{\mathbf{E}}$, it follows that $j_0(H)$ is the closure in $\overline{\mathbf{E}}$ of the set $j_0(H')$. Then $x \in \mathbf{E}$ belongs to $j_0(H')$ if and only if $j_0^{-1}(x) = j_0(x) \in H'$, i.e., the condition $x \in j_0(H')$ is equivalent to the condition

$$\lambda \frac{1}{|x|^2} - 2\langle \xi, \frac{x}{|x|^2} \rangle + \mu = 0 \Leftrightarrow x \neq 0 \& \lambda - 2\langle \xi, x \rangle + \mu |x|^2 = 0.$$

The equation $\lambda - 2\langle \xi, x \rangle + \mu |x|^2 = 0$ determines a sphere in $\overline{\mathbf{E}}$, and this proves the theorem.

THEOREM 3.5. *Let $\varphi: \overline{\mathbf{E}} \to \overline{\mathbf{E}}$ be a bijective mapping such that for any sphere H in $\overline{\mathbf{E}}$ the set $\varphi(H)$ is a sphere in $\overline{\mathbf{E}}$. Then φ is a Möbius transformation.*

PROOF. Assume that φ satisfies the conditions of the theorem. Suppose first that $\varphi(\infty) = \infty$. We prove that in this case φ is a similarity.

Let P be any hyperplane in $\mathbf{E}$. The set $H = P \cup \{\infty\}$ is a sphere in $\overline{\mathbf{E}}$. By a condition of the theorem, $\varphi(H)$ is also a sphere in $\overline{\mathbf{E}}$. Since $\varphi(\infty) = \infty$, it follows that $\infty \in \varphi(H)$, and hence $\varphi(H) = Q \cup \{\infty\}$, where Q is a hyperplane. Obviously, φ maps P onto Q. It follows next from the condition of the theorem that every sphere in $\mathbf{E}$ is transformed under φ into a sphere. Since φ is bijective, for any two sets A, $B \subset \overline{\mathbf{E}}$ with $A \cap B = \varnothing$ we have that $\varphi(A) \cap \varphi(B) = \varnothing$. This gives us, in particular, that if the hyperplanes P and Q are parallel, then so are the planes $\varphi(P)$ and $\varphi(Q)$.

Let x and y be arbitrary points in $\mathbf{E}$ such that $x \neq y$, and let S be the sphere constructed with the segment $[x, y]$ as a diameter. The center of S is the point $(x + y)/2$, and its radius is $|x - y|/2$. Let P and Q be hyperplanes tangent to S and passing through x and y, respectively. The planes P and Q are parallel, and each has a single point in common with S. This implies that the hyperplanes $\varphi(P)$ and $\varphi(Q)$ are also parallel, and each has a single point in common with $\varphi(S)$. The common point of $\varphi(P)$ and the sphere $\varphi(S)$ is $\varphi(x)$, and the common point of $\varphi(Q)$ and $\varphi(S)$ is $\varphi(y)$. We see that the sphere $\varphi(S)$ lies between two parallel hyperplanes and is tangent to them at the points $\varphi(x)$ and $\varphi(y)$. From this it follows that $\varphi(x)$ and $\varphi(y)$ are the endpoints of a diameter of $\varphi(S)$. We thus get that if the points $x, y \in S$ are the endpoints of a diameter of the sphere S in E, then $\varphi(x)$ and $\varphi(y)$ are the endpoints of a diameter of $\varphi(S)$.

Let us prove that for any $x, y \in \mathbf{E}$

$$|\varphi(x) - \varphi(y)| = \lambda(|x - y|), \tag{3.12}$$

where λ is a function on $[0, \infty)$. It clearly suffices to show that

$$|\varphi(x_1) - \varphi(y_1)| = |\varphi(x_2) - \varphi(y_2)| \tag{3.13}$$

for any pairs x_1, y_1 and x_2, y_2 of points in $\mathbf{E}$ such that $|x_1 - y_1| = |x_2 - y_2|$. If $|x_1 - y_1| = |x_2 - y_2| = 0$, then (3.13) obviously holds. Assume that $x_1 \neq y_1$ and $x_2 \neq y_2$. Let S_1 and S_2 be the spheres constructed with $[x_1, y_1]$ and $[x_2, y_2]$ as diameters. Since $|x_1 - y_1| = |x_2 - y_2|$, the radii of S_1 and S_2 are equal. This implies that there exist two parallel hyperplanes P and Q, each tangent to both spheres S_1 and S_2, such that these spheres lie in the strip bounded by the planes. The planes $\varphi(P)$ and $\varphi(Q)$ are also parallel, and the spheres $\varphi(S_1)$ and $\varphi(S_2)$ are tangent to each of them. This lets us conclude that the radii of $\varphi(S_1)$ and $\varphi(S_2)$ are equal. By what was proved, the points $\varphi(x_1)$ and $\varphi(y_1)$ are the endpoints of a diameter of $\varphi(S_2)$. Similarly, $\varphi(x_2)$ and $\varphi(y_2)$ are the endpoints of a diameter of $\varphi(S_2)$. Consequently, we get that $|\varphi(x_1) - \varphi(y_1)| = |\varphi(x_2) - \varphi(y_2)|$, which proves (3.13), and with it (3.12).

Let us now show that the function λ is linear, $\lambda(r) = kr$, where $k > 0$. Note first of all that λ is nonnegative. We prove that for any $r_1, r_2 > 0$

$$\lambda(r_1 + r_2) = \lambda(r_1) + \lambda(r_2). \tag{3.14}$$

Take arbitrary $r_1, r_2 > 0$, let x, y, and z be points on a single line such that $|x - y| = r_1, |y - z| = r_2$, and $|x - z| = r_1 + r_2$, and let $x' = \varphi(x)$, $y' = \varphi(y)$, and $z' = \varphi(z)$. Then $|x' - y'| = \lambda(r_1)$, $|y' - z'| = \lambda(r_2)$, and $|x' - z'| = \lambda(r_1 + r_2)$. We prove that the points x', y', and z' lie on a single line, with y' between x' and z'. Let S_1 and S_2 be the spheres constructed with $[x, y]$ and $[y, z]$ as diameters. Then $[x', y']$ is a diameter of $\varphi(S_1)$, and $[y', z']$ is a diameter of $\varphi(S_2)$. The spheres $\varphi(S_1)$ and $\varphi(S_2)$ have a unique common point, namely, y'. This implies that x', y', and z' all lie on a single line. To finish the proof it is necessary to see that neither of the spheres $\varphi(S_1)$ or $\varphi(S_2)$ lies interior to the other. Indeed, since S_1 and S_2 lie on different sides of a plane tangent to both, there exists a hyperplane P tangent to S_1 at some point a and to S_2 at a point $b \neq a$. Then the hyperplane $\varphi(P)$ is tangent to $\varphi(S_1)$ at $\varphi(a)$, and to $\varphi(S_2)$ at the point $\varphi(b) \neq \varphi(a)$. If one of the spheres $\varphi(S_1)$ or $\varphi(S_2)$ were interior to the other, then this would be impossible. Consequently, y' lies between x' and z', and thus

$$|x' - z'| = |x' - y'| + |y' - z'|,$$

i.e., $\lambda(r_1 + r_2) = \lambda(r_1) + \lambda(r_2)$, and (3.14) is established.

Since $\lambda(r)$ is nonnegative, (3.14) implies that the function λ is linear, $\lambda(r) = kr$ for all r, where $k \geq 0$. Since $\lambda(r) > 0$ for $r > 0$, it follows that $k > 0$.

Hence there exists a number $k > 0$ such that

$$|\varphi(x) - \varphi(y)|/|x - y| = k$$

for any $x, y \in \mathbf{E}, x \neq y$. As noted in §3.1, this implies that φ is a similarity, and hence $\varphi \in \mathbf{M}(\overline{\mathbf{E}})$.

In our arguments we assumed that $\varphi(\infty) = \infty$. Suppose that $\varphi(\infty) = a \neq \infty$. Let α be the inversion transformation with respect to the sphere $S(a, 1)$, and set $\psi = \alpha \circ \varphi$. Since each of φ and α maps every sphere in $\overline{\mathbf{E}}$ into a sphere, while ψ is bijective and satisfies $\psi(\infty) = \infty$, it follows from what was proved that ψ is a similarity. We have that $\varphi = \alpha \circ \psi$, and thus $\varphi \in M(\overline{E})$. This proves the theorem.

THEOREM 3.6. *Let H be a sphere in $\overline{E}$, and let $\varphi \in M(\overline{E})$ be such that $\varphi(x) = x$ for every $x \in H$. Assume that φ is not the identity mapping. Then φ is inversion with respect to H as a sphere in $\overline{E}$ in the case when $\infty \notin H$. But if $\infty \in H$, then φ is the mirror symmetry with respect to H.*

PROOF. Let us first consider the case when H is the closure in $\overline{E}$ of a hyperplane $P = \{x \in E | \langle e, x \rangle = 0\}$. Without loss of generality it can be assumed that e is a unit vector. Let $\varphi \in M(\overline{E})$ be such that $\varphi(x) = x$ for all $x \in H$. In particular, we get that $\varphi(\infty) = \infty$, and hence φ is a similarity, $\varphi(x) = k + Q(x)$, where Q is a general orthogonal transformation. Since $0 \in P$, it follows that $\varphi(0) = 0$, which leads to the conclusion that $k = 0$, so that $\varphi(x) = Q(x)$ for all $x \in E$. Let $\lambda > 0$ be the similarity coefficient with respect to φ. For any $x_1, x_2 \in P$ we have that

$$|x_1 - x_2| = |\varphi(x_1) - \varphi(x_2)| = \lambda|x_1 - x_2|,$$

and hence $\lambda = 1$. If $x \in P$, then

$$0 = \langle e, x \rangle = \langle Q(e), Q(x) \rangle = \langle Q(e), x \rangle.$$

From this, $Q(e) = \mu e$. Since Q is an orthogonal transformation, $\mu = \pm 1$. Let $x \in E$ be arbitrary. We have that $y = x - \langle x, e \rangle e \in P$; hence $\varphi(y) = y = x - \langle x, e \rangle e$. On the other hand, $\varphi(y) = Q(y) = Q(x) - \langle x, e \rangle Q(e)$, and we get that

$$\varphi(x) = Q(x) = x - \langle x, e \rangle e + \mu\langle x, e \rangle e = x - \langle x, e \rangle(1 - \mu)e.$$

It follows from this that φ is either the identity mapping (in the case $\mu = 1$) or the symmetry mapping $x \mapsto x - 2\langle x, e \rangle e$ with respect to the plane P.

Let H be an arbitrary sphere in $\overline{E}$. Two different Möbius transformations are known which leave the points of H fixed: one is the identity mapping, and the other is inversion with respect to the sphere when H is a sphere in E, and the symmetry with respect to H' when H is the closure of a hyperplane H'. It is required to prove that there is no third mapping having the same property. This has been established in the case when H is the closure of a hyperplane $\{x | \langle x, e \rangle = 0\}$. Let $\sigma \in M(\overline{E})$ be a Möbius transformation carrying H into the sphere H_0 which is the closure of $\{x | \langle x, e \rangle = 0\}$, and let $\varphi \in M(\overline{E})$ be such that $\varphi(x) = x \; \forall x \in H$. Define $\psi = \sigma \circ \varphi \circ \sigma^{-1}$. It is then obvious that $\psi_0(x) = x$ for all $x \in H_0$. We know two mappings φ such that $\varphi(x) = x$ for all $x \in H$. If there were a third mapping with the same property, then we would get three different mappings ψ such that $\psi(x) = x$ for all $x \in H_0$. As follows from what was proved above, this is impossible, and the theorem is proved.

THEOREM 3.7. *Let* $\varphi \in M(\overline{E})$. *Assume that there exists a sphere* $S(a, r)$ *such that its image is a sphere* $S(b, \rho)$, *and the center of* $S(a, r)$ *is carried by* φ *into the center of its image. Then* φ *is a similarity.*

PROOF. Assume that the conditions of the theorem hold. It is required to show that $\varphi(\infty) = \infty$. Suppose, on the contrary, that $\varphi(\infty) = c \neq \infty$. Let p and q be the points where $S(b, \rho)$ intersects the line bc, and let s and t be the points of $S(a, r)$ such that $\varphi(s) = p$ and $\varphi(t) = q$. We have that

$$\langle s, a, t, \infty \rangle = \frac{|s - a|}{|a - t|} = 1.$$

On the other hand,

$$\langle s, a, t, \infty \rangle = \langle \varphi(s), \varphi(a), \varphi(t), \varphi(\infty) \rangle = \langle p, b, q, c \rangle$$
$$= \frac{|p - b||q - e|}{|p - c||q - b|} = \frac{|q - c|}{|p - c|} \neq 1,$$

since the lengths of the segments $[q, c]$ and $[p, c]$ are clearly different. Thus, the assumption that $\varphi(\infty) \neq \infty$ leads to a contradiction. The theorem is proved.

3.5. The hypersphere bundle and linear representations of Möbius transformations. In the study of Möbius transformations a certain geometric construction which we call the hypersphere bundle turns out to be useful. The use of it enables us to describe Möbius transformations in the language of linear algebra.

As above, E will stand for a Euclidean vector space. Also as before, no restrictions are placed on the dimension of E; in particular, it can be infinite-dimensional. Consider the direct product $E \times R^2$. An arbitrary point $u \in E \times R^2$ will be understood as a triple (ξ, η, ζ), where $\xi \in E$, and η and ζ are real numbers. For $u_1 = (\xi_1, \eta_1, \zeta_1)$ and $u_2 = (\xi_2, \eta_2, \zeta_2)$ let

$$\langle u_1, u_2 \rangle_1 = \langle \xi_1, \xi_2 \rangle + \eta_1 \eta_2 - \zeta_1 \zeta_2.$$

A linear mapping $L: E \times R^2 \to E \times R^2$ is said to be *pseudo-orthogonal* if it is bijective and

$$\langle Lu_1, Lu_2 \rangle_1 = \langle u_1, u_2 \rangle_1$$

for any $u_1, u_2 \in E \times R^2$. The collection of all pseudo-orthogonal transformations of the space $E \times R^2$ will be denoted by $O_1(E \times R^2)$. Obviously, $O_1(E \times R^2)$ is a group.

Denote by K_E the set of all vectors $u \in E \times R^2$ such that $u \neq 0$ and $\langle u, u \rangle_1 = 0$. This set is a cone in $E \times R^2$ with vertex at 0. If l is the line passing through the points 0 and u, where $u \in K_E$, then $l \setminus \{0\} \subset K_E$. In this case we call $l \setminus \{0\}$ a *generator* of the cone $\mathbf{K_E}$. Denote by $\mathbf{K_E^+}$ the collection of all vectors $u = (\xi, \eta, \zeta) \in \mathbf{K_E}$ for which $\zeta > 0$, and let $\mathbf{K_E^-}$ be the set of all $u = (\xi, \eta, \zeta) \in \mathbf{K_E}$ for which $\zeta < 0$. The closure of the set $\mathbf{K_E^+}$ is the collection of all $u = (\xi, \eta, \zeta)$ such that $|\xi|^2 + \eta^2 - \zeta^2 = 0$ and $\zeta \geq 0$, and, similarly,

$$\overline{\mathbf{K_E^-}} = \{(\xi, \eta, \zeta) \in \mathbf{E} \times \mathbf{R}^2 \mid |\xi|^2 + \eta^2 - \zeta^2 = 0 \,\&\, \zeta \leq 0\}.$$

If $u = (\xi, \eta, \zeta)$ belongs to each of the sets $\overline{\mathbf{K_E^-}}$ and $\overline{\mathbf{K_E^+}}$, then $|\xi|^2 + \eta^2 - \zeta^2 = 0$ and $\zeta = 0$, which implies that $\eta = 0$ and $\xi = 0$, so that the intersection $\overline{\mathbf{K_E^-}} \cap \overline{\mathbf{K_E^+}}$ consists of a single element—the point 0. If $u = (\xi, \eta, \zeta) \in \mathbf{K_E}$, then $\zeta \neq 0$, for otherwise we would get that $\eta = 0$ and $\xi = 0$. However, the point 0 is excluded from the cone $\mathbf{K_E}$. Consequently,

$$\mathbf{K_E} = \mathbf{K_E^+} \cap \mathbf{K_E^-}.$$

Further, $\mathbf{K_E^+} \cap \mathbf{K_E^-} = \varnothing$ and $\overline{\mathbf{K_E^+}} \cap \overline{\mathbf{K_E^-}} = \{0\}$.

If $u \in \mathbf{K_E^+}$, then $-u \in \mathbf{K_E^-}$, and if $u \in \mathbf{K_E^-}$, then $-u \in \mathbf{K_E^+}$. We introduce some notation. Let e_1 and e_2 be the vectors $e_1 = (0, 1, 0)$ and $e_2 = (0, 0, 1)$ in $\mathbf{E} \times \mathbf{R}^2$.

We single out a certain subset $\mathbf{O_1^+}(\mathbf{E} \times \mathbf{R}^2)$ of $\mathbf{O_1}(\mathbf{E} \times \mathbf{R}^2)$. Let $L \in \mathbf{O_1}(\mathbf{E} \times \mathbf{R}^2)$ and $u = (\xi, \eta, \zeta) = Le_2$. We have that

$$|\xi|^2 + \eta^2 - \zeta^2 = \langle u, u \rangle_1 = \langle e_2, e_2 \rangle_1 = -1.$$

From this, $\zeta^2 = |\xi|^2 + \eta^2 + 1$, and hence $|\zeta| \geq 1$. Denote by $\mathbf{O_1^+}(\mathbf{E} \times \mathbf{R}^2)$ the collection of all $L \in \mathbf{O_1}(\mathbf{E} \times \mathbf{R}^2)$ such that the coordinate ζ of the vector $u = L(e_2)$ is positive.

LEMMA 3.1. *If $L \in \mathbf{O_1^+}(\mathbf{E} \times \mathbf{R}^2)$, then $L(\mathbf{K_{n+1}^+}) \subset \mathbf{K_{n+1}^+}$. Conversely, if $L \in \mathbf{O_1}(\mathbf{E} \times \mathbf{R}^2)$ is such that $L(\mathbf{K_E^+}) \subset \mathbf{K_E^+}$, then $L \in \mathbf{O_1^+}(\mathbf{E} \times \mathbf{R}^2)$.*

PROOF. Let $h = (p, q, r) = L(e_2)$ and $v = (\alpha, \beta, \gamma) = L(u)$, where $u = (\xi, \eta, \zeta) \in \mathbf{K_E^+}$. Then

$$\langle v, v \rangle_1 = \langle u, u \rangle_1 = 0 \quad \text{and} \quad \langle v, h \rangle_1 = \langle u, e_2 \rangle_1 = -\zeta < 0.$$

Further, $\langle v, v \rangle_1 = |\alpha|^2 + \beta^2 - \gamma^2$, and hence $|\gamma| = \sqrt{|\alpha|^2 + \beta^2}$. We have that

$$\langle h, h \rangle_1 = |p|^2 + q^2 - r^2 = \langle e_2, e_2 \rangle_1 = -1;$$

thus, $|p|^2 + q^2 = r^2 - 1$. It is required to prove that $\gamma > 0$. Assume, on the contrary, that $\gamma \leq 0$. We have that

$$0 > \langle v, h \rangle_1 = \langle \alpha, p \rangle + \beta q - \gamma z = |\gamma| r + \langle \alpha, p \rangle + \beta q$$

$$\geq |\gamma| r - |\alpha||p| - |\beta||q| \geq |\gamma| r - \sqrt{|\alpha|^2 + \beta^2}\sqrt{|p|^2 + q^2}$$

$$= |\gamma| r - |\gamma|\sqrt{r^2 - 1} = |\gamma|(r - \sqrt{r^2 - 1}) \geq 0.$$

The assumption that $\gamma \leq 0$ thus leads to a contradiction, and hence $\gamma > 0$. Accordingly, we get that if $u \in \mathbf{K}_\mathbf{E}^+$, then also $v = L(e) \in \mathbf{K}_\mathbf{E}^+$.

Assume that $L \in \mathbf{O}(\mathbf{E} \times \mathbf{R}^2)$ is such that $L(\mathbf{K}_\mathbf{E}^+) \subset \mathbf{K}_\mathbf{E}^+$. The vectors $a = (0, 1, 1)$ and $b = (0, -1, 1)$ belong to $\mathbf{K}_\mathbf{E}^+$; hence $L(a) = (\xi_1, \eta_1, \zeta_1) \in \mathbf{K}_\mathbf{E}^+$ and $L|b| = (\xi_2, \eta_2, \zeta_2) \in \mathbf{K}_\mathbf{E}^+$. We have that $e_2 = \frac{1}{2}(a + b)$. Hence,

$$(p, q, r) = L(e) = \frac{1}{2}(L(a) + L(b)).$$

In view of the condition $L(\mathbf{K}_\mathbf{E}^+) \subset \mathbf{K}_\mathbf{E}^+$ we have $\zeta_1, \zeta_2 > 0$, which give us that $r = \frac{1}{2}(\zeta_1 + \zeta_2) > 0$, i.e., $L \in \mathbf{O}_1^+(\mathbf{E} \times \mathbf{R}^2)$. The lemma is proved.

Let $L \in \mathbf{O}_1(\mathbf{E} \times \mathbf{R}^2)$. It follows from the definition that $L(\mathbf{K}_\mathbf{E}) \subset \mathbf{K}_\mathbf{E}$. We show that, actually, $L(\mathbf{K}_\mathbf{E}) = \mathbf{K}_\mathbf{E}$. Let $u \in \mathbf{K}_\mathbf{E}$ be arbitrary. Since L is bijective, there exists a $v \in \mathbf{E}$ such that $L(v) = u$. We have that

$$\langle v, v \rangle_1 = \langle Lv, Lv \rangle_1 = \langle u, u \rangle_1 = 0,$$

i.e., $v \in \mathbf{K}_\mathbf{E}$. Accordingly, for every $u \in \mathbf{K}_\mathbf{E}$ there exists a $v \in \mathbf{K}_\mathbf{E}$ such that $L(v) = u$, and hence $L(\mathbf{K}_\mathbf{E}) \supset \mathbf{K}_\mathbf{E}$.

If $L \in \mathbf{O}_1^+(\mathbf{E} \times \mathbf{R}^2)$, then $L(\mathbf{K}_\mathbf{E}^+) \subset \mathbf{K}_\mathbf{E}^+$. Further, if $u \in \mathbf{K}_\mathbf{E}^-$, then $-u \in \mathbf{K}_\mathbf{E}^+$, which implies that $-L(u) = L(-u) \in \mathbf{K}_\mathbf{E}^+$, and thus $L(u) \in \mathbf{K}_\mathbf{E}^-$, i.e., we get that $L(\mathbf{K}_\mathbf{E}^-) \subset \mathbf{K}_\mathbf{E}^-$. Since $L(\mathbf{K}_\mathbf{E}) = \mathbf{K}_\mathbf{E}$ and $\mathbf{K}_\mathbf{E} = \mathbf{K}_\mathbf{E}^+ \cup \mathbf{K}_\mathbf{E}^-$, and since $\mathbf{K}_\mathbf{E}^+ \cap \mathbf{K}_\mathbf{E}^- = \varnothing$, it follows from what has been proved that $L(\mathbf{K}_\mathbf{E}^+) = \mathbf{K}_\mathbf{E}^+$ and $L(\mathbf{K}_\mathbf{E}^-) = \mathbf{K}_\mathbf{E}^-$, i.e., a transformation of the class $\mathbf{O}_1(\mathbf{E} \times \mathbf{R}^2)$ carries each of the halves $\mathbf{K}_\mathbf{E}^+$ and $\mathbf{K}_\mathbf{E}^-$ of the cone $\mathbf{K}_\mathbf{E}$ into itself.

We now construct certain special mappings of $\mathbf{K}_\mathbf{E}$ onto $\overline{\mathbf{E}}$.

Let $u = (\xi, \eta, \zeta) \in \mathbf{E} \times \mathbf{R}^2$. In the space $\mathbf{E}$ consider the equation

$$k\langle x, x \rangle - 2\langle \xi, x \rangle + l = 0, \tag{3.15}$$

where $k = -\eta + \zeta$ and $l = \eta + \zeta$. We determine conditions under which the set of solutions of (3.15) consists of a unique point. Obviously, for this it is necessary that k be nonzero. Let $k \neq 0$. Then

$$k\langle x, x \rangle - 2\langle \xi, x \rangle - l = k\left(\langle x, x \rangle - 2\left\langle \frac{\xi}{k}, x \right\rangle + \frac{l}{k}\right)$$
$$= k\left(\left\langle x - \frac{\xi}{k}, x - \frac{\xi}{k} \right\rangle - \frac{\langle \xi, \xi \rangle - lk}{k^2}\right). \tag{3.16}$$

From this it is clear that (3.15) is uniquely solvable if and only if $k \neq 0$ and $\langle \xi, \xi \rangle - l_k = 0$. We have that

$$\langle \xi, \xi \rangle - lk = \langle \xi, \xi \rangle + \eta^2 - \zeta^2 = \langle u, u \rangle_1;$$

consequently, (3.15) is uniquely solvable in $\mathbf{E}$ if and only if $u \in \mathbf{K}_\mathbf{E}$, and $\eta \neq \zeta$.

We now define a certain mapping $\tau\colon \mathbf{K_E} \to \overline{\mathbf{E}}$. Let $u = (\xi, \eta, \zeta) \in \mathbf{K_E}$. If $\eta = \zeta$, then the condition $\langle u, u \rangle_1 = 0$ implies that $\xi = 0$, and hence $u = (0, \eta, \eta)$. In this case we set $\tau(u) = \infty$. Suppose that $\eta \neq \zeta$. Then the equation

$$(-\eta + \zeta)\langle x, x \rangle - 2\langle \xi, x \rangle + \zeta + \eta = 0$$

has a unique solution $x \in \mathbf{E}$, which is taken to be the point $\tau(u)$. It is not hard to see from (3.16) that

$$\tau(u) = \frac{\xi}{k} = (1/(-\eta + \zeta))\xi.$$

We show that τ is a mapping of $\mathbf{K_E}$ onto $\overline{\mathbf{E}}$. Take an arbitrary point $x_0 \in \overline{\mathbf{E}}$. It is required to prove that there is then a point $u \in \mathbf{K_E}$ such that $\tau(u) = x_0$. If $x_0 = \infty$, then the existence of the desired u follows immediately from the definition of τ. Suppose that $x_0 \neq \infty$. We construct an equation of the form (3.15) whose solution set consists of the single point x_0. Obviously,

$$\langle x - x_0, x - x_0 \rangle = \langle x, x \rangle - 2\langle x_0, x \rangle + \langle x_0, x_0 \rangle = 0$$

is such an equation. We let $\xi = x_0$ and determine η and ζ from the equations

$$-\eta + \zeta = 1, \qquad \eta + \zeta = \langle x_0, x_0 \rangle.$$

From this, $\eta = (\langle x_0, x_0 \rangle - 1)/2$ and $\zeta = (\langle x_0, x_0 \rangle + 1)/2$. Let

$$u = (\xi, \eta, \zeta) = (x_0, (\langle x_0, x_0 \rangle - 1)/2, (\langle x_0, x_0 \rangle + 1)/2). \qquad (3.17)$$

It is easily verified that $u \in \mathbf{K_E}$ and $\tau(u) = x_0$, which is what was required to prove.

If $x = \tau(u_1) = \tau(u_2)$, then $u_2 = \alpha u_1$, where $\alpha \neq 0$. Indeed, if $x = \infty$, then $u_1 = (0, \eta_1, \eta_1)$ and $u_2 = (0, \eta_2, \eta_2)$, where $\eta_1, \eta_2 \neq 0$, and hence $u_2 = \alpha u_1$ for $\alpha = \eta_2/\eta_1$. Suppose that $x \neq \infty$. Then

$$x = \frac{\xi_1}{-\eta_1 + \zeta_1} = \frac{\xi_2}{-\eta_2 + \zeta_2},$$

from which $\xi_2 = \alpha \xi_1$, where $\alpha = (\zeta_2 - \eta_2)(\zeta_1 - \eta_1)$. It then follows from the equalities

$$\langle \xi_1, \xi_1 \rangle + (\eta_1 - \zeta_1)(\eta_1 + \zeta_1) = 0,$$
$$\langle \xi_2, \xi_2 \rangle + (\eta_2 - \zeta_2)(\eta_2 + \zeta_2) = 0$$

that also $\eta_2 + \zeta_2 = \alpha(\eta_1 + \zeta_1)$, which gives us that $\eta_2 = \alpha \eta_1$ and $\zeta_2 = \alpha \zeta_1$, as required.

We get that for every $x \in \overline{\mathbf{E}}$ the set $\tau^{-1}(x)$ is a generator of the cone $\mathbf{K_E}$. Every point $u \in \tau^{-1}(x)$, $u = (\xi, \eta, \zeta)$, will be called a *vector of hyperspherical coordinates* of the point x, and we say correspondingly that ξ, η, ζ are *hyperspherical coordinates* of x.

The triple $(\mathbf{K_E}, \tau, \mathbf{E})$ is called the *hypersphere bundle* over $\overline{\mathbf{E}}$; $\overline{\mathbf{E}}$ is called the *base* of the bundle, $\mathbf{K_E}$ its *space*, and the mapping τ the *projection*. We remark that in older references the phrase "system of coordinates" is used instead of the word "bundle". However, according to the latest ideas a system of coordinates is always a one-to-one mapping. On the other hand, there is the general concept of a bundle, and a hypersphere bundle is a special case of this. In cases when the dimension n of $\mathbf{E}$ is finite, the given bundle is called a *quadrisphere bundle* or a *pentasphere bundle* if n is equal to 2 or 3, respectively, so in the general case we should call it an $(n + 2)$-*sphere bundle*.

Let H be a sphere in $\overline{\mathbf{E}}$. We determine the set $\tau^{-1}(H)$. Assume that H is given by

$$k|x|^2 - 2\langle a, x\rangle + l = 0, \tag{3.18}$$

where $|\alpha|^2 - kl > 0$, and $k \neq 0$. Let $u = (\xi, \eta, \zeta)$ be a point of the cone $\mathbf{K_E}$ such that $x = \tau(u) \in H$. Then $|\xi|^2 + \eta^2 - \zeta^2 = 0$, and since $\infty \notin H$, it follows that $\tau(u) \neq \infty$, and hence $\eta \neq \zeta$ and $\tau(u) = \xi/(-\eta+\zeta)$. Substituting the value $x = \tau(u)$ into (3.18), we get that

$$0 = k\frac{|\xi|^2}{|\zeta - \eta|^2} - 2\left\langle a, \frac{\xi}{\zeta - \eta}\right\rangle + l = k\frac{\zeta + \eta}{\zeta - \eta} - 2\left\langle a, \frac{\xi}{\zeta - \eta}\right\rangle + l.$$

From this,

$$\langle a, \xi\rangle + b\eta - c\zeta = 0, \tag{3.19}$$

where

$$b = \frac{-k + l}{2}, \quad c = \frac{k + l}{2}, \quad |a|^2 + b^2 - c^2 = |a|^2 - kl > 0. \tag{3.20}$$

Conversely, if (3.19) holds for the point $u = (\xi, \eta, \zeta) \in \mathbf{K_E}$, then in view of the conditions $k \neq 0$, $b - c \neq 0$, and $\eta \neq \zeta$, we get by reversing the computations that $\tau(u) \in H$. This allows us to conclude that $\tau^{-1}(H)$ is the section of the cone $\mathbf{K_E}$ by the plane (3.19) in this case. Let us now consider the case when H is the closure of the hyperplane

$$\{x \in \mathbf{E} | \langle a, x\rangle - l/2 = 0\},$$

where $a \neq 0$. Let the point $u = (\xi, \eta, \zeta) \in \mathbf{K_E}$ be such that $\tau(u) \in H$. If $\eta \neq \zeta$, then, as in the preceding case, we get that ξ, η, and ζ satisfy (3.19) with values of b and c determined according to (3.20) with $k = 0$, and, conversely, if $(\xi, \eta, \zeta) \in \mathbf{K_E}$ satisfies condition (3.19) and $\eta \neq \zeta$, then $\tau(u) \in H$. But if $\eta = \zeta$, then $\zeta = 0$, and $\tau(u) = \infty \in H$. In the given case $b - c = 0$, and the point $u(0, \eta, \eta)$ satisfies (3.19).

Accordingly, if H is a sphere in $\overline{\mathbf{E}}$, then the set $\tau^{-1}(H)$ is the section of $\mathbf{K_E}$ by the plane given by (3.19), where the vectors $a \in \mathbf{E}$ and the numbers $b, c \in \mathbf{R}$ are such that

$$|a|^2 + b^2 - c^2 > 0. \tag{3.21}$$

Conversely, suppose that a section of $\mathbf{K_E}$ by a plane of the form (3.19) is given, where a, b, and c satisfy (3.21). Let $k = -(b+c)$ and $l = b-c$, and let H be the sphere in $\overline{\mathbf{E}}$ determined by the equation $k|x|^2 - 2\langle a, x\rangle + l = 0$. Then, as follows from the preceding arguments, the set $\tau^{-1}(H)$ coincides with the given section of $\mathbf{K_E}$.

LEMMA 3.2. *If* $\varphi \in \mathbf{O}_1^+(\mathbf{E} \times \mathbf{R}^2)$ *carries every generator of the cone* $\mathbf{K_E}$ *into itself, then* φ *is the identity transformation.*

PROOF. Assume that φ satisfies the condition of the lemma. Then for every vector $u \in \mathbf{K_E^+}$ we have that $\varphi(u) = \lambda(u)u$, where $\lambda(u) \in \mathbf{R}$. Further, it is obvious that $\lambda(u) > 0$. Let $p = (0, 1, 1)$ and $q = (0, -1, 1)$. The vectors p and q belong to the cone $\mathbf{K_E^+}$. By assumption, $\varphi(p) = \lambda_1 p$ and $\varphi(q) = \lambda_2 q$, where $\lambda_1, \lambda_2 > 0$. Take an arbitrary vector $u \in \mathbf{K_E}$ which is not collinear to one of the vectors p and q. We have that $u = (\xi, \eta, \zeta)$, where $\xi \in \mathbf{E}$ and $\xi \neq 0$. Without loss of generality ξ can be assumed to be a unit vector. Then $\eta^2 - \zeta^2 = -1$. Let η_1 and ζ_1 be another pair of numbers such that $\eta_1^2 - \zeta_1^2 = -1$, and suppose that $\eta_1 + \zeta_1 \neq \eta + \zeta$. The vector $v = (\xi, \eta_1, \zeta_1)$ also belongs to $\mathbf{K_E}$ and is a linear combination of the vectors p, q, and u, namely, $v = \alpha p + \beta q + u$, where

$$\alpha = \frac{1}{2}[\eta_1 + \zeta_1 - (\eta + \zeta)],$$

$$\beta = \frac{1}{2}[\zeta_1 - \eta_1 - (\zeta - \eta)].$$

Since $\eta^2 - \zeta^2 = \eta_1^2 - \zeta_1^2 \neq 0$, the fact that $\eta_1 + \zeta_1 \neq \eta + \zeta$ implies that also $\eta_1 - \zeta_1 \neq \eta - \zeta$, and hence $\alpha, \beta \neq 0$. Let $\varphi(u) = \lambda_3 u$. Since $v \in \mathbf{K_E}$, it follows that $\varphi(v) = \mu v$. We have that

$$\varphi(v) = \varphi(\alpha p + \beta q + u) = \alpha\varphi(p) + \beta\varphi(q) + \varphi(u)$$
$$= \alpha\lambda_1 p + \beta\lambda_2 q + \lambda_3 u.$$

On the other hand, $\varphi(v) = \mu(\alpha p + \beta q + u)$. Since the vectors p, q, and u are linearly independent, this implies that $\mu\alpha = \alpha\lambda_1$, $\mu\beta = \beta\lambda_2$, and $\mu = \lambda_3$. And since $\alpha, \beta \neq 0$, this implies that $\lambda_1 = \lambda_2 = \lambda_3 = \lambda$. In particular, $\varphi(p) = \lambda p$, $\varphi(q) = \lambda q$, and $\varphi(u) = \lambda u$. Consequently, the arbitrariness of $u \in \mathbf{K_E}$ gives us that $\varphi(u) = \lambda u$ for all $u \in \mathbf{K_E}$, where $\lambda = \text{const}$. Let $x = (\xi, \eta, \zeta)$ be an arbitrary vector in $\mathbf{E} \times \mathbf{R}^2$. The vector $u = (\xi, 0, |\zeta|)$

belongs to $\mathbf{K_E}$, and the vector x can be represented as a linear combination of the vectors p, q, and u: $x = \alpha p + \beta q + \gamma u$. Since φ is linear, this implies that

$$\varphi(x) = \alpha\varphi(p) + \beta\varphi(q) + \gamma\varphi(u) = \lambda(\alpha p + \beta q + \gamma u) = \lambda x.$$

Thus, $\varphi(x) = \lambda x$ for all $x \in \mathbf{E} \times \mathbf{R}^2$. It follows from the condition $\langle x, x \rangle_1 = \langle \varphi(x), \varphi(x) \rangle_1$ for any $x \in \mathbf{E} \times \mathbf{R}^2$ that $\lambda^2 = 1$, and hence $\lambda = 1$, because $\lambda > 0$. We thus get that φ is the identity mapping. The lemma is proved.

THEOREM 3.8. *For every $\varphi \in \mathbf{M}(\overline{\mathbf{E}})$ there exists a transformation $T_\varphi \in \mathbf{O}_1^+(\mathbf{E} \times \mathbf{R}^2)$ such that the mapping diagram*

$$
\begin{array}{ccc}
\mathbf{K_E} & \xrightarrow{T_\varphi} & \mathbf{K_E} \\
\tau \downarrow & & \downarrow \tau \\
\overline{\mathbf{E}} & \xrightarrow{\varphi} & \overline{\mathbf{E}}
\end{array}
\qquad (3.22)
$$

is commutative, i.e., $\varphi(\tau(u)) = \tau[T_\varphi(u)]$ for every $u \in \mathbf{K_E}$. Such a mapping T_φ is unique, and the correspondence

$$\varphi \mapsto T_\varphi$$

is an isomorphism of the groups $\mathbf{M}(\overline{\mathbf{E}})$ and $\mathbf{O}_1^+(\mathbf{E} \times \mathbf{R}^2)$.

REMARK 1. Intuitively, the commutativity of the diagram (3.22) means the following. Let $x \in \overline{\mathbf{E}}$ and let $u = (\xi, \eta, \zeta)$ be a vector of polyspherical coordinates of x. Applying to u the transformation T_φ, we get a vector of hyperspherical coordinates of the point $\varphi(x)$.

REMARK 2. The isomorphism $\varphi \mapsto T_\varphi$ of $\mathbf{M}(\overline{\mathbf{E}})$ and $\mathbf{O}_1^+(\mathbf{E} \times \mathbf{R}^2)$ to be established in Theorem 3.8 will be called the *canonical isomorphism* in what follows.

PROOF. We first establish the uniqueness of a transformation T_φ satisfying the condition in the theorem if it exists. Let $\varphi \in \mathbf{M}(\overline{\mathbf{E}})$ and let T_1 and T_2 be transformations in $\mathbf{O}_1^+(\mathbf{E} \times \mathbf{R}^2)$ such that $\varphi \circ \tau = \tau \circ T_1$ and $\varphi \circ \tau = \tau \circ T_2$. Take an arbitrary generator of the cone $\mathbf{K_E}$, let u be a point on this generator, and let $x = \varphi(u)$. Then $\varphi(x) = \tau(T_1(u)) = \tau(T_2(u))$, and hence $T_1(u)$ and $T_2(u)$ lie on a single generator of $\mathbf{K_E}$. This implies that T_1 and T_2 carry the generator of $\mathbf{K_E}$ passing through a point $u \in \mathbf{K_E}$ into the same generator, which allows us to conclude that the mapping $T_2^{-1} \circ T_1$ carries every generator of $\mathbf{K_E}$ into itself, and thus $T_2^{-1} \circ T_1 = I$, by Lemma 3.2, i.e., $T_1 = T_2$.

To construct the required transformation T_φ we consider first the case when φ is one of the following transformations:

1)$\pi_a: x \mapsto x + a$;   2)$h_\lambda: x \mapsto \lambda x$, $\lambda \neq 0$;

3)$\varphi_A: x \mapsto Ax$, $A \in \mathbf{O_E}$;   4)$j: x \mapsto x/|x|^2$.

Suppose that φ is one of the transformations π_a, h_λ, or φ_A. Take an arbitrary vector $u \in \mathbf{K_E}, u = (\xi, \eta, \zeta)$. Assume first that $\eta \neq \zeta$ and let $k = -\eta + \zeta$, $l = \eta + \zeta$, and $x = (1/k)\xi = \tau(u)$. The point x is the unique solution of the equation

$$k\langle x, x \rangle - 2\langle \xi, x \rangle + l = 0. \tag{3.23}$$

Let $y = \varphi(x)$. Then $x = \varphi^{-1}(y)$. Substituting the expression for x into (3.22), we get after the obvious transformations that

$$k_1 \langle y, y \rangle - 2\langle \xi_1, y \rangle + l_1 = 0, \tag{3.24}$$

where k_1, ξ_1, and l_1 are expressed in terms of k, ξ, and l as follows. If $\varphi = \pi_a$, then

$$k_1 = k, \quad \xi_1 = \xi + ka, \quad l_1 = k|a|^2 + 2\langle \xi, a \rangle + l, \tag{3.25}$$

if $\varphi = h_\lambda$, then

$$k_1 = \frac{k}{\lambda}, \quad \xi_1 = \xi, \quad l_1 = \lambda l \tag{3.26}$$

and, finally, if $\varphi = \varphi_A$, then

$$k_1 = k, \quad \xi_1 = A\xi, \quad l_1 = l. \tag{3.27}$$

The transformation T_φ is now constructed as follows. Let $u = (\xi, \eta, \zeta) \in \mathbf{E} \times \mathbf{R}^2$ (u is not necessarily in $\mathbf{K_E}$). For u we define a vector $\xi_1 \in \mathbf{E}$ and numbers k_1 and l_1 by using (3.25) if $\varphi = \pi_a$, (3.26) if $\varphi = h\lambda$, and (3.27) if $\varphi = \varphi_A$. Let $\eta_1 = (-k_1 + l_1)/2$ and $\zeta_1 = (k_1 + l_1)/2$. We get that $T_\varphi(u) = (\xi_1, \eta_1, \zeta_1)$. The mapping T_φ is linear.

In all cases equalities (3.25)–(3.27) allow us to conclude that

$$|\xi_1|^2 + \eta_1^2 - \zeta_1^2 = |\xi_1|^2 - l_1 k_1 = |\xi|^2 - lk = |\xi|^2 + \eta^2 - \zeta^2,$$

which gives us that $T_\varphi \in \mathbf{O_1}(\mathbf{E} \times \mathbf{R}^2)$. Further, if $u = (\xi, \eta, \zeta) \in \mathbf{K_E}$ is such that $\zeta > 0$, then in view of the equality $|\xi|^2 = \zeta^2 - \eta^2$ we get that $\zeta^2 - \eta^2 \geq 0$, and hence $\zeta \geq |\eta|$. By using this inequality it is not hard to show that in each of the three cases $\varphi = \pi_a$, $\varphi = h\lambda$, and $\varphi = \varphi_A$ the component ζ of the vector $T_\varphi(u)$ is positive. Accordingly, T_φ transforms the upper half of the cone $\mathbf{K_E}$ into itself, and thus $T_\varphi \in \mathbf{O_1^+}(\mathbf{E} \times \mathbf{R}^2)$.

We show that the mapping T_φ we have constructed satisfies the condition $\tau \circ T_\varphi = \varphi \circ \tau$. Indeed, (3.24) is uniquely solvable, since (3.23) is uniquely solvable, and its solution is $y = \tau(u_1) = \tau[T_\varphi(u)]$. On the other hand, $y = \varphi(x) = \varphi[\tau(u)]$, i.e., $\tau[T_\varphi(x)] = \varphi[\tau(u)]$. This argument is valid only if $\eta \neq \zeta$ for the vector u. But if $\eta = \zeta$, then $u = (0, \eta, \eta)$, and $\tau(u) = \infty$.

In this case $\varphi(\infty) = \infty$. It is easy to verify that $T_\varphi(u) = u$ for the given u if φ is any one of the mappings π_a, $h\lambda$, or φ_A; hence $\tau[T_\varphi(u)] = \varphi[\tau(u)]$ also in this case.

Let $\varphi = j_0$. For a vector $u = (\xi, \eta, \zeta) \in \mathbf{E} \times \mathbf{R}^2$ let $T_\varphi(u) = (\xi, -\eta, \zeta)$. Then $T_\varphi \in \mathbf{O}_1^+(\mathbf{E} \times \mathbf{R}^2)$. It is easy to verify that $\varphi \circ \tau = \tau \circ T_\varphi$.

Let φ be an arbitrary Möbius transformation. Then φ can be represented in the form

$$\varphi = \varphi_1 \circ \varphi_2 \circ \cdots \circ \varphi_m,$$

where φ_i is a transformation of one of the following types for each i: $\pi_a, h\lambda, \varphi_A$, or j. Let $T_\varphi = T_{\varphi_1} \circ \cdots \circ T_{\varphi_m}$. It is easy to verify that $\varphi \circ \tau = \tau \circ T_\varphi$; hence T_φ is the desired mapping.

We now prove that the correspondence $\varphi \in \mathbf{M}(\overline{\mathbf{E}}) \mapsto T_\varphi$ is a homomorphism. Let $\varphi \in \mathbf{M}(\overline{\mathbf{E}})$ be arbitrary. For every $u \in \mathbf{K}_\mathbf{E}$ we have that $\varphi[\tau(u)] = \tau[T_\varphi(u)]$. From this, $\tau(u) = \varphi^{-1}[\tau(T_\varphi(u))]$. Replacing u by $T_\varphi^{-1}(u)$ here, we get that for any $u \in \mathbf{K}_\mathbf{E}$

$$\varphi^{-1}[\tau(u)] = \tau[T_\varphi^{-1}(u)],$$

and hence, since T_φ is unique,

$$T_\varphi^{-1} = T_{\varphi^{-1}}. \tag{3.28}$$

Let $\varphi, \psi \in \mathbf{M}(\overline{\mathbf{E}})$. We have that $\varphi[\tau(u)] = \tau[T_\varphi(u)]$, and so

$$\psi[\varphi(\tau(u))] = \psi(\tau[T_\varphi(u)]) = \tau[T_\psi(T_\varphi(u))],$$

i.e., $(\psi \circ \varphi) \circ \tau = \tau \circ (T_\psi \circ T_\varphi)$. On the other hand,

$$(\psi \circ \varphi)[\tau(u)] = \tau[T_{\psi \circ \varphi}(u)].$$

Comparing the last two equalities, we conclude that

$$T_{\psi \circ \varphi} = T_\psi \circ T_\varphi. \tag{3.29}$$

Equalities (3.28) and (3.29) mean that the correspondence $\varphi \mapsto T_\varphi$ is a homomorphism.

We now prove that the correspondence $\varphi \mapsto T_\varphi$ is one-to-one.

Let $\varphi_1 \in \mathbf{M}(\overline{\mathbf{E}})$ and $\varphi_2 \in \mathbf{M}(\overline{\mathbf{E}})$ be such that $T_{\varphi_1} = T_{\varphi_2} = T$. Let $x \in \overline{\mathbf{E}}$ be arbitrary, and let $u \in \mathbf{K}_\mathbf{E}$ be such that $\tau(u) = x$. Then

$$\varphi_1(x) = \varphi_1[\tau(u)] = \tau[T_{\varphi_1}(u)] = \tau[T(u)]$$

and

$$\varphi_2(x) = \varphi_2[\tau(u)] = \tau[T_{\varphi_2}(u)] = \tau[T(u)],$$

and hence $\varphi_1(x) = \varphi_2(x)$. Thus, $\varphi_1(x) = \varphi_2(x)$ for all $x \in \overline{\mathbf{E}}$, i.e., $\varphi_1 \equiv \varphi_2$. Accordingly, if $T_{\varphi_1} = T_{\varphi_2}$, then $\varphi_1 = \varphi_2$, and it is proved that the mapping $\varphi \mapsto T_\varphi$ is one-to-one.

We show that $\varphi \in M(E) \mapsto T_\varphi \in O_1^+(E \times R^2)$ is "onto". Let $T \in O_1^+(E \times R^2)$ be arbitrary and let $x \in \overline{E}$. We find the generator $\tau^{-1}(x)$ of the cone K_E. This generator is carried by T into some other generator λ of the cone. The set $\tau(\lambda)$ consists of a unique point y which we denote by $\varphi(x)$. Since $x \in \overline{E}$ was arbitrary, this defines a mapping $\varphi: \overline{E} \to E$. Further, as follows from the construction of φ, $\varphi(\tau(u)) = \tau[T(u)]$ for every $u \in K_E$. We show that φ is a Möbius mapping. If $x_1 \neq x_2$, then the generators $\alpha_1 = \tau^{-1}(x)$ and $\alpha_2 = \tau^{-1}(x)$ of K_E are distinct, and hence so are the generators $\beta_1 = T(\alpha_1)$ and $\beta_2 = T(\alpha_2)$. We have that $\varphi(x_1) = \tau(\beta_1)$ and $\varphi(x_2) = \tau(\beta_2)$, which implies that $\varphi(x_1) \neq \varphi(x_2)$. Accordingly, φ is injective. Let us prove that $\varphi(\overline{E}) = \overline{E}$. Indeed, take any $y \in \overline{E}$ and let $\beta = \tau^{-1}(y)$ be the generator of K_E corresponding to y; let $\alpha = T^{-1}(\beta)$. If $x = \tau(\alpha)$, then $\varphi(x) = y$. Thus, φ is a bijective mapping, and $\varphi \circ \tau = \tau \circ T$.

We prove that $\varphi \in M(\overline{E})$. For this it suffices to establish that φ satisfies the conditions of Theorem 3.5. Let H be an arbitrary sphere in $\overline{E}$. We have that $H = \tau(K_E \cap S)$, where S is the subspace of $E \times R^2$ determined by the equation $\langle p, u \rangle_1 = 0$, and the vector $p \in E \times R^2$ is such that $\langle p, p \rangle_1 > 0$. Let $q = Tu$. We have that

$$u \in T(S) \Leftrightarrow T^{-1}u \in S \Leftrightarrow \langle p, T^{-1}u \rangle_1 = 0.$$

But $\langle p, T^{-1}u \rangle_1 = \langle T^{-1}q, T^{-1}u \rangle_1 = \langle q, u \rangle$, and hence

$$T(S) = \{u \in E \times R^2 \langle q, u \rangle = 0\}.$$

For the vector q we have that

$$\langle q, q \rangle_1 = \langle Tp, Tp \rangle_1 = \langle p, p \rangle_1 > 0.$$

Let $\Gamma = T(S) \cap K_E$. The set $\tau(\Gamma)$ is a sphere in $\overline{E}$. It is easy to verify that $\tau(\Gamma) = \varphi(H)$. Thus, φ is a bijective mapping of $\overline{E}$ into $\overline{E}$ which transforms every sphere into a sphere. By Theorem 3.5, this implies that φ is a Möbius transformation. The mapping $T \in O_1^+(E \times R^2)$ was arbitrary, and we have that $T = T_\varphi$; hence it is established that $\varphi \mapsto T_\varphi$ is a mapping of $M(\overline{E})$ onto $O_1^+(E \times R^2)$. The theorem is proved.

§4. Definition of a mapping with bounded distortion

4.1. Orthogonal invariants of linear mappings of Euclidean spaces. A measure of nonorthogonality for a linear mapping. The symbols E and F below denote arbitrary finite-dimensional Euclidean spaces. The inner product of arbitrary vectors x, $y \in E$ is denoted by $\langle x, y \rangle$, and $|x| = \sqrt{\langle x, x \rangle}$ is the length of a vector $x \in E$. Analogous notation is used for the space F. The symbol I_E denotes the identity mapping of E onto itself. (The index E is omitted whenever no confusion is possible.)

Denote by $L(\mathbf{E}, \mathbf{F})$ the collection of all linear mappings of $\mathbf{E}$ into $\mathbf{F}$.

Let $\varphi: \mathbf{E} \to \mathbf{E}$ be a general orthogonal transformation. Then φ carries every sphere in $\mathbf{E}$ into another sphere. The purpose of this section is to construct certain quantities characterizing the degree of nonorthogonality of an arbitrary linear mapping.

Let $A: \mathbf{E} \to \mathbf{F}$ be a linear mapping, and let

$$\|A\| = \sup_{|x| \le 1} |A(x)|.$$

If $A \in L(\mathbf{E}, \mathbf{E})$, then $\det A$ denotes the determinant of A.

Let $A: \mathbf{E} \to \mathbf{F}$ be a given linear mapping. Then there exists a unique linear mapping $B: \mathbf{F} \to \mathbf{E}$ such that $\langle Ax, y \rangle_{\mathbf{F}} = \langle x, By \rangle_{\mathbf{E}}$ for any $x \in \mathbf{E}$ and $y \in \mathbf{F}$. This mapping B is called the *adjoint* of A and denoted by A^*. If Cartesian orthogonal coordinate systems are given in $\mathbf{E}$ and $\mathbf{F}$, then the matrix of A^* with respect to these coordinate systems is the transposed matrix of A.

As above, $\mathbf{O}(\mathbf{E})$ denotes the collection of all orthogonal transformations of $\mathbf{E}$. For every $P \in \mathbf{O}(\mathbf{E})$ we have that $\det P = \pm 1$, $\|P\| = 1$, and $P^* = P^{-1}$.

For an arbitrary linear mapping $A: \mathbf{E} \to \mathbf{F}$ let $\operatorname{Im} A = A(\mathbf{E})$ and $\operatorname{Ker} A = A^{-1}(0)$.

Mappings $A \in L(\mathbf{E}, \mathbf{F})$ and $B \in L(\mathbf{E}, \mathbf{F})$ are said to be *orthogonally equivalent* if there exist transformations $\varphi \in \mathbf{O}(\mathbf{E})$ and $\psi \in \mathbf{O}(\mathbf{F})$ such that $B = \psi \circ A \circ \varphi$. It is easy to verify that the relation of orthogonal equivalence is reflexive, symmetric, and transitive.

Denote the *Kronecker symbol* by δ_{ij}, where $i, j = 1, \ldots, s$; that is, $\delta_{ij} = 1$ for $i = j$, and $\delta_{ij} = 0$ for $i \ne j$. The numbers δ_{ij} are the elements of the $s \times s$ identity matrix.

A system of vectors $u_1, \ldots, u_k$ in $\mathbf{E}$ is said to be *orthogonal* if $\langle u_i, u_j \rangle = \delta_{ij}$ for any $i, j = 1, \cdots, k$.

LEMMA 4.1. *Let $\mathbf{E}$ and $\mathbf{F}$ be finite-dimensional Euclidean spaces, and let $A \in L(\mathbf{E}, \mathbf{F})$, $A \ne 0$. Then the subspaces $\operatorname{Ker} A$ and $\operatorname{Im} A^*$ are completely orthogonal, and their direct sum coincides with $\mathbf{E}$. There exist orthonormal systems of vectors $u_1, \ldots, u_k$ in $\mathbf{E}$ and $v_1, \ldots, v_k$ in $\mathbf{F}$ and numbers $\lambda_i > 0$ ($i = 1, \ldots, k$) such that $(u_1, \ldots, u_k)$ is a basis in $\operatorname{Im} A^*$, $(v_1, \ldots, v_k)$ is a basis in $\operatorname{Im} A$, and*

$$Au_i = \lambda_i v_i, \qquad A^* v_i = \lambda_i u_i \tag{4.1}$$

for each $i = 1, \ldots, k$.

PROOF. Consider the quadratic form

$$Q(x) = \langle Ax, Ax \rangle = \langle A^* Ax, x \rangle \tag{4.2}$$

on E. This quadratic form is nonnegative, and is not identically zero because $A \neq 0$ by assumption. Let $u_1, \ldots, u_n$ be an orthogonal basis in E such that $Q(x)$ can be expressed in terms of the coordinates of x with respect to this basis as follows:

$$Q(x) = \mu_1 x_1^2 + \mu_2 x_2^2 + \cdots + \mu_n x_n^2.$$

Since Q is nonnegative, $\mu_i \geq 0$ for each $i = 1, \ldots, n$, and since $Q \not\equiv 0$, at least one of the coefficients μ_i is nonzero. Without loss of generality it can be assumed that $\mu_i > 0$ for $i \leq k$, where $k \leq n$, and $\mu_i = 0$ for $i > k$. Let $\lambda_i = \sqrt{\mu_i}$, and let

$$v_i = (1/\lambda_i)Au_i, \qquad i = 1, 2, \ldots, k.$$

For arbitrary $x, y \in E$ we have that

$$\langle Ax, Ay \rangle = \frac{1}{2}[Q(x+y) - Q(x) - Q(y)]$$
$$= \lambda_1^2 x_1 y_1 + \lambda_2^2 x_2 y_2 + \cdots + \lambda_k^2 x_k y_k. \qquad (4.3)$$

For any $i, j = 1, \ldots, k$,

$$\langle v_i, v_j \rangle = \frac{1}{\lambda_i \lambda_j} \langle Au_i, Au_j \rangle.$$

Using (4.3), we get from this that $\langle v_i, v_j \rangle = \delta_{ij}$, $i, j = 1, 2, \ldots, k$; that is, the system of vectors $v_1, \ldots, v_k$ is orthogonal.

The set Ker A coincides with the set $\{x \in E | Q(x) = 0\}$, and hence Ker A is the subspace of E spanned by the vectors $u_{k+1}, \ldots, u_n$. In particular, $Au_i = 0$ for $i > k$.

Let us prove that the vectors $v_1, \ldots, v_k$ form a basis for Im A. Take an arbitrary vector $y \in \text{Im } A$. Then $y = Ax$ for some $x \in E$, $x = \sum_1^n x_i u_i$. From this, $y = \sum_1^n x_i Au_i = \sum_1^k \lambda_i x_i v_i$, and hence y belongs to the linear span of the system of vectors $v_1, \ldots, v_k$. Since $y \in \text{Im } A$ was arbitrary and $v_i \in \text{Im } A$ for each $i = 1, \ldots, k$, this implies that the linear span of the system of vectors $v_1, \ldots, v_k$ coincides with Im A.

We now prove that the vectors $u_1, \ldots, u_k$ form a basis for Im A^*. Indeed, let $v_{k+1}, \ldots, v_m$ be such that $v_1, \ldots, v_k, v_{k+1}, \ldots, v_m$ is an orthonormal basis for F. For any $i = 1, \ldots, n$ and $j = 1, \ldots, m$ we have that

$$\langle u_i, A^* v_j \rangle = \langle Au_i, v_j \rangle. \qquad (4.4)$$

For $j > k$ the vector v_j is orthogonal to Im A, and hence $\langle Au_i, v_j \rangle = 0$ in this case. By (4.4), $\langle u_i, A^* v_j \rangle = 0$ for each $i = 1, \ldots, n$ for $j > k$; that is, $A^* v_j = 0$. For $j \leq k$ it follows from (4.4) that $\langle A^* v_j, u_i \rangle = 0$ in the case $i > k$, since then $Au_i = 0$. For $i < k$,

$$\langle A^* v_j, u_i \rangle = \langle v_j, Au_i \rangle = \lambda_i \langle v_i, v_j \rangle = \lambda_i \delta_{ij}.$$

This implies that $A^*v_j = \lambda_j u_j$. Therefore, the vectors $u_1, \ldots, u_k$ belong to Im A^*. For an arbitrary $x \in$ Im A^* we have that $x = A^*y$ for some $y \in \mathbf{F}$, $y = y_1 v_1 + \cdots + y_m v_m$. From this, $x = \sum_1^m y_i A^* v_i = \sum_1^k \lambda_i y_i u_i$. We get that every vector $y \in$ Im A^m is a linear combination of the vectors $u_1, \ldots, u_k$; hence they form a basis for Im A^*.

It remains to observe that Im A^* is the subspace spanned by $u_1, \ldots, u_k$, and Ker A is the subspace spanned by $u_{k+1}, \ldots, u_n$. This implies that these subspaces are completely orthogonal and their direct sum is $\mathbf{E}$. The lemma is proved.

The next lemma enables us to give a precise description of the arbitrariness with which the systems of vectors $u_1, \ldots, u_k$ and $v_1, \ldots, v_k$ in Lemma 4.1 are determined, and establishes that the collection of numbers $\lambda_1, \ldots, \lambda_k$ is uniquely determined by A.

LEMMA 4.2. *Let $A \in L(\mathbf{E}, \mathbf{F})$, and let $\{u_1, \ldots, u_k\}$ and $\{v_1, \ldots, v_k\}$ be orthogonal systems of vectors in $\mathbf{E}$ and $\mathbf{F}$, respectively, such that $\{u_1, \ldots, u_k\}$ is a basis for Im $A^* \subset \mathbf{E}$, $\{v_1, \ldots, v_k\}$ is a basis for Im A, and for every $i = 1, \ldots, k$*

$$Au_i = \lambda_i v_i,$$

where $\lambda_i > 0$. Then $A^*Au_i = \lambda_i^2 u_i$ and $AA^*v_i = \lambda_i^2 v_i$ for each $i = 1, \ldots, k$, and all the remaining eigenvalues of the mappings A^*A and AA^* are equal to zero.

PROOF. Suppose that the systems of vectors $u_1, \ldots, u_k$ and $v_1, \ldots, v_k$ and the numbers λ_i, $i = 1, \ldots, k$, satisfy all the conditions of the lemma. To each of these systems of vectors we add new vectors in order to get an orthonormal basis in the corresponding space. Let $u_1, \ldots, u_n$ and $v_1, \ldots, v_m$ be bases constructed in this way. Let $j > k$; we have that $A^*v_j \in$ Im A^*. The vectors $u_{k+1}, \ldots, u_n$ are orthogonal to Im A^*, and hence $\langle A^*v_j, u_i \rangle = 0$ for $i > k$. For $i \leq k$ we get that

$$\langle A^*v_j, u_i \rangle = \langle v_j, Au_i \rangle = \lambda_i \langle v_j, v_i \rangle = 0.$$

Thus $\langle A^*v_j, u_i \rangle = 0$ for all $i = 1, \ldots, k$, and hence $A^*v_j = 0$ for $j > k$. Let $1 \leq j \leq k$. We have that $A^*v_j = \alpha_1 u_1 + \cdots + \alpha_k u_k$. From this,

$$\alpha_i = \langle u_i, A^*v_j \rangle = \langle Au_i, v_j \rangle = \lambda_i \langle v_i, v_j \rangle = \lambda_i \delta_{ij},$$

and hence $A^*v_j = \lambda_j u_j$. This leads us to conclude that

$$AA^*v_j = \lambda_j Au_j = \lambda_j^2 v_j$$

for $1 \leq j \leq k$. For $j > k$ we have

$$AA^*v_j = A(A^*v_j) = 0.$$

This, $v_1, \ldots, v_k$ are eigenvectors of the mapping AA^*, λ_i^2 is the eigenvalue of this mapping corresponding to v_i, and the remaining eigenvalues of AA^* are equal to zero.

For $i > k$ and $j > k$, $i \neq j$, the quantity $\langle v_j, Au_i \rangle$ vanishes in view of the fact that v_j is orthogonal to Im A. For $i < k$ and $j \leq k$ we have that

$$\langle v_j, Au_i \rangle = \langle A^* v_j, u_i \rangle = \lambda_j \langle u_j, u_i \rangle = 0,$$

and this leads to the conclusion that $Au_i = 0$ for $i = k + 1, \ldots, n$. By assumption, $Au_i = \lambda_i v_i$ for $i \leq k$. In view of what has been proved, this implies that $A^* Au_i = \lambda_i^2 u_i$ for $i = 1, \ldots, k$, and $A^* Au_i = 0$ for $i > k$.

The lemma is proved.

Let $A \in L(\mathbf{E}, \mathbf{F})$, $A \neq 0$, let $\lambda_1, \ldots, \lambda_k$, $\lambda_i > 0 \; \forall i$, be the numbers determined from A according to Lemma 4.1, and let $u_1, \ldots, u_k$ be the orthonormal system of vectors indicated in Lemma 4.1. The quantities $\lambda_1, \ldots, \lambda_k$ are called the *principal dilation coefficients* or *singular numbers* of A. We assume that they are numbered in such a way that $0 < \lambda_1 \leq \lambda_2 \leq \cdots \leq \lambda_k$. The vectors $u_i, i = 1, \ldots, k$, are called *principal vectors* of A. According to the definition, the vectors $v_i = (1/\lambda_i)Au_i, i = 1, \ldots, k$, are principal vectors of A^*. If $A = 0$, then it is convenient to assume that A does not have dilation coefficients, so that in this case the collection of dilation coefficients of A is empty.

Lemmas 4.1 and 4.2 yield a test for orthogonal equivalence of linear mappings.

COROLLARY. *Mappings A, $B \in L(\mathbf{E}, \mathbf{F})$ are orthogonally equivalent if and only if they have the same sets of principal dilation coefficients.*

The proof is left to the reader, since it is not used in what follows.

Let $A: \mathbf{E} \to \mathbf{F}$ be a linear mapping. Then $\|A\|$ is equal to the largest singular number of A. Indeed, let $\lambda_k \geq \cdots \geq \lambda_2 \geq \lambda_1 > 0$ be all the nonzero singular numbers of A, and let $u_1, \ldots, u_n$ be an orthonormal basis in $\mathbf{E}$ such that $u_1, \ldots, u_k$ are principal vectors of A. Let $x \in \mathbf{E}$ be such that $|x| \leq 1$ and $x = x_1 u_1 + \cdots + x_n u_n$. Then $Ax = \lambda_1 x_1 v_1 + \cdots + \lambda_k x_k v_k$, where $v_1, \ldots, v_k$ are principal vectors of A^*, and we have that

$$|Ax| = \langle Ax, Ax \rangle = \lambda_1^2 x_1^2 + \cdots + \lambda_k^2 x_k^2$$
$$\leq \lambda_k^2 (x_1^2 + \cdots + x_n^2) = \lambda_k^2 |x|^2 \leq \lambda_k^2.$$

Thus, if $|x| \leq 1$, then $|Ax| \leq \lambda_k$. If $x = u_k$, then $|Ax| = \lambda_k$, and hence

$$\|A\| = \sup_{|x| \leq 1} |Ax| = \lambda_k,$$

which was to be proved.

In particular, consider the case when $\mathbf{E} = \mathbf{F}$ and the mapping $A \in L(\mathbf{E}, \mathbf{E})$ is nonsingular, that is, $\det A \neq 0$. Then $\dim \operatorname{Im} A = \dim \mathbf{E} = n$, and thus all n principal dilation coefficients of A are nonzero. Let $\lambda_n \geq \lambda_{n-1} \geq \cdots \geq \lambda_1 > 0$ be the set of singular numbers of A, $u_1, \ldots, u_n$ principal vectors of A, and $v_i = (1/\lambda_i)Au_i$. Then $v_1, \ldots, v_n$ are principal vectors of A^*, and $Au_i = \lambda_i v_i$ for each i. Let $\varphi \in \mathbf{O}(\mathbf{E})$ be such that $\varphi(v_i) = u_i$. Then the mapping $B = A\varphi$ is such that $Bv_i = \lambda_i v_i$. Obviously, $\det B = \lambda_1 \lambda_2 \ldots \lambda_n$. On the other hand, $\det B = \det \varphi \det A$, and we get that

$$\lambda_1 \lambda_2 \ldots \lambda_n = |\det A| \tag{4.5}$$

because $\det \varphi = \pm 1$ and $\lambda_1 \cdots \lambda_n > 0$. If $\lambda_1 = \cdots = \lambda_n = \lambda$, then the mapping B constructed above is such that $Bv_i = \lambda v_i$ for all $i = 1, \ldots, n$. From this, $Bx = \lambda x$ for all $x \in \mathbf{E}$. We have that $A = B\varphi^{-1}$, and we get that if $\lambda_1 = \cdots = \lambda_n$, then A is the composition of an orthogonal transformation and a homothety; that is, A is a general orthogonal transformation in this case. It follows from the equalities $Au_i = \lambda_i v_i$, $i = 1, \ldots, n$, that

$$A^{-1} v_i = u_i / \lambda_i, \qquad i = 1, \ldots, n.$$

On the basis of Lemma 4.2 this enables us to conclude that the singular numbers of A^{-1} are $1/\lambda_1, \ldots, 1/\lambda_n$ $v_1, \ldots, v_n$, are principal vectors of A^{-1}, and $u_1, \ldots, u_n$ are principal vectors of $(A^*)^{-1}$.

For every general orthogonal transformation $A: \mathbf{E} \to \mathbf{E}$ we have that $|Ax| = \|A\| \cdot |x|$ for all $x \in \mathbf{E}$.

Let us determine the image of the sphere $S(0, 1)$ in $\mathbf{E}$ under a mapping $A \in L(\mathbf{E}, \mathbf{E})$ with $\det A \neq 0$. Let $u_1, \ldots, u_n$ and $v_1, \ldots, v_n$ be principal vectors of A and A^*, respectively, and let $\lambda_n \geq \lambda_{n-1} \geq \cdots \geq \lambda_1 > 0$ be the singular numbers of A. A point $x = x_1 v_1 + \cdots + x_n v_n$ belongs to $A[S(0, 1)]$ if and only if

$$A^{-1}(x) = \frac{x_1}{\lambda_1} u_1 + \frac{x_2}{\lambda_2} u_2 + \cdots + \frac{x_n}{\lambda_n} v_n \in S(0, 1);$$

that is, if and only if

$$\frac{x_1^2}{\lambda_1^2} + \frac{x_2^2}{\lambda_2^2} + \cdots + \frac{x_n^2}{\lambda_n^2} = 1. \tag{4.6}$$

We get that the set $A[S(0, 1)]$ is given by equation (4.6) in the Cartesian orthogonal coordinate system with basis $v_1, \ldots, v_n$; that is, $A[S(0, 1)]$ is an ellipsoid with semiaxes of lengths $\lambda_1, \ldots, \lambda_n$.

We now introduce some quantities characterizing the measure of how much an arbitrary nonsingular linear mapping $A: \mathbf{E} \to \mathbf{E}$ differs from a general orthogonal mapping. Let $0 < \lambda_1 \leq \cdots \leq \lambda_n$ be the singular numbers of the mapping. Then, as shown above, $\lambda_n = \|A\|$ and $\lambda_1 \lambda_2 \ldots \lambda_n = |\det A|$. In particular, this implies that

$$|\det| \leq \|A\|^n \qquad (4.7)$$

always, with equality if and only if $\lambda_1 = \cdots = \lambda_n$; that is, if and only if A is a general orthogonal transformation. Let

$$\|A\|^n / |\det A| = \lambda_n^n / \lambda_1 \lambda_2 \ldots \lambda_n = K(A). \qquad (4.8)$$

For every nonsingular linear mapping A we have that $K(A) \geq 1$, and $K(A) = 1$ if and only if A is a general orthogonal transformation.

We introduce some other characterizations. Let $A \colon \mathbf{E} \to \mathbf{E}$ be a nonsingular linear mapping, and let

$$K_0(A) = K(A^{-1}). \qquad (4.9)$$

Obviously,

$$K_0(A) = \lambda_1 \lambda_2 \ldots \lambda_n / \lambda_1^n, \qquad (4.10)$$

where $\lambda_1 \leq \cdots \leq \lambda_n$ are the principal dilation coefficients of A. For every A we have that $K_0(A) \geq 1$, and $K_0(A) = 1$ if and only if A^{-1}, hence also A, is a general orthogonal transformation.

The quantities $K(A)$ and $K_0(A)$ have the following geometric meaning. Let $\mathscr{D}$ be the ellipsoid into which the sphere $S(0, 1)$ is transformed by A, let B_I be the ball of smallest radius containing $\mathscr{D}$, and let B_0 be the ball of largest radius inscribed in $\mathscr{D}$. Then

$$K(A) = |B_I| / |\mathscr{D}|, \qquad K_0(A) = |\mathscr{D}| / |B_0|.$$

The quantity $K(A)$ is called the *outer dilation*, and $K_0(A)$ the *inner dilation* of the linear mapping A. (In the literature the notation $K(A) = K_O(A)$ (O for "outer") and $K_0(A) = K_I(A)$ (I for "inner") is used; to avoid errors we use here the notation in the Russian text of the book.)

We estimate $K(A)$ and $K_0(A)$ in terms of each other. Let $\alpha_i = \lambda_i / (\lambda_1 \ldots \lambda_n)^{1/n}$. Then $K_0(A) = 1/\alpha_1^n$, $K(A) = \alpha_n^n$, $\alpha_1 \leq \cdots \leq \alpha_n$, $\alpha_1 \ldots \alpha_n = 1$. Hence, $1/\alpha_1 = \alpha_2 \alpha_3 \cdots \alpha_n \leq \alpha_n^{n-1}$; that is,

$$K_0(A) \leq [K(A)]^{n-1}. \qquad (4.11)$$

Replacing A by A^{-1} here and considering that $K_0(A^{-1}) = K(A)$, we get that

$$K(A) \leq [K_0(A)]^{n-1}. \qquad (4.12)$$

Let A and B be arbitrary nonsingular linear mappings of the n-dimensional Euclidean space $\mathbf{E}$. Then

$$\det(AB) = \det A \det B.$$

Further, for any $x \in \mathbf{E}$ with $|x| \leq 1$

$$|A(Bx)| \leq \|A\| |Bx| \leq \|A\| \|B\|.$$

From this we see that

$$\|AB\| \leq \|A\| \|B\|, \tag{4.13}$$

and hence

$$K(AB) \leq K(A)K(B). \tag{4.14}$$

Applying the same argument to the mappings A^{-1} and B^{-1}, we get that

$$K_0(AB) = K(B^{-1}A^{-1}) \leq K(B^{-1})K(A^{-1}) = K_0(A)K_0(B). \tag{4.15}$$

If at least one of the given mappings A or B is a general orthogonal transformation, then equality holds in (4.13). This implies that here each of the relations (4.14) and (4.15) also becomes an equality.

Along with $K(A)$ and $K_0(A)$ it is useful to consider various kinds of other characterizations of the nonorthogonality of a linear mapping.

Let $\mathbf{E}$ be a finite-dimensional Euclidean space, and $A: \mathbf{E} \to \mathbf{E}$ a nonsingular linear mapping. Let λ be the smallest principal dilation coefficient of A, and Λ the largest. Define $\Lambda/\lambda = q(A)$. Obviously, $q(A) \geq 1$ always, and $q(A) = 1$ if and only if f is a general orthogonal mapping.

The quantity $q(A)$, as a measure of nonorthogonality of a linear mapping, is most natural. However, it turns out not to be very convenient analytically. The concept of a conformal norm of a linear mapping is used in [130] to define a whole class of nonorthogonality measures.

Let N be a norm in the vector space $L(\mathbf{E}, \mathbf{E})$. The norm N is said to be *conformal* if there exists a constant $\kappa_N > 0$ such that

$$[N(f)]^n \geq \kappa_N \det f$$

for any $f \in L(\mathbf{E}, \mathbf{E})$ such that $\det f > 0$, with equality if and only if f is a general orthogonal transformation.

We give examples of conformal norms. Let $f \in L(\mathbf{E})$, and let $\Lambda(f)$ be the largest dilation coefficient of f; this quantity is a norm in $L(\mathbf{E})$. It follows from what was proved above that Λ is a conformal norm in $L(\mathbf{E})$. The coefficient $\kappa_N = \kappa_\Lambda$ is equal to 1 in this case.

Fix a Cartesian orthogonal coordinate system and let $A = (a_{ik})$, $l, k = 1, \ldots, n$, be the matrix of a linear mapping $f \in L(\mathbf{E}, \mathbf{E})$ in this coordinate system. Let

$$M_p(f) = \left(\sum_{i=1}^n \left(\sum_{k=1}^n a_{ik}^2 \right)^{p/2} \right)^{1/p},$$

$$M_p^*(f) = \left(\sum_{k=1}^n \left(\sum_{i=1}^n a_{ik}^2 \right)^{p/2} \right)^{1/p},$$

where $p \geq 1$. Further, let

$$M_\infty(f) = \max_{i=1,2,\ldots,n} \left(\sum_{k=1}^n a_{ik}^2 \right)^{1/2},$$

$$M_\infty^*(f) = \max_{k=1,2,\ldots,n} \left(\sum_{i=1}^n a_{ik}^2 \right)^{1/2}.$$

By using the classical Hadamard inequality it is not hard to prove that M_p and M_p^*, $1 \leq p \leq \infty$, are conformal norms. The constant κ_N is equal to $n^{n/p}$ for $N = M_p$ and $N = M_p^*$.

Let N be a conformal norm in $L(\mathbf{E})$. The ratio

$$K_N(f) = \frac{[N(f)]^n}{\kappa_N \det f},$$

where $f \in L(\mathbf{E})$ is a nonsingular linear mapping such that $\det f > 0$, is called the *distortion* of f with respect to the conformal norm N.

Let $N: L(\mathbf{E}) \to \mathbf{R}$ be a conformal norm, and let

$$\tau_1(K; N) = \sup_{K(f) \leq K} K_N(f), \qquad \tau_2(K; N) = \sup_{K_N(f) \leq K} K(f).$$

It follows easily from continuity and compactness considerations that $\tau_1(K; N)$ and $\tau_2(K; N)$ are finite for any $K \geq 1$. The functions $K \mapsto \tau_1(K; N)$ and $K \mapsto \tau_2(K; N)$ are nondecreasing. Further, $\tau_1(K; N) \to 1$ and $\tau_2(K; N) \to 1$ as $K \to 1$ for every conformal norm N.

It follows from the definitions of the functions τ_1 and τ_2 that

$$K_N(f) \leq \tau_1[K(f); N], \qquad K(f) \leq \tau_2[K_N(f); N]$$

for every $f \in L(\mathbf{E})$ with $\det f > 0$. It is clear from these estimates that if $K(f)$ is close to 1 for a mapping f, then $K_N(f)$ is also close to 1, and, conversely, if $K_N(f)$ is close to 1, then so is $K(f)$.

4.2. Mappings with bounded distortion. Let $U \subset \mathbf{R}^n$ be an arbitrary open set, and let $f: U \to \mathbf{R}^n$ be a given mapping in $W_{p,\text{loc}}^1(U)$, where

$p \geq 1$. The linear mapping $f'(x)$ is defined for almost all $x \in U$. Its determinant $\det f'(x)$ is called the *Jacobian* of f at the point x, and is denoted by $\mathscr{J}(x, f)$.

We say that a function $u \colon U \to \mathbf{R}$ *does not change sign* on a set $A \subset U$ if either $u(x) \geq 0$ almost everywhere in A or $u(x) \leq 0$ almost everywhere in A.

A mapping $f \colon U \to \mathbf{R}^n$ is called a *mapping with bounded distortion* if it satisfies the following two conditions:

(C1) f is continuous.

(C2) f belongs to the class $W^1_{n,\mathrm{loc}}(U)$, the function $\mathscr{J}(x, f)$ does not change sign in U, and there exists a number $K \geq 1$ such that

$$\|f'(x)\|^n \leq K|\mathscr{J}(x, f)| \qquad (4.16)$$

for almost all $x \in U$.

A mapping f is said to be *quasiconformal* if instead of (C1) it satisfies the stronger condition

(C3) f is a homeomorphism.

The smallest constant K such that (1.1) holds almost everywhere in U is called the *distortion coefficient* of f and denoted by $K(f)$.

Condition (C2) is equivalent to the simultaneous satisfaction of the following three conditions:

α) $f \in W^1_{n,\mathrm{loc}}(U)$, and $\mathscr{J}(x, f)$ has constant sign in U.

β) There exists a set $E \subset U$ of measure zero such that if $x \in U \setminus E$ and $\mathscr{J}(x, f) = 0$, then the linear mapping $f'(x)$ is equal to zero.

γ) There exists a number $K \geq 1$ such that $K[f'(x)] \leq K$ at each point $x \in U \setminus E$ (E the set in β)) with $\mathscr{J}(x, f) \not\equiv 0$.

Quasiconformal mappings have been the subject of many investigations in the case $n = 2$. At present the theory of planar quasiconformal mappings is an extensive part of the theory of functions of a complex variable and has many important applications. A large role in the evolution of this theory has been played by the fundamental work of Lavrent'ev.

The concept of a quasiconformal mapping in n-space was introduced by Lavrent'ev as far back as the 1930's [80], but the beginning of intensive investigations in this area relates to 1960 (see [171], [33], [119], [145], [157], and [158]). Mappings with bounded distortion were considered by Lavrent'ev in the case $n = 2$. In the n-space case such mappings were introduced by the author in 1966 in [124]. There is a survey of investigations devoted to this subject in [175]. Some questions in the theory of quasiconformal mappings in space are considered in [15], [43], [74], [82], [83], [90], and [140]–[142]. A survey and an exhaustive bibliography of publications in the theory of n-space quasiconformal mappings up to 1967 is contained in the monograph [25].

4.3. Examples of mappings with bounded distortion.

1. Every nonsingular affine mapping of $\mathbf{R}^n$ is trivially a quasiconformal mapping. Let $f: U \to \mathbf{R}^n$ be a C^1-mapping of an open domain $U \subset \mathbf{R}^n$. In this case if there exist constants $\gamma > 0$ and $M < \infty$ such that $|f'(x)| \leq M$ and $|\mathscr{J}(x, f)| \geq \gamma$ for all $x \in U$, then f is clearly a mapping with bounded distortion. In particular, if $f: U \to \mathbf{R}^n$ is a C^1-mapping such that $\mathscr{J}(x, f) \neq 0$ for all $x \in U$, then the restriction of f to any open set V strictly inside U is a mapping with bounded distortion.

2. Let U be an arbitrary polyhedron in $\mathbf{R}^n$. A mapping $f: U \to \mathbf{R}^n$ is said to be *simplicial* if U can be partitioned into finitely many simplexes such that the restriction of f to each of these simplexes is affine. Every simplicial homeomorphism $f: U \to \mathbf{R}^n$ is a quasiconformal mapping. Indeed, let $f_1, \ldots, f_n$ be the components of the vector-valued function. The restriction of f_k to the intersection of U with an arbitrary line in $\mathbf{R}^n$ is a piecewise affine, hence absolutely continuous, function of one variable. The partial derivatives $\partial f_k / \partial x_i$ are constant on each simplex in the partition of U, and are hence bounded. This proves that f belongs to $W_n^1(U)$. The rest is obvious.

3. We give an example of a mapping having a peculiarity distinguishing an arbitrary mapping with bounded distortion from quasiconformal mappings. Take an arbitrary integer $m > 1$. Let $x = (x_1, \ldots, x_{n-2}, x_{n-1}, x_n) \in \mathbf{R}^n$. If $x_{n-1} = x_n = 0$, then let $f(x) = x$. Suppose that $x_{n-1}^2 + x_n^2 > 0$. Then $x_{n-1} = r \cos \varphi$ and $x_n = r \sin \varphi$, where $r = \sqrt{x_{n-1}^2 + x_n^2}$ and $0 \leq \varphi < 2\pi$. In this case let

$$f(x) = (x_1, \ldots, x_{n-2}, r \cos m\varphi, r \sin m\varphi).$$

This mapping is continuous, and all points of the plane

$$\mathbf{R}^{n-2} = \{x \in \mathbf{R}^n | x_{n-1} = x_n = 0\}$$

are carried into themselves. Every circle Γ with center in $\mathbf{R}^{n-2}$ and lying in the two-dimensional plane orthogonal to $\mathbf{R}^{n-2}$ is also transformed into itself under f. Further, if x makes a circuit of Γ in one direction, then $f(x)$ runs through Γ in the same direction m times. We call f a *twisting around an axis*.

The mapping f clearly belongs to C^1 on the open set $\mathbf{R}^n \setminus \mathbf{R}^{n-2}$. Every $(n-1)$-dimensional half-plane with boundary $\mathbf{R}^{n-2}$ is transformed isometrically by f into another such half-plane. This implies that the principal dilation coefficients of $f'(x)$ in the $n-1$ directions parallel to the indicated half-plane are equal to 1. The dilation coefficient of $f'(x)$ in the direction orthogonal to this half-plane is equal to m. This implies that at each point $x \notin \mathbf{R}^{n-2}$ we have that $\|f'(x)\| = m$, $\det f'(x) = m$, and

$$K[f'(x)] = m^{n-1}, \qquad K_0[f'(x)] = m.$$

The derivatives of f are bounded. Theorem 2.6 allows us to conclude that $f \in W^1_{n,\text{loc}}(\mathbf{R}^n)$. It follows from the foregoing that f is a mapping with bounded distortion. Further, $K(f) = m^{n-1}$ and $K_0(f) = m$.

The main peculiarity of the mapping in this example is that f is a homeomorphism in a sufficiently small neighborhood of every point $x \notin \mathbf{R}^{n-2}$ and is not a homeomorphism in any neighborhood of an arbitrary point $x \in \mathbf{R}^{n-2}$. It will be shown below (Chapter II, §10.4) that a peculiarity of the type considered in this example cannot occur for a mapping f with bounded distortion if $K(f)$ is sufficiently close to 1.

4. We give an example of a mapping with bounded distortion whose derivatives are not bounded in any neighborhood of some point a, in contrast to what happens in the examples already considered. Let $U = \mathbf{R}^n$ and $a = 0$, and take an arbitrary $\alpha > 0$. Let $f(0) = 0$, and let $f(x) = x|x|^{\alpha-1}$ for $x \neq 0$. This function f will be called a *nonlinear homothety*. Obviously, f is continuous on $\mathbf{R}^n$ and belongs to C^∞ on $\mathbf{R}^n \setminus \{0\}$. Every ray emanating from the point 0 is transformed by f into itself. The sphere $S(0, r)$ is transformed by f into the sphere $S(0, r^\alpha)$ by a similarity. Take an arbitrary point $x \neq 0$, and let $r = |x|$. The dilation coefficient of the linear mapping $f'(x)$ in the direction of the ray Ox is equal to $|dr^\alpha/dr| = \alpha r^{\alpha-1}$. The dilation coefficients of $f'(x)$ in the $n-1$ directions orthogonal to this ray are equal to $r^{\alpha-1}$. This implies that $\det f'(x) = \alpha r^{(\alpha-1)n}$. The quantity $\|f'(x)\|$ is equal to $\alpha r^{\alpha-1}$ in the case $\alpha \geq 1$, and $\|f'(x)\| = r^{\alpha-1}$ for $0 < \alpha < 1$, where $r = |x|$. Hence $K[f'(x)] = \alpha^{n-1}$ in the case $\alpha \geq 1$, and $K[f'(x)] = 1/\alpha$ for $0 < \alpha \leq 1$. We have that $\|f'(x)\| = |x|^{\alpha-1}$. It is clear from this that the derivatives of $f(x)$ are unbounded in a neighborhood of 0 in the case $0 < \alpha < 1$. Further, the function $x \mapsto |f'(x)|$ is integrable to any power $p < n/(1 - \alpha)$ in a neighborhood of 0. In particular, $f \in W^1_{p,\text{loc}}(\mathbf{R}^n)$ for any $p < n/(1 - \alpha)$. It follows from what has been proved that the mapping $f : \mathbf{R}^n \to \mathbf{R}^n$ constructed is quasiconformal. Further, $K(f) = \alpha^{n-1}$ and $K_0(f) = \alpha$ in the case $\alpha \geq 1$, while $K(f) = 1/\alpha$ and $K_0(f) = (1/\alpha)^{n-1}$ in the case $0 < \alpha < 1$.

5. Let Z be the right circular cylinder

$$\{(x_1, \ldots, x_n) \in \mathbf{R}^n \,|\, x_1^2 + \cdots + x_{n-1}^2 < \pi^2/4\}.$$

An arbitrary point $x = (x_1, \ldots, x_{n-1}, x_n) \in \mathbf{R}^n$ will be represented as a pair (y, z), where $y = (x_1, \ldots, x_{n-1})$ and $z = x_n$. Let $f(0, z) = (0, e^{-z}) = e^{-z}e_n$, and for $y \neq 0$ let

$$f(y, z) = \left(-\left(\frac{e^{-z}}{|y|}\sin|y|\right)y, e^{-z}\cos|y|\right).$$

The mapping f is obviously continuous and has derivatives of all orders at each point $x = (y, z)$ with $y \neq 0$. Consider an arbitrary plane Γ passing through the axis Ox_n. We introduce a Cartesian orthogonal coordinate system (ξ, η) in it, taking the line along which Γ intersects the hyperplane $x_n = 0$ as the axis $O\xi$, and the line Ox_n as the axis $O\eta$. Here the direction of the axis $O\eta$ is the same as for the axis Ox_n. The intersection $\Gamma \cap Z$ is the strip consisting of all points $x \in \Gamma$ such that $-\pi/2 < \xi < \pi/2$. The mapping f transforms $\Gamma \cap Z$ into a subset of the plane Γ. Further, if x has coordinates (ξ, η), then $f(x)$ has coordinates $(-e^{-\eta}\sin\xi, e^{-\eta}\cos\xi)$ in the plane Γ. In complex form the restriction of f to Γ is the mapping $\zeta \mapsto e^{i(\zeta+\pi/2)}$, where $\zeta = \xi + i\eta$, and it maps the strip $-\pi/2 < \operatorname{Re}\zeta < \pi/2$ conformally onto the upper half-plane $\operatorname{Im}\zeta > 0$. This implies that f is a homeomorphic mapping of the cylinder Z onto the half-plane $x_n > 0$. We prove that f is a quasiconformal mapping. The symbol $K(z, r)$ below denotes the $(n-2)$-dimensional sphere consisting of all the points $x = (y, z) \in \mathbf{R}^n$ such that $|y| = r$, while the z-component is fixed. The center of this sphere lies at the point $(0, z)$. The mapping f implements a similarity transformation of $K(z, r)$ into the sphere $K(e^{-z}\cos r, e^{-z}\sin r)$. We take an arbitrary point $x = (y, z) \in Z$ such that $y \neq 0$ and, as shown above, construct the two-dimensional plane Γ passing through x and the axis Ox_n. Consider also the $(n-2)$-dimensional sphere $K(z, |y|)$. The tangent plane T of $K(z, |y|)$ at x is completely orthogonal to the two-dimensional plane Γ. The mapping $f'(x)$ transforms the $(n-2)$-dimensional plane T and the plane Γ into two mutually orthogonal planes—the $(n-2)$-dimensional plane tangent to the sphere $K(e^{-z}\cos|y|, e^{-z}\sin|y|)$ at $f(x)$, and the plane Γ. The dilation coefficients of the mapping $f'(x)$ are equal to $e^{-z}\sin|y|/|y|$ on the plane T, and equal to e^{-z} on the plane Γ. We get that $n-2$ of the principal dilation coefficients of $f'(x)$ are equal to $e^{-z}(\sin|y|)/|y| \leq e^{-z}$, and two are equal to e^{-z}. From this,

$$|\det f'(x)| = e^{-nz}(\sin|y|/|y|)^{n-2}, \qquad \|f'(x)\| = e^{-z},$$

and hence

$$K(f'(x)) = (|y|/\sin|y|)^{n-2} \leq (\pi/2)^{n-2}.$$

It follows from the above that

$$K_0(f'(x)) = (|y|/\sin|y|)^2 \leq (\pi/2)^2.$$

We get that the mapping f is quasiconformal. Further,

$$K(f) = (\pi/2)^{n-2} \quad \text{and} \quad K_0(f) = (\pi/2)^2.$$

The composition $\varphi = g \circ h$ of a Möbius mapping h carrying $B(0, 1)$ into a half-plane and the mapping $g = f^{-1}$ inverse to f is a quasiconformal mapping of the ball onto the cylinder Z. Moreover,

$$K(\varphi) = (\pi/2)^2, \qquad K_0(\varphi) = (\pi/2)^{n-2}.$$

The closure of Z in the Möbius space $\mathbf{R}^n$ is homeomorphic to a closed ball. We have the following assertion ([141], [142], [168]). There exists a number $\varepsilon_0 > 0$ such that for every quasiconformal mapping $f: B(0, 1) \to \overline{\mathbf{R}}^n$ with $K(f) \le 1 + \varepsilon_0$, the closure of $f[B(0, 1)]$ is homeomorphic to the closed ball. This implies that a quasiconformal mapping of the ball onto the cylinder Z cannot have distortion coefficient $K(f)$ arbitrarily close to 1.

The existence of a quasiconformal mapping of a cylinder onto a ball was first shown by P. P. Belinskii; his example (which was different from that given) was not published.

One problem in the theory of n-space quasiconformal mappings is to describe domains that are quasiconformally equivalent to a ball, that is, which admit a quasiconformal mapping onto a ball. The domain with outward-directed peak formed by revolving the figure in Figure 1 in space about the OY-axis is quasiconformally equivalent to a ball. At the same time, this is not true for the domain with an inward-directed peak formed by revolving the figure in Figure 2 about the same axis.

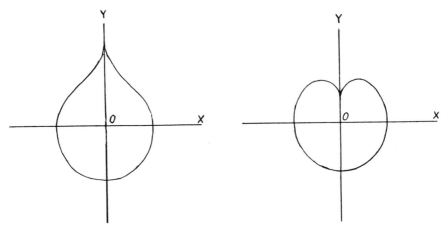

FIGURE 1   FIGURE 2

Let $\mathscr{D}$ be a domain in $\mathbf{R}^n$. If $\mathscr{D}$ is not quasiconformally equivalent to a ball, then let $K(\mathscr{D}) = K_0(\mathscr{D}) = \infty$. But if $\mathscr{D}$ is quasiconformally equivalent to a ball, then let

$$K(\mathscr{D}) = \inf K(f), \qquad K_0(\mathscr{D}) = \inf K_0(f),$$

where the infimum is over the set of all quasiconformal mappings of $\mathscr{D}$ onto a ball. It was shown in [60] (for the case $n = 3$) that for the cylinder Z

$$K(Z) = \frac{1}{2} \int_0^{\pi/2} \sqrt{\sin u}\, du = 1, 3 \ldots,$$

$$\sqrt[3]{2} \le K_0(Z) \le \sqrt{2}.$$

§5. Mappings with bounded distortion on Riemannian spaces

In what follows we have to do mainly with mappings of open subsets of $\mathbf{R}^n$ or of the Möbius space $\overline{\mathbf{R}}^n$. However, in some cases it is also convenient in studying mappings of sets in $\mathbf{R}^n$ to deal with mappings of Riemannian manifolds (given, for example, as surfaces in a Euclidean space of high dimension). It seems to us that the study of such mappings can itself be the source of many interesting problems.

5.1. Riemannian metrics in domains in $\mathbf{R}^n$. Let U be an open subset of $\mathbf{R}^n$. We say that a Riemannian metric is given in U if corresponding to each point $x \in U$ there is a positive-definite quadratic form of the variable $\xi \in \mathbf{R}^n$,

$$g(x, \xi) = \langle G(x)\xi, \xi \rangle = \sum_{i=1}^{n} \sum_{j=1}^{n} g_{ij}(x)\xi_i \xi_j. \tag{5.1}$$

Further, it is assumed that the functions $g_{ij}(x)$—the elements of the matrix $G(x)$—are continuous, and $g_{ij}(x) = g_{ji}(x)$ for any $i, j = 1, \ldots, n$. The quadratic form (5.1) is called the *linear element* of the given Riemannian metric. Let $dx_i \colon \mathbf{R}^n \to \mathbf{R}$ be the linear function defined by the condition $dx_i(\xi) = \xi_i$. Then the product $dx_i dy_j$ is the function whose value at the point $\xi = (\xi_1, \ldots, \xi_n) \in \mathbf{R}^n$ is equal to $\xi_i \xi_j$. Correspondingly, the quadratic form (5.1) can be written in the form

$$g(x) = g_{ij}(x)\, dx_i\, dx_j. \tag{5.2}$$

Here repeated indices are understood to be summed from 1 to n.

Our goal is to define the concept of a mapping with bounded distortion for the case of a mapping of one Riemannian space into another. It is expedient to consider initially the case when we have a mapping of a domain in $\mathbf{R}^n$, equipped with some Riemannian metric, into another such domain. We first make some remarks about quadratic forms.

Assume that $a(\xi) = \langle A\xi, \xi \rangle$ and $b(\xi) = \langle B\xi, \xi \rangle$ are given positive-definite quadratic forms on $\mathbf{R}^n$, where A and B are symmetric matrices. Let $\mathbf{E}_1$ and $\mathbf{E}_2$ be n-dimensional Euclidean spaces, each coinciding with $\mathbf{R}^n$ as a vector space, but with the inner product $\langle \xi, \eta \rangle_{\mathbf{E}_1} = \langle A\xi, \eta \rangle$ in $\mathbf{E}_1$ and $\langle \xi, \eta \rangle_{\mathbf{E}_2} = \langle B\xi, \eta \rangle$ in $\mathbf{E}_2$. Let $L : \mathbf{R}^n \to \mathbf{R}^n$ be a given linear mapping, and let $0 \le \lambda_1 \le \cdots \le \lambda_n$ be the principal dilation coefficients of L, as a mapping from $\mathbf{E}_1$ into $\mathbf{E}_2$. Assume that L is a nonsingular mapping. Then $\lambda_1 > 0$, and we let

$$K(a, b; L) = \lambda_n^n / \lambda_1 \lambda_2 \ldots \lambda_n,$$
$$K_0(a, b; L) = K(b, a; L^{-1}) = \lambda_1 \lambda_2 \ldots \lambda_n / \lambda_1^n.$$

It is useful to assign definite values to $K(a, b; L)$ and $K_0(a, b; L)$ in the case when $\det L = 0$. If $\det L = 0$ and L is identically equal to zero (that is, if $L\xi = 0$ for all $\xi \in \mathbf{R}^n$), then we let $K(a, b; L) = K_0(a, b; L) = 1$. But if $\det L = 0$ and $L \ne 0$ (that is, there exists a $\xi \in \mathbf{R}^n$ such that $L(\xi) \ne 0$), then we let $K(a, b; L) = K_0(a, b; L) = \infty$.

LEMMA 5.1. *Let the numbers $\alpha_0, \alpha_1, \beta_0, \beta_1$ be such that $0 < \alpha_0 \le \alpha_1 < \infty$, $0 < \beta_0 \le \beta_1 < \infty$, and*

$$\alpha_0^2 |\xi|^2 \le \langle A\xi, \xi \rangle \le \alpha_1^2 |\xi|^2 \quad and \quad \beta_0^2 |\xi|^2 \le \langle B\xi, \xi \rangle \le \beta_1^2 |\xi|^2$$

for any $\xi \in \mathbf{R}^n$. Then for every linear operator $L : \mathbf{R}^n \to \mathbf{R}^n$

$$K(a, b; L) \le \left(\frac{\alpha_1 \beta_1}{\alpha_0 \beta_0} \right)^n K(L), \qquad K(L) \le \left(\frac{\alpha_1 \beta_1}{\alpha_0 \beta_0} \right)^n K(a, b; L). \quad (5.3)$$

PROOF. The desired inequalities are obvious if $\det L = 0$, for then $(a, b; L)$ is either 0 or ∞. Below we assume that $\det L \ne 0$. As shown in §4.1, there exist systems of vectors $\{u_1, \ldots, u_n\}$ and $\{v_1, \ldots, v_n\}$ such that the first of them is orthogonal in $\mathbf{E}_1$, the second is orthogonal in $\mathbf{E}_2$, and $Lu_i = \lambda_i v_i$ for each $i = 1, \ldots, n$. For any $i, j = 1, \ldots, n$ we have that $\langle Lu_i, Lu_j \rangle_{\mathbf{E}_2} = \lambda_i \lambda_j \delta_{ij}$, where $\delta_{ij} = 1$ for $i = j$ and $\delta_{ij} = 0$ for $i \ne j$. By definition,

$$\langle Lu_i, Lu_j \rangle_{\mathbf{E}_2} = \langle Lu_i, BLu_j \rangle = \langle u_i, L^*BLu_j \rangle$$
$$= \langle u_i, A(A^{-1}L^*BL)u_j \rangle = \langle u_i, A^{-1}L^*BLu_j \rangle_{\mathbf{E}_1}.$$

Let $A^{-1}L^*BL = \tilde{L}$. We see that $\langle \tilde{L}u_j, u_i \rangle_{\mathbf{E}_1} = 0$ for $i \ne j$, and $\langle \tilde{L}u_j, u_j \rangle = \lambda^2 j$. From this, $\tilde{L}u_j = \lambda_j^2 u_j$. Accordingly, $\lambda_1^2, \ldots, \lambda_n^2$ are the eigenvalues of the mapping $\tilde{L} = A^{-1}L^*BL$, and $u_1, \ldots, u_n$ are eigenvectors. In particular, this implies that

$$\lambda_1\lambda_2 \cdots \lambda_n = \sqrt{\det \tilde{L}} = |\det L|\sqrt{(\det B)/\det A}.$$

The conditions of the lemma give us that all the eigenvalues of the matrix A lie between α_0^2 and α_1^2, while the eigenvalues of B are between β_0^2 and β_1^2, which implies that

$$\beta_0^{2r} \leq \det B \leq \beta_1^{2n}, \qquad \alpha_0^{2n} \leq \det A \leq \alpha_1^{2n}.$$

This allows us to conclude that

$$(\beta_0/\alpha_1)^n |\det L| \leq \lambda_1\lambda_2 \cdots \lambda_n \leq (\beta_1/\alpha_0)^n |\det L|. \tag{5.4}$$

The quantity λ_n is the norm of L, as a linear mapping of E_1 into E_2; that is, λ_n is the largest of the numbers λ such that $\|L\xi\|_{E_2} \leq \lambda\|\xi\|_{E_1}$ for $\xi \in \mathbf{R}^n$. In precisely the same way, $\|L\|$ is the smallest of the numbers λ such that $|L\xi| \leq \lambda|\xi|$. In particular, $\|L\xi\|_{E_2} \leq \lambda_n\|\xi\|_{E_1}$ for every ξ. From this,

$$\beta_0|L\xi| \leq \|L\xi\|_{E_2} \leq \lambda_n\alpha_1|\xi|,$$

and hence $|L\xi| \leq (\alpha_1/\beta_0)\lambda_n|\xi|$, for every $\xi \in \mathbf{R}^n$. Consequently,

$$\|L\| \leq \frac{\alpha_1}{\beta_0}\lambda_n. \tag{5.5}$$

The inequality $|L\xi| \leq \|L\|\,|\xi|$ implies that

$$\|L\xi\|_{E_2} \leq \beta_1|L\xi| \leq \beta_1\|L\|\,|\xi| \leq \frac{\beta_1}{\alpha_0}\|L\|\,\|\xi\|_{E_1}.$$

Since ξ is arbitrary, this leads to the conclusion that

$$\lambda_n \leq \frac{\beta_1}{\alpha_0}\|L\|. \tag{5.6}$$

The required result follows immediately from (5.4)–(5.6). The lemma is proved.

The characteristics of a linear mapping introduced here have the following invariance property. Let $\varphi: \mathbf{R}^n \to \mathbf{R}^n$ and $\psi: \mathbf{R}^n \to \mathbf{R}^n$ be nonsingular linear mappings, and let $(\varphi^*a)(\xi) = a[\varphi(\xi)]$ and $(\psi_*b)(\xi) = b[\psi^{-1}(\xi)]$. Then

$$K(\varphi^*a, \psi_*b, \psi L\varphi) = K(a, b, L), \tag{5.7}$$

$$K_0(\varphi^*a, \psi_*b, \psi L\varphi) = K_0(a, b, L). \tag{5.8}$$

For a proof we introduce Euclidean spaces E_1' and E_2', each coinciding with $\mathbf{R}^n$ as a vector space, with $\|\xi\|_{E_1'}^2 = (\varphi^*a)(\xi)$ and $\|\xi\|_{E_2'}^2 = (\psi_*b)(\xi)$. The mappings φ and ψ, regarded as mappings of the Euclidean spaces E_1'

and E_2 into E_1 and E_2', respectively, are orthogonal. This implies that the principal dilation coefficients of the mapping $\psi L \varphi : E_1' \to E_2'$ are the same as those of the mapping $L: E_1 \to E_2$. This proves (5.7) and (5.8).

Let U and V be open subsets of $\mathbf{R}^n$. Assume that U and V have Riemannian metrics given by the linear elements $g(x) = g_{ij}(x) \, dx_i \, dx_j$ and $h(y) = h_{ij}(y) \, dy_i \, dy_j$. A mapping $f: U \to V$ is called a *mapping with bounded distortion* with respect to the Riemannian metrics on U and V if the following conditions hold:

(A) f is continuous and belongs to the class $W^1_{n,\mathrm{loc}}(U)$.

(B) The Jacobian $\mathscr{J}(x, f)$ of f has a constant sign in U.

(C) There exists a number K, $1 \le K < \infty$, such that

$$K[g(x), h[f(x)], f'(x)] \le K \qquad (5.9)$$

for almost all $x \in G$.

The smallest number K for which (5.9) holds for almost all $x \in G$ is denoted by $K(g, h; f)$. If $f: U \to V$ satisfies conditions (A), (B), and (C), then there exists a $K < \infty$ such that $K_0[g(x), h(x), f'(x)] \le K$ for almost all $x \in U$, and we denote the smallest such K by $K_0(g, h; f)$.

If $f: U \to V$ is a mapping with bounded distortion with respect to Riemannian metrics $g(x)$ and $h(y)$ given on U and V, respectively, then for every open set G strictly inside U the restriction of f to G is a mapping with bounded distortion. Indeed, G lies strictly inside U. Then $\overline{G}$ is compact, and $U \supset \overline{G}$. Let $M = f(\overline{G})$. The set M is compact, and $V \supset M$. It follows from continuity considerations that there exist constants $\alpha_0, \alpha_1, \beta_0$, and β_1 such that $0 < \alpha_0 \le \alpha_1 < \infty$, $0 < \beta_0 \le \beta_1 < \infty$, and

$$\alpha_0^2 |\xi|^2 \le g_{ij}(x) \xi_i \xi_j \le \alpha_1^2 |\xi|^2$$

for all $x \in G$, while

$$\beta_0^2 |\xi|^2 \le h_{ij}(y) \xi_i \xi_j \le \beta_1^2 |\xi|^2$$

for all $y \in M$. Lemma 5.1 permits us to conclude that

$$K[f'(x)] \le \left(\frac{\alpha_1 \beta_1}{\alpha_0 \beta_0} \right)^n K[g(x), h[f(x)], f'(x)] \le \left(\frac{\alpha_1 \beta_1}{\alpha_0 \beta_0} \right)^n K(g, h; f)$$

for almost all $x \in G$. This establishes that the restriction of f to G is a mapping with bounded distortion.

Let $\varphi: U \to \mathbf{R}^n$ be a diffeomorphism, and $V = \varphi(U)$. Assume that a Riemannian metric is given on V by the linear element $h(y) = \langle H(y)\eta, \eta \rangle$. The diffeomorphism φ permits us in a certain sense to carry this metric over to U. For an arbitrary $x \in U$ let the symbol $(\varphi^* h)(x)$ denote the quadratic form $g(x)$ defined by

$$g(x;\xi) = \langle H[\varphi(x)]\varphi'(x)\xi, \varphi'(x)\xi \rangle$$
$$= \langle G(x)\xi, \xi \rangle,$$

where $G(x) = [\varphi'(x)]^* H[\varphi(x)]\varphi'(x)$. We have that $G(x) = \|g_{ij}(x)\|_{i,j=1,\dots,n}$. The elements of $G(x)$ are expressed by the equalities

$$g_{ij}(x) = h_{\alpha\beta}[\varphi(x)]\frac{\partial\varphi_\alpha}{\partial x_i}(x)\frac{\partial\varphi_\beta}{\partial x_j}(x). \tag{5.10}$$

Let U and V be open subsets of $\mathbf{R}^n$. Assume that U has a Riemannian metric given by the linear element $g(x) = g_{ij}(x)\,dx_i\,dx_j$, and V has a Riemannian metric given by the linear element $h(y) = h_{kl}(y)\,dy_k\,dy_l$. Let $f: U \to V$ be a mapping with bounded distortion with respect to the given Riemannian metrics. Assume now that $\varphi: U_1 \to \mathbf{R}^n$ and $\psi: V \to \mathbf{R}^n$ are given diffeomorphisms, with $\varphi(U_1) = U$, and let $\varphi(V) = V_1$. Then the mapping $\tilde{f} = \psi \circ f \circ \varphi: U_1 \to V_1$ is defined. The diffeomorphisms φ and ψ can be interpreted as different coordinate systems given on the respective sets U and V. Then $\tilde{f}$ should be understood as the representation of f in these coordinate systems. We define the Riemannian metrics

$$(\varphi^* g)(t) = g_{ij}[\varphi(t)]\,d\varphi_i\,d\varphi_j \quad \text{and} \quad (\psi_* h)(y) = [(\psi^{-1})^* h](u)$$

in the sets U_1 and V_1, and we show that f_1 is a mapping with bounded distortion with respect to these metrics, and that

$$K(\varphi^* g, \psi_* h; \tilde{f}) = K(g, h; f),$$
$$K_0(\varphi^* g, \psi_* h; \tilde{f}) = K_0(g, h; f). \tag{5.11}$$

Indeed, the mapping $f_1 = f \circ \varphi$ belongs to the class $W^1_{1,\text{loc}}$ in view of Theorem 2.8. Further, $f_1'(t) = f'(\varphi(t)) \circ \varphi'(t)$ for almost all $t \in U_1$. We have that

$$\|f_1'(t)\| \leq \|f'(\varphi(t))\| \cdot \|\varphi'(t)\|.$$

On every compact set $A \subset U_1$ we have that $\|\varphi'(t)\|^n \leq C|\mathscr{J}(t,\varphi)|$, which leads us to conclude that

$$\int_A \|f_1'(t)\|^n\,dt \leq C\int_A \|f'(\varphi(t))\|^n|\mathscr{J}(t,\varphi)|\,dt$$
$$\leq C\int_{\varphi(A)} \|f'(x)\|^n\,dx < \infty$$

for any compact set $A \subset U_1$. Consequently, f_1 is a mapping in $W^1_{n,\text{loc}}(U_1)$. Application of Theorem 2.9 gives us that $\tilde{f} = \psi \circ f_1$ is a mapping of class $W^1_{n,\text{loc}}$. Further,

$$\tilde{f}'(t) = \psi'(y) \circ f'(x) \circ \varphi'(t)$$

for almost all t, where $x = \varphi(t)$ and $y = f(x) = f[\varphi(t)]$. Let

$$a(\xi) = g_{ij}(x)\xi_i\xi_j, \qquad b(\xi) = h_{kl}(f(x))\xi_k\xi_l.$$

Then, by (5.7) and (5.8),

$$K((\varphi'(t))^*a, \psi(u)_*b; \psi'(y) \circ f'(x) \circ \varphi'(t)) = K(a, b; f'(x))$$

and similarly for K_0. The equality holds for almost all t, and this proves (5.11). The mappings φ and ψ are diffeomorphisms, and hence $\mathcal{J}(t, \varphi)$ and $\mathcal{J}(u, \psi)$ have constant sign on every connected component of the domain of φ and ψ. This implies that the Jacobian of $\tilde{f}$ has constant sign on each connected component of U_1. Consequently, $\tilde{f}$ is a mapping with bounded distortion.

5.2. Mappings with bounded distortion on Riemannian spaces. For the convenience of the reader we define certain concepts.

Let M be a Hausdorff topological space with a countable base. The space M is called an *n-dimensional manifold* if each point of it has a neighborhood homeomorphic to $\mathbf{R}^n$.

A *local coordinate system*, or *chart*, in an n-dimensional manifold M is defined to be a homeomorphism $\varphi: U \to \mathbf{R}^n$, where U is an open set in M. For $x \in U$ let $\varphi(x) = (t_1(x), \ldots, t_n(x))$. The numbers $t_1(x), \ldots, t_n(x)$ are called the *coordinates* of x with respect to the chart φ, and the real functions t_i are called the *coordinate functions* corresponding to this chart. The set $G = \varphi(U)$ is open in $\mathbf{R}^n$.

Let $\varphi: U \to \mathbf{R}^n$ and $\psi: V \to \mathbf{R}^n$ be two arbitrary charts of M such that $U \cap V \neq \varnothing$. In this case the charts φ and ψ are said to be *overlapping*. The sets $G = \varphi(U \cap V)$ and $H = \varphi(U \cup V)$ are open in $\mathbf{R}^n$. The functions $\theta = \psi \circ \varphi^{-1}$ and $\omega = \varphi \circ \psi^{-1}$ implement homeomorphisms of G onto H and H onto G, respectively. Further, $\theta = \omega^{-1}$ and $\omega = \theta^{-1}$. The charts φ and ψ are said to be *C^r-compatible*, where $1 \leq r \leq \infty$, if each of the functions θ and ω belongs to C^r.

We regard any two charts with disjoint domains as C^r-compatible for any r such that $1 \leq r \leq \infty$.

The functions $\theta = \psi \circ \varphi^{-1}$ and $\omega = \varphi \circ \psi^{-1}$, which are defined for a pair of overlapping charts $\varphi: U \to \mathbf{R}^n$ and $\psi: V \to \mathbf{R}^n$, have the following simple meaning. Let $x \in U \cap V$, $t = \varphi(x)$, and $u = \psi(x)$. Then $u = \theta(t)$ and $t = \omega(u)$; hence θ enables us to find the coordinates of the point x in the chart ψ from its coordinates in the chart φ. Conversely, the function ω enables us to compute the coordinates of x in φ from its coordinates in ψ.

An *n-dimensional differentiable manifold of class C^r* is an *n*-dimensional manifold M with a set $\mathfrak{A} = \{\varphi_\alpha, U_\alpha \to \mathbf{R}^n\}_{\alpha \in A}$ of charts such that the domains of these charts cover M, that is, $M = \bigcup_{\alpha \in A} U_\alpha$, and any two charts in this set are C^r-compatible. A set $\mathfrak{A}$ of charts satisfying this condition is called an *atlas of class C^r* on M. A chart $\varphi\colon U \to \mathbf{R}^n$ of M is said to be *admissible* if it is smoothly compatible with any chart $\varphi_\alpha\colon U_\alpha \to \mathbf{R}^n$ of the atlas $\mathfrak{A}$. The collection of all charts is itself an atlas of class C^r.

We define the concept of a Riemannian space. For this we first define the general concept of a tensor and a tensor field on a differentiable manifold. In what follows we need only certain special cases of these general concepts, but here it is expedient to define the general concept. Let $k \geq 0$ and $l \geq 0$ be integers such that $k + l > 0$. One says that *a k-fold covariant and l-fold contravariant tensor*, or, briefly, a *tensor of type (k, l)*, is given at a point p of a manifold M if to each admissible chart $\varphi\colon U \to \mathbf{R}^n$ defined in a neighborhood of p there corresponds a collection of n^{k+l} numbers $a^{i_1 \ldots i_l}_{j_1 \ldots j_k}$, where the indices $i_1, \ldots, i_l$ and $j_1, \ldots, j_k$ run independently from 1 to n. The numbers $a^{i_1 \ldots i_l}_{j_1 \ldots j_k}$ are the coordinates of the tensor with respect to the chart φ. The following condition must hold here. Let $\varphi\colon U \to \mathbf{R}^n$ and $\psi\colon V \to \mathbf{R}^n$ be arbitrary admissible charts, each defined in a neighborhood of p, let $t_1, \ldots, t_n$ be the coordinate functions of φ, and let $u_1, \ldots, u_n$ be the coordinate functions of ψ. Then the coordinates of the tensor $b^{i_1 \ldots i_l}_{j_1 \ldots j_k}$ with respect to ψ can be expressed in terms of its coordinates $a^{i_1 \ldots i_l}_{j_1 \ldots j_k}$ with respect to φ according to the formula

$$b^{i_1 \ldots i_l}_{j_1 \ldots j_k} = a^{\alpha_1 \ldots \alpha_l}_{\beta_1 \ldots \beta_k} \frac{\partial u_{i_1}}{\partial t_{\alpha_1}} \cdots \frac{\partial u_{i_l}}{\partial t_{\alpha_l}} \cdot \frac{\partial t_{\beta_1}}{\partial u_{j_1}} \cdots \frac{\partial t_{\beta_k}}{\partial u_{j_k}},$$

where repeated indices are summed from 1 to n, and the values of the partial derivatives are taken at the point p.

Let g be a tensor of type $(2, 0)$ at a point p of the manifold M. This tensor is said to be *symmetric* if for some admissible chart defined in a neighborhood of p the coordinates of g satisfy the relations $g_{ij} = g_{ji}$ $(i, j = 1, \ldots, n)$. If this condition holds for some admissible chart in a neighborhood of p, then it holds for any other admissible chart in a neighborhood of p. A symmetric tensor g is said to be *positive-definite* at a point if the quadratic form $g(\xi) = g_{ij}\xi_i\xi_j$ is positive-definite, where the g_{ij} are the coordinates of the tensor with respect to an admissible chart φ defined in a neighborhood of p.

A tensor of type $(0, 1)$ at a point p of M is called a *vector* at this point. The collection of all vectors at a point p of M is denoted by $T_M(p)$ and called the *tangent space* of the manifold at this point.

One says that a tensor field of class C^r and type (k, l) is given on a manifold M of class C^s ($0 \leq r \leq s - 1$) if to each point $p \in M$ there corresponds a tensor of type (k, l), and for every admissible chart in M the coordinates of the tensor with respect to this chart are functions of class C^r.

In studying functions on a differentiable manifold it proves expedient to endow the manifold with a Riemannian metric. For example, specifying the metric facilitates the introduction of a norm in function spaces analogous to the spaces W_p^1, and so on. We give the necessary definitions.

A *Riemannian space of class* C^r is a pair (M, g), where M is an n-dimensional differentiable manifold of class C^{r+1}, and g is a tensor field of class C^r and type $(2, 0)$ that is symmetric and positive-definite. The tensor g here is called the *metric tensor* of the Riemannian space.

Let M be an n-dimensional Riemannian space of class C, and let $\varphi\colon U \to \mathbf{R}^n$ and $\psi\colon U \to \mathbf{R}^n$ be two admissible charts in M that have the same domain $U \subset M$. Let $G = \varphi(U)$ and $H = \psi(U)$. Then G and H are open sets in $\mathbf{R}^n$, and the tensor g on M enables us to define certain Riemannian metrics in G and H. Namely, let $t \in G$ and let $g_{ij}(t)$ be the coordinates of the metric tensor g at the point $p = \varphi^{-1}(t)$ with respect to the chart φ. Similarly, if $u \in H$, then let $h_{ij}(u)$ be the coordinates of the metric tensor of the space at $p = \psi^{-1}(u)$ with respect to φ. We get the Riemannian metric $ds^2 = g_{ij}(t)\, dt_i dt_j$ in G, and the Riemannian metric $d\sigma^2 = h_{ij}(u)\, du_i\, du_j$ in H. By the general rule for transforming tensor coordinates we have that

$$g_{ij}(t) = h_{\alpha\beta}(u(t))\frac{\partial u_\alpha}{\partial t_i} \cdot \frac{\partial u_\beta}{\partial t_j};$$

that is, $ds^2 = (u^* d\sigma)^2$, where $u = \psi \circ \varphi^{-1}$ is the function implementing the coordinate change in passing from the chart φ to the chart ψ.

A certain measure called the volume of a set is defined in a natural way on every Riemannian space. Let $\varphi\colon U \to \mathbf{R}^n$ be an arbitrary admissible chart in M, let $G = \varphi(U)$, and let $g_{ij}(t), i, j = 1, \ldots, n, t \in G$, be the coordinates of the metric tensor of the space in this chart. For an arbitrary Borel set $E \subset U$ we let

$$v_g(E) = \int_{\varphi(E)} \sqrt{\det \|g_{ij}(t)\|}\, dt.$$

By what was proved in §5.1, $v_g(E)$ does not depend on the choice of the chart φ defined on an open set containing E. Let E be an arbitrary Borel set in M. Then E can be represented in the form

$$E = \bigcup_{i=1}^{\infty} E_i \qquad (5.12)$$

where $E_i \cap E_j = \varnothing$ for $i \neq j$, E_i is a Borel set for each i, and E_i is contained in the domain of some admissible chart $\varphi_i : U_i \to \mathbf{R}^n$. The number $v_g(E_i)$ is defined for each $i = 1, 2, \ldots$. The sum $\sum_{i=1}^{\infty} v_g(E_i)$ does not depend on the choice of the representation (5.12), and by definition it is set equal to $v_g(E)$.

Let M_1 and M_2 be Riemannian spaces of class C^1, and let $f : M_1 \to M_2$ be a continuous mapping. Then we say that f is a *mapping with bounded distortion* if there exists a constant K, $1 \leq K < \infty$, such that the following condition holds. Let $\varphi : U \to \mathbf{R}^n$ and $\psi : V \to \mathbf{R}^n$ be arbitrary admissible charts in M_1 and M_2, respectively, such that $f(U) \subset V$. Then the open sets $G = \varphi(U)$ and $H = \psi(V)$ are defined in $\mathbf{R}^n$, and the mapping $f^* = \psi \circ f \circ \varphi^{-1} : G \to H$ is the coordinate representation of f by means of the charts φ and ψ. In the set G we have a Riemannian metric $ds^2 = g_{ij}(t)\, dt_i\, dt_j$— the coordinate representation of the metric of the Riemannian space M_1 by means of the chart φ—and in H the metric $d\sigma^2 = h_{ij}(u)\, du_i\, du_j$ is defined—the coordinate representation of the metric of the Riemannian space M_2 by means of the chart ψ. The condition to be satisfied by f is as follows. For any charts φ and ψ the mapping $f^* : G \to H$ is a mapping with bounded distortion with respect to the metrics ds^2 in G and $d\sigma^2$ in H, and

$$K(f; g, h) \leq K.$$

The smallest number K for which this condition holds is called the *distortion coefficient* of f and denoted by $K(f, M_1, M_2)$. If $K(f, M_1, M_2) = 1$, then f is said to be *conformal*.

Let M_1 and M_2 be differentiable manifolds. We say that $f : M_1 \to M_2$ is a *mapping of class C^r*, where $r \geq 1$, if f is continuous, and the mapping $\psi \circ f \circ \varphi^{-1}$ belongs to the class C^r for any admissible charts $\varphi : U \to \mathbf{R}^n$ in M_1 and $\psi : V \to \mathbf{R}^m$ in M_2 such that $f(U) \subset V$.

We give special consideration to the case of manifolds imbedded in $\mathbf{R}^n$. Let M be a subset of $\mathbf{R}^n$. The set M is called a *k-dimensional manifold of class C^r* in $\mathbf{R}^n$ if it is a k-dimensional manifold and it has an atlas $\mathfrak{A} = \{\varphi_\alpha : U_\alpha \to \mathbf{R}^k\}$ of class C^r such that the following condition holds. For every admissible chart $\varphi : U \to \mathbf{R}^k$ the inverse mapping $x = \varphi^{-1} : G = \varphi(U) \to \mathbf{R}^n$ belongs to the class C^r, and the vectors $f(\partial x / \partial t_i)(t)$, $i = 1, \ldots, k$, are linearly independent at each point $t \in G$. The mapping $x = \varphi^{-1}$ is called a *parametrization* of M. Let M be a k-dimensional manifold of class C^r imbedded in $\mathbf{R}^n$, and let $x : G \to M$ be an arbitrary parametrization of it. Take a $p \in M$, $p = x(t)$. The set R_p

of all points $z \in \mathbf{R}^n$ of the form

$$z = x(t) + \sum_{i=1}^{k} \lambda_i \frac{\partial x}{\partial t_i}(t),$$

where $\lambda_i \in \mathbf{R}$, $i = 1, \ldots, k$, is a k-dimensional plane passing through z. The plane R_p does not depend on the choice of the parametrization x with $p \in x(G)$, and it is called the *tangent plane of M at the point p*. Denote by T_p the k-dimensional subspace of $\mathbf{R}^n$ consisting of all vectors of the form

$$u = \sum_{i=1}^{k} u^i \frac{\partial x}{\partial t_i}(t).$$

The numbers u^i, $i = 1, \ldots, k$, will be called the *coordinates of the vector u with respect to the chart $\varphi = x^{-1}$*. The numbers u^i transform like the coordinates of a tensor of type $(0, 1)$ in passing from one chart to another. In view of this, the vector u can be identified naturally with a vector at the point p of M. Setting

$$g_{ij}(t) = \left\langle \frac{\partial x}{\partial t_i}(t), \frac{\partial x}{\partial t_j}(t) \right\rangle$$

$(i, j = 1, \ldots, k)$ for an arbitrary parametrization $x\colon G \to M$ of M, we get in M a tensor of type $(2, 0)$ that is symmetric and positive-definite; that is, a Riemannian metric in M.

To conclude this section we give some examples of conformal mappings of manifolds.

1. *Stereographic projection*. The following notation will be used. We consider here the $(n + 1)$-dimensional space $\mathbf{R}^{n+1}$. Let $z \in \mathbf{R}^{n+1}$, $z = (x_1, \ldots, x_n, x_{n+1})$. Take $(x_1, \ldots, x_n) = x$ and $x_{n+1} = y$, and regard the vector z here as the pair (x, y). Then $|z|^2 = |x|^2 + y^2$. The space $\mathbf{R}^n$ is identified with the set of all points $z \in \mathbf{R}^{n+1}$ of the form $z = (x, 0)$.

Let S^n be the sphere $\{z \in \mathbf{R}^{n+1} | |z| = 1\}$, and let $N = (0, 1)$ (the north pole) and $S = (0, -1)$ (the south pole). The sphere S^n is a C^∞-differentiable submanifold of $\mathbf{R}^{n+1}$. We define a mapping σ of the sphere S^n onto $\overline{R}^n$. Let $\sigma(N) = \infty$. If $z = (x, y) \neq N$, then $y < 1$, and in this case we let $\sigma(z) = x/(1 - y)$; $\sigma(z)$ is the point where the line passing through N and z intersects the plane $\mathbf{R}^n = \{(x, y) \in \mathbf{R}^{n+1} | y = 0\}$. The mapping σ is called the *stereographic projection* of S^n into $\overline{R}^n$. Let $x \in \overline{R}^n$, $x \neq \infty$. We find a point $z = (u, v) \in S^n$ such that $\sigma(z) = x$. We have the equalities $x = u/(1 - v)$ and $|u|^2 + v^2 = 1$. Solving this system with respect to u and v, we find that

$$u = \frac{2x}{1 + |x|^2}, \qquad v = \frac{|x|^2 - 1}{|x|^2 + 1}. \tag{5.13}$$

The point $z = (u, v)$ with u and v determined by (5.13) does belong to S^n, and $\sigma(z) = x$. This proves that σ is a mapping of S^n onto $\overline{R}^n$. The mapping σ is one-to-one. Each of the mappings $\sigma: S^n \setminus \{N\} \to R^n$ and $\sigma^{-1}: R^n \to S^n \setminus \{N\}$ belongs to the class C^∞.

We introduce the structure of a differentiable manifold in $\overline{R}^n$ by agreeing that the mapping $\sigma: S^n \to \overline{R}^n$ is a diffeomorphism.

Let us show that the stereographic projection maps $S^n \setminus \{N\}$ conformally onto R^n.

Take an arbitrary point $z = (u, v) \in S^n$, $z \neq N$, and let $x = \sigma(z) = u/(1 - v)$. Consider the n-dimensional plane P passing through z parallel to the plane R^n. Projection onto R^n from the point N is a similarity mapping of P onto R^n with similarity coefficient $1/(1 - v)$. It is clear from this that the dilation coefficients of σ are equal to $1/(1 - v)$ in directions lying in the section $S^n \cap P$. We compute the dilation coefficients of σ in the directions orthogonal to this section. To do this we construct the two-dimensional plane H through the diameter SN and the point z. We introduce an orthogonal coordinate system in H, taking the line $O\sigma(z)$ as the axis OX. In this coordinate system the circle $H \cap S^n$ is given by the parametric equations $x = \cos t$, $y = \sin t$, and the restriction of σ to $H \cap S^n$ is the mapping

$$\varphi: t \mapsto \frac{\cos t}{1 - \sin t}.$$

We have that $\varphi'(t) = 1/(1 - \sin t)$, from which we conclude that the dilation coefficient of σ in a direction lying in H is also equal to $1/(1 - v)$, and it is proved that σ is conformal.

It is also possible to consider the stereographic projection with respect to the south pole S of S^n. In this case the image of a point $z = (u, v) \in S^n$ is the point $x = u/(1 + v)$, and the inverse mapping is determined by the formulas

$$u = \frac{2x}{1 + |x|^2}, \qquad v = \frac{1 - |x|^2}{|x|^2 + 1}.$$

2. *The Mercator projection.* By analogy with the well-known cartographical projection we define the Mercator projection to be a mapping of R^n with deleted point O onto a right circular cylinder. Denote by C^n the set

$$\{(x_1, \ldots, x_n, y) \in R^{n+1} | x_1^2 + \cdots + x_n^2 = 1\}$$

in R^{n+1}. Let x be an arbitrary point in R^n, $x = (x_1, \ldots, x_n) \neq 0$. We set

$$f(x) = \left(\frac{x_1}{|x|}, \frac{x_2}{|x|}, \ldots, \frac{x_n}{|x|}, \ln \frac{1}{|x|} \right).$$

Obviously, f is a mapping of $\mathbf{R}^n \setminus \{0\}$ onto the cylinder C^n. Here every sphere $S(0, r) = \{x \in \mathbf{R}^n \,||x| = r\}$ is mapped by a similarity onto the unit sphere obtained in the section of C^n by the plane $y = \ln 1/r$. The dilation coefficients of f at a point $x \in S(0, r)$ in the directions tangent to the sphere are thus equal to $1/r$—the ratio of the radii of the image and inverse image spheres. The rays emanating from O are carried under the mapping f into generators of the cylinder C^n, i.e., into curves orthogonal to the images of $S(0, r)$. The dilation coefficient along any such ray is equal to

$$\left| \frac{d}{dr} \ln \left(\frac{1}{r} \right) \right| = \frac{1}{r}.$$

We see that the dilation coefficients of $f'(x)$ in n mutually orthogonal directions are equal to $1/r$. Consequently, f is conformal.

CHAPTER II

Main Facts in the Theory of Mappings with Bounded Distortion

§1. Estimates of the moduli of continuity and differentiability almost everywhere of mappings with bounded distortion

§1.1. Some auxiliary facts. Let U be an arbitrary open domain in $\mathbf{R}^n$, and let $f: U \to \mathbf{R}^n$ be a continuous mapping. The mapping f is said to *satisfy Hölder condition with exponent α on compact subsets of U*, where $0 < \alpha \leq 1$, if for every compact set $A \subset U$ there is a number $M(A)$, $0 \leq M(A) < \infty$, such that for any $x_1, x_2 \in A$

$$|f(x_1) - f(x_2)| \leq M(A)|x_1 - x_2|^\alpha. \tag{1.1}$$

It will be shown below that every mapping with bounded distortion satisfies a Hölder condition with exponent $\alpha = 1/K$ on compact subsets of its domain, where $K = K(f)$. We remark that an estimate of the modulus of continuity of quasiconformal mappings was first established by Kreĭnes [73]. The Hölder property was first proved for a mapping with bounded distortion by the author ([119], [124]) and simultaneously by Callender [24]. The proof that a mapping with bounded distortion satisfies a Hölder condition is based on two facts. The first is a lemma due to Morrey [102]. The second is an inequality which is a corollary of a known isoperimetric inequality.

Let U be an open set in $\mathbf{R}^n$. For any point $x \in U$ let $\rho_U(x) = \rho(x, \partial U)$, with the subscript U omitted whenever no confusion can result.

LEMMA 1.1 (Morrey's lemma). *Let $U \subset \mathbf{R}^n$ be an open set in $\mathbf{R}^n$, and $f: U \to \mathbf{R}^k$ a function of the class $W_m^1(U)$, where $1 \leq m \leq n$. Assume that there exist numbers α $(0 < \alpha \leq 1)$, $M < \infty$, and $\delta > 0$ such that*

$$\int_{B(a,r)} |f'(x)|^m \, dx \leq M r^{n-m+m\alpha}$$

for every ball $B(a, r) \subset U$ with radius at most δ. Then there exists a continuous function $\tilde{f}$ such that $f(x) = \tilde{f}(x)$ almost everywhere, and the oscillation of $\tilde{f}$ on any ball $B(x, r) \subset U$ with $r \leq \delta/3$ and $r < \rho(x)/3$ does not exceed $CM^{1/m}r^\alpha$, where $C < \infty$ is a constant.

A proof of the lemma will be given in Chapter III, §4.2.

If a continuous function f satisfies the conditions of Lemma 1.1, then it satisfies a Hölder condition with exponent α on compact subsets of U. Indeed, let A be a compact subset of U, and let γ be the smaller of the numbers $\delta/3$ and $d/3$, where δ is the constant in Lemma 1.1 and $d = \text{dist}(A, \partial U)$. We consider the function h defined as follows on the product $A \times A$: $h(x, y) = |f(x) - f(y)|/|x - y|^\alpha$ for $x \neq y$, and $h(x, x) = 0$. Let H be the set of pairs $(x, y) \in A \times A$ such that $|x - y| \geq \gamma$, and let $G = (A \times A) \backslash H$. The set H is compact, and thus h is bounded on H by continuity. The conclusion of the lemma enables us to deduce that h is bounded also on G. Consequently, h is bounded on $A \times A$, and thus $|f(x) - f(y)| \leq M|x - y|^\alpha$ for any $x, y \in A$.

LEMMA 1.2. *Suppose that $U \subset \mathbf{R}^n$ and the mapping $f: U \to \mathbf{R}^n$ is in the class $W_n^1(U)$. Then for any $a \in U$ and almost all $t \in (0, \rho(a))$*

$$\int_{B(a,t)} \det f'(x)\, dx \leq \frac{t}{n} \int_{S(a,t)} \|f'(x)\|^n\, d\sigma_x,$$

where $d\sigma$ is the area element of the sphere $S(a, t)$.

PROOF. We get the lemma as a corollary to a known isoperimetric inequality. To formulate the latter we give some necessary definitions. Let $A: \mathbf{R}^n \to \mathbf{R}^n$ be a linear mapping, ν a unit vector in $\mathbf{R}^n$, P_ν the $(n - 1)$-dimensional subspace of $\mathbf{R}^n$ orthogonal to ν, and $Q_\nu = A(P_\nu)$. The restriction of A to P_ν is a linear mapping of an $(n - 1)$-dimensional Euclidean space P_ν onto the Euclidean space Q_ν. The determinant of this mapping is denoted by $\Delta_\nu(A)$. The norm of the linear mapping $A|P_\nu$ clearly does not exceed $\|A\|$. We have $\Delta_\nu(A) \leq \|A|P_\nu\|^{n-1}$, which implies that

$$\Delta_\nu(A) \leq \|A\|^{n-1}. \tag{1.2}$$

Suppose that U is an arbitrary domain in $\mathbf{R}^n$, $a \in U$, and $f: U \to \mathbf{R}^n$ is a mapping of class W_n^1. Let

$$V(t, f) = \int_{B(a,t)} \det f'(x)\, dx, \tag{1.3}$$

where $0 < t < \rho_U(a)$, and let

$$F(t, f) = \int_{S(a,t)} \Delta_{\nu(x)}[f'(x)]\, d\sigma_x. \tag{1.4}$$

Here $d\sigma_x$ is the area element of the sphere $S(a, t)$, and $\nu(x)$ is the unit vector normal to $S(a, t)$ at the point x. The integral (1.4) is obviously defined for almost all $t \in (0, \rho(a))$.

In the case when f is a mapping of class C^1 and the Jacobian of f is nonzero everywhere, $F(t, f)$ is clearly the area of the $(n - 1)$-dimensional surface into which $S(a, t)$ is transformed by f, and $V(t, f)$ is the oriented volume bounded by this surface. For a suitable definition of the concept of the area of a surface it is possible to interpret the integral (1.4) as the area of a surface also in the case when f is a mapping of the class W_n^1. We remark that, by (1.2), $\Delta_\nu[f'(x)] \geq \|f'(x)\|^{n-1}$ for all x such that $f'(x)$ is defined, and this gives us that

$$F(t, f) \leq \int_{S(a,t)} \|f'(x)\|^{n-1} \, d\sigma_x. \tag{1.5}$$

For almost all $t \in (0, \rho(a))$

$$V(t, f) \leq (1/n\beta_n)[F(t, f)]^{n/(n-1)}, \tag{1.6}$$

where $\beta_n = \omega_n^{1/(n-1)}$. This is the isoperimetric inequality. It expresses the fact that among all closed surfaces in $\mathbf{R}^n$ with given area a sphere bounds the largest volume. We do not give a proof of (1.6), but refer the reader to the author's paper [123].

It follows from (1.5) and (1.6) that for almost all $t \in (0, \rho(a))$

$$\int_{B(a,t)} \det f'(x) \, dx \leq (1/n\beta_n) \left(\int_{S(a,t)} \|f'(x)\|^{n-1} \, d\sigma_x \right)^{n/(n-1)}. \tag{1.7}$$

By Hölder's inequality,

$$\int_{S(a,t)} \|f'(x)\|^{n-1} \, d\sigma_x \leq \left(\int_{S(a,t)} \|f'(x)\|^n \, d\sigma_x \right)^{1-1/n} \left(\int_{S(a,t)} d\sigma_x \right)^{1/n}.$$

The second integral on the right-hand side is equal to $\omega_n t^{n-1}$. Substituting the last expression into (1.7), we get that

$$\int_{B(a,t)} \det f'(x) \, dx \leq \frac{t}{n} \int_{S(a,t)} \|f'(x)\|^n \, d\sigma_x,$$

as required.

§1.2. An estimate of the modulus of continuity of a mapping with bounded distortion.

LEMMA 1.3. *Suppose that* $U \subset \mathbf{R}^n$ *is an open set in* $\mathbf{R}^n$, $f: U \to \mathbf{R}^n$ *is a mapping of class* $W^1_{n,\mathrm{loc}}(U)$ *whose Jacobian has positive sign in* U, *and* $\|f'(x)\|^n \le K|\mathcal{J}(x, f)|$ *for almost all* $x \in U$, *where* $1 \le K < \infty$. *For* $x \in U$ *and* $r < \rho(x)$ *let*

$$v(x, r) = (1/r^{n/K}) \int_{B(x,r)} \|f'(x)\|^n \, dx.$$

Then the function $r \to v(x, r)$ *is nondecreasing.*

PROOF. For $r < \rho(x)$

$$\int_{B(x,r)} \|f'(x)\|^n \, dx \le K \int_{B(x,r)} |\mathcal{J}(x, f)| \, dx = K \left| \int_{B(x,r)} \mathcal{J}(x, f) \, dx \right| \quad (1.8)$$

because $\mathcal{J}(x, f)$ has constant sign in U. On the basis of Lemma 1.2

$$\left| \int_{B(x,r)} \det f'(x) \, dx \right| \le \frac{r}{n} \int_{S(a,r)} \|f'(x)\|^n \, d\sigma_x \quad (1.9)$$

for almost all $r \in (0, \rho(x))$. From (1.8) and (1.9) we get

$$\int_{B(x,r)} \|f'(x)\|^n \, dx \le \frac{Kr}{n} \int_{S(x,r)} \|f'(x)\|^n \, d\sigma_x. \quad (1.10)$$

Let

$$\int_{B(x,r)} \|f'(x)\|^n \, dx = w(r), \qquad \int_{S(x,r)} \|f'(x)\|^n \, d\sigma_x = s(r).$$

Applying Fubini's theorem, we get that $w(r) = \int_0^r s(t) \, dt$ for all $r \in (0, \rho(x))$. This leads us to conclude that the function w is absolutely continuous and $w'(r) = s(r)$ for almost all $r \in (0, \rho(x))$. From (1.10) we have that $w(r) \le Krw'(r)/n$ for almost all r. Multiplying both sides of this inequality by $r^{-(n/K)-1}$, we get after obvious transformations that $(w(r)/r^{n/K})' \ge 0$. Consequently, the function $w(r)/r^{n/K}$ is nondecreasing, and the lemma is proved.

THEOREM 1.1. *Suppose that* $U \subset \mathbf{R}^n$ *is an open set,* $f: U \to \mathbf{R}^n$ *is a mapping of class* $W^1_n(U)$ *such that the function* $x \to \mathcal{J}(x, f)$ *has constant sign in* U, *i.e., is either nonnegative almost everywhere or nonpositive almost everywhere, and* $\|f'(x)\|^n \le K|\mathcal{J}(x, f)|$ *for almost all* $x \in U$. *Let* $\int_U \|f'(x)\|^n \, dx = M^n$. *Then the vector-valued function* f *is equivalent, in the sense of the theory of the integral, to some continuous function* f^*. *Further, for every set* V *lying strictly inside* U *the oscillation of* f^* *on any ball*

$B(a, r)$ *of radius* $r < 2d/3$ *about an* $a \in V$ *does not exceed* $(CM/d^{1/K})r^{1/K}$, *where* $d = \text{dist}(V, \partial U)$.

PROOF. Let $a \in V$, and let

$$w(a, r) = \int_{B(a,r)} \|f'(x)\|^n \, dx \le M^n.$$

According to Lemma 1.3, the function $v(r) = w(a, r)r^{-n/K}$ is nondecreasing, and thus

$$w(a, r)r^{-n/K} \le v(d) \le M^n d^{-n/K}$$

for all $r \in (0, d)$; hence $w(a, r) \le (M^n/d^{n/K})r^{n/K}$. The required result follows directly from Lemma 1.1.

COROLLARY 1. *Let* $f: U \to \mathbf{R}^n$ *be a mapping with bounded distortion. Assume that* $\int_U \|f'(x)\|^n \, dx = M < \infty$. *Then the function* f *satisfies a Hölder condition with exponent* $\alpha = 1/K$ *on compact subsets of* U. *Further, if* V *is contained strictly inside* U, *then for any* $x, y \in V$

$$|f(x) - f(y)| \le L|x - y|^{1/K}, \tag{1.11}$$

where the constant L *depends only on* V, *the distance from* V *to the boundary of* U, *and the constant* M.

REMARK. The example of the mapping with $f(0) = 0$ and $f: x \to x|x|^{\alpha-1}$ for $x \ne 0$, where $\alpha = 1/K$ and $K \ge 1$, shows that the exponent $1/K$ in (1.11) cannot be decreased. For this f we have that $K(f) = K$ and $|f(x) - f(0)| = |x|^{\alpha}$.

COROLLARY 2. *Let* U *be an open domain in* $\mathbf{R}^n$, *and* $F(U, K, M)$ *the collection of all mappings* f *with bounded distortion of* U *such that* $\int_U \|f'(x)\|^n \, dx \le M$. *Then the set of functions* f *is equi-uniformly continuous on every compact subset of* U.

§1.3. Differentiability almost everywhere of mappings with bounded distortion.

Let U be an open set in $\mathbf{R}^n$. As we know, a mapping $f: U \to \mathbf{R}^n$ is said to be differentiable at a point $a \in U$ if there exists a linear mapping $L: \mathbf{R}^n \to \mathbf{R}^n$ such that $f(x) = f(a) + L(x - a) + \alpha(x)|x - a|$ for all $x \in U$, where $\alpha(x) \to 0$ as $x \to a$. The mapping L is called the differential of f at the point a. Our immediate goal is to prove that every mapping with bounded distortion is differentiable almost everywhere in its domain. The proof of this is based on a certain general proposition about differentiability almost everywhere of functions with generalized derivatives. We present formulation of this proposition here. Its proof will be given in Chapter III, §4.1. Let us first define some concepts.

Let $\mathfrak{R}$ be a topological vector space whose elements are functions with values in a finite-dimensional vector space V and defined on the ball $B_1 = \overline{B}(0, 1) \subset \mathbf{R}^n$. (Only the case when $V = \mathbf{R}^k$ is considered in what follows.) Let $U \subset \mathbf{R}^n$ be an open set. We say that the function $f: U \to V$ is differentiable in the sense of convergence in $\mathfrak{R}$ (briefly, $\mathfrak{R}$-differentiable) at the point $a \in U$ if there exists a linear mapping $L: \mathbf{R}^n \to V$ such that the function $r_h: X \in B_1 \to (1/h)[f(a + hX) - f(a) - hL(X)]$ is defined on B_1 and belongs to $\mathfrak{R}$ for sufficiently small h, $0 < h \leq h_0$, and $r_h \to 0$ in the topology of $\mathfrak{R}$ as $h \to 0$. The linear mapping L is called the $\mathfrak{R}$-differential of f at the point a.

Let $\mathfrak{R}$ be the Banach space L_∞ of bounded functions $f: B_1 \to V$, with the norm in L_∞ defined by $\|f\|_{L_\infty} = \operatorname{ess\,sup}_{x \in B_1} |f(x)|$. Assume that the function $f: U \to V$ is differentiable at a point $a \in U$ in the sense of convergence in L_∞. Let $\Delta(h) = \|r_h\|_{L_\infty}$. Then it is obvious that $|r_h(X)| \leq \Delta(h)$ for every $X \in B_1$. Take an arbitrary vector ξ such that $|\xi| < h_0$. Substituting $h = |\xi|$ and $X = \xi/|\xi|$ in the expression for $r_h(X)$, we get that

$$\frac{|f(a + \xi) - f(n) - A(\xi)|}{|\xi|} \leq \Delta(|\xi|),$$

and hence $|f(a + \xi) - f(a) - A(\xi)|/|\xi| \to 0$ as $\xi \to 0$. Thus, L_∞-differentiability at the point a implies differentiability in the classical sense. It is not hard to verify that, conversely, if f is differentiable at a in the classical sense, then it is L_∞-differentiable at this point.

LEMMA 1.4. *Suppose that* $f: U \to \mathbf{R}^k$ *is a function of class* $W_p^1(U)$, *where* $U \subset \mathbf{R}^n$ *is an open set. Then* f *is differentiable in the sense of convergence in* $W_p^1(U)$ *at almost all points* $x \in U$, *and its formal differential* $f'(x)$ *is the* W_p^1-*differential almost everywhere in* U.

This is a special case of a general theorem in [131]. A proof of the lemma will be given in Chapter III, §4.1.

THEOREM 1.2. *Let* $f: U \to \mathbf{R}^n$ *be a mapping with bounded distortion. Then for almost all* $x \in U$ *the linear mapping* $f'(x)$ *is the differential of* f *at the point* x.

PROOF. Let $f: U \to \mathbf{R}^n$ be a mapping with bounded distortion. Then $f \in W_n^1$, and thus, according to Lemma 3, the linear mapping $f'(x)$ is the differential in the sense of convergence in W_n^1 for almost all $x \in U$. Suppose that the point $x \in U$ is such that $L = f'(x)$ is the W_n^1-differential of f at x. Let $f_h(X) = (1/h)[f(x + hX) - f(x)]$. Then $\|f_h - L\|_{1,n,B_1} \to 0$ as $h \to 0$. The vector-valued function f_h belongs to W_n^1, like f, and its derivatives can be expressed in terms of the derivatives of f by the same formulas as in the case of functions of class C^1. This gives us that $f_h'(X) = f'(x + hX)$ for almost all $X \in B_1$, which enables us to conclude that f_h is a mapping with bounded distortion. Further, it is not hard to see that $K(f_h) \leq K(f)$. Let $h_0 > 0$ be such that for $0 < h < h_0$ the mapping f_h is defined everywhere on the ball B_1. Since f_h converges in $W_n^1(B_1)$ to the linear mapping $f'(a)$ as $h \to 0$, it can be assumed that the norms $\|f_h\|_{1,n,B_1}$ are bounded for $0 < h \leq h_0$, $\|f_h\|_{1,n,B_1} < M < \infty$ for all such h. Hence, by Corollary 1 to Theorem 1.1, the family of mappings f_h, $0 < h < h_0$, is equi-uniformly continuous on every ball B_r, where $0 < r < 1$. Since $f_h \to L$ in $W_n^1(B_1)$ as $h \to 0$, $f_h \to f'(x)$ in $L_n(B_1)$ in view of the Sobolev imbedding theorem. Since the family of functions f_h is equi-uniformly continuous, this implies that $f_h \to L$ uniformly on every ball B_r, where $0 < r < 1$. We have the relation $f_h(X) = 2f_{2h}(X/2)$. For $|X| \leq 1$ the point $X/2$ is in $B_{1/2}$. Since $f_{2h}(X) \to L(X)$ uniformly on $B_{1/2}$, this implies that $2f_{2h}(X/2) \to L(X)$ uniformly on B_1, and hence $f_h \to L$ uniformly on B_1. This proves that f is differentiable at x and $L = f'(x)$ is the differential of f at x. The theorem is proved.

§2. Some facts about continuous mappings on $\mathbf{R}^n$

§2.1. The degree of a mapping.
The main topological concept used in studying n-space mappings is the degree of a mapping. Its definition does not require a cumbersome topological apparatus. Here we present some necessary facts. A set $G \subset \mathbf{R}^n$ is called a *compact domain* if G is compact, its interior G^0 is connected, and it is the closure of its interior.

Let $G \subset \mathbf{R}^n$ be a compact domain, and $f: G \to \mathbf{R}^n$ a continuous mapping. A point $y \in \mathbf{R}^n$ is said to be (f, G)-*admissible* if $y \notin f(\partial G)$. With every triple (f, G, y) with G a compact domain in $\mathbf{R}^n$, $f: G \to \mathbf{R}^n$ a continuous mapping, and y an (f, G)-admissible point we can associate a number $\mu(y, f, G)$ called the *topological index* or *degree* of the mapping $f: G \to \mathbf{R}^n$ at the point y.

We first define the concept of degree for the case when f is sufficiently smooth. Let $G \subset \mathbf{R}^n$ be a compact domain. The mapping $f: G \to \mathbf{R}^n$ is said to belong to the class C^k, where $1 \leq k \leq \infty$, if the restriction of f to G^0 belongs to C^k, and each of its derivatives of order at most k is uniformly continuous on G^0.

Let $f: G \to \mathbf{R}^n$ be a mapping of class C^k, $k \geq 1$, on a compact domain $G \subset \mathbf{R}^n$. We say that f is *regular* with respect to a point $y \in \mathbf{R}^n$ if the following conditions hold:

1) y is (f, G)-admissible, i.e., $y \notin f(\partial G)$.
2) The set $f^{-1}(y)$ is finite.

3) At each point $x \in f^{-1}(y)$ the Jacobian of f is nonzero.

In particular, f is regular with respect to y if $f^{-1}(y)$ is empty.

Let $f: G \to \mathbf{R}^n$ be a mapping of class C^k which is regular with respect to a point $y \in \mathbf{R}^n$. The total number of points in $f^{-1}(y)$ is denoted by N. Let N^+ be the number of points $x \in f^{-1}(y)$ at which $\mathscr{J}(x, f) > 0$, and N^- the number of points $x \in f^{-1}(y)$ at which $\mathscr{J}(x, f) < 0$. Obviously, $N = N^+ + N^-$. The number $\mu = N^+ - N^-$ is called the *degree* or *topological index* of f with respect to the point y and the domain G, and is denoted by $\mu(y, f, G)$. If $f^{-1}(y)$ is empty, then each of the numbers N, N^+, and N^- is equal to zero, and hence $\mu(y, f, G) = 0$ in this case.

Let A and B be arbitrary sets in $\mathbf{R}^n$. They are said to form a *pair* if $A \supset B$. Assume that pairs (A_1, B_1) and (A_2, B_2) of sets are given. A mapping of the pair (A_1, B_1) into the pair (A_2, B_2) is defined to be a mapping $f: A_1 \to A_2$ such that $f(B_1) \subset B_2$.

We define the concept of a homotopy of continuous mappings of pairs. Let (A_1, B_1) and (A_2, B_2) be two arbitrary pairs of sets in $\mathbf{R}^n$, and let f and g be continuous mappings of the first pair into the second; f and g are said to be *homotopic* if there is a family of mappings φ_t, $t \in [0, 1]$, of (A_1, B_1) into (A_2, B_2) such that: 1) $\varphi_0 \equiv f$ and $\varphi_1 \equiv g$; and 2) the mapping $\Phi: (x, t) \to \varphi_t(x)$ of $A_1 \times [0, 1] \subset \mathbf{R}^{n+1}$ into $\mathbf{R}^n$ is continuous.

The mapping Φ is called a *homotopy* of the mappings f and g. If f is homotopic to g, then we write $f_{\bar{H}} g$. Homotopy of mappings is an equivalence relation, i.e., the following hold for any continuous mappings f, g, and h of the pair (A_1, B_1) into (A_2, B_2): 1) $f_{\bar{H}} f$; 2) if $f_{\bar{H}} g$, then $g_{\bar{H}} f$; and 3) if $f_{\bar{H}} g$ and $g_{\bar{H}} h$, then $f_{\bar{H}} h$. The proofs of these properties are obvious and are left to the reader.

The following simple remark will be used repeatedly below. Let $G \subset \mathbf{R}^n$ be a compact domain, $f: G \to \mathbf{R}^n$ a continuous mapping, and $y \in \mathbf{R}^n$ an (f, G)-admissible point. Let $(f_\nu: G \to \mathbf{R}^n)$, $\nu = 1, 2, \ldots$, be a sequence of continuous mappings uniformly convergent to f. Then there is a ν_0 such that for $\nu > \nu_0$ the point y is (f_ν, G)-admissible, and f_ν and f are homotopic as mappings of the pair $(G, \partial G)$ into the pair $(\mathbf{R}^n, \mathbf{R}^n \setminus \{y\})$. Indeed, $f(\partial G)$ is compact and $y \notin f(\partial G)$, and thus $\delta = \rho(y, f(\partial G)) > 0$. Let ν_0 be such that $|f_\nu(x) - f(x)| < \delta$ for $\nu > \nu_0$ for all $x \in \partial G$. This is the desired ν_0. The homotopy of f and f_ν for $\nu > \nu_0$ can be taken to be the function

$$\varphi_t(x) = (1 - t)f(x) + tf_\nu(x), \qquad t \in [0, 1].$$

LEMMA 2.1. *Let G be a compact domain in $\mathbf{R}^n$, and let $f: G \to \mathbf{R}^n$ and $g: G \to \mathbf{R}^n$ be mappings of class C^k, $k \geq 1$, which are regular with respect* *to y. If the pair mappings*

$$f: (G, \partial G) \to (\mathbf{R}^n, \mathbf{R}^n \backslash \{y\}) \quad and \quad g: (G, \partial G) \to (\mathbf{R}^n, \mathbf{R}^n \backslash \{y\})$$

are homotopic, then $\mu(y, f, G) = \mu(y, g, G)$.

LEMMA 2.2. *Let G be a compact domain in $\mathbf{R}^n$, and let $f: G \to \mathbf{R}^n$ be a continuous mapping. Then for every (f, G)-admissible point y there exists a sequence of mappings $f_\nu: G \to \mathbf{R}^n$ of class C^∞, $\nu = 1, 2, \ldots$, such that f_ν converges to f uniformly on G as $\nu \to \infty$, and each of the mappings f_ν is regular with respect to y.*

Proofs of Lemmas 2.1 and 2.2 are given in Chapter III, §5.

Using Lemmas 2.1 and 2.2, we define the concept of degree for continuous mappings. Let G be a compact domain in $\mathbf{R}^n$, $f: G \to \mathbf{R}^n$ a continuous mapping, and y an arbitrary (f, G)-admissible point.

We construct a sequence of mappings $f_\nu: G \to \mathbf{R}^n$ satisfying all the conditions of Lemma 2.2. For each ν let $\mu(y, f_\nu, G) = \mu_\nu$. Since $f_\nu \to f$ uniformly on G, there is a ν_0 such that f_ν and f are homotopic as mappings of the pair $(G, \partial G)$ into the pair $(\mathbf{R}^n, \mathbf{R}^n \backslash \{y\})$ for $\nu > \nu_0$. If $\nu' > \nu_0$ and $\nu'' > \nu_0$, then $f_{\nu'}$ and $f_{\nu''}$ are also homotopic as mappings of $(G, \partial G)$ into $(\mathbf{R}^n, \mathbf{R}^n \backslash \{y\})$. Hence, by Lemma 2.1, $\mu_{\nu'} = \mu_{\nu''}$ for any ν' and ν'' larger than ν_0. This shows, in particular, that the limit $\mu_0 = \lim_{\nu \to \infty} \mu_\nu$ exists. This limit does not depend on the choice of the sequence $(f_\nu)_{\nu \in \mathbf{N}}$. Indeed, let (g_ν) be any other sequence of C^k-mappings of G convergent uniformly to f. Then $g_\nu \sim_H f_\nu$ for sufficiently large $\nu \geq \nu_1$, and hence $\mu(y, f_\nu, G) = \mu(y, g_\nu, G)$ for all $\nu \geq \nu_1$. Consequently,

$$\lim_{\nu \to \infty} \mu(y, f_\nu, G) = \lim_{\nu \to \infty} \mu(y, g_\nu, G),$$

which is what was to be proved. The limit $\mu_0 = \lim_{\nu \to \infty} \mu(y, f_\nu, G)$ is called the *degree* of the mapping $f: G \to \mathbf{R}^n$ at the point y, and will be denoted by $\mu(y, f, G)$.

For the case of mappings of class C^k this definition leads to the same result as the original definition. Indeed, if $f_\nu: G \to \mathbf{R}^n$, $\nu = 1, 2, \ldots$, is a sequence of mappings of class C^∞ which converges uniformly to $f: G \to \mathbf{R}^n$ as $\nu \to \infty$, then f_ν and f are homotopic as mappings of $(G, \partial G)$ into $(\mathbf{R}^n, \mathbf{R}^n \backslash \{y\})$ beginning with some $\nu = \nu_0$, and thus $\mu(y, f_\nu, G) = \mu(y, f, G)$ for $\nu > \nu_0$, i.e.,

$$\mu(y, f, G) = \lim_{\nu \to \infty} \mu(y, f_\nu, G).$$

We make note of some properties of the degree of a mapping which follow immediately from the definition and the main Lemmas 2.1 and 2.2.

I. *Let G be a compact domain in $\mathbf{R}^n$, and $f_0: G \to \mathbf{R}^n$, and $f_1: G \to \mathbf{R}^n$ continuous mappings. Assume that the point $y \in \mathbf{R}^n$ is simultaneously (f_0, G) admissible and (f_1, G)-admissible. If f_0 and f_1 are homotopic as mappings of the pair $(G, \partial G)$ into the pair $(\mathbf{R}^n, \mathbf{R}^n \setminus \{y\})$, then*

$$\mu(y, f_0, G) = \mu(y, f_1, G).$$

PROOF. By Lemma 2.2, we construct sequences of mappings $f_\nu^{(0)}: G \to \mathbf{R}^n$ and $f_\nu^{(1)}: G \to \mathbf{R}^n$ of class C^∞ which are uniformly convergent to f_0 and f_1, respectively, as $\nu \to \infty$. There is a ν_0 such that $f_\nu^{(0)}$ is homotopic to f_0 and $f_\nu^{(1)}$ is homotopic to f_1 for $\nu > \nu_0$. Then for $\nu > \nu_0$

$$\mu(y, f_\nu^{(0)}, G) = \mu(y, f_0, G),$$
$$\mu(y, f_\nu^{(1)}, G) = \mu(y, f_1, G).$$

Since f_0 and f_1 are homotopic, $f_\nu^{(0)}$ is homotopic to $f_\nu^{(1)}$ for $\nu > \nu_0$, which implies that $\mu(y, f_\nu^{(0)}, G) = \mu(y, f_\nu^{(1)}, G)$ for these values of ν, and thus $\mu(y, f_0, G) = \mu(y, f_1, G)$, which is what was to be proved.

II (The additive property of the degree of a mapping). *Let $G_0, \ldots, G_m$ be compact domains such that $G_i \subset G_0^0$ for all $i = 1, \ldots, m$ and G_i and G_j have no common interior points for $i \neq j$. Let $f: G_0 \to \mathbf{R}^n$ be continuous. Assume that the point $y \in \mathbf{R}^n$ is such that $f^{-1}(y) \subset \bigcup_1^m G_i$ and y is (f, G_i)-admissible for $i = 0, \ldots, m$. Then*

$$\mu(y, f, G_0) = \sum_{i=1}^m \mu(y, f, G_i).$$

PROOF. Suppose that $(f_\nu: G_0 \to \mathbf{R}^n)$ is a sequence of mappings of class C^∞ which are regular with respect to y and converge uniformly to f on G_0. Then there is a ν_0 such that $\mu(y, f_\nu, G_i) = \mu(y, f, G_i)$ for $\nu > \nu_0$ and $i = 0, \ldots, m$. Further, since f_ν converges uniformly to f, the set $f_\nu^{-1}(y)$ is contained in $\bigcup_1^m G_i^0$ for sufficiently large ν, $\nu \geq \nu_1 \geq \nu_0$. It is obvious from the definition of the degree for mappings of class C^k that in this case

$$\mu(y, f_\nu, G_0) = \sum_{i=1}^m \mu(y, f_\nu, G_i).$$

Passing to the limit as $\nu \to \infty$, we get the required result.

III. *Let $f: G \to \mathbf{R}^n$ be the restriction of a nonsingular affine mapping of $\mathbf{R}^n$. Then for every (f, G)-admissible point $y \in \mathbf{R}^n$*

$$\mu(y, f, G) = \begin{cases} 0 & \text{if } y \notin f(G), \\ \operatorname{sgn}(\det f) & \text{if } y \in f(G). \end{cases}$$

This proposition is an obvious consequence of the definition of degree for mappings of class C^k.

IV. *The degree $\mu(y, f, G)$ is 0 for every mapping $f: G \to \mathbf{R}^n$ identically constant on G and for any (f, G)-admissible point y.*

PROOF. Suppose that $f \equiv \text{const}$ on G, $f(x) = b$ for all $x \in G$. Then a point y is (f, G)-admissible if and only if $y \neq b$. Then $f^{-1}(y)$ is empty, and hence $\mu(y, f, G) = 0$, as required.

V. *Suppose that $f: G \to \mathbf{R}^n$ is continuous and $y \notin f(G)$. Then $\mu(y, f, G) = 0$.*

The assertion is obvious from the definition of degree in the case when $f \in C^k$. In the general case it can easily be established by passing to the limit.

VI. *Let $f: G \to \mathbf{R}^n$ be a continuous mapping. Then the function $y \to \mu(y, f, G)$ of the variable y is constant on each connected component of $\mathbf{R}^n \backslash f(\partial G)$.*

PROOF. We first make the following simple observation. Let $f: G \to \mathbf{R}^n$ be continuous, y an (f, G)-admissible point, and a an arbitrary vector in $\mathbf{R}^n$. Then

$$\mu(y + a, f + a, G) = \mu(y, f, G). \tag{2.1}$$

Indeed, assume that f is in C^∞ and is regular with respect to y. Then the mapping $g = f + a$ is also in C^∞. Further, $g^{-1}(y + a) = f^{-1}(y)$, and at each point $x \in g^{-1}(y + a)$ the Jacobian of g is equal to that of f at x. Hence, g is regular with respect to y, and $\mu(y + a, g, G) = \mu(y, f, G)$. In the general case (2.1) can be established by an obvious passage to the limit.

Let $f: G \to \mathbf{R}^n$ be continuous, and Δ an arbitrary connected component of the open set $\mathbf{R}^n \backslash f(\partial G)$. Let $y_0, y_1 \in \Delta$. Since Δ is connected, y_0 and y_1 can be joined by a continuous path in Δ. Let $y: [0, 1] \to \Delta$ be a continuous path such that $y(0) = y_0$ and $y(1) = y_1$. Let $\varphi_t(x) = f(x) - y(t)$, $t \in [0, 1]$. For each $t \in [0, 1]$, φ_t is a mapping of the pair $(G, \partial G)$ into $(\mathbf{R}^n, \mathbf{R}^n \backslash \{0\})$. Indeed, if $f(x) - y(t) = 0$, then $x \notin \partial G$, since $y(t) \notin \partial G$ for all $t \in [0, 1]$, and hence if $x \in \partial G$, then $\varphi_t(x) \neq 0$. The function φ_t is a homotopy of the mappings $\varphi_0 = f - y_0$ and $\varphi_1 = f - y_1$, which implies that

$$\mu(0, f - y_0, G) = \mu(0, f - y_1, G).$$

We can conclude from the foregoing that

$$\mu(y_0, f, G) = \mu(y_1, f, G).$$

Since y_0 and y_1 are arbitrary points of Δ, this proves the given statement.

VII. *Let* $f: G \to \mathbf{R}^n$ *be continuous and* Δ *the unbounded component of the set* $\mathbf{R}^n \backslash f(\partial G)$. *Then* $\mu(y, f, G) = 0$ *for every* $y \in \Delta$.

Indeed, $f(G)$ is compact, and if Δ is unbounded, then there is a point $y \in \Delta$ such that $y \notin f(G)$. Then $\mu(y, f, G) = 0$. Since $\mu(y, f, G)$ is continuous on Δ, $\mu(y, f, G) = 0$ for any $y \in \Delta$.

§2.2. The degree of a mapping and exterior differential forms. We describe a certain analytic technique used to study the degree of a mapping. Here we assume that the reader is familiar with concepts involving exterior differential forms in $\mathbf{R}^n$, in particular, concepts such as the product of exterior forms, the differential of an exterior form, transformations of forms under differentiable mappings, etc. Only exterior forms of degrees $n - 1$ and n will be considered. Every exterior form of degree $n - 1$ in a domain U in the space $\mathbf{R}^n$ has an expression

$$\omega(x) = \sum_{h=1}^{n} (-1)^{k-1} u_k(x) \, dx_1 \ldots dx_{k-1} \, dx_{k+1} \ldots dx_n, \qquad (2.2)$$

where $u_1, \ldots, u_n$ are functions defined in U. An exterior differential form of degree n is an expression like

$$\omega(x) = u(x) \, dx_1 \, dx_2 \ldots dx_n. \qquad (2.3)$$

The differential $d\omega(x)$ of the form (2.2) is the form

$$\left(\sum_{k=1}^{n} \frac{du_k}{dx_k}(x) \right) dx_1 \, dx_2 \ldots dx_n. \qquad (2.4)$$

Finally, if $\theta(x) = \theta_1(x) \, dx_1 + \cdots + \theta_n(x) \, dx_n$ is a form of the first degree, and $\omega(x)$ is a form like (2.2), then their exterior product is

$$\theta(x) \wedge \omega(x) = \left(\sum_{i=1}^{n} \theta_i(x) u_i(x) \right) dx_1 \ldots dx_n. \qquad (2.5)$$

LEMMA 2.3. *Suppose that* ω *is an exterior differential form like* (2.2), *where the* u_i *are functions of class* C_0^1 *in an open subset* U *of* $\mathbf{R}^n$. *Then* $\int_U d\omega(x) = 0$.

PROOF. Let us extend the functions u_i to the whole of $\mathbf{R}^n$ by setting $u_i(x) = 0$ for $x \notin U$. Then

$$\int_U d\omega(x) = \int_{\mathbf{R}^n} \left[\sum_{i=1}^{n} \frac{du_i}{dx_i}(x) \right] dx_1 \, dx_2 \ldots dx_n.$$

Since each u_i has compact support,

$$\int_{-\infty}^{\infty} \frac{du_i}{dx_i}(x_1, \ldots, x_i, \ldots, x_n) \, dx_i = 0,$$

which gives us that $\int_U d\omega(x) = 0$, as was to be proved.

Let a be an arbitrary point in $\mathbf{R}^n$. Denote by θ_a the exterior differential form of degree $n - 1$ given by

$$\theta_a(x) = \sum_{k=1}^{n}(-1)^{k-1}\frac{(x_k - a_k)\,dx_1 \ldots dx_{k-1}\,dx_{k+1}\ldots dx_n}{|x - a|^n}, \quad (2.6)$$

defined in $\mathbf{R}^n\backslash\{a\}$. Simple computations show that $d\theta_a(x) \equiv 0$.

Let G be an arbitrary compact domain in $\mathbf{R}^n$, and $f\colon G \to \mathbf{R}^n$ a mapping of class C^∞. Suppose that the point y is (f, G)-admissible and the mapping f is regular with respect to y. Let $A = f^{-1}(y)$. The set A is finite and contained in G^0. The exterior differential form θ_y is defined in $\mathbf{R}^n\backslash\{y\}$. We use it to construct a certain exterior form $f^*\theta_y$ defined on $G^0\backslash A$ by

$$f^*\theta_y(x)$$
$$= \sum_{k=1}^{n}(-1)^{k-1}\frac{[f_k(x) - y_k]\,df_1 \wedge \cdots \wedge df_{k-1} \wedge df_{k+1} \wedge \cdots \wedge df_n}{|f(x) - y|^n}. \quad (2.7)$$

Formally, $f^*\theta_y$ is obtained from θ_y if in the expression for the latter the quantity x_i is replaced by $f_i(x)$, $i = 1,\ldots, n$. By known results in the theory of exterior forms, the differential of the form $f^*\theta_y$ is zero.

LEMMA 2.4. *For every function $\zeta \in C_0^\infty(\mathbf{R}^n)$ equal to 1 in a neighborhood of A and with support contained in G^0,*

$$\int_{G^0\backslash A} d\zeta(x) \wedge f^*\theta_y(x) = -\omega_n\mu(y, f, G). \quad (2.8)$$

PROOF. First of all we show that the integral on the left-hand side of (2.8) does not depend on the choice of the function ζ. Indeed, let ζ_1 and ζ_2 be two functions of class $C_0^\infty(\mathbf{R}^n)$ equal to 1 in a neighborhood of the set $A = f^{-1}(y)$ and with supports contained in G^0. Let $\zeta = \zeta_1 - \zeta_2$. The function ζ belongs to $C_0^\infty(\mathbf{R}^n)$ and has support $G^0\backslash A$. We have that

$$\int_{G^0\backslash A} d\zeta_1 \wedge f^*\theta_y - \int_{G^0\backslash A} d\zeta_2 \wedge f^*\theta_y = \int_{G^0\backslash A} d\zeta \wedge f^*\theta_y$$

$$= \int_{G^0\backslash A} d(\zeta f^*\theta_y).$$

The coefficients of the form $\omega = \zeta f^* \theta_y$ are functions of class C_0^∞ with supports in $G^0 \backslash A$. Hence, on the basis of Lemma 2.3 we have that $\int_{G^0 \backslash A} d\omega = 0$, which implies that

$$\int_{G^0 \backslash A} d\zeta_1 \wedge f^* \theta_y = \int_{G^0 \backslash A} d\zeta_2 \wedge f^* \theta_y.$$

If $A = f^{-1}(y)$ is empty, then $\mu(y, f, G) = 0$. In this case the coefficients of the exterior form $f^* \theta_y$ are functions of class C^∞ with supports in G^0, and so by Lemma 2.3

$$\int_G d\zeta \wedge f^* \theta_y = \int_G d(\zeta f^* \theta_y) = 0,$$

i.e., the equality to be proved is true in this case.

Let us now consider the case when $f^{-1}(y)$ is nonempty. Then $f^{-1}(y)$ consists of finitely many points $x_1, \ldots, x_m$, and $\mathscr{F}(x, f) \neq 0$ at each of them. There is a $\delta > 0$ such that the balls $B(x_i, \delta)$ are disjoint and f maps each of them homeomorphically into $\mathbf{R}^n$, with the inverse mapping also in C^∞. Let $V_i = f[B(x_i, \delta)]$. There is an $\varepsilon > 0$ such that $B(y, \varepsilon) \subset V_i$ for all $i = 1, \ldots, m$. Take a function $\eta \in C^\infty$ such that $S(\eta) \subset B(y, \varepsilon)$ and $\eta(x) = 1$ in a neighborhood of y, and let $\zeta(x) = \eta[f(x)]$. Then $\zeta \in C^\infty$, and $\zeta(x) = 1$ in some neighborhood of the set A. The support of ζ is obviously contained in the union of the balls $B(x_i, \delta)$. The restriction of ζ to $B(x_i, \delta)$ is denoted by ζ_i, and the restriction of f itself to $B(x_i, \delta)$ by f_i. We have that

$$\int_{G^0 \backslash A} d\zeta \wedge f^* \theta_y = \sum_{i=1}^m \int_{B(x_i, \delta)} d\zeta_i \wedge f_i^* \theta_y.$$

Let $d\eta \wedge \theta_y = u(x) \, dx_1 \ldots dx_n$. Then, by known results in the theory of exterior forms,

$$d\zeta \wedge f^* \theta_y = f^*(d\eta \wedge \theta_y) = u[f(x)] \mathscr{F}(x, f) \, dx_1 \, dx_2 \ldots dx_n,$$

and thus

$$\int_{B(a_i, \delta)} d\zeta_i \wedge f_i^* \theta_y = \int_{B(a_i, \delta)} u[f(x)] \mathscr{F}(x, f) \, dx.$$

On the basis of a classical change of variables formula for multiple integrals, the last integral is equal to

$$\sigma_i \int_{B(y, \varepsilon)} u(x) \, dx = \sigma_i \int_{B(y, \varepsilon)} d\eta(x) \wedge \theta_y(x),$$

where $\sigma_i = \operatorname{sgn} \mathscr{F}(x, f)$. To compute the last integral, observe that it does not depend on the choice of the function η (this follows formally from what has been proved when $G = \overline{B}(y, \varepsilon)$ and f is the identity mapping). We take $\eta(x) = \varphi(|x - y|)$, where φ is a function of a single variable. Then

$$d\eta(x) = \varphi'(|x - y|)|x - y|^{-1} \sum_{i=1}^{n} (x_i - y_i)\, dx_i.$$

From this,

$$d\eta(x) \wedge \theta_y(x) - \varphi'(|x - y|)|x - y|^{1-n}\, dx_1\, dx_2 \ldots dx_n.$$

Introducing spherical coordinates with center y in $\mathbf{R}^n$, we find that

$$\int_{B(y,\varepsilon)} d\eta \wedge \theta_y(x) = \omega_n \int_0^\infty \varphi'(r)\, dr = -\varphi(0)\omega_n = -\omega_n.$$

Finally,

$$\int_{G^0 \backslash A} d\zeta(x) \wedge f^*\theta_y = -\omega_n \sum_{i=1}^{m} \operatorname{sgn} \mathscr{J}(x_i, f) = -\omega_n \mu(y, f, G).$$

The lemma is proved.

The proof of Lemma 2.1 is based on the use of Lemma 2.4. We omit this proof, referring the reader to [139].

§2.3. Change of variables in a multiple integral. Let U be an open set in $\mathbf{R}^n$, and $f: U \to \mathbf{R}^n$ a continuous mapping. The mapping f is said to have *property N* if the image of every set $E \subset U$ of measure zero is a set of measure zero.

THEOREM 2.1. *Suppose that U is an open set in $\mathbf{R}^n$, and $f: U \to \mathbf{R}^n$ a continuous mapping. If f has property N, then the image under f of every measurable set $A \subset U$ is a measurable set.*

PROOF. Let A be any measurable subset of U. Then there exists a sequence $(A_m)_{m \in \mathbf{N}}$ of compact sets such that $A \supset A_m$ for each m and $E = A \backslash \bigcup_1^\infty A_m$ is a set of measure zero. We have that $A = E \cup \bigcup_1^\infty A_m$, which implies that

$$f(A) = f(E) \cup \bigcup_{m=1}^{\infty} f(A_m). \tag{2.14}$$

Since f is continuous, each of the sets $f(A_m)$ is compact, and hence measurable. Since f has property N, $f(E)$ is a set of measure zero and is thus measurable. By (2.14), this implies that $f(A)$ is measurable. The theorem is proved.

Let U be an open set in $\mathbf{R}^n$. A mapping $f: U \to \mathbf{R}^m$ is said to be *locally Lipschitz* if for every compact set $A \subset U$ there exists a constant $L_A < \infty$

such that

$$|f(x) - f(y)| \leq L_A|x - y|$$

for any $x, y \in A$.

Every C^1-mapping $f: U \to \mathbf{R}^m$, where U is an open set in $\mathbf{R}^n$, is locally Lipschitz. Indeed, let $A \subset U$ be compact, and let

$$L_A = \sup_{x,y \in A} \frac{|f(x) - f(y)|}{|x - y|}.$$

It is required to prove that L_A is finite. Let (x_ν, y_ν), $\nu = 1, 2, \ldots$, be a sequence of pairs of points in A such that

$$\frac{|f(x_\nu) - f(y_\nu)|}{|x_\nu - y_\nu|} \to L_A \quad \text{as } \nu \to \infty.$$

Since A is compact, it can be assumed without loss of generality that $x_\nu \to x_0$ and $y_\nu \to y_0$ as $\nu \to \infty$, where $x_0, y_0 \in A$. If $x_0 \neq y_0$, then by passing to the limit we get that

$$L_A = \frac{|f(x_0) - f(y_0)|}{|x_0 - y_0|} < \infty.$$

Suppose that $x_0 = y_0$. Since $f \in C^1$, it follows that

$$f(x') - f(x'') = f'(x_0)(x' - x'') + \alpha(x', x'')|x' - x''|,$$

where $\alpha(x', x'') \to 0$ as $x', x'' \to x_0$. From this,

$$\frac{|f(x_\nu) - f(y_\nu)|}{|x_\nu - y_\nu|} \leq |f'(x_0)| + |\alpha(x_\nu, y_\nu)|,$$

and we get by passing to the limit that $L_A \leq |f'(x_0)| < \infty$, which is what was to be proved.

THEOREM 2.2. *Let U be an open set in $\mathbf{R}^n$. Every locally Lipschitz mapping $f: U \to \mathbf{R}^n$ has property N.*

PROOF. Let $f: U \to \mathbf{R}^n$ be a locally Lipschitz mapping, and let $A \subset U$ be a set of measure zero. It must be proved that $|f(A)| = 0$. Let (U_m), $m = 1, 2, \ldots$, be a sequence of open sets such that U_m is strictly interior to U for each m and $U = \bigcup_1^\infty U_m$. Since $\overline{U}_m$ is compact, there exists a constant $L_m < \infty$ such that $|f(x) - f(y)| \leq L_m|x - y|$ for any $x, y \in \overline{U}_m$.

Let $A_m = A \cap U_m$. Then $|A_m| = 0$ and $A = \bigcup_1^\infty A_m$. Let $\varepsilon > 0$ be arbitrary. For ε there exists an open set $V \subset U_m$ such that $A_m \subset V$ and $|V| < \varepsilon$. Let $V = \bigcup_1^\infty Q_\nu$ be a cube subdivision of V, and r_ν the length of an edge of the cube Q_ν. Then $f(A_m) \subset \bigcup_1^\infty f(Q_\nu)$. The diameter of Q_ν is equal to $\sqrt{n}r_\nu$, and hence the diameter of $f(Q_\nu)$ does not exceed $L_m\sqrt{n}r_\nu$. This implies that

$$|f(Q_\nu)| \le \sigma_n(L_m\sqrt{n})^n r_\nu^n = Cr_\nu^n,$$

and thus

$$|f(A_m)| \le \sum_{\nu=1}^\infty |f(Q_\nu)| \le C\sum_{\nu=1}^\infty r_\nu^n < C\varepsilon.$$

Since $\varepsilon > 0$ is arbitrary, it thus follows that $|f(A_m)| = 0$. It remains to note that $f(A) = \bigcup_1^\infty f(A_m)$, which gives us that $|f(A)| = 0$. The theorem is proved.

COROLLARY. *Let* $U \subset \mathbf{R}^n$ *be an open set. Then every C^1-mapping $f\colon U \to \mathbf{R}^n$ has property N.*

The following result was established in [147].

THEOREM 2.3. *Let U be an open set in $\mathbf{R}^n$, and $f\colon U \to \mathbf{R}^n$ a mapping in the class $W_{p,\mathrm{loc}}^1(U)$, where $p > n$. Then f has property N.*

PROOF. Let $f \in W_{p,\mathrm{loc}}^1(U)$, and let the set $A \subset U$ be such that $|A| = 0$. It is required to prove that $|f(A)| = 0$. We construct a sequence (U_m), $m = 1, 2, \ldots$, of open sets such that U_m is strictly interior to U for each m and $U = \bigcup_1^\infty U_m$. Let $A_m = A \cap U_m$. It suffices to show that $|f(A_m)| = 0$ for each m. Let $\varepsilon > 0$ be arbitrary, and for it choose an open set $V \supset A_m$ such that $|V| < \varepsilon$ and $U_m \supset V$. Let $V = \bigcup_1^\infty Q_\nu$ be a cube subdivision of V, and r_ν the length of an edge of Q_ν. We have that

$$f(A_m) \subset \bigcup_{m=1}^\infty f(Q_\nu).$$

In view of the estimate (2.15) in §2 of Chapter I, the diameter d_ν of the set $f(Q_\nu)$ does not exceed

$$Cr_\nu^{1-(n/p)}\left(\int_{Q_\nu} |f'(x)|^p\,dx\right)^{1/p}.$$

We have that $|f(Q_\nu)| \le \sigma_n d_\nu^n$, from which

$$|f(A_m)| \le \sum_{\nu=1}^\infty |f(Q_\nu)| \le C^n \sigma_n \sum_{\nu=1}^\infty r_\nu^{n(1-(1/p))}\left(\int_{Q_\nu} |f'(x)|^p\,dx\right)^{n/p}.$$

Let us apply the Hölder inequality to the sum on the right-hand side, obtaining

$$|f(A_m)| \leq C^n \sigma_n \left(\sum_{\nu=1}^{\infty} r_\nu^n \right)^{1-(n/p)} \left(\sum_{\nu=1}^{\infty} \int_{Q_\nu} |f'(x)|^p \, dx \right)^{n/p}$$

$$= C^n \sigma_n |V|^{1-(n/p)} \left(\int_V |f'(x)|^p \, dx \right)^{n/p}$$

$$\leq C^n \sigma_n \varepsilon^{1-(n/p)} \left(\int_{U_m} |f'_{(x)}|^p \, dx \right)^{n/p}.$$

Since $\varepsilon > 0$ is arbitrary, this gives us that $|f(A_m)| = 0$, and the theorem is proved.

In §6.2 of this chapter we consider the question of whether property N holds for mappings in the class $W^1_{p,\text{loc}}$ when $p \leq n$.

LEMMA 2.7. *Let $Q = \overline{Q}(a, r)$ be a closed cube in $\mathbf{R}^n$, and let $f \colon Q \to \mathbf{R}^n$ be a continuous mapping. Let $Q_\tau = \overline{Q}(a, r\tau)$, where $\tau > 0$. Assume that there exists a nonsingular linear mapping $L \colon \mathbf{R}^n \to \mathbf{R}^n$ such that if $x \in Q$ and $x \neq a$, then*

$$|f(x) - f(a) - L(x - a)| < \delta|x - a|,$$

where $\delta = \text{const} > 0$. Let $\varphi(x) = f(a) + L(x-a)$, and let $\varepsilon = \delta\sqrt{n}|L^{-1}| < 1$. Then

$$G_{1-\varepsilon} = \varphi(Q_{1-\varepsilon}) \subset f(Q), \qquad G_{1-\varepsilon} \cap f(\partial Q) = \varnothing,$$

and for every $y \in G_{1-\varepsilon}$

$$\mu(y, f, Q) = \text{sgn} \det L.$$

PROOF. Take an arbitrary $x \in \partial Q$ and let $x' \in Q_{1-\varepsilon}$. Let $y = \varphi(x)$ and $y' = \varphi(x')$. We have that $y - y' = L(x - x')$, from which $x - x' = L^{-1}(y - y')$, and hence

$$|x - x'| \leq |L^{-1}||y - y'|.$$

Obviously, $|x - x'| \geq \varepsilon r/2$, which gives us that

$$|\varphi(x) - \varphi(x')| \geq \varepsilon r/2|L^{-1}| = \delta\sqrt{n}r/2.$$

We have that $y' = \varphi(x') \in G_{1-\varepsilon}$. Since $x' \in Q_{1-\varepsilon}$ was arbitrary, we get that for every $x \in \partial Q$ the distance from the point $\varphi(x)$ to the set $G_{1-\varepsilon}$ is at least $\delta\sqrt{n}r/2$. For every $x \in \partial Q$ we have that

$$|f(x) - \varphi(x)| < \delta|x - a| \leq \delta\sqrt{n}r/2.$$

This implies that if $x \in \partial Q$, then $f(x) \notin G_{1-\varepsilon}$, and hence $G_{1-\varepsilon} \cap f(\partial Q) = \varnothing$.

Let us now consider the function $f(t, x) = \varphi(x) + t[f(x) - \varphi(x)]$, where $0 \le t \le 1$. For $x \in \partial Q$ we have that

$$|f(t, x) - \varphi(x)| \le t|f(x) - \varphi(x)| < \delta\sqrt{n}r/2,$$

and thus $f(t, x) \notin G_{1-\varepsilon}$ for $x \in \partial Q$. Obviously, $f(0, x) = \varphi(x)$, $f(1, x) = f(x)$, and $f(t, x)$ is continuous on $[0, 1] \times Q$. We get that $f(t, x)$ is a homotopy between φ and f, as mappings of the pair $(Q, \partial Q)$ into the pair $(\mathbf{R}^n, \mathbf{R}^n \backslash G_{1-\varepsilon})$. From this,

$$\mu(y, f, Q) = \mu(y, \varphi, Q) = \operatorname{sgn} \det L$$

for all $y \in G_{1-\varepsilon}$. In particular, $\mu(y, f, Q) \neq 0$ for any $y \in G_{1-\varepsilon}$. This implies that $G_{1-\varepsilon} \subset f(Q)$. The lemma is proved.

LEMMA 2.8. *Let $f: U \to \mathbf{R}^n$ be a continuous mapping, and $G \subset U$ a compact region. Assume that the point $y \in \mathbf{R}^n$ is such that $y \notin f(\partial G)$, the set $f^{-1}(y) \cap G$ is finite, and the mapping f is differentiable at each point $x \in f^{-1}(y) \cap G$, with $\mathscr{J}(x, f) \neq 0$. Let N^+ be the number of points $x \in f^{-1}(y) \cap G$ at which $\mathscr{J}(x, f) > 0$, and N^- the number of such points at which $\mathscr{J}(x, f) < 0$. Then*

$$\mu(y, f, G) = N^+ - N^-.$$

PROOF. Assume all the conditions of the lemma. If the set $f^{-1}(y) \cap G$ is empty, then $y \notin f(G)$, and hence $\mu(y, f, G) = 0$, so there is nothing to prove in this case. Assume that $f^{-1}(y) \cap G \neq \varnothing$, and let $x_1, \ldots, x_N$ be all the elements in the set $f^{-1}(y) \cap G$. The mapping f is differentiable at each of the points x_i, and thus for each $i = 1, \ldots, N$

$$f(x) - f(x_i) - f'(x_i)(x - x_i) = \delta_i(x)|x - x_i|,$$

where $\delta_i(x) \to 0$ as $x \to 0$. The linear mappings $f'(x_i)$ are nonsingular. By Lemma 2.7, there is an r_0 such that

$$\mu[y, f, \overline{Q}(x_i, r)] = \operatorname{sgn} \mathscr{J}(x_i, f)$$

for $r < r_0$ for each $i = 1, \ldots, N$. We choose $r < r_0$ such that the cubes $\overline{Q}_i = \overline{Q}(x_i, r)$ are contained in G^0 and are disjoint. Then, by property II of the degree of a mapping,

$$\mu(y, f, G) = \sum_{i=1}^{N} \mu(y, f, Q_i) = \sum_{i=1}^{N} \operatorname{sgn} \mathscr{J}(x_i, f)$$
$$= N^+ - N^-.$$

The lemma is proved.

LEMMA 2.9. *Let U be an open set in $\mathbf{R}^n$, and $f: U \to \mathbf{R}^n$ a continuous mapping. Assume that f is differentiable at the point $a \in U$, and let $Q_r = Q(a, r)$. Then*

$$|f(Q_r)|/r^n \to |\det f'(a)|$$

as $r \to 0$.

PROOF. Let $L = f'(a)$, and for $r > 0$ and $x \in \mathbf{R}^n$ let

$$F_r(X) = \frac{f(a + rX) - f(a)}{r}.$$

Then $F_r(X) \to L(X)$ as $r \to 0$, and the convergence is uniform on the cube $\overline{Q}(0, 1)$. The mapping $X \to a + rX$ transforms $\overline{Q}(0, 1)$ into the cube Q_r. This implies that the set $F_r[\overline{Q}(0, 1)]$ is obtained from $f(Q_r)$ by a parallel translation and a similarity dilation with similarity coefficient equal to $1/r$, and hence

$$|F_r[\overline{Q}(0, 1)]| = \frac{|f(Q_r)|}{r^n}.$$

Let $G = L[\overline{Q}(0, 1)]$. If $\det L \neq 0$, then G is an n-dimensional parallelepiped whose volume is equal to $|\det L|$. But if $\det L = 0$, then G lies in some hyperplane, and thus $|G| = 0 = |\det L|$ in this case. Let

$$\delta_r = \sup_{x \in \overline{Q}(0,1)} |F_r(X) - L(X)|.$$

Then $\delta_r \to 0$ as $r \to 0$. For each r the set $F_r[\overline{Q}(0, 1)]$ is contained in the closed δ_r-neighborhood $\overline{U}_{\delta_r}(G)$ of G. We have that $|\overline{U}_{\delta_r}(G)| \to |G|$ as $r \to 0$. For each r

$$|F_r[\overline{Q}(0, 1)]| \le |\overline{U}_{\delta_r}(G)|,$$

which implies that

$$\varlimsup_{r \to 0} \frac{|f(Q_r)|}{r^n} = \varlimsup_{r \to 0} |F_r[\overline{Q}(0, 1)]| \le |G| = |\det L|. \tag{2.15}$$

In particular, this gives us the assertion of the lemma for the case when $\det f'(a) = 0$. We assume below that $\det f'(a) \neq 0$. Let $\varphi(x) = f(a) + L(x - a)$, where $L = f'(a)$ as above. Since f is differentiable at the point a, it follows that $|f(x) - \varphi(x)| \le \alpha(x)|x - a|$, where $\alpha(x) \to 0$ as $x \to a$. Let $0 < \varepsilon < 1$, and let $r_0 > 0$ be such that $|\alpha(x)| < \varepsilon/\sqrt{n}|L^{-1}|$ for all $x \in Q_{r_0}$. By Lemma 2.7, if $r < r_0$, then $f(Q_r) \supset \varphi[Q_{(1-\varepsilon)r}]$, and hence $|f(Q_r)| \ge (1 - \varepsilon)^n|Q_r||\det L|$. From this,

$$\varliminf_{r \to 0} \frac{|f(Q_r)|}{r^n} \ge (1 - \varepsilon)^n|\det L|.$$

Since $\varepsilon > 0$ was arbitrary, this implies that

$$\lim_{r \to 0} \frac{|f(Q_r)|}{r^n} \geq |\det L|. \tag{2.16}$$

The lemma is proved in view of (2.15).

Let U be an open set in $\mathbf{R}^n$, $f: U \to \mathbf{R}^n$ a continuous mapping, and A a subset of U. Assume that $v: U \to \mathbf{R}^n$ is a given nonnegative function. For any point $y \in \mathbf{R}^n$ denote by $N_A(y, f, v)$ the sum of the values of v at the points of $f^{-1}(x) \cap A$ (values $N_A(y, f, \infty) = \infty$ are allowed). In the case when $A = U$ the index A is omitted in the notation: $N(y, f, v) \equiv N_u(y, f, v)$. Further, we simply write $N_A(y, f)$ in place of $N_A(y, f, v)$ when $v \equiv 1$. The quantity $N_A(y, f)$ is equal to the number of elements in the set $f^{-1}(y) \cap A$.

THEOREM 2.2 [116]. *Let U be an open set in $\mathbf{R}^n$, and $f: U \to \mathbf{R}^n$ a continuous mapping. Assume that f has property N and is differentiable almost everywhere in U, with the function $x \mapsto \mathscr{J}(x, f)$ locally integrable on U. Then for every nonnegative measurable function $v: U \to \mathbf{R}$ the function $y \mapsto N(y, f, v)$ is measurable on $\mathbf{R}^n$, and*

$$\int_{\mathbf{R}^n} N(y, f, v)\, dy = \int_U v(x)|\mathscr{J}(x, f)|\, dx. \tag{2.17}$$

Further, if $G \subset U$ is a compact region whose boundary has measure zero, then, for every nonnegative measurable function $u: \mathbf{R}^n \to \mathbf{R}$ such that the function $y \mapsto N_G(y, f)u(y)$ is integrable, the function

$$x \mapsto u[f(x)]\mathscr{J}(x, f)$$

is integrable over G, and

$$\int_G u[f(x)]|\mathscr{J}(x, f)|\, dx = \int_{\mathbf{R}^n} u(y)N_G(y, f)\, dy, \tag{2.18}$$

$$\int_G u[f(x)]\mathscr{J}(x, f)\, dx = \int_{\mathbf{R}^n} u(y)\mu(y, f, G)\, dy. \tag{2.19}$$

PROOF. Suppose that the mapping $f: U \to \mathbf{R}^n$ satisfies the conditions of the theorem. We first consider the case when U is bounded, the function $x \mapsto \mathscr{J}(x, f)$ is integrable, and $v(x) \equiv 1$, so that $N(y, f, v) \equiv N(y, f)$. Let E' be the set of all points $x \in U$ such that f is not differentiable at x, and let E'' be the collection of points $x \in U$ which are not Lebesgue points of $\mathscr{J}(x, f)$. By the conditions of the theorem $|E'| = 0$ and $|E''| = 0$. Let $E_0 = E' \cup E''$ and $U_0 = U \backslash E_0$. Obviously, $|U_0| = |U|$.

Let m be an arbitrary positive integer. Let $x \in U_0$. Then $x \notin E'$, and hence f is differentiable at x. On the basis of Lemma 2.9, this implies

that
$$|f(\overline{Q}(x,r)|/r^n \to |\mathscr{J}(x,f)|$$
as $r \to 0$. In the given case it is also true that $x \notin E''$, and thus x is a Lebesgue point of $\mathscr{J}(x,f)$, i.e.,
$$\frac{1}{r^n} \int_{Q(x,r)} |\mathscr{J}(t,f) - \mathscr{J}(x,f)| \, dt \to 0$$
as $r \to 0$. Let $r_m(x) > 0$ be such that $r_m(x) \leq 1/m$ and such that if $0 < r < r_m(x)$, then $\overline{Q}(x,r) \subset U$ and
$$\left| \frac{|f(\overline{Q}(x,r)|}{r^n} - |\mathscr{J}(x,f)| \right| < \frac{1}{2m}, \tag{2.20}$$
$$\frac{1}{r^n} \int_{Q(x,r)} |\mathscr{J}(t,f) - \mathscr{J}(x,f)| \, dt < \frac{1}{2m}. \tag{2.21}$$

The set Q_m of cubes $\overline{Q}(x,r)$ such that $x \in U_0$ and $0 < r < r_m(x)$ covers U_0 in the Vitali sense. Hence, by Vitali's theorem, there exists a sequence $Q_\nu^m = \overline{Q}_\nu(x_\nu, r_\nu)$, $\nu = 1, 2, \ldots$, of disjoint cubes in Q_m such that $|U_0 \backslash \bigcup_\nu Q_\nu^m| = 0$. Let $\bigcup_\nu Q_\nu^m = V_m$. Obviously, $U \supset V_m$, and hence $E_m = U \backslash V_m$ is a set of measure zero. We carry out this construction for each $m = 1, 2, \ldots$, and let $E = \bigcup_0^\infty E_m$. Then $|E| = 0$, and then $A = f(E)$ is also a set of measure zero, because f has property N.

We define certain functions $\mathscr{J}_m$ and $\mathscr{N}_m$ for each $m = 1, 2, \ldots$. Let $\mathscr{J}_m(x) = 0$ for $x \in E_m$, and let $\mathscr{J}_m(x) = |f(Q_\nu^m)|/r_\nu^m$ if $x \in Q_\nu^m$. Take χ_ν to be the indicator function of the set $f(Q_\nu^m)$, and define $\mathscr{N}_m(y) = \sum_\nu \chi_\nu(y)$. For each m we obviously have that
$$\int_{\mathbf{R}^n} \mathscr{N}_m(y) \, dy = \sum_\nu |f(Q_\nu^m)| = \int_U \mathscr{J}_m(x) \, dx. \tag{2.22}$$

It follows from the definition of the functions $\mathscr{J}_m$ and from (2.19) and (2.20) that for each ν
$$\int_{Q_\nu^m} ||\mathscr{J}_m(x) - |\mathscr{J}(x,f)|| \, dx \leq \int_{Q_\nu^m} \left| \frac{|f(Q_\nu^m)|}{r_\nu^m} - \mathscr{J}(x_\nu, f) \right| \, dx$$
$$+ \int_{Q_\nu^m} |\mathscr{J}(x_\nu, f) - \mathscr{J}(x,f)| \, dx < \frac{|Q_\nu^m|}{m}.$$

Summing over ν, we get that
$$\int_U ||\mathscr{J}_m(x) - |\mathscr{J}(x,f)|| \, dx \leq \frac{|U|}{m},$$
and hence
$$\int_U \mathscr{J}_m(x) \, dx \to \int_U |\mathscr{J}(x,f)| \, dx \tag{2.23}$$
as $m \to \infty$.

Let $y \notin f(E)$ be arbitrary. The quantity $\mathscr{N}_m(y)$ is clearly equal to the number of cubes Q_ν^m containing points of the set $f^{-1}(y)$, and thus $\mathscr{N}_m(y) \leq \mathscr{N}_U(y, f)$, since these cubes are disjoint. We prove that $\mathscr{N}_m(y) \to \mathscr{N}_U(y, f)$ as $m \to \infty$. This follows from the preceding inequality in the case when $\mathscr{N}_U(y, f) = 0$. Suppose that $\mathscr{N}_U(y, f) > 0$. We specify a $K > 0$ by setting $K = \mathscr{N}_U(y, f)$ when the latter is finite and taking K arbitrarily otherwise. The set $f^{-1}(y)$ contains K distinct elements $x_1, \ldots, x_K$. Let $\delta > 0$ be the smallest of the distances $|x_i - x_j|$, $i, j = 1, \ldots, K$. Take m_0 such that $\sqrt{n}/m < \delta$ for $m > m_0$. Since $y \notin f(E_m) \subset A$, it follows that $f^{-1}(x)$ is contained in V_m. The diameter of each of the cubes Q_ν^m is at most $\sqrt{n}/m < \delta$, and hence none of them can contain more than one of the chosen points $x_1, \ldots, x_K$. Consequently, the cubes Q_ν^m containing points from $x_1, \ldots, x_K$ are disjoint, which implies that $\mathscr{N}_m(y) \geq K$. Accordingly, $\mathscr{N}_m(y) \geq K$ for each $m \geq m_0$, and it is thus proved that $\mathscr{N}_m(y) \to \mathscr{N}_U(y, f)$ for the given y.

Therefore, we get that $\mathscr{N}_m(y) \leq \mathscr{N}_U(y, f)$ and $\mathscr{N}_m(y) \to \mathscr{N}_U(y, f)$ as $m \to \infty$ for all y not in the set A of measure zero. This implies that

$$\int_{\mathbf{R}^n} \mathscr{N}_m(y)\, dy \to \int_{\mathbf{R}^n} \mathscr{N}_U(y, f)\, dy \tag{2.23$'$}$$

as $m \to \infty$. It follows from (2.21), (2.22), (2.23), and (2.23)$'$ that

$$\int_U |\mathscr{J}(x, f)|\, dx = \int_{\mathbf{R}^n} \mathscr{N}_U(y, f)\, dy. \tag{2.24}$$

The first assertion of the theorem is thus proved, at least under the assumption that U is bounded, the function $|\mathscr{J}(x, f)|$ is integrable, and $v(x) \equiv 1$. We now prove that under the same assumption on U and $\mathscr{J}(x, f)$ the equality (2.17) is true also when v is an arbitrary measurable function.

It follows from what has been proved that if U is bounded and $|\mathscr{J}(x, f)|$ is integrable, then $\mathscr{N}_U(y, f)$ is finite for almost all $y \in \mathbf{R}^n$.

Let (A_m), $m = 1, 2, \ldots$, be an arbitrary sequence of subsets of U such that $A_m \supset A_{m+1}$ for each m, and let $A = \bigcap_1^\infty A_m$. We show that $\mathscr{N}_{A_m}(y, f) \to \mathscr{N}_A(y, f)$ as $m \to \infty$ for every y such that $\mathscr{N}_U(y, f) < \infty$. Indeed, take such a y. If $\mathscr{N}_U(y, f) = 0$, then it is clear that $\mathscr{N}_E(y, f) = 0$ also for every $E \subset U$, and there is nothing to prove in this case. Let $\mathscr{N}_U(y, f) > 0$ and let $f^{-1}(y) = \{x_1, \ldots, x_{\mathscr{N}}\}$, where $\mathscr{N} = \mathscr{N}_U(y, f)$. Suppose that $x_1, \ldots, x_k$ are all the elements in $f^{-1}(y) \cap A$. If $i > k$, then $x_i \notin A$, and hence there is an index m_i such that $x_i \notin A_m$ for $m \geq m_i$. Consequently, for $m \geq \max_{i=k+1,\ldots,N} m_i$ the number of elements in $f^{-1}(y) \cap A_m$

is equal to $k = \mathcal{N}_A(y, f)$. This proves that $\mathcal{N}_{A_m}(y, f) \to \mathcal{N}_A(y, f)$ as $m \to \infty$.

Let $E \subset U$ be a measurable set strictly interior to U. Then there is a sequence of open sets $V_m \supset E$ such that V_m is strictly interior to U and $V_m \supset V_{m+1}$ for each m, and $|V_m| \to |E|$ as $m \to \infty$. Let $E_0 = \bigcap_{m=1}^{\infty} V$. Then $E_0 \supset E$ and $|E_0 \backslash E| = 0$. Each of the sets V_m is bounded, and the function $\mathcal{J}(x, f)$ is integrable over V_m. Hence,

$$\int_{V_m} |\mathcal{J}(x, f)| \, dx = \int_{\mathbf{R}^n} \mathcal{N}_{V_m}(y, f) \, dy,$$

by what was proved. The sequence $(\mathcal{N}_{V_m}(y, f))$, $m = 1, 2, \ldots$, of functions is decreasing, and each of them is finite almost everywhere. Passing to the limit, we get that

$$\int_{E_0} |\mathcal{J}(x, f)| \, dx = \int_{\mathbf{R}^n} \mathcal{N}_{E_0}(y, f) \, dy.$$

It is obvious that $\mathcal{N}_E(y, f) = \mathcal{N}_{E_0}(y, f)$ for every $y \notin f(E_0 \backslash E)$. Since f has property $\mathcal{N}$, it follows that $|f(E_0 \backslash E)| = 0$, and hence $\mathcal{N}_{E_0}(y, f) = \mathcal{N}_E(y, f)$ almost everywhere. As a result we get that

$$\int_E |\mathcal{J}(x, f)| \, dx = \int_{\mathbf{R}^n} \mathcal{N}_E(y, f) \, dy. \tag{2.25}$$

A function $v: U \to \mathbf{R}$ is called a *step function* if v is a linear combination of the indicator functions of finitely many measurable sets contained strictly interior to U. It follows from what was proved that (2.17) holds if v is a step function. For every nonnegative measurable function v there exists an increasing sequence of step functions such that $v_m(x) \to v(x)$ for all x. We have that

$$\mathcal{N}_U(y, f, v_m) \to \mathcal{N}_U(y, f, v)$$

as $m \to \infty$ for all $y \in \mathbf{R}^n$. Further, the sequence $(y \mapsto \mathcal{N}_U(y, f, v_m))$, $m = 1, 2, \ldots$, of functions is increasing, which implies that

$$\int_{\mathbf{R}^n} \mathcal{N}_U(y, f, v_m) \, dy \to \int_{\mathbf{R}^n} \mathcal{N}_U(y, f, v) \, dy.$$

For each m

$$\int_U v(x) |\mathcal{J}(x, f)| \, dx = \int_{\mathbf{R}} \mathcal{N}_U(y, f, v_m) \, dy.$$

Passing to the limit in this equality as $m \to \infty$, we get (2.17).

Assume now that $G \subset U$ is a given closed region whose boundary has measure zero. Since f has property $\mathcal{N}$, $f(\partial G)$ is a set of measure zero. Assume first that $u: \mathbf{R}^n \to \mathbf{R}$ is a continuous function. Then the function $v(x) = u[f(x)]$ is continuous, and hence bounded on G. Since G is compact, $\mathscr{J}(x, f)$ is integrable on G, and thus so is the function $u[f(x)]\mathscr{J}(x, f)$. By what was proved,

$$\int_G |\mathscr{J}(x, f)|\, dx = \int_{\mathbf{R}^n} \mathscr{N}_G(y, f)\, dy;$$

hence the function $\mathscr{N}_G(y, f)$ of the variable y is integrable, and, in particular, it is finite for almost all $y \in \mathbf{R}^n$. The function $v(x) = u(f(x))$ takes the same value $u(y)$ at all points of the set $f^{-1}(u)$; consequently, $\mathscr{N}_G(y, f, v) = \mathscr{N}_G(y, f)u(y)$ for every y with $\mathscr{N}_G(y, f) < \infty$. We thus get that (2.18) holds when $u: \mathbf{R}^n \to \mathbf{R}$ is a continuous function. The general case is derived from this by a standard passage to the limit.

Thus (2.18) is proved.

Choosing the v in (2.17) to be the indicator function of the set S of points $x \in U$ at which f is differentiable and $\mathscr{J}(x, f) = 0$, we get that

$$\int_{\mathbf{R}^n} \mathscr{N}_S(y, f)\, dy = 0,$$

which implies that $f(S)$ is a set of measure zero. Let $G \subset U$ be a compact region such that $|\partial G| = 0$, and let $u: \mathbf{R}^n \to \mathbf{R}$ be a measurable function such that the function $u(y)\mathscr{N}_G(y, f)$ is integrable. Suppose that $A_1 = f(\partial G)$, $A_2 = f(S \cap G)$, A_3 is the set of y such that $\mathscr{N}_G(y, f) = \infty$, and, finally, A_4 is the set of y such that $u(y)$ is not defined or is equal to ∞. The measure of each of the sets A_i is equal to zero. Let $A = A_1 \cup A_2 \cup A_3 \cup A_4$, and let

$$v(x) = u[f(x)]\, \text{sgn}\, \mathscr{J}(x, f).$$

Then the function $v(x)|\mathscr{J}(x, f)|$ is integrable, and

$$\int_G v(x)|\mathscr{J}(x, f)|\, dx = \int_{\mathbf{R}^n} \mathscr{N}_G(y, f, v)\, dy.$$

By Lemma 2.8, for the given choice of v and for $y \notin A$ we have that

$$\mathscr{N}_G(y, f, v) = u(y)\mu(y, f, G),$$

i.e., $\mathscr{N}_G(y, f, v) = u(y)\mu(y, f, G)$ almost everywhere. This establishes the equality (2.19).

The theorem is proved.

§3. Conformal capacity

§3.1. The capacity of a capacitor. A *capacitor* in $\mathbf{R}^n$ is defined to be any ordered pair $H = (A_0, A_1)$ of disjoint nonempty closed sets of which one is compact. Let a capacitor $H = (A_0, A_1)$ be given. The open set $U = \mathbf{R}^n \backslash (A_0 \cup A_1)$ is called the *field* of the capacitor H. Denote by $\tilde{W}_p(H) = \tilde{W}_p(A_0, A_1)$, where $p > 1$, the collection of all functions $v: \mathbf{R}^n \to \mathbf{R}$ of class $C^\infty(\mathbf{R}^n)$ for which there exist open sets U_0 and U_1 such that $A_0 \subset U_0$, $A_1 \subset U_1$, $v(x) = 0$ for $x \in U_0$, $v(x) = 1$ for $x \in U_1$, $0 \leq v(x) \leq 1$ for all x, and the integral

$$\int_{\mathbf{R}^n} |\nabla v(x)|^p \, dx \tag{3.1}$$

is finite. The infimum of the integral in (3.1) on the set $\tilde{W}_p(A_0, A_1)$ is called the *p-capacity* of the capacitor $H = (A_0, A_1)$ and denoted by $C_p(H) = C_p(A_0, A_1)$.

The concept of the capacity of a ring is a special case of the concept of the capacity of a capacitor. A *ring* in $\mathbf{R}^n$ is defined to be an open set U whose complement $\mathbf{R}^n \setminus U$ consists of two connected components A_0 and A_1, of which at least one (say A_0 for definiteness) is compact. Obviously, the pair (A_0, A_1) is a capacitor whose field is the set U. The set $\partial U \cap A_0$ is then called the *inner boundary* and $\partial U \cap A_1$ the *outer boundary* of the ring U. The capacity of the capacitor (A_0, A_1) is called *the capacity of the ring* U and denoted by $\operatorname{Cap}_p U$. In the case when $p = n$ the p-capacity is called the *conformal capacity*.

The concept of conformal capacity in $\mathbf{R}^n$ was introduced by Löwner [89], who also indicated a number of applications of the concept of conformal capacity to the theory of quasiconformal mappings. In particular, he used the concept of conformal capacity to prove that there is no quasiconformal mapping of $\mathbf{R}^n$ onto a subset of $\mathbf{R}^n$ whose boundary contains more than one point. Further progress in the investigation of the concept of the capacity of a ring has been made by Gehring, Väisälä, and others (see [174] and [34–37]). Closely connected with the concept of conformal capacity is the concept of the modulus of a family of curves or surfaces, but we shall not dwell on this.

Next, we need the concept of the capacity of a capacitor with respect to an open set U. Let (A_0, A_1) be a capacitor and U an open set such that A_0 and A_1 are contained in the closure of U. If this condition is satisfied, then we say that the capacitor (A_0, A_1) and the open set U are *joined*. We denote by $\tilde{W}_p(A_0, A_1, U)$ the collection of all functions $u \in C^\infty(U)$ such that there exist open sets $G_0 \supset A_0$ and $G_1 \supset A_1$ with $u(x) = 0$ for $x \in G_0 \cap U$, $u(x) = 1$ for $x \in G_1 \cap U$, and $0 \leq u(x) \leq 1$ for all $x \in U$. The infimum of the integral $\int_U |u'(x)|^p \, dx$ is called *the p-capacity of the capacitor* (A_0, A_1) *with respect to* U, and is denoted by $C_p(A_0, A_1, U)$. Obviously, $C_p(A_0, A_1) = C_p(A_0, A_1, \mathbf{R}^n)$. If the function $u(x)$ belongs to the class $\tilde{W}_p(A_0, A_1)$, then for every open set U joined with (A_0, A_1) the restriction of $u(x)$ to U belongs to $\tilde{W}_p(A_0, A_1, U)$. This implies that

$$C_p(A_0, A_1, U) \le \int_U |u'(x)|^p \, dx \le \int_{\mathbf{R}^n} |u'(x)|^p \, dx,$$

and the arbitrariness of $u \in \tilde{W}_p(A_0, A_1)$ gives us that

$$C_p(A_0, A_1, U) \le C_p(A_0, A_1).$$

Suppose that the capacitor (A_0, A_1) and the set U are joined. If $u \in \tilde{W}_p(A_0, A_1, U)$, then $v = 1 - u \in \tilde{W}_p(A_1, A_0, U)$, and, since $|\nabla u(x)| = |\nabla v(x)|$ for all $x \in U$, this implies that $C_p(A_0, A_1, U) = C_p(A_1, A_0, U)$.

Suppose that a connected open set $U \subset \mathbf{R}^n$ and a capacitor (A_0, A_1) joined with U are given. The collection of all functions $u \in C^\infty(U)$ vanishing in a neighborhood of A_0 and satisfying

$$\|u\|_{1,p,U} = \left(\int_U |u'(x)|^p \, dx \right)^{1/p} < \infty$$

is a vector space, and the functional $u \to \|u\|_{1,p,U}$ is a norm in this space. The closure of the set $\tilde{W}_p(A_0, A_1, U)$ of functions with respect to the norm $\|u\|_{1,p,U}$ is denoted by $W_p(A_0, A_1, U)$. According to this definition, $W_p(A_0, A_1, U)$ is the collection of all functions u on U for which there exist sequences (u_m) of functions in $\tilde{W}_p(A_0, A_1, U)$ converning to u in $L_{1,\text{loc}}(U)$ and such that $\|u_k - u_m\|_{1,p,U} \to 0$ as $k, m \to \infty$.

These definitions extend in a natural way to the case of the space $\overline{\mathbf{R}}^n$. A capacitor in $\overline{\mathbf{R}}^n$ is defined to be any pair (A_0, A_1) of disjoint closed sets. We say that a capacitor $H = (A_0, A_1)$ and an open set U in $\overline{\mathbf{R}}^n$ are joined if $A_0 \cup A_1$ is contained in $\overline{U}$. For an arbitrary capacitor (A_0, A_1) in $\overline{\mathbf{R}}^n$ let

$$C_n(A_0, A_1, U) = C_n(A_0 \backslash \{\infty\}, A_1 \backslash \{\infty\}, U \backslash \{\infty\}).$$

We compute the p-capacity of a certain special capacitor in $\mathbf{R}^n$ when $p > 1$. Let a and b be such that $0 < a < b$. Denote by $H(a, b)$ the capacitor (A_0, A_1), where $A_0 = \mathbf{R}^n \backslash B(0, b)$ and $A_1 = \overline{B}(0, a)$. Let φ be an arbitrary function in $\tilde{W}_p[H(a, b)]$. We specify in $\mathbf{R}^n$ a spherical coordinate system with origin at 0. Let $\mathbf{e}$ be an arbitrary unit vector in $\mathbf{R}^n$. We have that

$$1 = \varphi(a\mathbf{e}) - \varphi(b\mathbf{e}) = - \int_a^b < \varphi'(r\mathbf{e}), \qquad \mathbf{e} > dr \le \int_a^b |\varphi'(r\mathbf{e})| \, dr.$$

Hölder's inequality gives us that

$$1 \le \left(\int_a^b |\varphi'(r\mathbf{e})|^p r^{n-1} \, dr \right) \left(\int_a^b r^{-(n-1)/(p-1)} \, dr \right)^{p-1}. \tag{3.2}$$

From this, in the case $p = n$

$$\int_a^b |\varphi'(re)|^n r^{n-1} \, dr \geq 1/[\ln(b/a)]^{n-1}.$$

Integrating with respect to e, we find that

$$\int_{a \leq |x| \leq b} |\varphi'(x)|^n \, dx \geq \omega_n/[\ln(b/a)]^{n-1},$$

which implies that

$$C_n[H(a,b)] \geq \omega_n/[\ln(b/a)]^{n-1}. \tag{3.3}$$

In the case $p \neq n$ we arrive at the following estimate, in which $\gamma = (p-n)/(p-1)$:

$$C_p[H(a,b)] \geq \omega_n((b^\gamma - a^\gamma)/\gamma)^{p-1}. \tag{3.4}$$

We show that equality holds in each of the relations (3.3) and (3.4). To do this it suffices to construct a function φ of class $W_p(A_0, A_1)$ such that the integral

$$\int_{\mathbf{R}^n} |\varphi'(x)|^p \, dx \tag{3.5}$$

is equal to the right-hand side of (3.3) in the case $p = n$ and the right-hand side of (3.4) in the case $p \neq n$. The required function φ will be found in the form $\varphi(x) = \psi(|x|)$, where ψ is a function of a single variable, from the condition that equality holds in (3.2). This leads to the following differential equation for ψ: $[\psi'(r)]^p r^{p-1} = C r^{-(n-1)/(p-1)}$, where C is a constant. From this,

$$\psi(r) = C_1 r^{(p-n)/(p-1)} + C_2$$

in the case $p \neq n$, while

$$\psi(r) = C_1 \ln(1/r) + C_2$$

in the case $p = n$. The constants C_1 and C_2 are chosen so that $\psi(a) = 1$ and $\psi(b) = 0$. It is easy to verify that the function $\varphi(x) = \psi(|x|)$ belongs to the class $W_p[H(a,b)]$, and for it the integral (3.5) really is equal to the right-hand side of (3.4) in the case $p \neq n$ and the right-hand side of (3.3) in the case $p = n$. Thus,

$$C_p[H(a,b)] = \omega_n[(b^\gamma - a^\gamma)/\gamma]^{p-1}$$

for $p \neq n$, where $\gamma = (p-n)/(p-1)$, and $C_n[H(a,b)] = \omega_n/[\ln(b/a)]^{n-1}$.

Next, we use a certain special sequence of functions on $\mathbf{R}^n$. Let $\tilde{\varphi}_m(x) = 1$ for $|x| \leq m+1$, $\varphi_m(x) = 2 - (\ln x)/(\ln(m+1))$ for $m+1 < |x| \leq (m+1)^2$, and $\tilde{\varphi}_m(x) = 0$ for $|x| > (m+1)^2$. Let $\varphi_m(x)$ be the mean function for $\tilde{\varphi}_m(x)$ corresponding to the value $h = 1$ of the averaging parameter. Then $\varphi_m(x) = 1$ for $|x| \leq m$, $\varphi_m(x) = 0$ for $|x| \geq (m+1)^2 + 1$ (hence, φ_m has compact support), $0 \leq \varphi_m(x) \leq 1$ for all x, and

$$\int_{\mathbf{R}^n} |\varphi'_m(x)|^n \, dx \leq \int_{\mathbf{R}^n} |\tilde{\varphi}'_m(x)|^n \, dx = \omega_n / [\ln(m+1)]^{n-1} \to 0$$

as $m \to \infty$.

THEOREM 3.1. *Let* $U \subset \overline{\mathbf{R}}^n$ *be an open set, and* (A_0, A_1) *a capacitor joined to* U. *Then*

$$C_n[f(A_0), f(A_1), f(U)] = C_n(A_0, A_1, U)$$

for every Möbius transformation f.

PROOF. Let $u \in \tilde{W}_n(A_0, A_1, U)$ be an arbitrary function, and let $v(x) = u[f^{-1}(x)]$, $B_0 = f(A_0)$, $B_1 = f(A_1)$ and $V = f(U)$. The function v vanishes in a neighborhood of B_0 and takes the value 1 in a neighborhood of B_1. However, v can fail to belong to the class $C^\infty(V)$ here. This happens in the case when V contains the point z_0 such that $f(\infty) = z_0$. To avoid the difficulty arising in this connection we proceed as follows. It will be assumed that $\infty \notin A_1$. Since $C_n(A_0, A_1, U) = C_n(A_1, A_0, U)$, this clearly involves no loss of generality. The set A_1 is compact. Suppose that the integer k is such that $A_1 \subset B(0, k)$. Let $u_m = u\varphi_{m+k}$, where (φ_m), $m = 1, 2, 3, \ldots$ is the sequence constructed above. Obviously, $u_m \in \tilde{W}_n(A_0, A_1, U)$, and the function u_m is compactly supported. The latter property of u_m implies that the function $v_m = u_m \circ f^{-1}$ belongs to the class $C^\infty(V)$, and it is zero in a neighborhood of z_0. Consequently,

$$C_n(B_0, B_1, V) \leq \int_V |v'_m(y)|^n \, dy = \int_U |v'_m[f(x)]|^n |\mathcal{F}(x, f)| \, dx$$

$$= \int_U |u'_m(x)|^n \, dx.$$

Here we have used the fact that $|\mathcal{F}(x, f)| = |f'(x)|^n$, and hence

$$|v'_m[f(x)]|^n |\mathcal{F}(x, f)| = |v'_m[f(x)] f'^*(x)|^n = |u'_m(x)|^n.$$

It is easy to show that

$$\int_U |u'_m(x)|^n \, dx \to \int_U |u'(x)|^n \, dx \quad \text{as } m \to \infty.$$

From this,

$$C_n(B_0, B_1, V) \leq \int_U |u'(x)|^n \, dx,$$

and since $u \in \tilde{W}_n(A_0, A_1, U)$ is arbitrary, it is thereby proved that

$$C_n(B_0, B_1, V) \leq C_n(A_0, A_1, U).$$

Replacing f by f^{-1} in the arguments and interchanging the triples (A_0, A_1, U) and (B_0, B_1, V), we get that

$$C_n(A_0, A_1, U) \leq C_n(B_0, B_1, V),$$

and the theorem is proved.

The estimate of the conformal capacity in the next lemma will be needed later.

LEMMA 3.1. *Let U be an open subset of $\mathbf{R}^n$, and A_0 and A_1 closed sets, constructed as follows*:

1) *U is the domain in $\mathbf{R}^n$ consisting of all the points $x \in \mathbf{R}^n$ such that $0 < a < |x - x_a| < b$ and the vector $x - x_0$ forms an angle less than $\theta(|x - x_0|)$ with a certain unit vector $\mathbf{e}_0 \in \mathbf{R}^n$, where $0 < \theta(r) < \theta_0 \leq \pi$.*

2) *A_0 is the segment formed by the points $x_r = x_0 + r\mathbf{e}_0$, where $a \leq r \leq b$.*

3) *The set A_1 is contained in ∂U, and the sphere $S(x_0, r)$ contains points of A_1 for each $r \in (a, b)$.*

Then $C_n(A_0, A_1, U) \geq \gamma_n[\ln(b/a)]/\theta_0$, where $\gamma_n > 0$ is a constant.

PROOF. Let Σ_r be the spherical segment which is the intersection of U with $S(x_0, r)$. Let v be an arbitrary function of class $\tilde{W}(A_0, A_1, U)$. The function v is equal to 0 in a neighborhood of A_1, and to 1 in a neighborhood of A_1. We have that

$$\int_U |v'(x)|^n \, dx = \int_a^b \left(\int_{\Sigma_r} |v'(x)|^n \, d\sigma_x \right) dr,$$

where $d\sigma_x$ is the area element of the sphere $S(a, r)$. We get a lower estimate of the integral $\int_{\Sigma_r} |v'(x)|^n \, dx$ for each $r \in (a, b)$. Let ξ_r be a point of A_1 in $S(x_0, r)$. Obviously, ξ_r lies on the boundary of Σ_r. Let $x(t)$, $p \leq t \leq q$, be an arbitrary smooth curve on $S(x_0, r)$ such that $x(p) = x_r$, $x(q) = \xi_r$, and $x(t) \in \Sigma_r$ for $p < t < q$. Then

$$\int_p^q |v'[x(t)]||x'(t)| \, dt \geq \int_p^q \langle v'[x(t)], x'(t) \rangle \, dt$$
$$= \int_p^q \frac{d}{dt}(v[x(t)]) \, dt = v[x(q)] - v[x(p)] = 1. \tag{3.6}$$

We introduce a certain coordinate system on $S(x_0, r)$. Namely, let a plane P_0 orthogonal to the radius going to the point ξ_r be drawn through the point x_0. A Cartesian system with origin at x_0 is introduced in P_0. Let σ be the projection mapping of $S(x_0, r_0)$ onto P_0 with projection from the point ξ_r, and let τ be the inverse mapping. We take the coordinates of the point $\sigma(x) \in P_0$ as the coordinates of an arbitrary point $x \in S(x_0, r)$, $x \neq \xi_r$. The line element of the sphere in this stereographic coordinate system will have the form

$$ds^2 = [\lambda(t)]^2(dt_1^2 + \cdots + dt_{n-1}^2),$$

where $\lambda(t) = 2r^2/(r^2 + |t|^2)$. We regard x_r as having the coordinates $(\alpha, 0, \ldots, 0)$, where $\alpha > 0$. This can be made the case by a suitable rotation of the coordinate axes in the plane P_0. It is not hard to see that $\alpha = r\cot(\theta/2)$, where $\theta = \theta(r)$, and the segment Σ_r is defined by the inequality $t_1 \geq r\cot\theta$.

We now consider on $S(x_0, r)$ the family of circular arcs lying in Σ_r and joining x_r and ξ_r. Let $a_0 = \sigma(x_r)$. Every circle in this family has a parametrization $x(\rho) = \tau[a_0 + \rho\mathbf{e}]$, $0 \leq \rho < \infty$, where $\mathbf{e}$ is a unit vector in P_0 directed into the half-plane $t_1 \geq 0$. Let $h(t) = |v'[\tau(t)]|$. Then, using (3.6), we get that

$$1 \leq \int_0^\infty |v'[x(\rho)]||x'(\rho)|\,d\rho = \int_0^\infty h(a_0 + \rho\mathbf{e})\lambda(a_0 + \rho\mathbf{e})\,d\rho.$$

From this, by Hölder's inequality,

$$\begin{aligned}
1 \leq & \left(\int_0^\infty [h(a_0 + \rho\mathbf{e})]^n[\lambda(a_0 + \rho\mathbf{e})]^{n-1}\rho^{n-2}\,d\rho \right) \\
& \times \left(\int_0^\infty [\lambda(a_0 + \rho\mathbf{e})]^{1/(n-1)}\rho^{(1/(n-1))-1}\,d\rho \right)^{n-1}.
\end{aligned} \tag{3.7}$$

Further,

$$\begin{aligned}
\lambda(a_0 + \rho\mathbf{e}) &= 2r^2/(r^2 + |a_0 + \rho\mathbf{e}|^2) \leq 2r^2/(r^2 + |a_0|^2 + \rho^2) \\
&= 2r^2\sin^2(\theta/2)/[r^2 + \rho^2\sin^2(\theta/2)].
\end{aligned}$$

This gives us that the second integral on the right-hand side in (3.7) does not exceed

$$\int_0^\infty \left(\frac{2r^2\sin^2(\theta/2)}{r^2 + \rho^2\sin^2(\theta/2)} \right)^\alpha \rho^{\alpha-1}\,d\rho,$$

where $\alpha = 1/(n-1)$. Performing the change of variable $\rho = ur/\sin(\theta/2)$ in this integral, we find after obvious transformations that the integral is equal to $[\gamma_n r\sin(\theta/2)]^\alpha$, where $\gamma_n > 0$ is a constant. The foregoing gives us

$$\frac{1}{\gamma_n r\sin[\theta(r)/2]} \leq \int_0^\infty [h(a_0 + \rho\mathbf{e})]^n[\lambda(a_0 + \rho\mathbf{e})]^{n-1}\rho^{n-2}\,d\rho.$$

We integrate both sides of the last inequality with respect to the variable $\mathbf{e}$ over the unit $(n-2)$-dimensional hemisphere. On the right-hand side we get the integral

$$\int_{t_1 > r\cot(\theta/2)} [h(t)]^n [\lambda(t)]^{n-1}\, dt,$$

which is equal to the integral of $|v'(x)|^n$ over some segment contained in Σ_r. From this,

$$\int_{\Sigma_r} |v'(x)|^n\, d\tau_x \geq \omega_n/\gamma_n r \sin[\theta(r)/2],$$

and this proves the lemma.

§3.2. Sets of zero capacity. Let A be an arbitrary compact subset of $\mathbf{R}^n$. We say that A is a *set of zero p-capacity*, where $1 < p \leq n$, if there exists a closed set B disjoint from A such that the set $\mathbf{R}^n \backslash B$ is bounded and $C_p(A, B) = 0$. We investigate the properties of sets of zero p-capacity.

LEMMA 3.2. *For every compactly supported function $u(x)$ of class C^∞ on $\mathbf{R}^n$ and all $x \in \mathbf{R}^n$*

$$u(x) = \frac{1}{\omega_n} \int_{\mathbf{R}^n} \langle \nabla u(y), x - y \rangle \frac{dy}{|x - y|^n}.$$

PROOF. We fix a point x and a unit vector $\mathbf{e}$ in $\mathbf{R}^n$. Since $u(x)$ is compactly supported,

$$u(x) = -\int_0^\infty \frac{d}{dt} u(x + t\mathbf{e})\, dt = -\int_0^\infty \langle \nabla u(x + t\mathbf{e}), \mathbf{e} \rangle\, dt.$$

The required formula is obtained after obvious transformations by integrating both sides of this equality over the sphere in $\mathbf{R}^n$ and then setting $x + r\mathbf{e} = y$.

LEMMA 3.3. *Let u be a function of class $C_0^\infty(\mathbf{R}^n)$ with support in the ball $B(0, r)$. Then $\|u\|_p \leq 2r\|u\|_{1,p}$.*

PROOF. Using the integral representation of the previous lemma, we get

$$|u(x)| \leq \frac{1}{\omega_n} \int_{|y| \leq r} \frac{|\nabla u(y)|}{|x - y|^{n-1}}\, dy.$$

By Hölder's inequality, this gives us the required estimate. to simplify the expression let

$$|\nabla u(y)| = v(y), \quad \frac{1}{\omega_n |x - y|^{n-1}} = K(x, y), \quad B(0, r) = U.$$

We have

$$\int_U K(x,y)v(y)\,dy = \int_U [K(x,y)]^{1/p}v(y)[K(x,y)]^{1-1/p}\,dy$$

$$\leq \left(\int_U K(x,y)[v(y)]^p\,dy\right)^{1/p}\left(\int_U K(x,y)\,dy\right)^{1-1/p}.$$

From this,

$$\int_U |u(x)|^p\,dx \leq \int_U \left(\int_U K(x,y)\,dx\right)[v(y)^p]\,dy \left(\int_U K(x,y)\,dy\right)^{p-1}.$$

$$(3.8)$$

For every $x \in U = B(0,r)$

$$\int_U K(x,y)\,dy = \frac{1}{\omega_n}\int_U \frac{dy}{|x-y|^{n-1}} \leq \frac{1}{\omega_n}\int_{B(0,2r)} \frac{dz}{|z|^{n-1}}.$$

The last integral is easy to compute; it is equal to $2r$, and thus $\int_U K(x,y)\,dy \leq 2r$. Similarly, $\int_U K(x,y)\,dx \leq 2r$ for every $y \in U$. Substituting these estimates into (3.8), we get that

$$\int_U |u(x)|^p\,dx \leq (2r)^p \int_U |v(y)|^p\,dy,$$

and the lemma is proved.

LEMMA 3.4. *Let A be a compact subset of $\mathbf{R}^n$ of p-capacity zero. Then $C_p(B,A) = 0$ for every nonempty closed set B disjoint from A.*

PROOF. By definition, there is a closed set $B_0 \subset \mathbf{R}^n$ such that $\mathbf{R}^n \backslash B_0$ is bounded and contains A, and $C_p(B_0,A) = 0$. Take an arbitrary function $u \in \tilde{W}_p(B_0,A)$. Let $\eta \in \mathring{W}_p(B,A)$. Then it is clear that $v = u\eta \in \tilde{W}_p(B,A)$. We have that

$$|\nabla v(x)| \leq \eta(x)|\nabla u(x)| + u(x)|\nabla\eta(x)| \leq |\nabla u(x)| + Mu(x),$$

where $M = \sup_x |\nabla\eta(x)| < \infty$. From this,

$$\|\nabla v\|_{L_p} \leq \|\nabla u\|_{L_p} + M\|u\|_{L_p}.$$

Suppose that $r > 0$ is such that $\mathbf{R}^n \backslash B_0$ is contained in $B(0,r)$. By the preceding lemma, $\|u\|_{L_p} \leq 2r\|\nabla u\|_{L_p}$, which gives us that $\|\nabla v\|_{L_p} \leq (1 + 2rM)\|\nabla u\|_{L_p}$. This obviously implies that $C_p(B,A) = 0$, since the supremum of the norm $\|\nabla u\|_{L_p}$ on the set $\tilde{W}_p(B_0,A)$ is zero by assumption; the lemma is proved.

THEOREM 3.2. *Let $A \subset \mathbf{R}^n$ be a compact set, and $U \supset A$ a bounded open set in $\mathbf{R}^n$. The p-capacity of A is zero if and only if there exists a function $v \in L_p(\mathbf{R}^n)$ such that $v(y) = 0$ for $y \notin U$ and*

$$\int_{\mathbf{R}^n} \frac{v(y)}{|x - y|^{n-1}}\, dy = \infty$$

for all $x \in A$.

PROOF. *Necessity.* Suppose that A and U satisfy the condition of the theorem. Assume that $C_p(A) = 0$. Let $A_0 = \mathbf{R}^n \backslash U$. Then $C_p(A_0, A_1) = 0$ by Lemma 3.4. For every $m \in \mathbf{N}$ we construct a function u_m of class $\tilde{W}_p(A_0, A_1)$ such that $\|u_m\|_{1,p} < \omega_n/2^m$. By Lemma 3.2,

$$u_m(x) = \frac{1}{\omega_n} \int_{\mathbf{R}^n} \frac{\langle \nabla u_m(y), x - y \rangle}{|x - y|^n}\, dy,$$

which leads to the conclusion that

$$|u_m(x)| \leq \int_{\mathbf{R}^n} \frac{v_m(y)}{|x - y|^{n-1}}\, dy$$

for all $x \in \mathbf{R}^n$, where $v_m(y) = |\nabla u_m(y)|/\omega_n$. We have that $\|v_m\|_p < 1/2^m$. Let $v = v_1 + v_2 + \dots$. Obviously, $v \in L_p(\mathbf{R}^n)$, $\|v\|_p \leq \|v_1\|_p + \|v_2\|_p + \dots \leq 1$, v is nonnegative, and its support lies in $\overline{B}(0, r)$. For every $x \in A$ and for any m

$$\int_{\mathbf{R}^n} \frac{v(y)\, dy}{|x - y|^{n-1}} \geq \sum_{u=1}^{m} \int_{\mathbf{R}^n} \frac{v_k(y)\, dy}{|x - y|^{n-1}} \geq \sum_{n=1}^{m} u_k(x) \geq m.$$

Since m is arbitrary, this means that

$$\int_{\mathbf{R}^n} \frac{v(y)}{|x - y|^{n-1}}\, dy = \infty$$

for all $x \in A$. The necessity is proved.

Sufficiency. Let $A \subset \mathbf{R}^n$ be a compact set, and $U \supset A$ a bounded open subset of $\mathbf{R}^n$. Assume that there exists a nonnegative function $v \in L_p(\mathbf{R}^n)$ such that

$$u(x) = \int_{\mathbf{R}^n} v(y)|x - y|^{1-n}\, dy = \infty$$

for all $x \in A$, and $v(y) = 0$ for $y \notin U$. Let D be the 1-neighborhood of U. It is not hard to see that u is bounded above on $\mathbf{R}^n \backslash D$, namely,

$$u(x) \leq \int_{\mathbf{R}^n} v(y)\, dy \leq |U|^{1-1/p}\|v\|_p < \infty.$$

Let L be the supremum of u on $\mathbf{R}^n \backslash D$, and let $k > 0$ be an integer. Denote by G_k the set of all $x \in \mathbf{R}^n$ such that $u(x) > k + L$. Obviously,

$A \subset G_k \subset D$, and G_k is open. Indeed, let x_0 be an arbitrary point in G_k, and assume that x_0 is not an interior point. Then there exists a sequence (x_m), $m = 1, 2, \ldots$, such that $x_m \to x_0$ as $m \to \infty$, and $u(x_m) \leq k + L$ for all m. We have that

$$v(y)|x_m - y|^{1-n} \to v(y)|x_0 - y|^{1-n} \quad \text{as } m \to \infty$$

for all y such that the expression on the right-hand side is defined, i.e., for almost all y, and thus, by Fatou's lemma,

$$u(x_0) = \int_{\mathbf{R}^n} v(y)|x_0 - y|^{1-n} \, dy \leq \varliminf_{m \to \infty} \int_{\mathbf{R}^n} v(y)(x_m - y)^{1-n} \, dy$$
$$= \varliminf_{m \to \infty} u(x_m) \leq k + L.$$

This contradicts the fact that $u(x_0) > k + L$ by assumption.

Let $\delta_k > 0$ be the smallest of the distances from the points of A to the boundary of G_{k+1}. Let $h < \delta_k/2$ and $h > 0$. We consider the averaging u_h of u with averaging parameter h. Obviously, $u_h(x) > k+L+1$ everywhere in an h-neighborhood of A. Then

$$u_h(x) = \int_{\mathbf{R}^n} \omega(z)u(x + hz)\, dz$$
$$= \int_{\mathbf{R}^n} \omega(z) \left(\int_{\mathbf{R}^n} v(y)|x + hz - y|^{1-n}\, dy \right) dz$$
$$= \int_{\mathbf{R}^n} \omega(z) \left(\int_{\mathbf{R}^n} v(y + hz)|x - y|^{1-n}\, dy \right) dz$$
$$= \int_{\mathbf{R}^n} v_h(y)|x - y|^{1-n}\, dy.$$

Further, $\|v_h\|_p \leq \|v\|_p$. By a theorem of Calderón and Zygmund (see [23 and 99]), $\|\nabla u_h(x)\|_p \leq C\|v_h\|_p \leq C\|v\|_p$, where C is a constant.

We now revise u_h with the object of getting a function equal to zero outside D and equal to k in a neighborhood of A. To do this let $\zeta_0 \colon \mathbf{R} \to \mathbf{R}$ be a function of class C^∞ such that $0 \leq \zeta_0'(x) \leq 1$ for all x, $\zeta_0(x) = 0$ for $x < L$, and $\zeta_0(x) = k$ for $x > L + k + 1$. Such a function can clearly be obtained by smoothing the function $\overline{\zeta}$ with $\overline{\zeta}(x) = 0$ for $x \leq L + 1/2$, $\overline{\zeta}(x) = x - L - 1/2$ for $L + 1/2 < x < L + k + 1/2$, and $\overline{\zeta}(x) = k$ for $x \geq L + k + 1/2$. Let $w_k(x) = \zeta_0[u_h(x)]$. Then $w_k \in C_0^\infty(\mathbf{R}^n)$, and the support of w_k is contained in $\overline{D}$. For $x \in G_{k+1}$ we have that $w_k(x) \geq k$. Further, $\|\nabla w_k\|_p \leq \|\nabla u_h\|_p \leq C\|v\|_p$. The function $(1/k)w_k$ belongs to the class $\mathring{W}(\mathbf{R}^n \backslash D, A)$, and $\|(1/k)w_k\|_{1,p} \leq M/k$, where M is a constant. Since k is arbitrary here, this implies that $C_p(A) = 0$. The theorem is proved.

§**3.3. The concept of a Hausdorff measure. Cartan's lemma.** We introduce some characteristics of massiveness for subsets of $\mathbf{R}^n$. Let h be a real function defined on $[0, \infty)$ and satisfying the following conditions:

1) h is a nondecreasing continuous function on $[0, \infty)$.
2) $h(0) = 0$, and $h(r) > 0$ for $r > 0$.
3) $h(r) \to \infty$ as $r \to \infty$.

This function h is called a *gauge function*.

Let $E \subset \mathbf{R}^n$ be an arbitrary set, and let $B_1 = B(x_1, r_1)$, $B_2 = B(x_2, r_2), \ldots$. The sequence (B_ν) of balls is said to *cover* the set E if $E \subset \bigcup_1^\infty B_\nu$. If the radius r_ν of each ball B_ν is $\leq \varepsilon$, where $\varepsilon > 0$, then the sequence (B_ν) is said to form an *ε-covering* of E. For each sequence of balls $(B_\nu = B(x_\nu, r_\nu))$, $\nu = 1, 2, \ldots$, covering E we form the sum

$$\sum_1^\infty h(r_\nu) \qquad\qquad (*).$$

The infimum of this sum over the set of all sequences of balls covering E is called the *h-content* of E and denoted by $\gamma_h(E)$.

It should be noted that h-content is a comparatively coarse characteristic of sets. Therefore, we introduce another characteristic. Let $\varepsilon > 0$. The infimum of the sum $(*)$ on the set of all sequences (B_ν), $\nu = 1, 2, \ldots$, forming an ε-covering of E is denoted by $m_h(E, \varepsilon)$. It is not hard to see that $m_h(E, \varepsilon)$ is a nonincreasing function of the variable $\varepsilon > 0$, and $m_h(E, \varepsilon) \to \gamma_h(E)$ as $\varepsilon \to \infty$. The limit $\lim_{\varepsilon \to 0} m_h(E, \varepsilon) = m_h(E)$ is called the *Hausdorff h-measure* of E.

In the case when $h(r) = r^\alpha$, where $\alpha > 0$, the quantities $\gamma_h(E)$ and $m_h(E)$ are denoted by $\gamma_\alpha(E)$ and $m_\alpha(E)$, respectively.[*] It is clear from the definition that the h-content and the Hausdorff h-measure always satisfy the inequality

$$m_h(E) \geq \gamma_h(E). \qquad\qquad (3.9)$$

In one important special case the equality sign can be substituted here. Namely, we have the following statement.

LEMMA 3.5. *The Hausdorff h-measure of a set E is equal to zero if and only if its h-content is equal to zero.*

PROOF. The necessity follows at once from (3.9).

Sufficiency. Assume that $\gamma_h(A) = 0$. Let ν be an arbitrary positive integer, and let ρ_ν be the largest number r such that $h(r) = 1/\nu$. Let $B(a, r_1), B(a, r_2), \ldots$ be an arbitrary sequence of balls covering E and such

---

[*]*Editor's note.* For this special choice of h, Hausdorff h-measure is called α-dimensional Hausdorff measure.

that $h(r_1) + h(r_2) + \cdots < 1/\nu$. Then $h(r_k) < 1/\nu$ for all k, and hence $r_k < \rho_\nu$, so that the given sequence of balls forms an ε-covering of A for $\varepsilon = \rho_\nu$. This implies that $m_h(A, \rho_\nu) \leq \sum_k h(r_k) < 1/\nu$. As $\nu \to \infty$ we have that $\rho_\nu \to 0$; thus,

$$m_h(A) = \lim_{\varepsilon \to 0} m_h(A, \varepsilon) = \lim_{\nu \to \infty} m_h(A, \rho_\nu) = 0.$$

The lemma is proved.

LEMMA 3.6. *For every integer $n > 1$ there exists a constant $C_n < \infty$ having the property that if $u_1, \ldots, u_m, \ldots$ is an arbitrary system of nonzero vectors in $\mathbf{R}^n$ such that the angle formed by any two of them is at least $\pi/3$, then the number of vectors in the system does not exceed C_n.*

The proof reduces to the following. On the unit sphere in $\mathbf{R}^n$ let there be given points $A_1, \ldots, A_m, \ldots$ such that the distance between any two of them, measured on the sphere, is at least $\pi/3$. The geodesic balls of radius $\pi/3$ about $A_1, \ldots, A_m, \ldots$ on the sphere are disjoint. Let τ_n be the area ($(n-1)$-dimensional) of such a ball. Obviously, the number of points $A_1, A_2, \ldots$ does not exceed the ratio ω_n/τ_n. The desired constant is $C_n = \omega_n/\tau_n$.

Estimates concerning the connection between capacity and Hausdorff measures are essentially based on a certain general lemma given by H. Cartan in the case $n = 2$. The following statement is an immediate extension of Cartan's lemma to the case $n > 2$. Suppose that μ is a measure, i.e., a nonnegative completely additive function on the σ-algebra of Borel subsets of $\mathbf{R}^n$. We define $\mu(x, r) = \mu[B(x, r)]$.

LEMMA 3.7 (Cartan's lemma). *Let μ be an arbitrary measure on $\mathbf{R}^n$ such that $\mu(\mathbf{R}^n) < \infty$, and let $h: [0, \infty) \to \mathbf{R}$ be a gauge function. Let $A \subset \mathbf{R}$ be the set of $x \in \mathbf{R}^n$ such that $\mu(x, r) \leq h(r)$ for all $r > 0$. Then the h-content of $\mathbf{R}^n \backslash A$ does not exceed $C_n \mu(\mathbf{R}^n)$, where C_n is the constant in Lemma 1.*

PROOF. The following arguments amount to a direct extension of the proof given by Ahlfors for $n = 2$ (see [109]) to the general case. We define by induction a finite or infinite sequence (E_m), $m = 0, 1, \ldots$, of sets. Let $E_0 = \mathbf{R}^n$. Assume that the set E_{m-1} has been defined for some $m \geq 1$. Let

$$\lambda_m(r) = \sup_{x \in E_{m-1}} \mu(x, r),$$

where $r > 0$. Obviously, $\lambda_m(r)$ is a monotone nondecreasing function. Furthermore,

$$\lim_{r \to \infty} \lambda_m(r) \leq \mu(\mathbf{R}^n) < \infty.$$

Since, on the other hand, $h(r) \to \infty$ as $r \to \infty$,

$$\lambda_m(r) \le h(r) \tag{3.10}$$

for all sufficiently large r. It can turn out that (3.10) holds for all r. In this case the construction is complete.

Assume that $\lambda_m(r) > h(r)$ for some $r > 0$, and denote by r_m the supremum of such values r. Then $0 < r_m < \infty$. Each interval $(r_m - \varepsilon, r_m)$ contains a value r such that $\lambda_m(r) > h(r)$, and this gives us by passage to the limit that

$$\lambda_m(r_m - 0) \ge h(r_m). \tag{3.11}$$

On the other hand, $\lambda_m(r) < h(r)$ for all $r > r_m$, which implies that

$$\lambda_m(r_m + 0) \le h(r_m). \tag{3.12}$$

Comparing (3.11) and (3.12), we get that the function λ_m is continuous at r_m, and $\lambda_m(r_m) = h(r_m)$. Let $\varepsilon > 0$ be arbitrary, and let x_m be a point of E_{m-1} such that

$$\mu(x_m, r_m) > \lambda_m(r_m) - \mu(\mathbf{R}^n)\varepsilon/2^m.$$

Define $B_m = B(x_m, r_m)$ and $E_m = E_{m-1} \backslash B_m$. It is obvious that in this case $E_m \subset E_{m-1}$ and $E_m \ne E_{m-1}$.

The construction just described obviously defines a finite or infinite sequence of balls $B_m = B(x_m, r_m)$, $m = 1, 2, \ldots$, and a sequence of sets E_m. Further, $E_m = \mathbf{R}^n \backslash \bigcup_1^m B_k$ for each $m \ge 1$, and the center of B_m is in E_{m-1} and hence does not belong to any of the balls $B_1, \ldots, B_{m-1}$. Moreover, for each m

$$\mu(B_m) \ge \lambda_m(r_m) - \varepsilon \frac{\mu(\mathbf{R}^n)}{2^m} = h(r_m) - \frac{\varepsilon \mu(\mathbf{R}^n)}{2^m}. \tag{3.13}$$

Finally, note that, since $E_m \subset E_{m-1}$,

$$\lambda_m(r) = \sup_{x \in E_m} \mu(x, r) \le \sup_{x \in E_{m-1}} \mu(x, r) = \lambda_{m-1}(r).$$

It follows from the definition of r_m that $r_m \le r_{m-1}$ (r_m is the supremum of the set S_m of all values of r such that $\lambda_m(r) > h(r)$, and the inequality $\lambda_m(r) \le \lambda_{m-1}(r)$ clearly implies that $S_m \subset S_{m-1}$, whence $r_m \le r_{m-1}$). The sequence (r_m) of radii is thus decreasing.

Let x be an arbitrary point of $\mathbf{R}^n$. We prove that it cannot belong to more than C_n balls in the sequence (B_m), where C_n is the constant in Lemma 3.6. Indeed, let $B_k = B(x_k, r_k)$ and $B_l = B(x_l, r_l)$ be two balls in our sequence containing a point x. Assume for definiteness that $k > l$. Then $r_k \le r_l$ and $x_k \notin B_l$. We have that $|x - x_k| < r_k$ and $|x - x_l| < r_l$. Since $x_k \notin B_l$, it follows that $x \ne x_k$ and $|x_k - x_l| \ge r_l$. Further, $x \ne x_l$, for otherwise we would have that $|x_l - x_k| = |x - x_k| < r_k \le r_l$, which is impossible. In the triangle with vertices x, x_k, and x_l the side $x_k x_l$ is the largest. Consequently, the angle of this triangle at the vertex x is the largest of its angles, and hence has measure greater than $\pi/3$. We thus get that the angles between any two vectors emanating from the point x and directed toward the centers of those balls $B_1, \ldots, B_m, \ldots$ containing x is greater than $\pi/3$. By Lemma 3.6, this implies that x belongs to at most C_n balls in the sequence (B_m).

Denoting by χ_m the indicator function of B_m, we get that $\sum_m \chi_m(x) \le C_n$ for all $x \in \mathbf{R}^n$. From this,

$$\sum_m \mu(B_m) = \sum_m \int_{\mathbf{R}^n} \chi_m(x)\mu(dx) \le C_n \mu(\mathbf{R}^n).$$

By (3.13),

$$\sum_m h(r_m) \le \sum_m \mu(B_m) + \varepsilon\mu(\mathbf{R}^n) \le (C_n + \varepsilon)\mu(\mathbf{R}^n). \tag{3.14}$$

We now prove that the set $\mathbf{R}^n \backslash A$ is contained in the union of the balls $B_1, \ldots, B_m, \ldots$. If this sequence is finite, it follows from the construction that $\mu(x, r) \le h(r)$ for every x not belonging to any of these balls, and this gives us the required inclusion. Assume that the sequence (B_m) is infinite. By (3.14), the series $\sum_1^\infty h(r_m)$ converges, and hence $r_m \to 0$ as $m \to \infty$. Take any

$$x \in E = \mathbf{R}^n \backslash \bigcup_{m=1}^\infty B_m,$$

and let $r > 0$. Choose m such that $r_m < r$. The point x is in E_{m-1}, and $\mu(x, r) \le h(r)$, because $r > r_m$. Since x is an arbitrary point of E and $r > 0$ is arbitrary, this implies that $E \subset A$, and hence the balls $B_1, \ldots, B_m, \ldots$ form a covering of $\mathbf{R}^n \backslash A$.

Thus a sequence (B_m), $m = 1, 2, \ldots$, covering $\mathbf{R}^n \backslash A$ has been constructed. The radii of these balls satisfy (3.14), which implies that $\gamma_h(\mathbf{R}^n \backslash A) \le (C_n + \varepsilon)\mu(\mathbf{R}^n)$. Since $\varepsilon > 0$ is arbitrary, this proves the lemma.

COROLLARY. *Suppose that μ is an arbitrary measure on $\mathbf{R}^n$ such that $\mu(\mathbf{R}^n) < \infty$, and $h(r)$ is an arbitrary gauge function. Denote by A_λ, where $\lambda > 0$, the set of all $x \in \mathbf{R}^n$ such that $\mu(x, r) \le h(r)/\lambda$ for all r. Then*

$$\gamma_h(\mathbf{R}^n \backslash A_\lambda) \le C_n \lambda \mu(\mathbf{R}^n),$$

where C_n is the constant in Lemma 3.6, $C_n < \infty$.

For a proof it suffices to apply Lemma 3.7 to the measure $\mu_1 = \lambda\mu$.

§3.4. Capacity and Hausdorff measures.

LEMMA 3.8. *Let* $F: (0, \infty) \to \mathbf{R}$ *be a monotonically decreasing absolutely continuous function such that* $F(r) \to \infty$ *as* $r \to 0$ *and* $F(r) \to 0$ *as* $r \to \infty$. *Let* $0 < \alpha < \beta$, *and define*

$$F_\alpha(r) = \begin{cases} F(\alpha) & \text{for } 0 \leq r \leq \alpha, \\ F(r) & \text{for } \alpha < r < \infty; \end{cases}$$

$$F_{\alpha,\beta}(r) = \begin{cases} F(\alpha) - F(\beta) & \text{for } 0 \leq r < \alpha, \\ F(r) - F(\beta) & \text{for } \alpha \leq r \leq \beta, \\ 0 & \text{for } \beta < r < \infty. \end{cases}$$

Let μ *be a nonnegative measure on* $\mathbf{R}^n$, *and define* $\mu(x, r) = \mu[B(x, r)]$. *Then*

$$\int_{\mathbf{R}^n} F_{\alpha,\beta}(|x - y|)\mu(dy) = -\int_\alpha^\beta F'(r)\mu(x, r)\, dr, \qquad (3.15)$$

$$\int_{\mathbf{R}^n} F_\alpha(|x - y|)\mu(dy) = -\int_\alpha^\infty F'(r)\mu(x, r)\, dr, \qquad (3.16)$$

$$\int_{\mathbf{R}^n} F(|x - y|)\mu(dy) = -\int_0^\infty F'(r)\mu(x, r)\, dr. \qquad (3.17)$$

PROOF. Define $\chi(r, l) = 1$ for $r > l$, and $\chi(r, l) = 0$ for $r \leq l$. Then for every $x \in \mathbf{R}^n$

$$\mu(x, r) = \int_{\mathbf{R}^n} \chi(r, |x - y|)\mu(dy).$$

Substituting this in the right-hand side of (3.15) and then using Fubini's theorem, we get

$$-\int_\alpha^\beta F'(r)\mu(x, r)\, dr = -\int_{\mathbf{R}^n} \left(\int_\alpha^\beta F'(r)\chi(r, |x - y|)\, dr \right) \mu(dy).$$

After uncomplicated transformations we have that

$$-\int_\alpha^\beta F'(r)\chi(r, l)\, dr = F_{\alpha\beta}(l)$$

for any $l > 0$. This gives us (3.15).

The equalities (3.16) and (3.17) are obtained from (3.15) by passing to the limit as $\beta \to \infty$ and as $\alpha \to 0$ and $\beta \to \infty$, respectively. The lemma is proved.

THEOREM 3.3 [132]. *Let*

$$u(x) = \int_{B(0,r)} \frac{v(y)}{|x - y|^{n-\lambda}}\, dy,$$

where $0 < \lambda < n$, $v(y) \geq 0$, v is integrable to the power $p \geq 1$ in the ball $B(0, R)$, and $\lambda p \leq n$. Suppose that $h(r)$ is a gauge function such that

$$\int_0^1 [h(r)]^{1/p} r^{-(n/p)+\lambda-1} \, dr < \infty.$$

Then there exist constants K_1 and K_2 depending only on λ, n, p, h, and R such that for any $\delta > 0$ the h-content of the set of all $x \in \mathbf{R}^n$ with

$$u(x) > \frac{K_1}{\delta} + K_2 \|v\|_{p,B(0,R)}$$

does not exceed $C_n(\delta\|v\|_p)^p$, where C_n is the constant in Lemma 3.6.

PROOF. Denote by $\theta(v, x, r)$ the integral of v over $B(x, r)$, and by $\theta(v^p, x, r)$ the integral of v^p over the same ball. It will be assumed that v is defined for all $x \in \mathbf{R}^n$, with $v(x) = 0$ for $x \notin B(0, R)$.

By Hölder's inequality,

$$\theta(v, x, r) \leq \sigma_n^{1-1/p} r^{n-n/p} [\theta(v^p, x, r)]^{1/p}. \tag{3.18}$$

Setting $F(r) = r^{\lambda-n}$ in Lemma 3.7, we get that

$$u(x) = (n - \lambda) \int_0^\infty \frac{\theta(v, x, r)}{r^{n-\lambda+1}} \, dr$$

$$= (n - \lambda) \int_0^R \frac{\theta(v, x, r)}{r^{n-\lambda+1}} \, dr + (n - \lambda) \int_R^\infty \frac{\theta(v, x, r)}{r^{n-\lambda+1}} \, dr.$$

Let us estimate separately each of the integrals on the right-hand side. For the first integral

$$\int_0^R \frac{\theta(v, x, r)}{r^{n-\lambda+1}} \, dr \leq \sigma_n^{1-1/p} \int_0^R \frac{\theta(v^p, x, r)]^{1/p} \, dr}{r^{(n/p)-\lambda+1}}$$

by (3.18). To estimate the second integral we note that, since $v(x) = 0$ for $|x| \geq r$, $\theta(v, x, r)$ is equal to the integral of v over the intersection of $B(0, R)$ with $B(x, r)$, and hence

$$\theta(v, x, r) \leq \theta(v, 0, R) \leq \sigma_n^{1-1/p} R^{n-n/p} [\theta(v^p, 0, R)]^{1/p}$$
$$= \sigma_n^{1-1/p} R^{n-n/p} \|v\|_{p,B(0,R)}.$$

Comparing the above inequalities, we get that

$$u(x) \leq (n - \lambda)\sigma_n^{1-1/p} \int_0^R \frac{[\theta(v^p, x, r)]^{1/p}}{r^{n/p-\lambda+1}} \, dr + K_2 \|u\|_p, \tag{3.19}$$

where $K_2 = \sigma_n^{1-1/p} R^{n-n/p}$.

Let $\delta > 0$ be arbitrary, and denote by B_δ the collection of all $x \in \mathbf{R}^n$ such that $\theta(v^p, x, r) \leq h(r)/\delta^p$ for any $r > 0$. By Lemma 3.7, the h-content of $A_\delta = \mathbf{R}^n \backslash B_\delta$ does not exceed

$$\delta^p C_n \int_{\mathbf{R}^n} [v(x)]^p \, dx = C_n (\delta \|v\|_{p,B(0,R)})^p.$$

By (3.19), $u(x) \leq \frac{K_1}{\delta} + K_2 \|v\|_{p,\mathbf{R}^n}$ for all $x \in B_\delta$, where the constant K_1 is equal to the integral

$$(n - \lambda)\sigma_n^{1-1/p} \int_0^R [h(r)]^{1/p} r^{-(n/p)+\lambda-1} \, dr.$$

The set of all x such that $v(x) > K_1/\delta + K_2 \|u\|_p$ is contained in A_δ, and hence its h-content does not exceed $C_n(\delta \|u\|_{p,\mathbf{R}^n})^p$. The theorem is proved.

COROLLARY 1. *Let A be a compact subset of $\mathbf{R}^n$. If the p-capacity of A is equal to zero, then for every gauge function h with*

$$\int_0^1 [h(r)]^{1/p} r^{-n/p} \, dr < \infty \tag{3.20}$$

the h-content of A is equal to zero.

PROOF. Let $R > 0$ be such that $B(0, R) \supset A$. Suppose that the gauge function h satisfies (3.20). By Theorem 3.3, there exists a function $v \in L_p(\mathbf{R}^n)$ such that $v(x) = 0$ for $|x| > R$, and for all $x \in A$

$$u(x) = \int_{\mathbf{R}^n} v(y)|x - y|^{1-n} \, dy = \infty.$$

Let A_δ be the set of all $x \in \mathbf{R}^n$ such that $u(x) > K_1/\delta + K_2 \|v\|_p$, where K_1 and K_2 are the constants in Theorem 3.3 corresponding to the case $\lambda = 1$. The set A is contained in A_δ for any $\delta > 0$; hence $\gamma_h(A) \leq \gamma_h(A_\delta)$ for any $\delta > 0$. According to the lemma, $\gamma_h(A_\delta) \leq C_n(\delta \|v\|_{p,\mathbf{R}^n})^n$. Since $\delta > 0$ is arbitrary, this implies that $\gamma_h(A) = 0$, and the corollary is proved.

Setting $h(r) = r^\alpha$, we get

COROLLARY 2. *If A is a compact set of p-capacity zero, then for every $\alpha > n - p$ the α-dimensional Hausdorff measure of A is equal to zero. In particular, if the conformal capacity of A is equal to zero, then for any $\alpha > 0$ its α-dimensional Hausdorff measure is equal to zero.*

§3.5. Estimates of the capacity of certain capacitors.

We consider some special capacitors in $\mathbf{R}^n$ and show that they have definite extremal properties. The result in this section is due to Gehring [34], [35].

Let $t \in (0, 1)$. Denote by $K_G(t)$ the capacitor (A_0, A_1) with A_0 the exterior of the ball $B(0, 1)$, $A_0 = \mathbf{R}^n \backslash B(0, 1)$, and A_1 the segment consisting of all points $x = \lambda e_n$, where $0 \leq \lambda \leq t$; $K_G(t)$ is called a *Grötzsch capacitor*. Let $t > 0$. The symbol $K_T(t)$ denotes the capacitor (A_0, A_1) with A_0 the ray consisting of all points $x = \lambda e_n$, $\lambda \geq t$, and A_1 the segment formed by the points $x = \mu e_n$, $-1 \leq \mu \leq 0$. This capacitor is called a *Teichmüller capacitor*.

The capacitors $K_G(t)$ and $K_T(t)$ have extremal properties expressed by the following theorems.

THEOREM 3.4. *Let (A_0, A_1) be a capacitor in $\mathbf{R}^n$ such that A_0 contains the exterior of some ball $B(a, r)$, while A_1 has a connected component joining the center a of this ball to a point $x_0 \in B(a, r)$, $x_0 \neq a$. Then*

$$C_n(A_0, A_1) \geq C_n[K_G(|x_0 - a|/r)].$$

THEOREM 3.5. *Let (A_0, A_1) be a capacitor in $\mathbf{R}^n$. Assume that A_0 contains an unbounded connected component E_0, while A_1 has a connected component E_1 containing two distinct points x_0 and y_0. Let ρ be the distance from x_0 to E_0. Then*

$$C_n(A_0, A_1) \geq C_n[K_T(\rho/|x_0 - y_0|)].$$

We introduce some notation. The collection of all capacitors in $\mathbf{R}^n$ satisfying the conditions of Theorem 3.4 and such that the ratio $|x_0 - a|/r$ is equal to some number $t \in (0, 1)$ is denoted by $K_n(t)$. The set of all capacitors such that the conditions of Theorem 3.5 hold with $\rho/|x_0 - y_0| = t > 0$ is denoted by $H_n(t)$.

It is not hard to see that $K_G(t) \in K_n(t)$ for every $t \in (0, 1)$, and hence $K_G(t)$ has the smallest capacity in the class $K_n(t)$, by Theorem 3.4. Similarly, it follows from Theorem 3.5 that $K_T(t)$ is a capacitor of smallest capacity in $H_n(t)$.

The proof of Theorems 3.4 and 3.5 is our immediate problem. It is based on a certain transformation of functions on $\mathbf{R}^n$ called *symmetrization*. We give without proof the properties of symmetrization we need. These properties can be found, for example, in [115]. We remark that the type of symmetrization used here is simpler than in [35]. The proofs of the necessary properties of symmetrization in our case do not require using the Brunn-Minkowski theorem.

Symmetrization of functions of a single variable. Let I denote either the whole set $\mathbf{R}$ or an interval $[-h, h]$ with $0 < h < \infty$. Nonnegative measurable functions $f(x)$ and $g(x)$ defined on I are said to be *equimeasurable* if the measures of the sets $E_f(t) = \{x \in I | f(x) \geq t\}$ and $E_g(t) = \{x \in I | g(x) \geq t\}$ coincide for every $t > 0$. In this case f and g are said to be *rearrangements* of each other.

Let $f(x)$ be a nonnegative integrable function. The *symmetrization,* or symmetrically decreasing rearrangement of f, is defined to be the function $\sigma f \geq 0$ defined on I and such that the following conditions hold:

1) σf is equimeasurable with f.

2) $(\sigma f)(x) = (\sigma f)(-x)$ for every $x \in I$.

3) σf is a nonincreasing left-continuous function on $(0, \infty) \cap I$, and $(\sigma f)(0) = \lim_{x \to 0} \sigma f(x)$.

A function σf satisfying all these conditions exists and is unique, as is easy to show. We mention an important special case. Let A be an arbitrary measurable subset of I with finite measure, and let $f = \chi_A$ be the indicator function of A, i.e., the function with $f(x) = 1$ for $x \in A$ and $f(x) = 0$ for $x \notin A$. Then the symmetrization of f is the indicator function of the closed interval symmetric with respect to 0 whose measure is equal to that of A. We call this interval the *symmetrization* of A.

Symmetrization with respect to a plane in $\mathbf{R}^n$. We hold to the following notation. Let $x = (x_1, \ldots, x_{n-1}, x_n)$ be a point in $\mathbf{R}^n$. We write $x = (y, z)$, where $y = (x_1, \ldots, x_{n-1}) \in \mathbf{R}^{n-1}$ and $z = x_n$. The space $\mathbf{R}^{n-1}$ is assumed to be imbedded in $\mathbf{R}^n$ by identifying the point $y = (x_1, \ldots, x_{n-1}) \in \mathbf{R}^{n-1}$ with the point $(y, 0) \in \mathbf{R}^n$. Take a point $y \in \mathbf{R}^{n-1}$. We denote by $l(y)$ the line in $\mathbf{R}^n$ through y orthogonal to the plane $\mathbf{R}^{n-1}$ and thus consisting of all the points of the form (y, z), where y is the given point.

Denote by G^n a domain in $\mathbf{R}^n$ which is the whole of $\mathbf{R}^n$, or the strip formed by the points $x = (y, z)$ such that $|z| \leq h$, or, finally, the half-strip consisting of the points $(x_1, \ldots, x_{n-1}, x_n)$ such that $x_{n-1} \geq 0$ and $|x_n| \leq h$. Let G_0^n denote $G^n \cap \mathbf{R}^{n-1}$. If G^n is $\mathbf{R}^n$ or a strip, then G_0^n is obviously $\mathbf{R}^{n-1}$. If G^n is a half-strip, then G_0^n is a half-plane. If G^n is a strip or a half-strip, then the number $2h$ is called the *width* of G^n.

Let $f: G^n \to \mathbf{R}$ be a nonnegative integrable function. Then for almost all $y \in G_0^n$ the function $f_y: z \in I \to f(y, z)$ is integrable. For every such y we construct the symmetrization σf_y of f_y. As a result we get a function $\sigma f: G^n \to \mathbf{R}$ defined for almost all $y \in G_0^n$ and for almost all z, and for each y the function $z \to (\sigma f)(y, z)$ is the symmetrization of the function $z \to f(y, z)$.

We next describe a certain form of symmetrization needed later. For an arbitrary point $x = (x_1, \ldots, x_{n-2}, x_{n-1}, x_n) \in \mathbf{R}^n$ we set $x = (t, x_{n-1}, x_n)$, where $t = (x_1, \ldots, x_{n-2}) \in \mathbf{R}^{n-2}$. Every point x of the form $x = (t, 0, 0)$ in $\mathbf{R}^n$ will be identified with the point $t \in \mathbf{R}^{n-2}$, and the plane $\mathbf{R}^{n-2}$ will be identified with the plane $x_{n-1} = x_n = 0$ in $\mathbf{R}^n$. Finally, $\mathbf{R}_+^{n-1}$ denotes the set of all points x of the form $x = (t, x_{n-1}, 0)$, where $x_{n-1} > 0$. Let D_n be the set of all points $(t, r, \theta) \in \mathbf{R}^n$ with $t = (x_1, \ldots, x_{n-2}) \in \mathbf{R}^{n-2}$, $r \geq 0$

and $-\pi \le \theta \le \pi$, and let

$$\zeta(t, r, \theta) = (t, r \cos \theta, r \sin \theta).$$

We get a certain mapping ζ of D_n into $\mathbf{R}^n$. Obviously, $\zeta(D_n) = \mathbf{R}^n$, and ζ is one-to-one on the set D_n^0 of all points $(t, r, \theta) \in D_n$ such that $r > 0$ and $-\pi < \theta < \pi$. Further, $\zeta(D_n^0) = \mathbf{R}^n \setminus \mathbf{R}_+^{n-}$. For $t \in \mathbf{R}^{n-2}$ and $r > 0$ denote by $C(t, r)$ the subset $\{\zeta(t, r, \theta) \mid -\pi \le \theta \le \pi\}$ of $\mathbf{R}^n$. Obviously, $C(t, r)$ is a circle of radius r about $(t, 0, 0) \in \mathbf{R}^{n-2}$ whose plane is completely orthogonal to $\mathbf{R}^{n-2}$.

Let f be a nonnegative integrable function on $\mathbf{R}^n$. We define from it a new function $\varphi \colon D_n \to \mathbf{R}$ by setting

$$\varphi(t, r, \theta) = f(t, r \cos \theta, r \sin \theta).$$

Let $\sigma\varphi$ be the symmetrization of φ. Obviously,

$$(\sigma\varphi)(t, r, \pi) = (\sigma\varphi)(t, r, -\pi)$$

for any $t \in \mathbf{R}^{n-2}$ and $r \ge 0$, and $(\sigma\varphi)(t, 0, \theta)$ does not depend on θ. This implies the existence on $\mathbf{R}^n$ of a function zf such that

$$(zf)(t, r \cos \theta, r \sin \theta) = (\sigma\varphi)(t, r, \theta)$$

for any $t \in \mathbf{R}^{n-2}$, $r \ge 0$ and $\theta \in [-\pi, \pi]$. The function zf is called the *cylindrical symmetrization* of f with respect to the plane $\mathbf{R}^{n-2}$ and the half-plane $\mathbf{R}_+^{n-1}$.

It is useful to see how to look at the cylindrical symmetrization geometrically. For any $t \in \mathbf{R}^{n-2}$ and $r > 0$ we have the circle $C(t, r)$. Under cylindrical symmetrization the restriction of f to any such circle is replaced by its symmetrization with respect to the parameter θ, $-\pi \le \theta \le \pi$.

THEOREM 3.6. *Let f be a nonnegative compactly supported function of class $W_p^1(G^n)$ in the domain G^n, where $p \ge 1$. Then its symmetrization σf also belongs to $W_p^1(G^n)$, and*

$$\int_{G^n} |(\sigma f)'(x)|^p \, dx \le \int_{G^n} |f'(x)|^p \, dx.$$

THEOREM 3.7. *Let f be a nonnegative compactly supported function of class $W_p^1(\mathbf{R}^n)$ in $\mathbf{R}^n$, where $p \ge 1$. Then its cylindrical symmetrization zf also belongs to $W_p^1(\mathbf{R}^n)$, and*

$$\int_{\mathbf{R}^n} |(zf)'(x)|^p \, dx \le \int_{\mathbf{R}^n} |f'(x)|^p \, dx.$$

See [115] for proofs of Theorems 3.6 and 3.7.

PROOF OF THEOREM 3.4. Let $K = (A_0, A_1)$ be a capacitor of class $K_n(t)$, where $0 < t < 1$. It will be assumed that $A_0 \supset \mathbf{R}^n \backslash B(0, 1)$, and A_1 is a closed connected set joining the points 0 and $t\mathbf{e}_1$. The general case clearly can be reduced to this by the successive application of a similarity dilation and a motion of $\mathbf{R}^n$. Let $A_0' = \mathbf{R}^n \backslash B(0, 1)$. Since $A_0 \supset A_0'$, it follows that $C_n(A_0, A_1) \geq C_n(A_0', A_1)$. We assume below that $A_0 = \mathbf{R}^n \backslash B(0, 1)$.

Consider the orthogonal projection of A_1 on the plane $\mathbf{R}^{n-1}$. The projection is a connected set A_1' lying in $\mathbf{R}^{n-1}$ and joining the points O and $t\mathbf{e}_1$. Let us prove that

$$C_n(A_0, A_1) \geq C_n(A_0, A_1'). \qquad (3.21)$$

Indeed, take an arbitrary function $f \in \tilde{W}(A_0, A_1)$, and let σf be its symmetrization. We prove that $\sigma f(x)$ is equal to 1 in a neighborhood of A_1' and to 0 in a neighborhood of A_0. There is a $\delta > 0$ such that $f(x) = 1$ on the ball $B(x, \delta)$ for every $x \in A_1$ and the support of f is contained in the ball $B(0, 1 - \delta)$. Then the support of σf is clearly also contained in $B(0, 1 - \delta)$, since σf vanishes in a neighborhood of A_0. Denote by l_y the line orthogonal to $\mathbf{R}^{n-1}$ and intersecting $\mathbf{R}^{n-1}$ at the point $(y, 0)$. If $(y, 0) \in A_1'$, then l_y contains points of A_1. Since $f(x) \leq 1$ and $f(x)$ is equal to 1 in the δ-neighborhood of every point $x_0 \in A_1$, this implies that for $(y, 0) \in A_1'$ the function $z \to f(y, z)$, and with it also $(\sigma f)(y, z)$, is equal to 1 on some segment of length 2δ. This enables us to conclude that σf is equal to 1 in a neighborhood of the set A_1'. Therefore, if $f \in \tilde{W}(A_0, A_1)$, then $\sigma f \in \tilde{W}(A_0, A_1')$. From this,

$$\int_{\mathbf{R}^n} |f'(x)|^n \, dx \geq \int_{\mathbf{R}^n} |\tau f'(x)|^n \, dx \geq C_n(A_0, A_1').$$

Since $f \in \tilde{W}(A_0, A_1)$ is arbitrary, this proves (3.21). By considering the invariance of the capacity with respect to motions it is not hard to show that the same thing is valid when A_1 is replaced by the projection on an arbitrary hyperplane passing through the points O and $\mathbf{e}_1$. Let $P_1, \ldots, P_{n-1}$ be a system of $n - 1$ mutually orthogonal hyperplanes intersecting in the line $O\mathbf{e}_1$. We construct a sequence of sets $A_1^{(1)}, \ldots, A_1^{(n-1)}$, where $A_1^{(1)}$ is the orthogonal projection of A_1 on P_1, and $A_1^{(i+1)}$ is the orthogonal projection of $A_1^{(i)}$ on P_{i+1} for each $i < n - 1$. Obviously,

$$C_n(A_0, A_1) \geq C_n(A_0, A_1^{(1)}) \geq \cdots \geq C_n(A_0, A_1^{(n-1)}).$$

It remains to note that $A_1^{(n-1)}$ is a connected set in the line passing through the points O and $t\mathbf{e}_1$ and contains these points, hence it also contains $[0, t\mathbf{e}_1]$. This gives us that

$$C(A_0, A_1) \geq C(A_0, [0, t\mathbf{e}_1]) = \gamma(t),$$

and Theorem 3.4 is proved.

PROOF OF THEOREM 3.5. Let (A_0, A_1) be a capacitor of class $H_n(t)$, where $t > 0$, and A_0 and A_1 are closed connected sets with A_1 bounded and A_0 unbounded. It will be assumed that A_1 contains the points O and $\mathbf{e}_{n-1}$, the distance from O to A_0 is equal to t, and the point x_0 of A_0 closest to O lies in the plane of the vectors $\mathbf{e}_{n-1}$ and $\mathbf{e}_n$, $x_0 = \alpha \mathbf{e}_{n-1} + \beta \mathbf{e}_n$. An arbitrary capacitor of class $H_n(t)$ can always be made into a capacitor satisfying these conditions by a motion and a similarity dilation. The capacity of the capacitor is not changed by this. We now construct another capacitor (A_0', A_1'). Let $x = (y, x_{n-1} x_n)$. Define

$$r(x) = \sqrt{x_{n-1}^2 + x_n^2}, \qquad \zeta_1(x) = (y, r(x), 0), \quad \zeta_2(x) = (y, -r(x), 0).$$

The mappings ζ_1 and ζ_2 are continuous. Let $A_0' = \zeta_2(A_0)$ and $A_1' = \zeta_1(A_1)$. The sets A_0' and A_1' are connected and closed, and lie in the plane $\mathbf{R}^{n-1}$. Further, A_1' contains the points O and $\mathbf{e}_{n-1}$. The set A_0' lies in the half-plane $x_n = 0, x_{n-1} \leq -t$, and contains the point $-t\mathbf{e}_{n-1} = \zeta_2(x_0)$.

We prove that

$$C_n(A_0, A_1) \geq C_n(A_0', A_1'). \tag{3.22}$$

Let $f \in \tilde{W}(A_0, A_1)$ be an arbitrary function, and consider its cylindrical symmetrization zf. For $t \in \mathbf{R}^{n-2}$ and $r > 0$ denote by $C(t, r)$ the circle formed by the points $(t, r\cos\theta, r\sin\theta)$, where $-\pi \leq \theta \leq \pi$. A point (t, r, θ) with $r > 0$ belongs to the set A_1' if and only if $C(t, r)$ intersects A_1. Similarly, a point $(t, -r, 0)$ with $r > 0$ belongs to A_0' if and only if $C(t, r) \cap A_0$ is nonempty. Let

$$\varphi(t, r, \theta) = f(t, r\cos\theta, r\sin\theta),$$
$$\overline{\varphi}(t, r, \theta) = (zf)(t, r\cos\theta, r\sin\theta).$$

Assume that $(t, r, 0) \in A_1'$. Then the function $\theta \to (t, r, \theta)$ is equal to 1 on some interval contained in $[-\pi, \pi]$. Since $0 \leq f(x) \leq 1$ for all x, this implies that $\overline{\varphi}(t, r, \theta) \equiv 1$ on some interval $-\delta < \theta < \delta$, and thus $\overline{\varphi}(t, r, \theta)$ is equal to 1 on some arc of $C(t, r)$. The middle of this arc is precisely the point $(t, r, 0)$. We show that $(zf)(x) \equiv 1$ in some neighborhood of $(t, r, 0)$. Let (x_m), $m = 1, 2, \ldots,$ be an arbitrary sequence of points in $\mathbf{R}^n$ converging to $(t, r, 0)$, and let $x_m = (t_m, x_{n-1,m}, x_{n,m})$. The point x_m lies on the circle $C(t_m, r_m)$, where $r_m = r(x_m) \to r$ as $m \to \infty$. Since f is equal to 1 in a neighborhood of every point $x \in A_1$, the length of the arc of the circle $C(t_m, r_m)$ on which $f(x) = 1$ is not less than some $\varepsilon > 0$ for sufficiently large m. It is not hard to deduce from this that $(zf)(x_m) = 1$ for sufficiently large m, and since (x_m), $m = 1, 2, \ldots,$ is an arbitrary sequence convergent to $x = (t, r, 0)$, it is thereby proved that $zf \equiv 1$ in some neighborhood of the given point x. We now show that $(zf)(x) \equiv 0$ in a neighborhood of every point $(t, -r, 0) \in A'_0$. Suppose $(t, -r, 0) \in A'_0$. Then the function $\theta \to \varphi(t, r, \theta)$ vanishes on certain intervals contained in $[-\pi, \pi]$. This implies that the function $\tilde{\varphi}$ vanishes in certain intervals $[-\pi, -\pi + \delta]$ and $[\pi - \delta, \pi]$ symmetric with respect to the point O. The points of $C(t, r)$ corresponding to the values of θ in these intervals form an arc whose midpoint is precisely the point $(t, -r, 0)$. The function zf vanishes on this arc. Arguing again as in the preceding case, we get that zf vanishes in a neighborhood of every point $x \in A'_0$.

The function zf has compact support. It follows from what has been proved that $zf \in \tilde{W}(A'_0, A'_1)$. By Theorem 3.8,

$$\int_{\mathbf{R}^n} |f'(x)|^n \, dx \geq \int_{\mathbf{R}^n} |(zf)'(x)|^n \, dx \geq C_n(A_0, A_1).$$

Since $f \in \tilde{W}(A_0, A_1)$ is arbitrary, inequality (3.22) is established.

To conclude the proof we use the usual symmetrization. Assume that (A_0, A_1) is a capacitor constructed as follows: A_1 is a bounded connected set containing O and $\mathbf{e}_{n-1}$; A_0 is an unbounded set lying in the half-plane $x_{n-1} \leq -t$; A_0 contains the point $-t\mathbf{e}_{n-1}$, and $\inf_{A_0} x_{n-1} = -\infty$. The collection of all capacitors satisfying these conditions is denoted by $\tilde{H}_n(t)$. Take an arbitrary function $f \in \tilde{W}(A_0, A_1)$, and let σf be its symmetrization. Denote by A'_0 and A'_1 the orthogonal projections of A_0 and A_1 on the plane $\mathbf{R}^{n-1}$. The function σf is equal to zero in a neighborhood of A'_0 and to 1 in a neighborhood of A'_1; hence, $\sigma f \in \tilde{W}(A'_0, A'_1)$. We have that

$$\int_{\mathbf{R}^n} |f'(x)|^n \, dx \geq \int_{\mathbf{R}^n} |(\sigma f)'(x)|^n \, dx \geq C_n(A'_0, A'_1).$$

Since $f \in \tilde{W}(A_0, A_1)$ is arbitrary, this proves that

$$C_n(A_0, A_1) \geq C_n(A'_0, A'_1).$$

Thus, if (A_0, A_1) is a capacitor of class $\tilde{H}_n(t)$, then the capacity of (A_0, A_1) is not increased if A_0 and A_1 are replaced by their orthogonal projections on $\mathbf{R}^{n-1}$. It is clear that instead of $\mathbf{R}^{n-1}$ we can take any hyperplane passing through O and $\mathbf{e}_{n-1}$. The capacitor (A_0, A_1) is made into the capacitor $H(t)$ by successive projections. This gives us that $C_n(A_0, A_1) \geq C_n[H(t)] = \eta(t)$, and the proof of Theorem 3.5 is thus complete.

We introduce some notation. Let

$$C_n[K_G(t)] = \omega_n / [\ln \Psi_n(1/t)]^{n-1},$$
$$C_n[K_T(t)] = \omega_n / [\ln \Phi_n(t)]^{n-1}.$$

THEOREM 3.8 [35]. *The function* $\Psi_n(t)/t$ *is nondecreasing in* $(1,\infty)$, *and there exists a constant* $\lambda_n > 1$ *such that* $t \leq \Psi_n(t) \leq \lambda_n t$ *for all* $t \in (1,\infty)$. *The functions* Ψ_n *and* Φ_n *are connected as follows*: $\Phi_n(t) = (\Psi_n[(t+1)^{1/2}])^2$ *for all* $t > 0$.

We shall not present the proof of Theorem 3.8.

The so-called *Thompson principle*, which gives means for constructing a lower estimate for the capacity of a capacitor, is well known in potential theory. Here we present a certain modification of this principle that can be used for getting a lower estimate of the p-capacity of a capacitor in $\mathbf{R}^n$ for arbitrary $p > 1$.

Let $H = (A, B)$ be a given capacitor in $\mathbf{R}^n$, and let $U = \mathbf{R}^n \backslash (A \cup B)$ be its field. Also, let $p > 1$ be a given number, and let $q = p/(p - 1)$.

For a given vector-valued function $f : U \to \mathbf{R}^n$ of class $L_q(U)$ we say that f is *divergence-free (solenoidal)* if

$$\int_U \langle \nabla\varphi(x), f(x) \rangle \, dx = 0 \tag{3.23}$$

for every $\varphi \in C_0^\infty(U)$. If f satisfies this condition, then we also write $\operatorname{div} f(x) = 0$ in U. It is easy to establish by passing to the limit that if $f(x)$ is divergence-free, then (3.23) holds for any function $\varphi \in \overset{\circ}{W}{}^1_p(U)$.

Suppose that $f : U \to \mathbf{R}^n$ is divergence-free, and φ_1 and φ_2 are two different functions of class $\tilde{W}^1_p(A, B)$. Their difference $\varphi_1 - \varphi_2$ belongs to the class $\overset{\circ}{W}{}^1_p(U)$, and hence

$$\int_U \langle \nabla\varphi_1(x) - \nabla\varphi_2(x), f(x) \rangle \, dx = 0.$$

From this we conclude that the quantity

$$\gamma(f, A, B) = \int_U \langle \nabla\varphi(x), f(x) \rangle \, dx \tag{3.24}$$

does not depend on the choice of the function $\varphi \in \tilde{W}^1_p(A, B)$. We call $\gamma(f, A, B)$ the *flow* of the divergence-free vector-valued function $f(x)$ in the capacitor H. If A and B are bounded by smooth $(n - 1)$-dimensional surfaces and $f(x)$ is continuous, then

$$\gamma(f, A, B) = \int_{\partial U \cap B} \langle f(x), \nu(x) \rangle \, d\sigma = - \int_{\partial U \cap A} \langle f(x), \nu(x) \rangle \, d\sigma, \tag{3.25}$$

where $\nu(x)$ is the outward normal vector at the point $x \in \partial U$ and $d\sigma$ is the area element of the surface ∂U. Formula (3.25) can have a definite heuristic value.

THEOREM 3.9. *Let $H = (A, B)$ be a capacitor, let $U = \mathbf{R}^n \backslash (A \cup B)$ be its field, and let $f: U \to \mathbf{R}^n$ be a vector-valued function of class $L_q(U)$. Assume that f is divergence-free, and $\gamma(f, A, B) = 1$. Then*

$$C_p(A, B) \geq \left(\int_U |f(x)|^{p/(p-1)} \, dx \right)^{1-p}. \tag{3.26}$$

PROOF. Let $\varphi \in \tilde{W}_p^1(A, B)$ be arbitrary. Then

$$1 = \int_U \langle \nabla \varphi(x), f(x) \rangle \, dx \leq \int_U |\nabla \varphi(x)| |f(x)| \, dx$$

$$\leq \left(\int_U |\nabla \varphi(x)|^p \, dx \right)^{1/p} \left(\int_U |f(x)|^q \, dx \right)^{1/q}.$$

From this,

$$\int_U |\nabla \varphi(x)|^p \, dx \geq \left(\int_U |f(x)|^q \, dx \right)^{-p/q}.$$

Since $\varphi \in \tilde{W}_p^1(A, B)$ was arbitrary, the theorem is proved.

REMARK. The estimate of the lemma is sharp in the sense that there is equality for some vector-valued function $f: U \to \mathbf{R}^n$ such that $\operatorname{div} f = 0$. Namely, suppose that $u(x) \in W_p^1(A, B)$ supplies the smallest values of the functional

$$\int_U |\nabla u(x)|^p \, dx \tag{3.27}$$

in the class $W_p^1(A, B)$. Then $u(x)$ is extremal for the functional (3.27), and hence

$$\int_U \langle \nabla \eta(x), |\nabla u(x)|^{p-2} \nabla u(x) \rangle \, dx = 0 \tag{3.28}$$

for every function $\eta \in \overset{\circ}{W}{}_p^1(U)$, with compact support, by Corollary 1 to Lemma 5.9 of Chapter II. It is easy to establish by passing to the limit that (3.28) holds for every function $\eta \in \overset{\circ}{W}{}_p^1(U)$, and hence the vector-valued function $g = |\nabla u|^{p-2} \nabla u$ is divergence-free. We have that

$$\gamma(g, A, B) = \int_U \langle \nabla \eta(x), |\nabla u(x)|^{p-2} \nabla u(x) \rangle \, dx,$$

where η is an arbitrary function in $W_p^1(A, B)$. Setting $\eta = u$ here, we get that

$$\gamma(g, A, B) = C_p(A, B).$$

For brevity of notation we set $C_p(A, B) = C$ and let $f = (1/C)g$. We have that

$$\left(\int_U |f(x)|^q \, dx \right)^{p-1} = \frac{1}{C^p} \left(\int_U |\nabla u(x)|^p \, dx \right)^{p-1} = \frac{1}{C},$$

and so equality holds in (3.26) for this function f.

§4. The concept of the generalized differential of an exterior form

§4.1. General facts about exterior forms. We assume that the reader is familiar with the concept of an exterior differential form defined in a domain in $\mathbf{R}^n$, along with the main properties of exterior differential forms, in particular, the properties of the operations of addition, exterior multiplication, and differentiation of exterior forms. Expositions of these topics can be found in, for example, [146], [139], and [164].

We introduce some notation. Let Γ_n^m, where $1 \leq m \leq n$, be the collection of all m-element finite sequences $I = (i_1, \ldots, i_m)$, where $i_1, \ldots, i_m$ are integers such that $1 \leq i_1 < i_2 < \cdots < i_m \leq n$. Let $K = (k_1, \ldots, k_m) \in \Gamma_n^m$ be a given m-tuple of indices. The exterior form $dx_{k_1} \ldots dx_{k_m}$ of degree m is denoted by dx_K. An arbitrary exterior form $\omega(x)$ of degree m on a set U can be written in the form

$$
\begin{aligned}
\omega(x) &= \sum_{1 \leq i_1 < i_2 < \cdots < i_m \leq n} f_{i_1 i_2 \ldots i_m}(x)\, dx_{i_1}\, dx_{i_2} \ldots dx_{i_m} \\
&= \sum_{I \in \Gamma_n^m} f_I(x)\, dx_I,
\end{aligned}
\tag{4.1}
$$

where the $f_I(x)$ are real functions defined on U. The functions $f_I(x)$ are called the *coefficients* of the form $\omega(x)$. The equality (4.1) is called the *canonical representation* of the exterior differential form $\omega(x)$.

The degree of a form $\omega(x)$ is denoted by $\deg \omega$. A form ω will be said to belong to the class $C^\infty(U)$ $(L_{p,\text{loc}}(U), W_{p,\text{loc}}^1(U)$, etc.) if all its coefficients belong to $C^\infty(U)$ (respectively, to $L_{p,\text{loc}}(U)$, $W_{p,\text{loc}}^1(U)$, etc.). A form ω is said to have *compact support in U* if all its coefficients have compact support in U.

We introduce the notation

$$
|\omega|(x) = \sqrt{\sum_{I \in \Gamma_n^m} |f_I(x)|^2},
$$

where f_I $(I \in \Gamma_n^m)$ are the coefficients of ω. If $E \subset U$, then let

$$
\|\omega\|_{p,E} = \left[\int_E (|\omega|(x))^p\, dx \right]^{1/p}.
$$

If $\deg \omega = n$, then the form ω has an expression $\omega(x) = u(x)\, dx_1 \ldots dx_n$. If $E \subset U$ is a measurable set, then by the integral of a form ω of degree n over E we mean the quantity $\int_E \omega = \int_E u(x)\, dx$.

An exterior form of degree equal to zero on an open set U is defined to be any real function defined on U. The product of exterior forms ω and

φ is denoted by $\omega \wedge \varphi$, and $d\omega$ denotes the differential of an exterior form ω of class C^1.

Let φ and ω be forms of degrees l and k, respectively, and let $\theta = \omega \wedge \varphi$. Then it is not hard to show that for all $x \in U$

$$|\theta|(x) \leq C|\varphi|(x)|\omega|(x), \tag{4.2}$$

where C is a constant depending only on k, l, and n.

Let $\omega \in L_{p,\mathrm{loc}}(U)$ and $\psi \in L_{q,\mathrm{loc}}(U)$ be exterior forms on U of degrees k and l, respectively, with $k + l \leq n$, and let $1/p + 1/q = 1/r \leq 1$. Then their exterior product $\omega \wedge \varphi$ is in $L_{r,\mathrm{loc}}(U)$, and for every compact set $A \subset U$

$$\|\omega \wedge \varphi\|_{r,A} \leq C\|\omega\|_{p,A}, \tag{4.3}$$

where C is a constant depending only on k, l, p, and n. Inequality (4.3) is an obvious consequence of (4.2) and Hölder's inequality.

Let (ω_ν), $\nu = 1, 2, \ldots$, be a sequence of exterior forms of degree k defined on U. Then the forms ω_ν will be said to converge as $\nu \to \infty$ to an exterior form ω in $L_{p,\mathrm{loc}}(U)(W^l_{p,\mathrm{loc}}(U)$, locally uniformly) if all the coefficients of ω_ν converge in the indicated sense to the corresponding coefficients of ω.

Let $\omega_\nu \in L_{p,\mathrm{loc}}(U)$ and $\varphi_\nu \in L_{q,\mathrm{loc}}(U)$, $\nu = 1, 2, \ldots$, be sequences of exterior forms converging to forms ω and φ in $L_{p,\mathrm{loc}}$ and $L_{q,\mathrm{loc}}$, respectively, as $\nu \to \infty$. Assume that $1/p + 1/q = 1/r \leq 1$. We have that

$$\|\omega_\nu \wedge \varphi_\nu - \omega \wedge \varphi\|_{r,A} \leq \|(\omega_\nu - \omega) \wedge \varphi_\nu\|_{r,A} + \|\omega \wedge (\varphi_\nu - \varphi)\|_{r,A}$$
$$\leq \gamma(\|\omega_\nu - \omega\|_{p,A}\|\varphi_\nu\|_{q,A} + \|\omega\|_{p,A}\|\varphi_\nu - \varphi\|_{q,A}),$$

which implies that $\omega_\nu \wedge \varphi_\nu \to \omega \wedge \varphi$ in $L_{r,\mathrm{loc}}$.

LEMMA 4.1. *Let ω be an exterior form of class $L_{1,\mathrm{loc}}(U)$ with degree $r \geq 0$ defined in an open set U. If $\int_U \eta(x) \wedge \omega(x) = 0$ for every exterior form $\eta \in C^\infty(U)$ of degree $n - r$ with compact support in U, then $\omega(x) = 0$ almost everywhere in U.*

PROOF. Let

$$\omega(x) = \sum_{\mathscr{J} \in \Gamma^r_n} \omega_{\mathscr{J}}(x)\, dx_{\mathscr{J}}.$$

Fix arbitrarily an r-tuple $\mathscr{J}_0 \in \Gamma^r_n$, $\mathscr{J}_0 = (j_1, \ldots, j_r)$, of indices, and let $I_0 = (i_1, \ldots, i_{n-r}) \in \Gamma^{n-r}_n$ be an $(n-r)$-tuple of indices complementary to $\mathscr{J}_0$, i.e., such that $(j_1, \ldots, j_r, i_1, \ldots, i_{n-r})$ is a permutation of $(1, 2, \ldots, n)$. Let $\varphi \in C_0^\infty(U)$ be arbitrary, and let $\eta(x) = \varphi(x)\, dx_{I_0}$. Then

$$\eta(x) \wedge \omega(x) = \sigma \varphi(x)\omega_{\mathscr{J}_0}(x)\, dx_1 \ldots dx_n,$$

where $\sigma = \pm 1$, and we see from the condition of the lemma that $\int_U \varphi(x)\omega_{\mathcal{I}_0}(x)\,dx = 0$ for every function $\varphi \in C^\infty$ with compact support in U. This implies that $\omega_{\mathcal{I}_0}(x) = 0$ almost everywhere, and the lemma is proved.

COROLLARY. *Let ω_1 and ω_2 be rth-degree forms of class $L_{1,\mathrm{loc}}(U)$ such that for every form $\eta \in C^\infty$ of degree $n - r$ with compact support in U*

$$\int_U \eta(x) \wedge \omega_1(x) = \int_U \eta(x) \wedge \omega_2(x).$$

Then $\omega_1(x) = \omega_2(x)$ almost everywhere in U.

§4.2. The concept of generalized differential of an exterior form.   Let $U \subset \mathbf{R}^n$ be an arbitrary open set, and let ω and φ be exterior forms of class $L_{1,\mathrm{loc}}(U)$, with $\deg\varphi = \deg\omega + 1$ and $\deg\omega = r \geq 0$. Then we say that φ is the *generalized exterior differential* of the form ω if for every form $\eta(x)$ of class C^∞ with degree $n - r - 1$ that is compactly supported in U

$$\int_U \eta \wedge \varphi = (-1)^{n-r} \int_U d\eta \wedge \omega.$$

The generalized exterior differential of ω is denoted by $d\omega$, as in the case of forms with smooth coefficients. By the corollary to Lemma 4.1, the generalized exterior differential of a form is unique if it exists.

Let ω be an exterior form of class C^1. Then its ordinary differential is also the generalized differential. Indeed, let η be a form of class C^∞ with degree $n - r - 1$, where $r = \deg\omega$, and with compact support in U. Then the form $\eta \wedge \omega$ is compactly supported in U, and $d(\eta \wedge \omega) = d\eta \wedge \omega + (-1)^{n-r-1}\eta \wedge d\omega$, where the differential on the right-hand side is understood in the classical sense. Since the form $\eta \wedge \omega$ has compact support by Lemma 2.3, we have that $\int_U d(\eta \wedge \omega) = 0$, which implies that

$$\int_U \eta \wedge d\omega = (-1)^{n-r} \int_U d\eta \wedge \omega.$$

Since the form η of class C^∞ here is arbitrary, this means that $d\omega$ is also the generalized differential of ω.

LEMMA 4.2. *Let ω and φ be exterior forms of class $L_{1,\mathrm{loc}}(U)$, where U is an open set in $\mathbf{R}^n$, and $\deg\varphi = \deg\omega + 1$. Assume that for every set V lying strictly inside U there exists a sequence (ω_ν) of forms such that $\omega_\nu \to \omega$ in $L_1(V)$ as $\nu \to \infty$, ω_ν has a generalized differential for each ν, and $d\omega_\nu \to \varphi$ in $L_1(V)$. Then φ is the generalized differential of ω.*

PROOF. Let η be an arbitrary exterior form of class $C^\infty(U)$ with compact support in U and with degree $n - r - 1$, where $r = \deg\omega$. Let V be an open set lying strictly inside U and such that $\eta(x) = 0$ for $x \notin V$. We construct a sequence (ω_ν), $\nu = 1, 2, \ldots$, of exterior forms of class $L_1(V)$ such that the forms ω_ν converge in $L_1(V)$ to the form ω, and the $d\omega_\nu$ converge in $L_1(V)$ to the form φ. For each ν

$$\int_V \eta \wedge d\omega_\nu = (-1)^{n-r} \int_V d\eta \wedge \omega_\nu. \qquad (4.4)$$

Since η is in C^∞ and has compact support, the coefficients of the forms η and $d\eta$ are bounded functions. Passing to the limit as $\nu \to \infty$ in (4.4), we get that

$$\int_V \eta \wedge \varphi = (-1)^{n-r} \int_V d\eta \wedge \omega.$$

Obviously, the equality is not violated if V is replaced by U. Since η is an arbitrary form of class C^∞ with compact support in U, it follows from what has been proved that $\varphi = d\omega$ by definition. The lemma is proved.

We make some more remarks about approximation of exterior forms by smooth functions. Take an arbitrary averaging kernel α, and let ω be an arbitrary exterior form of degree $r \geq 1$ and of class $L_{1,\text{loc}}(U)$,

$$\omega(x) = \sum_I \omega_I(x)\, dx_I.$$

Let

$$(\alpha_h * \omega)(x) = \sum_I (\alpha_h * \omega_I)(x)\, dx_I.$$

Denote by $\tilde{\alpha}$ the function $x \to \alpha(-x)$. Obviously, $\tilde{\alpha}$ is also an averaging kernel. By using Fubini's theorem it is not hard to show that if u is locally integrable and v is a bounded measurable function with compact support in U (U an open set in $\mathbf{R}^n$), then

$$\int_U (\alpha_h * u)(x)v(x)\, dx = \int_U u(x)(\tilde{\alpha}_h * v)(x)\, dx.$$

Applying this result to exterior forms, we get that if ω and η are exterior forms of class $L_{1,\text{loc}}(U)$, η is compactly supported and has bounded functions as coefficients, and $\deg \omega + \deg \eta = n$, then

$$\int_U \eta(x) \wedge (\alpha_h * \omega)(x) = \int_U (\tilde{\alpha}_h * \eta)(x) \wedge \omega(x).$$

Note further that $d(\alpha_h * \omega)(x) = (\alpha_h * d\omega)(x)$ for every exterior form ω of class C^1.

LEMMA 4.3. *Let ω and φ be forms of class $L_{1,\mathrm{loc}}(U)$, where U is an open set in $\mathbf{R}^n$ and φ is the generalized differential of ω in U. Then $\alpha_h * d\varphi = d(\alpha_h * \omega)$ for every $h > 0$.*

PROOF. As above, let $\hat{U}_h$ be the set of all points $x \in U$ whose distance to the boundary of U is greater than h. Let $\eta \in C^\infty(U)$ be any exterior form with compact support in $\hat{U}_h$ and with $\deg \eta = n - r - 1$, where $r = \deg \omega$. We have the following chain of equalities:

$$\int_{U_h} d\eta \wedge (\alpha_h * \omega) = \int_U (\tilde{\alpha}_h * d\eta) \wedge \omega = \int_U d(\tilde{\alpha}_h * \eta) \wedge \omega$$

$$= (-1)^{n-r} \int_U (\tilde{\alpha}_h * \eta) \wedge \varphi = (-1)^{n-r} \int_{\hat{U}_h} \eta \wedge (\alpha_h * \varphi).$$

Since $\eta \in C^\infty$ is an arbitrary form with compact support in $\hat{U}_h$, this implies that $\alpha_h * \varphi = d(\alpha_h * \omega)$, which is what was required.

COROLLARY. *Let U be an open set in $\mathbf{R}^n$ and ω and φ exterior forms in the respective classes $L_{p_1,\mathrm{loc}}(U)$, $p_1 \geq 1$, and $L_{p_2,\mathrm{loc}}(U)$, $p_2 \geq 1$, and suppose that φ is the generalized exterior differential of ω. Then there exist sequences (ω_m) and (φ_m) of forms of class C^∞ such that $\varphi_m = d\omega_m$ for each m, and for every measurable set A lying strictly inside U the forms ω_m and φ_m are defined on A beginning with some $m = m_0(A)$ and satisfy the convergence relations $\omega_m \to \omega$ in $L_{p_1}(A)$ and $\varphi_m \to \varphi$ in $L_{p_2}(A)$.*

Obviously, if (h_m), $m = 1, 2, \ldots$, is an arbitrary sequence such that $h_m > 0$ for all m and $h_m \to 0$ as $m \to \infty$, then the sequences of exterior forms $\omega_m = \alpha_{h_m} * \omega$ and $\varphi_m = \alpha_{h_m} * \varphi$ are the desired sequences.

§4.3. **Properties of the generalized differential of an exterior form.** The linearity of a generalized differential follows in an obvious way from its definition: if exterior forms ω_1 and ω_2 of degree k in the class $L_{1,\mathrm{loc}}(U)$ have generalized differentials $d\omega_1$ and $d\omega_2$, then for any $\alpha_1, \alpha_2 \in \mathbf{R}$ the form $\alpha_1 \omega_1 + \alpha_2 \omega_2$ has a generalized differential, and $d(\alpha_1 \omega_1 + \alpha_2 \omega_2) = \alpha_1 d\omega_1 + \alpha_2 d\omega_2$.

LEMMA 4.4. *Suppose that ω and φ are exterior forms in an open set $U \subset \mathbf{R}^n$ with $\deg \omega = r$, $\deg \varphi = s$, $r + s \leq n$, $\omega \in L_{p,\mathrm{loc}}(U)$ and $\varphi \in L_{q,\mathrm{loc}}(U)$, where p and q are such that $1/p + 1/q \leq 1$. Assume that the generalized differentials $d\omega$ and $d\varphi$ exist, and that $d\omega \in L_{p,\mathrm{loc}}(U)$ and $d\varphi \in L_{q,\mathrm{loc}}(U)$. Then the form $\omega \wedge \varphi$ has a generalized differential, and*

$$d(\omega \wedge \varphi) = d\omega \wedge \varphi + (-1)^r \omega \wedge d\varphi.$$

PROOF. We construct sequences (ω_m) and (φ_m), $m = 1, 2, \ldots$, of exterior forms of class C^∞ such that $\omega_m \to \omega$ and $d\omega_m \to d\omega$ in $L_{p,\mathrm{loc}}(U)$, and

$\varphi_m \to \varphi$ and $d\varphi_m \to d\varphi$ in $L_{q,\text{loc}}(U)$. Let $t \geq 1$ be such that $1/t = 1/p + 1/q$. Then $\omega_m \wedge \varphi_m \to \omega \wedge \varphi$ in $L_{t,\text{loc}}(U)$. For each m we have that

$$d(\omega_m \wedge \varphi_m) = d\omega_m \wedge \varphi_m + (-1)^r \omega_m \wedge d\varphi_m.$$

The right-hand side of this equality converges in $L_{t,\text{loc}}(U)$ to the form $\theta = d\omega \wedge \varphi + (-1)^r \omega \wedge d\varphi$ as $m \to \infty$. It follows from this by Lemma 4.2 that θ is the generalized differential of $\omega \wedge \varphi$, and the lemma is proved.

LEMMA 4.5. *Let ω be an exterior form of class $L_{1,\text{loc}}(U)$. Assume that ω is the generalized exterior differential of some form of class $L_{1,\text{loc}}(U)$ with degree $n-1$ and with compact support in U. Then $\int_U \omega = 0$.*

PROOF. The required result is contained in Lemma 2.3 in the case where ω is a form of class C^1. Let α be an arbitrary averaging kernel, and let $\omega = d\theta$. We set $\alpha_h * \omega = \omega_h$ and $\alpha_h * \theta = \theta_h$. The forms θ_h are compactly supported in U and $\omega_h = d\theta_h$ for sufficiently small h. We have that $\int_U \omega_h = 0$. Passage to the limit as $h \to 0$ gives the required result.

LEMMA 4.6. *If a form ω of class $L_{1,\text{loc}}(U)$ with degree $r \geq 1$ is the generalized differential of some form θ, then the generalized differential of ω is equal to zero.*

PROOF. Let $\eta \in C^\infty$ be an arbitrary form with degree $n-r-1$ and with compact support in U. The form $d\eta \wedge \theta$ has a generalized differential, by Lemma 4.4. Further, $d(d\eta \wedge \theta) = (-1)^r d\eta \wedge \omega$. By the lemma, this means that $\int_U d\eta \wedge \omega = 0$. Since $\eta \in C^\infty$ is arbitrary, this means by definition that $d\omega = 0$. The lemma is proved.

§4.4. The homomorphism induced on the algebra of exterior forms by a mapping of the domain. Let $U \subset \mathbf{R}^n$ and $V \subset \mathbf{R}^n$ be open sets in the corresponding spaces. Assume that $f: U \to V$ is a given mapping of class C^∞. Then for every exterior form ω of class C^∞ defined on V we can construct a certain exterior form φ on U denoted by $f^*\omega$ and defined as follows. Let $f(x) = (f_1(x), \ldots, f_m(x))$. If $\omega(y)$ is a form of zero degree, i.e., ω is simply a real function defined on V, then $(f^*\omega)(x) = \omega[f(x)]$. Assume that $\deg \omega(y) > 0$. Then

$$\omega(y) = \sum_{1 \leq j_1 < j_2 < \cdots < j_k \leq m} \omega_{j_1 j_2 \ldots j_k}(y) \, dy_{j_1} \, dy_{j_2} \ldots dy_{j_k},$$

$$(f^*\omega)(x) = \sum_{1 \leq j_1 < j_2 < \cdots < j_k \leq m} \omega_{j_1 j_2 \ldots j_k}[f(x)] \, df_{j_1}(x)$$
$$\wedge df_{j_2}(x) \wedge \cdots \wedge df_{j_k}(x).$$

Here df_j, $j = 1, \ldots, m$, is an exterior form of the first degree, namely,

$$df_j = \sum_{i=1}^{n} \frac{\partial f_j}{\partial x_i}(x)\, dx_i.$$

We present some properties of the operation f^*, assuming that $f \in C^\infty$ and all the forms considered are of class C^∞.

1. $f^*(\omega_1 + \omega_2) = f^*\omega_1 + f^*\omega_2$ for any forms ω_1 and ω_2 of the same degree.

2. $f^*(u\omega) = (f^*u)(f^*\omega)$ for every form ω and any function $u \colon V \to \mathbf{R}$.

3. $f^*(\omega \wedge \theta) = (f^*\omega) \wedge (f^*\theta)$ for any forms ω and θ.

4. Let U be an open set in $\mathbf{R}^n$, and i_U the identity mapping of U. Then $i_U^*\omega = \omega$ for every exterior form ω defined on U.

5. Suppose that $U \subset \mathbf{R}^n$, $V \subset \mathbf{R}^m$, and $W \subset \mathbf{R}^k$ are sets open in the corresponding spaces, and that $f \colon U \to V$ and $g \colon V \to W$ are mapping of class C^∞, and let $h = g \circ f$. Then $h^* = f^* \circ g^*$, i.e., $h^*\omega = f^*(g^*\omega)$ for every form ω on W.

6. $f^*(d\omega) = d(f^*\omega)$ for every form ω.

See, for example, [139] for proofs of these properties.

Our goal is to extend the operation f^* to cases when f satisfies minimal smoothness conditions. Let $U \subset \mathbf{R}^n$ and $G \subset \mathbf{R}^m$ be open sets in $\mathbf{R}^n$ and $\mathbf{R}^m$, respectively, and let $f \colon U \to G$ be a continuous mapping of class $W_{p,\mathrm{loc}}^1(U)$, $p \geq 1$. Let $f(x) = (f_1(x), \ldots, f_m(x))$. The exterior forms

$$df_i(x) = \sum_{j=1}^{n} \frac{\partial f_i}{\partial x_j}(x)\, dx_j$$

are defined on U, where the derivatives are understood as generalized derivatives. Let $\omega(y)$ be an arbitrary exterior form on G. The symbol $f^*\omega(x)$ denotes the form on U defined by the same formulas as in the case when f is a mapping of class C^∞. Namely, if ω is a form of degree zero, then $(f^*\omega)(x) = \omega[f(x)]$. But if $\deg \omega > 0$ and

$$\omega(y) = \sum_{1 \leq j_1 < j_2 < \cdots < j_k \leq m} \omega_{j_1 j_2 \ldots j_k}(y)\, dy_{j_1}\, dy_{j_2} \ldots dy_{j_k},$$

then we set

$$f^*\omega(x) = \sum_{1 \leq j_1 < j_2 < \cdots < j_k \leq m} \omega_{j_1 j_2 \ldots j_k}[f(x)]\, df_{j_1}(x)$$
$$\wedge df_{j_2}(x) \wedge \cdots \wedge df_{j_k}(x).$$

If the degree k of ω exceeds n, then clearly $f^*\omega(x) = 0$.

We remark that certain difficulties can arise almost everywhere in the general case when we consider arbitrary forms with measurable coefficients defined on G. For example, it can happen that the form $f^*\omega(x)$ is not defined on a set of positive measure, etc. In this connection we consider here only forms with continuous coefficients.

Since f is in $W^1_{p,\mathrm{loc}}(U)$, it follows that $df_i \in L_{p,\mathrm{loc}}(U)$. Further, $\|df_i\|_{p,A} \leq \|f\|_{1,p,A}$ for every compact set $A \subset U$. On the basis of (1.3), it is easy to establish from this by induction that the form $\mathscr{J} = (j_1, \ldots, j_k)$, $k \leq p$ is in $L_{p/k,\mathrm{loc}}(U)$ for $f^* dy_{\mathscr{J}} = df_{j_1} \wedge \cdots \wedge df_{j_k}$. Further, for every compact set $A \subset U$

$$\|f^* dy_{\mathscr{J}}\|_{p,k,A} \leq \gamma_k(\|f\|_{1,p,A})^k,$$

where γ_k is a constant that depends only on n and k.

Let V be an arbitrary open set contained strictly inside U, and let $f_\nu : G \to D$, $\nu = 1, 2, \ldots$, be a sequence of mappings of class C^∞ such that $f_\nu \to f$ in $W^1_r(V)$ as $\nu \to \infty$. Then $df_{i,\nu} \to df_i$ in $L_r(V)$ as $\nu \to \infty$, and hence for $k \leq r$

$$f_\nu^* dy_{\mathscr{J}} = df_{j_1,\nu} \wedge df_{j_2,\nu} \wedge \cdots \wedge df_{j_k,\nu}$$

converges to the form $f^* dy_{\mathscr{J}}$ in $L_p(V)$ as $\nu \to \infty$, where $p = r/k$.

It follows, in particular, that if $f_h = \alpha_h * f$, where α is an averaging kernel on $\mathbf{R}^n$, then the forms $f_h^* dy_{\mathscr{J}}$ converge to the form $f^* dy_{\mathscr{J}}$ in $L_{p,\mathrm{loc}}(U)$ as $h \to 0$, where $p = r/k$, $k = \deg dy_{\mathscr{J}}$.

We remark that the particular properties 1–4 above follow algebraically from the definition of the operation f^* and thus remain true also when f is an arbitrary continuous mapping of class W^1_p. Property 5 is also an algebraic consequence of the definition of f^* and the rule for differentiating a composite function. Therefore, it is true also for the case of mappings determined by functions with generalized derivatives under the condition that f, g, and $h = g \circ f$ belong to the corresponding function classes, and the usual rule for differentiating a composite function remains in force also in this case.

LEMMA 4.7. *Let $U \subset \mathbf{R}^n$ and $G \subset \mathbf{R}^m$ be open sets in the corresponding spaces, and $f : U \to G$ a continuous mapping of class $W^1_{r,\mathrm{loc}}(U)$. Then for every exterior form $\omega(y)$ of degree k, $0 \leq k \leq r$, and class C on G the form $f^*\omega$ is of class C in the case when $k = 0$, and of class $L_{p,\mathrm{loc}}(U)$ in the case when $k > 0$, where $p = r/k$. If the form ω has a generalized differential $d\omega$, the form $d\omega$ belongs to the class C, and $k + 1 \leq r$, then $f^* d\omega$ is the generalized differential of $f^*\omega$.*

PROOF. Let $\omega(y)$ be a form of degree k and class C on G. If $k = 0$, then $(f^*\omega)(x) = \omega[f(x)]$, and hence $(f^*\omega)(x)$ is a continuous function on U. We consider the case $k > 0$. Let

$$\omega(y) = \sum_{\mathscr{J} \in \Gamma_m^k} \omega_{\mathscr{J}}(y) \, dy_{\mathscr{J}}.$$

Then

$$(f^*\omega)(x) = \sum_{\mathscr{J} \in \Gamma_m^k} \omega_{\mathscr{J}}[f(x)] f^* \, dy_{\mathscr{J}}.$$

The functions $\omega_{\mathscr{J}}[f(x)]$ are all continuous, the form $f^* \, dy_{\mathscr{J}}$ belongs to the class $L_{r/k,\mathrm{loc}}(U)$, and, hence, $f^*\omega \in L_{r/k,\mathrm{loc}}(U)$.

Let $\varphi = d\omega$, and suppose that φ belongs to the class C. Then, applying the constructions to the form φ, we get that $f^*\varphi \in L_{r/(k+1),\mathrm{loc}}(U)$. Take an arbitrary open set V lying strictly inside U, and let $f_\nu : V \to \mathbf{R}^m$ be a sequence of mappings of class C^∞ such that $f_\nu \to f$ uniformly as $\nu \to \infty$ and, moreover, $f_\nu \to f$ in $W_r^1(V)$. These conditions are satisfied by any sequence $\alpha_{h_\nu} * f$, where $h_\nu \to 0$ as $\nu \to \infty$, and $\overline{V} \subset \hat{U}_{h_\nu}$ for all ν, so that the forms $\alpha_{h_\nu} * f$ are defined on V.

The set $f(\overline{V}) \subset G$ is compact, since f is continuous, and hence there exists a $\delta > 0$ such that $\rho(y, \partial G) \geq \delta$ for all $y \in f(\overline{V})$. It will be assumed that $|f_\nu(x) - f(x)| < \delta/2$ for all $x \in \overline{V}$. Since $f_\nu \to f$ uniformly on $\overline{V}$, this assumption obviously involves no loss of generality. By this condition, $f_\nu(\overline{V})$ is contained in the closed $\delta/2$-neighborhood (denote it by H) of $f(\overline{V})$. Let $\mu(\delta)$, $\delta > 0$, be the modulus of continuity of $\omega_{\mathscr{J}}$ on the compact set H. Then for each ν

$$|\omega_{\mathscr{J}}[f_\nu(x)] - \omega_{\mathscr{J}}[f(x)]| \leq \mu(\|f_\nu - f\|_{C(\overline{V})}).$$

This implies that $\omega_{\mathscr{J}}[f_\nu(x)] \to \omega_{\mathscr{J}}[f(x)]$ uniformly on $\overline{V}$ as $\nu \to \infty$.

Further, $f_\nu^* \, dy_{\mathscr{J}} \to f^* \, dy_{\mathscr{J}}$ in $L_{r/k}(U)$, and hence $(f_\nu^*\omega)(x) \to (f^*\omega)(x)$ in $L_{r/k}(\overline{V})$ as $\nu \to \infty$. Similarly, $(f_\nu^*\varphi)(x) \to (f^*\varphi)(x)$ in $L_{r/(k+1)}(\overline{V})$ as $\nu \to \infty$.

Assume first that ω is a form of class C^∞. Then $f_\nu^*\omega$ is also a form of class C^∞, and $d(f_\nu^*\omega) = f_\nu^* \, d\omega = f_\nu^*\varphi$. Since $f_\nu^*\omega \to f^*\omega$ and $d(f_\nu^*\omega) = f_\nu^* \, d\omega \to f^* \, d\omega$ in $L_{1,\mathrm{loc}}(V)$ as $\nu \to \infty$, and the set V lying strictly inside U was taken arbitrarily, this proves that $f^* \, d\omega = d(f^*\omega)$.

We now consider the case when ω and φ are forms of class C, and $\varphi = d\omega$, understood as the generalized differential. Again, let V be an open set lying strictly inside U. Let H be an open set in $\mathbf{R}^m$ lying strictly inside G and such that $f(\overline{V}) \subset H$. Let α be an arbitrary regularizer* on $\mathbf{R}^m$, and let $\omega_\nu = \alpha_{h_\nu} * \omega$ and $\varphi_\nu = \alpha_{h_\nu} * \varphi$, where the sequence (h_ν), $\nu = 1, 2, \ldots$, is such that $h_\nu > 0$ for all ν, $h_\nu \to 0$ as $\nu \to \infty$, and $H \subset \hat{G}_{h_\nu}$

---

*Editor's note. Called an averaging kernel above.

for all ν. For each ν the forms ω_ν and φ_ν belong to the class C^∞, and since ω and φ belong to the class C, it follows that $\omega_\nu \to \omega$ and $\varphi_\nu \to \varphi$ uniformly on H as $\nu \to \infty$. For each ν we have that $\varphi_\nu = d\omega_\nu$. Consider the forms $f^*\omega_\nu$ and $f^*\varphi_\nu = f^*(d\omega_\nu)$. Then $f^*\varphi_\nu = d(f^*\omega_\nu)$ for each ν, where the differential is understood as the generalized differential. For each ν

$$(f^*\omega_\nu)(x) = \sum_{\mathscr{I}} \omega_{\nu,\mathscr{I}}[f(x)] f^* \, dy_{\mathscr{I}},$$

$$(f^*\omega)(x) = \sum_{\mathscr{I}} \omega_{\mathscr{I}}[f(x)] f^* \, dy_{\mathscr{I}},$$

and $\omega_{\nu,\mathscr{I}}[f(x)] \to \omega_{\mathscr{I}}[f(x)]$ uniformly as $\nu \to \infty$. Since $f^* \, dy_{\mathscr{I}} \in L_1(\overline{V})$, this implies that $f^*\omega_\nu(x) \to f^*\omega(x)$ in $L_1(\overline{V})$. We conclude similarly that $(f^*\varphi_\nu)(x) \to (f^*\varphi)(x)$ in $L_1(\overline{V})$.

The set V lying strictly inside U was chosen arbitrarily. We thus get that for every V lying strictly inside U there is a sequence of forms η_ν converging to $f^*\omega$ in $L_1(V)$ such that for each ν the generalized differential $d\eta_\nu$ exists, and $d\eta_\nu \to f^*(d\omega)$ in $L_1(V)$. By Lemma 4.2, this implies that $f^*(d\omega) = d(f^*\omega)$. The proof of the lemma is complete.

We note the special case when the degree k of the form ω in the formulation of Lemma 4.7 is equal to zero. This leads to the following statement.

COROLLARY. *Suppose that G is an open set in $\mathbf{R}^m$, $v: G \to \mathbf{R}$ is a function of class C^1, and U is an open set in $\mathbf{R}^n$. Then for every continuous mapping $f: U \to G$ of class $W^1_{r,\mathrm{loc}}(U)$, where $r \geq 1$, the function $u = v \circ f$ is in $W^1_{r,\mathrm{loc}}(U)$, and its derivatives can be expressed in terms of the derivatives of v and f according to the same formulas as in the case of sufficiently smooth functions.*

§4.5. **Weak convergence of sequences of exterior forms.** Below, U is an open subset of $\mathbf{R}^n$. Let f_m, $m = 0, 1, \ldots$, be functions of class $L_{p,\mathrm{loc}}(U)$, where $p \geq 1$. We say that the sequence f_m converges to f_0 weakly in $L_{p,\mathrm{loc}}(U)$ if it is locally bounded in $L_p(U)$, and for every function $\varphi \in C_0^\infty(U)$

$$\int_U f_m(x)\varphi(x) \, dx \to \int_U f_0(x)\varphi(x) \, dx$$

as $m \to \infty$. Let (ω_m), $m = 1, 2, \ldots$, be some sequence of exterior forms in $L_{p,\mathrm{loc}}(U)$ of degree k, where $1 \leq k \leq n$. Then we say that the sequence of forms ω_m converges to a form ω_0 weakly in $L_{p,\mathrm{loc}}(U)$ as $m \to \infty$ if the coefficients of the forms ω_m converge weakly in $L_{p,\mathrm{loc}}(U)$ to the corresponding coefficients of ω_0.

LEMMA 4.8. *Let (ω_m) be a sequence of exterior forms in $L_{p_1,\mathrm{loc}}(U)$ of degree k that converges weakly in $L_{p_1,\mathrm{loc}}(U)$ to a form ω_0 as $m \to \infty$. Assume that each of the exterior forms ω_m, $m = 0, 1, \ldots$, has in U a generalized differential, and that the sequence $(d\omega_m)$, $m = 1, 2, \ldots$, is locally bounded in $L_{p_2,\mathrm{loc}}(U)$. Then the sequence $d\omega_m$ converges to $d\omega_0$ weakly in $L_{p_2,\mathrm{loc}}(U)$ as $m \to \infty$.*

PROOF. Let α be an arbitrary exterior form of degree $n - k - 1$ in the class $C_0^\infty(U)$. For each m we have

$$\int_U \omega_m \wedge d\alpha = (-1)^{k-1} \int_U d\omega_m \wedge \alpha.$$

Since $\omega_m \to \omega_0$ in $L_{p_1,\mathrm{loc}}$ as $m \to \infty$, it follows that

$$\int_U \omega_m \wedge d\alpha \to \int_U \omega_0 \wedge d\alpha = (-1)^{k-1} \int_U d\omega_0 \wedge \alpha$$

as $m \to \infty$. This leads to the conclusion that for any exterior form $\alpha \in C_0^\infty$ of degree $n - k - 1$

$$\lim_{m \to \infty} \int_U d\omega_m \wedge \alpha = \int_U d\omega_0 \wedge \alpha.$$

The lemma is proved.

LEMMA 4.9. *Let U be an open set in $\mathbf{R}^n$, and let $f_m = (f_{m1}, \ldots, f_{mk})$, $1 \le k \le n$, $m = 1, 2, \ldots$, be a sequence of vector-valued functions of class $W_{p,\mathrm{loc}}^1(U)$, where $p \ge k$, that is locally bounded in $W_p^1(U)$. Assume that f_m converges in $L_{1,\mathrm{loc}}$ to a vector-valued function $f_0 = (f_{01}, \ldots, f_{0k})$ as $m \to \infty$, and let $\omega_m = df_{m1} \wedge \ldots \wedge df_{mk}$. Then the sequence of forms ω_m converges weakly in $L_{p/k,\mathrm{loc}}(U)$ to the form ω_0.*

PROOF. We prove the lemma by induction on k. The lemma is an immediate consequence of Lemma 4.8 in the case $k = 1$. Assume that the lemma has been proved for some k, and let $f_m: U \to \mathbf{R}^{k+1}$ be a sequence of vector-valued functions of class $W_{p,\mathrm{loc}}^1(U)$, $p \ge k + 1$, which is locally bounded in W_p^1 and locally convergent in L_1 to a function f_0. By the Sobolev imbedding theorem, $f_m \to f_0$ also in $L_{p,\mathrm{loc}}$. Let

$$f_m(x) = (f_{m1}(x), f_{m2}(x), \ldots, f_{m,k+1}(x)).$$

In $\mathbf{R}^{k+1}$ we consider the exterior forms

$$u = dy_1\, dy_2 \ldots dy_k, \qquad v = (-1)^k y_{k+1} u,$$
$$w = u \wedge dy_{k+1} = dy_1 \ldots dy_k\, dy_{k+1}.$$

Then $w = dv$ for each k. Let

$$\tilde{\omega}_m = f_m^* u = df_{m1} \wedge df_{m2} \wedge \cdots \wedge df_{mk},$$

$$\psi_m = f_m^* v = (-1)^k f_{m,k+1} \tilde{\omega}_m, \qquad \omega_m = f_m^* w.$$

Then $\omega_m = d\psi_m$ for each m, by Lemma 4.7. The induction hypothesis gives us that $\tilde{\omega}_m \to \tilde{\omega}_0$ in $L_{p/k,\mathrm{loc}}$ for each k as $m \to \infty$. On the basis of the Sobolev imbedding theorem, $f_{m,k+1} \to f_{0,k+1}$ in $L_{s,\mathrm{loc}}$ for $s < np/(n-p)$. Let θ be an arbitrary C^∞-form of degree $n-k$ with compact support in U. We have that $\tilde{\omega}_n \to \tilde{\omega}_0$ weakly in $L_{p/k,\mathrm{loc}}$ as $m \to \infty$, and $f_{m,k+1}\theta \to f_{0,k+1}\theta$ in L_s. Let us show that as $m \to \infty$

$$\int_U f_{m,k+1}\tilde{\omega}_m \wedge \theta \to \int_U f_{0,k+1}\tilde{\omega}_0 \wedge \theta. \qquad (4.5)$$

Indeed,

$$\left| \int_U f_{m,k+1}\tilde{\omega}_m \wedge \theta - \int_U f_{0,k+1}\tilde{\omega}_m \wedge \theta \right|$$

$$\leq C\|\tilde{\omega}_m\|_{L_{p/k}(A)}\|f_{m,k+1} - f_{0,k+1}\|_{L_s(A)},$$

where A is the support of the form θ and $s = p/(p-k)$, $1 < s < p$. From this,

$$\int_U f_{m,k+1}\tilde{\omega}_m \wedge \theta - \int_U f_{0,k+1}\tilde{\omega}_m \wedge \theta \to 0 \qquad (4.6)$$

as $m \to \infty$. Let $f \in C_0^\infty$ be such that $\|f - f_{0,k+1}\|_{L_s} < \varepsilon$. Then

$$\left| \int_U f_{0,k+1}\tilde{\omega}_m \wedge \theta - \int_U f_{0,k+1}\tilde{\omega}_0 \wedge \theta \right|$$

$$\leq \left| \int_U (f_{0,k+1} - f)\tilde{\omega}_m \wedge \theta \right| + \left| \int_U f(\tilde{\omega}_m \wedge \theta - \tilde{\omega}_0 \wedge \theta) \right|$$

$$+ \left| \int_U (f - f_{0,k+1})\tilde{\omega}_0 \wedge \theta \right|.$$

The first and third integrals on the right-hand side are each less than $C\varepsilon$, and the second tends to 0 as $m \to \infty$. Since $\varepsilon > 0$ is arbitrary, this implies that

$$\int_U f_{0,k+1}\tilde{\omega}_m \wedge \theta \to \int_U f_{0,k+1}\tilde{\omega}_0 \wedge \theta. \qquad (4.7)$$

It is obvious from (4.6) and (4.7) that (4.5) holds. Since the form $\theta \in C_0^\infty(U)$ is arbitrary, this implies that the sequence of forms $\psi_m = f_{m,k+1}\tilde{\omega}_m$, $m = 1, 2, \ldots$, converges weakly in $L_{1,\mathrm{loc}}(U)$ to the form $\psi_0 = f_{0,k+1}\tilde{\omega}_0$ as $m \to \infty$. By the condition of the lemma, the sequence of forms $\omega_m = d\psi_m$ is locally bounded in $L_{p/(k+1)}$. This implies on the basis of Lemma 4.8 that $\omega_m \to \omega_0$ weakly in $L_{p/(k+1),\mathrm{loc}}(U)$, and the lemma is proved.

COROLLARY. *Let* $f_m : U \to \mathbf{R}^n$ *be a sequence of mappings of class* $W^1_{n,\text{loc}}(U)$. *Assume the following conditions:*

a) (f_m) *is locally bounded in* $W^1_n(U)$.

b) (f_m) *converges in* L_1 *to some mapping* f_0 *as* $m \to \infty$.

Then the limit mapping f_0 *belongs to* $W^1_{n,\text{loc}}(U)$, *and for every continuous real function* $\varphi : U \to \mathbf{R}$ *with compact support in* U

$$\int_U \varphi(x) \mathscr{F}(x, f_m)\, dx \to \int_U \varphi(x) \mathscr{F}(x, f_0)\, dx$$

as $m \to \infty$.

PROOF. The fact that the limit mapping f_0 is in $W^1_{n,\text{loc}}(U)$ follows immediately from the general properties of functions with generalized derivatives. Let $f_m = (f_{m1}, \ldots, f_{mn})$, $m = 0, 1, 2, \ldots$, and let $\omega_m = df_{m1} \wedge \cdots \wedge df_{mn}$, $m = 0, 1, 2, \ldots$. Then $\omega_m \to \omega_0$ weakly in $L_{1,\text{loc}}(U)$, by the lemma. This means that $\int_U \varphi \omega_m \to \int_U \varphi \omega_0$ for every function $\varphi \in C_0(U)$. It remains to see that

$$\omega_m(x) = \mathscr{F}(x, f_m) \times dx_1\, dx_2 \ldots dx_n,$$

and the corollary is proved.

§5. Mappings with bounded distortion and elliptic differential equations

§5.1. A description of a certain class of functionals of the calculus of variations.
The real and imaginary parts of an analytic function of a single complex variables are well known to be solutions of the Laplace equation. Here we establish that an analogous fact holds also for arbitrary mappings with bounded distortion. We first give some facts about differential equations in the form we need. Proofs of the statements about the relevant equations are also presented in part in Chapter III. Some theorems establishing a connection between mappings with bounded distortion and extremal functionals of the calculus of variations are proved.

Let U be an open set in $\mathbf{R}^n$. We consider functionals defined on the set $W^1_{1,\text{loc}}(U)$ of functions and representable in the form

$$I_F(u, A) = \int_A F[x, u'(x)]\, dx, \tag{5.1}$$

where $A \subset U$ is a measurable set, and $F(x, q)$ is a nonnegative function of the variables (x, q), defined for almost all $x \in U$ and any $q \in \mathbf{R}^n$. Some restrictions will be imposed on $F(x, q)$. First of all, we assume $F(x, q)$ is such that $f[x, u'(x)]$ is measurable for any $u \in W^1_{1,\text{loc}}(U)$. For studying mappings with bounded distortion the most important case is that when $F(x, q)$ has the form

$$F(x, q) = \left(\sum_{i=1}^{n} \sum_{j=1}^{n} \theta_{ij}(x) q_i q_j \right)^{n/2}, \tag{5.2}$$

where the quadratic form on the right-hand side satisfies the following conditions: the functions $\theta_{ij}(x)$ are defined and finite almost everywhere and are measurable in U, and there exist constants λ_1 and λ_2 such that $0 < \lambda_1 \leq \lambda_2 < \infty$ and

$$\lambda_1^2 |q|^2 \leq \sum_{i=1}^{n} \sum_{j=1}^{n} \theta_{ij}(x) q_i q_j \leq \lambda_2^2 |q|^2 \tag{5.3}$$

for any x such that all the quantities $\theta_{ij}(x)$ are defined and finite, and for all $q \in \mathbf{R}^n$.

Although we shall need only the case when $F(x, q)$ has the special structure (5.2) in what follows, it is more convenient to carry out the arguments concerning the properties of the functional (5.1) in general form. We list the conditions on $F(x, q)$ which ensure the properties we need for the functional (5.1).

A_1. There exists a set $E \subset U$ with $\operatorname{meas} E = 0$ such that $F(x, q)$ is defined and continuous as a function of q in the whole space $\mathbf{R}^n$ for every $x \notin E$, and $F(x, q) \geq 0$ for any x and q with $x \notin E$.

A_2. $F(x, q)$ has partial derivatives $F_{q_i} = \partial F / \partial q_i$, $i = 1, \ldots, n$, continuous in the variables $q_1, \ldots, q_n$ for $x \notin E$.

Let

$$F_q(x, q) = (F_{q_1}(x, q), F_{q_2}(x, q), \ldots, F_{q_n}(x, q)).$$

A_3. For any function $u \in W_1^1(U)$ the functions $x \to F[x, u'(x)]$ and $x \to F_q[x, u'(x)]$ are measurable in U.

B. For all $x \notin E$ (where E is the set of measure zero indicated in condition A_1) the function $F(x, q)$ is essentially convex with respect to the variable q, i.e.,

$$F(x, \lambda q_1 + \mu q_2) \leq \lambda F(x, q_1) + \mu F(x, q_2)$$

for any $q_1, q_2 \in \mathbf{R}^n$ and any numbers λ and μ such that $\lambda, \mu > 0$ and $\lambda + \mu = 1$, with equality if and only if $q_1 = q_2$.

C. There exists a number $p > 1$ such that for $x \notin E$ and any q

$$a_1 |q|^p \leq F(x, q) \leq a_2 |q|^p + c(x), \tag{5.4}$$

where a_1 and a_2 are positive constants, and $c(x)$ is an integrable function.

D. For $x \notin E$

$$|F_q(x, q)| \leq a_3 |q|^{p-1}, \tag{5.5}$$

$$\langle q, F_q(x, q) \rangle \geq a_4 |q|^p, \tag{5.6}$$

where $p > 1$ is the same as in C, and a_3 and a_4 are positive constants.

E. $F(x, q) \geq F(x, 0)$ for $x \notin E$ and for any q.

In what follows we assume that F satisfies all the conditions A, B, C, D, and E here, though some of the statements below are true for the case when F satisfies only part of these conditions. For brevity a function F satisfying all the conditions A–E simultaneously will be called a *normal kernel*. Conditions A–E hold with $p = n$ in the special case when F is defined by (5.2), where the matrix function θ satisfies (5.3).

Let $F(x, q)$ be a normal kernel, and V an open set lying strictly inside U. The number

$$I_F(u, V) = \int_V F[x, u'(x)] \, dx$$

is defined for every function $u \in W_p^1(V)$.

We consider the following boundary value problem for the functional I_F: among all the functions taking given values on the boundary of V, find the one which gives I_F the smallest possible value. A function satisfying this condition is called an *extremal function* for the functional I_F. However, it is first necessary to specify what it means to say that some function $u \in W_p^1(U)$ (it can even be discontinuous in general) takes given values on the boundary of V, which can in turn to be a fairly bad set. The necessary precise definitions will be given in §5.2.

Let F be an arbitrary normal kernel. A function $u \in W_{p,\mathrm{loc}}^1(U)$ is called a *generalized solution* of the equation

$$\sum_{i=1}^n \frac{\partial}{\partial x_i} \left\{ \frac{\partial F}{\partial q_i} [x, u'(x)] \right\} = 0 \tag{5.7}$$

in U if the equality

$$\int_U \sum_{i=1}^n \frac{\partial F}{\partial q_i} [x, u'(x)] \frac{\partial \eta}{\partial x_i}(x) \, dx = \int_U \langle F_q[x, u'(x)], \eta'(x) \rangle \, dx = 0 \tag{5.8}$$

holds for every function $\eta \in W_p^1(U)$ with compact support in U.

Equation (5.7) is none other than the Euler equation for the functional (5.1). As will be shown below, every extremal function for (5.1), where F is a normal kernel, is a generalized solution of (5.7). The converse assertion is also true. Equation (5.7) will be written briefly as follows:

$$\operatorname{div} F_q[x, u'(x)] = 0. \tag{5.9}$$

If $u \in W_p^1(V)$ is a generalized solution of (5.7), then we say that u is a *stationary function* for the functional I_F.

We mention a special case of functionals of the form (5.1) that is important for what follows. By the use of concepts introduced in §4, the condition that u be a generalized solution of (5.7) can be represented as follows. We introduce an exterior form $\omega_F(u, x)$ of degree $n - 1$ by setting

$$\omega_F(u, x) = \sum_{k=1}^{n} (-1)^{k-1} F_{q_k}[x, u'(x)] \, dx_1 \qquad (5.10)$$
$$\ldots dx_{k-1} \, dx_{k+1} \ldots dx_n.$$

Then

$$\langle F_q[x, u'(x)], \eta'(x) \rangle \, dx_1 \, dx_2 \ldots dx_n = d\eta(x) \wedge \omega_F(u, x).$$

Therefore, satisfaction of (5.8) for every $\eta \in C_0^\infty(U)$ means that the generalized differential of the form $\omega_F(u, x)$ is equal to zero. Thus, u is a generalized solution of (5.7) if and only if $d\omega_F(u, x) = 0$.

We mention a special case of functionals of the form (5.1) that is important for what follows. Let $F(x, q) = |q|^n$. All the conditions A–E are satisfied here. Equation (5.7) takes the form

$$n \operatorname{div}[|u'(x)|^{n-2} u'(x)] = 0. \qquad (5.11)$$

We have the following particular solutions of (5.11):

$$u_1(x) = \sum_{i=1}^{n} a_i x_i + a_0, \qquad u_2(x) = C_1 \ln \frac{1}{|x - a|} + C_2,$$

where $a_0, a_1, \ldots, a_n, C_1, C_2 \in \mathbf{R}$ and $a \in \mathbf{R}^n$ are constants.

§5.2. Variational properties of mappings with bounded distortion.
It is known that if h is a harmonic function on the plane of the complex variable w and $f(z)$ is a holomorphic function, then $h[f(z)]$ is also harmonic. The goal of this subsection is to establish an analogue of this property for mappings with bounded distortion.

We first present some heuristic considerations. Let $f: U \to \mathbf{R}^n$ be a diffeomorphism, and let $V = f(U)$. Assume that $u \in W_n^1(V)$ minimizes the functional

$$\int_V |u'(y)|^n \, dy \qquad (5.12)$$

among all functions coinciding with $u(x)$ on the boundary of V. Let $v(x) = u[f(x)]$. For each $i = 1, \ldots, n$

$$\frac{\partial v}{\partial x_i}(x) = \sum_{\alpha=1}^{n} \frac{\partial u}{\partial y_\alpha}[f(x)] \frac{\partial f_\alpha}{\partial x_i}(x).$$

This equality can be written as follows in matrix form:

$$v'(x) = f'^*(x)u'[f(x)].$$ (5.13)

Making the change of variable $y = f(x)$ in (5.12), we get that

$$\int_V |u'(y)|^n \, dy = \int_U |u'[f(x)]|^n |\mathscr{J}(x, f)| \, dx.$$

Let us express $u'[f(x)]$ in terms of $v'(x)$ by using (5.13). We get that $u'[f(x)] = [f'^*(x)]^{-1}v'(x)$. From this,

$$|u'(f(x))|^2 = \langle u'[f(x)], u'[f(x)] \rangle = \langle [f'^*(x)]^{-1}v'(x), [f'^*(x)]^{-1}v'(x) \rangle$$
$$= \langle [f'(x)]^{-1}(f'^*(x))^{-1}v'(x), v'(x) \rangle.$$

Setting

$$\theta(x, f) = |\mathscr{J}(x, f)|^{2/n}[f'(x)]^{-1}[f'^*(x)]^{-1},$$ (5.14)

we find that the integral in (5.12) is equal to

$$\int_U \langle \theta(x, f)v'(x), v'(x) \rangle^{n/2} \, dx.$$ (5.15)

Assigning to a $v \in W_n^1(V)$ the function $v \circ f$, we get a one-to-one correspondence between the spaces $W_n^1(V)$ and $W_n^1(U)$ under certain natural assumptions about the diffeomorphism f. (As follows from what was proved earlier, this condition holds if, for example, $\sup |f'(x)| < \infty$ and $\inf |\mathscr{J}(x, f)| > 0$; in this case the mapping f is quasiconformal.) Therefore, if u is extremal for the functional (5.12), then $v = u \circ f$ is extremal for (5.15). Equation (5.9) has the following form for the functional (5.15):

$$\operatorname{div}(\langle \theta(x, f)v'(x), v'(x) \rangle^{n/2-1}\theta(x, f)v'(x)) = 0.$$ (5.16)

This is a relation between the derivatives, and the supposition naturally arises that it can be obtained by formal transformations, without resorting to the fairly stringent assumptions about f that were made above. It is the purpose of the following arguments to prove this supposition.

Let $f: U \to \mathbf{R}^n$ be a mapping with bounded distortion, where U is an open subset of $\mathbf{R}^n$. The matrix $f'(x)$ is defined for almost all $x \in U$, and

$$|f'(x)|^n \leq K\mathscr{J}(x, f).$$ (5.17)

We construct a matrix function $\theta(x) \equiv \theta(x, f) \equiv \theta f(x)$. Let $x \in U$ be such that $f'(x)$ is defined and (5.17) holds. If $\mathscr{J}(x, f) = 0$, then let $\theta(x) = I$ (the $n \times n$ identity matrix). But if $\mathscr{J}(x, f) \neq 0$, then $\theta(x, f)$ is determined by (5.14). The matrix-valued function $\theta(x)$ is defined for almost all $x \in U$. Let $\theta(x) = (\theta_{ij}(x))$, $i, j = 1, \ldots, n$. The elements θ_{ij} of the matrix $\theta(x)$ are measurable functions on U.

We establish certain properties of the matrix-valued function $\theta(x) \equiv \theta(x; f)$ constructed from an arbitrary mapping $f: U \to \mathbf{R}^n$ with bounded distortion.

First, a simple observation about matrices. Let X be a nonsingular $n \times n$ matrix, and let $Y = X^{-1}$. Then $XY = I$, which implies that $I = (XY)^* = Y^*X^*$, and hence $Y^* = (X^*)^{-1}$, so that $(X^{-1})^* = (X^*)^{-1}$.

We show that the matrix $\theta(x)$ is symmetric and positive-definite, and we find bounds for its eigenvalues. If $\mathcal{J}(x, f) = 0$ at x, then $\theta(x) = I$, and in this case $\theta(x)$ is a positive-definite symmetric matrix with all eigenvalues equal to 1. Suppose that $\mathcal{J}(x, f) \neq 0$, and let $[\mathcal{J}(x, f)]^{2/n} = \Delta$ and $[f'(x)]^{-1} = L$. Then $[f'(x)^*]^{-1} = L^*$ and $\theta(x) = \Delta L L^*$. From this, $\theta(x)^* = \Delta(LL^*)^* = \Delta LL^* = \theta(x)$, and the symmetry of $\theta(x)$ is proved. For an arbitrary vector $\xi \neq 0$

$$\langle \theta(x)\xi, \xi \rangle = \Delta \langle LL^*\xi, \xi \rangle = \Delta \langle L^*\xi, L^*\xi \rangle = \Delta |L^*\xi|^2.$$

Let $0 < k_1 \leq \cdots \leq k_n$ be the principal dilations of the mapping $f'(x)$. Then the principal dilations of the mapping $L^* = (f'(x)^*)^{-1}$ are equal to $1/k_1 \geq \cdots \geq 1/k_n$, and hence $|\xi|/k_1 \geq |L^*\xi| \geq |\xi|/k_n$ for every ξ. We have that $\Delta = (k_1 \ldots k_n)^{2/n}$. This leads to the conclusion that for every ξ

$$\Delta |\xi|^2/k_1^2 \geq \langle \theta(x)\xi, \xi \rangle \geq \Delta |\xi|^2/k_n^2.$$

The quadratic form $\langle \theta(x)\xi, \xi \rangle$ is thus positive-definite. We now remark that

$$\left(\frac{\Delta}{k_1^2} \right)^{n/2} = (k_1 k_2 \ldots k_n)/k_1^n \leq K_0(f),$$

$$(\Delta/k_n^2)^{n/2} = (k_1 k_2 \ldots k_n)/k_n^n \geq (K(f))^{-1}.$$

From this,

$$\forall \xi \in \mathbf{R}^n [K_0(f)]^{2/n} |\xi|^2 \geq \langle \theta(x)\xi, \xi \rangle \geq [K(f)]^{-2/n} |\xi|^2. \qquad (5.14)$$

In particular, the eigenvalues of the matrix $\theta(x)$ lie in the interval

$$[K(f)^{-2/n}, K_0(f)^{2/n}].$$

THEOREM 5.1. *Suppose that U and G are open subsets of $\mathbf{R}^n$, and $f: U \to G$ is a mapping with bounded distortion. Assume that the function $v \in C^1(G)$ is a generalized solution in G of the equation*

$$\operatorname{div}[|v'(y)|^{n-2} v'(y)] = 0. \qquad (5.15)$$

Then the function $u = v \circ f$ belongs to $W^1_{n, \mathrm{loc}}(U)$ and is a generalized solution in U of the equation

$$\operatorname{div}[\langle \theta(x, f)u'(x), u'(x) \rangle^{n/2-1} \theta(x, f)u'(x)]. \qquad (5.16)$$

PROOF. The fact that $u = v \circ f$ belongs to the class $W^1_{p,\mathrm{loc}}(U)$ follows from Theorem 2.8 of Chapter I. This theorem gives us that for almost all $x \in U$

$$\frac{\partial u}{\partial x_i}(x) = \sum_{k=1}^n \frac{\partial v}{\partial y_k}[f(x)]\frac{\partial f_k}{\partial x_i}(x).$$

This means that the vectors $u'(x)$ and $v'[f(x)]$ are connected as follows:

$$u'(x) = [f'(x)]^* v'[f(x)]. \tag{5.17}$$

Denote by ξ the vector $\langle \theta(x)u'(x), u'(x)\rangle^{n/2-1}\theta(x)u'(x)$, and by ω the exterior form

$$\sum_{j=1}^n (-1)^{j-1}\xi_j\, dx_1 \ldots dx_{j-1}\, dx_{j+1} \ldots dx_n.$$

It is required to prove that the generalized differential form ω is equal to zero. Let $\sigma(y) = |v'(y)|^{n-2}v'(y)$, $\sigma(y) = (\sigma_1(y), \ldots, \sigma_n(y))$, where

$$\sigma_j(y) = |v'(y)|^{n-2}\frac{\partial v}{\partial y_j}(y).$$

The form

$$\tau(y) = \sum_{i=1}^n (-1)^{i-1}\sigma_i(y)\, dy_1 \ldots dy_{i-1}\, dy_{i+1} \ldots dy_n$$

is defined in G. The condition that v is a generalized solution of (5.15) means that $d\tau = 0$. The theorem will be proved if we establish that

$$\omega(x) = (f^*\tau)(x). \tag{5.18}$$

We have

$$(f^*\sigma)(x) = \sum_{i=1}^n (-1)^{i-1}\sigma_i[f(x)]\, df_1 \wedge \cdots \wedge df_{i-1} \wedge df_{i+1} \wedge \cdots \wedge df_n. \tag{5.19}$$

Suppose that the point $x \in U$ is such that the derivative $f'(x)$ is defined there and (5.12) holds. If $\mathscr{J}(x, f) = 0$, then $f'(x) = 0$. It follows from the preceding formula that $(f^*\sigma)(x) = 0$. By (5.17), $u'(x) = 0$ in this case, and hence $\omega(x) = 0$. Thus, (5.18) holds if $\mathscr{J}(x, f) = 0$. Next, assume that $\mathscr{J}(x, f) \neq 0$. Let F be the matrix with element F_{ij} equal to the cofactor of the element $(\partial f_i/\partial x_j)(x)$ of the matrix $f'(x)$. Then $F = \mathscr{J}(x, f)(f'^*)^{-1}$. Let $\lambda = (\lambda_1, \ldots, \lambda_n)$ be an arbitrary vector, and let μ be the exterior form $\lambda_1\, dx_1 + \cdots + \lambda_n\, dx_n$.

Then

$$\mu \wedge (f^*\sigma)(x) = \sum_{i=1}^{n}(-1)^{i-1}\sigma_i[f(x)]\mu \wedge df_1 \wedge \cdots \wedge df_{i-1} \wedge df_{i+1} \wedge \cdots \wedge df_n$$

$$= \sum_{i=1}^{n} \sigma[f(x)]\, df_1 \wedge \cdots \wedge df_{i-1} \wedge \mu \wedge df_{i+1} \wedge \cdots \wedge df_n.$$

(5.20)

We introduce the notation $f_{i,j} = \frac{\partial f_i}{\partial x_j}$. Then

$$df_1 \wedge \cdots \wedge df_{i-1} \wedge \mu \wedge df_{i+1} \wedge \cdots \wedge df_n$$

$$= \begin{vmatrix} f_{1,1}, f_{1,2}, \ldots, f_{1,n} \\ \cdots\cdots\cdots\cdots\cdots\cdots \\ f_{i-1,1}, f_{i-1,2}, \ldots, f_{i-1,n} \\ \lambda_1, \lambda_2, \ldots, \lambda_n \\ f_{i+1,1}, f_{i+1,2}, \ldots, f_{i+1,n} \\ \cdots\cdots\cdots\cdots\cdots\cdots \\ f_{n,1}, f_{n,2}, \ldots, f_{n,n} \end{vmatrix} dx_1\, dx_2 \ldots dx_n.$$

Expanding the determinant here with respect to the minors of the ith row and substituting the result into (5.20), we find that

$$\mu \wedge (f^*\sigma)(x) = \sum_{i=1}^{n}\sum_{j=1}^{n} F_{ij}(x)\sigma_i[f(x)]\lambda_j\, dx_1\, dx_2 \ldots dx_n$$

(5.21)

$$= \langle F^*(x)\sigma[f(x)], \lambda \rangle\, dx_1\, dx_2 \ldots dx_n.$$

Let

$$(f^*\sigma)(x) = \sum_{i=1}^{n}(-1)^{i-1}\alpha_i(x)\, dx_1 \ldots dx_{i-1}\, dx_{i+1} \ldots dx_n$$

be the canonical representation of the form $(f^*\sigma)(x)$, and let $\beta(x) = (\alpha_1(x), \ldots, \alpha_n(x))$. Then

$$\mu \wedge (f^*\sigma)(x) = \langle \beta(x), \lambda \rangle\, dx_1\, dx_2 \ldots dx_n.$$ (5.22)

Comparing (5.22) and (5.21), we get that

$$\beta(x) = F^*(x)\sigma[f(x)] = \mathscr{J}(x, f)(f')^{-1}\sigma[f(x)].$$

We have

$$\sigma(y) = |v'(y)|^{n-2}v'(y), \qquad u'(x) = (f')^{-1}v'[f(x)],$$

which implies that

$$\beta(x) = \langle f'(f'^*)^{-1}u'(x), u'(x)\rangle^{(n-2)/2} \mathscr{J}(x, f)(f')^{-1}(f'^*)^{-1}u'(x)$$

$$= \langle \theta(x)u'(x), u'(x)\rangle^{n/2-1}\theta(x)u'(x).$$

This establishes (5.18), and the theorem is proved.

§**5.3. The classes** $W_p^1(U/A)$ **and** $\overset{+}{W}{}_p^1(U/A)$. We define what it means for two functions u and v of class $W_p^1(U)$ to coincide on some closed set $A \subset \overline{U}$, and also what it means to say that $u \geq v$ on A. Throughout this subsection U denotes a bounded open set, $\overline{U}$ (as usual) is the closure of U, and ∂U is the boundary of U.

Let $A \subset \overline{U}$ be a given nonempty closed set, and let u be a function of class $W_p^1(U)$. We say that u belongs to the class $\tilde{W}_p^1(U/A)$ if there exists an open set $V \supset A$ such that $u(x) = 0$ almost everywhere on $U \cap V$. The closure of $\tilde{W}_p^1(U/A)$ in the space $W_p^1(U)$ is denoted by $W_p^1(U/A)$. Note that a linear combination of two functions in $\tilde{W}_p^1(U/A)$ is a function in $\tilde{W}_p^1(U/A)$. This implies that $W_p^1(U/A)$ is a subspace of the vector space $W_p^1(U)$.

Let u and v be two functions in $W_p^1(U)$, and let $A \subset \overline{U}$ be a closed set. We say that $u = v$ on A and write $u = v|A$ if the difference $u - v$ belongs to $W_p^1(U/A)$. We mention an important particular case. Let $A = \partial U$. In this case we set $W_p^1(U/A) = W_p^1(U/\partial U)$ is denoted by $\overset{\circ}{W}{}_p^1(U)$.

We remark that if a function u is in $\tilde{W}_p^1(U/\partial U)$, then there exists a sequence (u_m), $m = 1, 2, \ldots$, of functions in $C_0^\infty(U)$ convergent to u in $W_p^1(U)$. Indeed, let V be an open set containing ∂U and such that $u(x) = 0$ almost everywhere on $U \cap V$, and let $G = U \backslash V$. The set G is compact. Let $\delta = \mathrm{dist}\{G, \mathbf{R}^n \backslash U\} > 0$. The *averaging* u_h of u is a function of class $C^\infty(\hat{U}_h)$ with compact support in $\hat{U}_h$ for sufficiently small h. Let us extend u_h to the whole of U by setting $u_h(x) = 0$ for $x \notin \hat{U}_h$. Let (h_m), $m = 1, 2, \ldots$, be a sequence of values of $h > 0$ such that $h_m \to 0$ as $m \to \infty$. Then the sequence of functions $u_m = u_{h_m}$ is obviously the desired sequence.

We thus get that the set $C_0^\infty(U)$ is dense in $\tilde{W}_p^1(U/\partial U)$, hence also in $W_p^1(U/\partial U) = \overset{\circ}{W}{}_p^1(U)$. This lets us conclude that u is in $\overset{\circ}{W}{}_p^1(U)$ if and only if there exists a sequence (u_m) of functions in $C_0^\infty(U)$ such that $\|u_m - u\|_{1,p,U} \to 0$ as $m \to \infty$. Let U be a bounded open set in $\mathbf{R}^n$, and u a function of class $\overset{\circ}{W}{}_p^1(U)$. We extend u to $\mathbf{R}^n$ by setting $u_0(x) = 0$ for $x \notin U$, and $u_0(x) = u(x)$ for $x \in U$. The function u_0 will be called the *zero extension* of u. We prove that u_0 is a function in $W_p^1(\mathbf{R}^n)$. Indeed, if $u \in C_0^\infty(U)$, then by setting $u(x) = 0$ for $x \notin U$ we obviously obtain a function in $C_0^\infty(\mathbf{R}^n)$ whose norm in $W_p^1(\mathbf{R}^n)$ coincides with that of u in $W_p^1(U)$. This implies that if a sequence of functions u_ν in $C_0^\infty(U)$ converges in $W_p^1(U)$ to a function u, then a sequence of functions converging in $W_p^1(\mathbf{R}^n)$ is obtained by extending the functions u_ν to $\mathbf{R}^n$ by setting $u_\nu(x) = 0$ for

$x \notin U$. Its limit is some function $\tilde{u}_0$. Clearly, $\tilde{u}_0(x) = 0$ for $x \notin U$, and $\tilde{u}_0(x) = u_0(x)$ for $x \in U$, i.e., $\tilde{u}_0$ is the desired extension of u.

Let u be a function in $C_0^\infty(\mathbf{R}^n)$ whose support is contained in a ball of radius $r > 0$. Then, as shown above,

$$\|u\|_{p,\mathbf{R}^n} \leq 2r\|u'(x)\|_{p\mathbf{R}^n}.$$

This implies that if the set $U \subset \mathbf{R}^n$ is bounded, with U contained in a ball of radius $r > 0$, then for every $u \in \overset{\circ}{W}{}^1_p(U)$

$$\|u\|_{p,U} \leq 2r\|u'(x)\|_{p,U}, \tag{5.18}$$

which implies that for $u \in \overset{\circ}{W}{}^1_p(U)$

$$\|u\|_{1,p,U} \leq C\|u'\|_{p,U} \tag{5.19}$$

where the constant C depends only on the radius of the smallest ball containing U.

LEMMA 5.1. *Let U be an open subset of $\mathbf{R}^n$. Then every sequence (u_ν), $\nu = 1, 2, \ldots$, of functions in $W^1_{p,\mathrm{loc}}(U)$ which is locally bounded in W^1_p, where $p \leq n$, has a subsequence which converges locally in L_r, where $r < np/(n-p)$, to some function $u_0 \in W^1_{p,\mathrm{loc}}(U)$.*

PROOF. Let (B_k), $k = 1, 2, \ldots$, be a sequence of balls such that the closure of B_k is contained in U for each k and $\bigcup_1^\infty B_k = U$. We use induction to construct subsequences $(u_{k,\nu})$, $\nu = 1, 2, \ldots$, of the original sequence (u_ν), $\nu = 1, 2, \ldots$. Let $(u_{0,\nu})$, $\nu = 1, 2, \ldots$, be the original sequence (u_ν). Assume that the sequence $(u_{k,\nu})$, $\nu = 1, 2, \ldots$, has been defined. Then the norm sequence $(\|u_{k,\nu}\|_{1,p,B_{k+1}})$, $\nu = 1, 2, \ldots$, is bounded, by a condition of the lemma. Hence, by the Sobolev-Kondrashov theorem on compactness of an imbedding, $(u_{k,\nu})$ has a subsequence convergent in $L_r(B_{k+1})$. Let $(u_{k+1,\nu})$ be this subsequence. The diagonal sequence $u_{1,1}, \ldots, u_{\nu,\nu}, \ldots$ is a subsequence of (u_ν), $\nu = 1, 2, \ldots$, converging in L_r on each of the balls B_k, and thus on any compact subset of U, because every compact set $A \subset U$ can be covered by finitely many balls in the sequence B_k, $k = 1, 2, \ldots$. The lemma is proved.

The definition of what it means for a function $u \in W^1_p$ to be nonnegative on the boundary of U in the case when this boundary is a fairly "bad" set is based on Lemmas 5.2 and 5.3.

LEMMA 5.2. *Suppose that u is a function of class $W^1_{p,\mathrm{loc}}(U)$, where $p \geq 1$, and $\varphi: \mathbf{R} \to \mathbf{R}$ is a C^1-function with bounded derivative φ'. Then the* *function $v(x) = \varphi[u(x)]$ is also in $W^1_{p,\mathrm{loc}}(U)$. Further, the derivatives of v can be expressed in terms of those of u by the formula*

$$\frac{\partial v}{\partial x_i}(x) = \varphi'[u(x)]\frac{\partial u}{\partial x_i}(x). \tag{5.20}$$

PROOF. Suppose that $u \in W^1_{p,\mathrm{loc}}(U)$. We construct a sequence (u_ν), $\nu = 1, 2, \ldots$, of functions of class $C^\infty(U)$ that converges to u in $W^1_{p,\mathrm{loc}}(U)$ as $\nu \to \infty$. Let V be an arbitrary open set lying strictly inside U. Then

$$\|u_\nu - u\|_{p,V} \to 0, \qquad \left\|\frac{\partial u_\nu}{\partial x_i} - \frac{\partial u}{\partial x_i}\right\|_{p,V} \to 0$$

as $\nu \to \infty$.

Define $v_\nu(x) = \varphi[u_\nu(x)]$. Let $K < \infty$ be such that $|\varphi'(u)| \leq K$ for all $u \in \mathbf{R}$. Then

$$|\varphi(u_1) - \varphi(u_2)| \leq K|u_1 - u_2|$$

for any $u_1, u_2 \in \mathbf{R}$. This implies that

$$|v_\nu(x) - v(x)| \leq K|u_\nu(x) - u(x)|,$$

and hence $\|v_\nu \to v\|_{p,V} \to 0$ as $\nu \to \infty$. Each of the functions v_ν belongs to the class C^1, and for every ν

$$\frac{\partial v_\nu}{\partial x_i}(x) = \varphi'[u_\nu(x)]\frac{\partial u_\nu}{\partial x_i}(x), \qquad i = 1, 2, \ldots, n.$$

This leads to the conclusion that

$$\left|\frac{\partial v_\nu}{\partial x_i}(x) - \varphi'[u(x)]\frac{\partial u}{\partial x_i}(x)\right| \leq |\varphi'[u_\nu(x)]|\left|\frac{\partial u_\nu}{\partial x_i}(x) - \frac{\partial u}{\partial x_i}(x)\right|$$
$$+ |[\varphi'[u_\nu(x)] - \varphi'[u(x)]|\left|\frac{\partial u}{\partial x_i}(x)\right|.$$

As a result we get the estimate

$$\left\|\frac{\partial v_\nu}{\partial x_i} - \varphi'(u)\frac{\partial u}{\partial x_i}\right\|_{p,V} \leq K\left\|\frac{\partial u_\nu}{\partial x_i} - \frac{\partial u}{\partial x_i}\right\|_{p,V}$$
$$+ \left(\int_V |\varphi'[u_\nu(x)] - \varphi'[u(x)]|^p\left|\frac{\partial u}{\partial x_i}(x)\right|^p dx\right)^{1/p}. \tag{5.21}$$

The first term tends to zero as $\nu \to \infty$. The functions u_ν converge to u in measure on V as $\nu \to \infty$. It follows from this that also $\varphi' \circ u_\nu \to \varphi' \circ u$ in measure. We have that $|\varphi'[u_\nu(x)] - \varphi'[u(x)]| \leq 2K$ for all $x \in U$. This enables us to conclude that the second term in (5.21) also tends to zero as $\nu \to \infty$.

The set V lying strictly inside U was chosen arbitrarily; hence we have proved that the functions $\partial v_\nu/\partial x_i$ converge in $L_{p,\mathrm{loc}}$ to the function

$\varphi'(u)(\partial u/\partial x_i)$ as $\nu \to \infty$. This allows us to deduce that the function $(\varphi' \circ u)\partial u/\partial x_i$ is the generalized derivative $\partial v/\partial x_i$ of v. It is thereby established that the function $v = \varphi \circ u$ belongs to $W^1_{p,\text{loc}}(U)$. Further,

$$\frac{\partial v}{\partial x_i}(x) = \varphi'[u(x)]\frac{\partial u}{\partial x_i}(x)$$

for all $x \in U$, and the lemma is proved.

LEMMA 5.3. *Suppose that u is a function of class $W^1_{p,\text{loc}}(U)$, $p \geq 1$. Let*

$$u^+(x) = \max\{u(x), 0\} = \tfrac{1}{2}\{|u(x)| + u(x)\}.$$

Then u^+ belongs to $W^1_{p,\text{loc}}(U)$, and its partial derivatives can be expressed for almost all $x \in U$ in terms of the derivatives of u according to the formula

$$\frac{\partial u^+}{\partial x_i}(x) = \tau[u(x)]\frac{\partial u}{\partial x_i}(x),$$

where $\tau(u) = 1$ for $u > 0$, and $\tau(u) = 0$ for $u \leq 0$.

PROOF. Let θ be a nonnegative C^∞-function on $\mathbf{R}$ such that $\theta(x) = 0$ for $x \notin [0, 1]$ and $\int_{-\infty}^\infty \theta(x)\,dx = 1$. Let $h > 0$, and define

$$\tau_h(u) = \frac{1}{h}\int_{-\infty}^u \theta\left(\frac{t}{h}\right)dt, \qquad \sigma_h(u) = \int_{-\infty}^u \tau_h(t)\,dt.$$

Obviously, $0 \leq \tau_h(u) \leq 1$ for all u, $\tau_h(u) = 0$ for $u \leq 0$, $\tau_h(u) = 1$ for $u \geq h$, and the function σ_h is increasing. We have that $\tau_h(u) \to \tau(u)$ as $h \to 0$ for any u, where τ is the function indicated in the statement of the lemma: $\tau(u) = 0$ for $u \leq 0$, and $\tau(u) = 1$ for $u > 0$. Note, further, that $\sigma_h(u) = 0$ for $u \leq 0$, and $0 \leq \sigma_h(u) \leq u$ for $u > 0$; $\sigma_h(u) \to u^+$ as $h \to 0$. Then $u^+ - h \leq \sigma_h(u) \leq u^+$ and $\sigma_h(u) \geq 0$ for all u.

By Lemma 5.2, the functions $\sigma_h[u(x)]$ belongs to $W^1_{p,\text{loc}}(U)$. Further,

$$\frac{\partial(\sigma_h \circ u)}{\partial x_i}(x) = \tau_h[u(x)]\frac{\partial u}{\partial x_i}(x)$$

for all x. We have that $\sigma_h(u) \to u^+$ and $\tau_h(u) \to \tau(u)$ as $h \to 0$ for all u. This implies that the function $\sigma_h \circ u$ converge in $L_{p,\text{loc}}(U)$ to u^+, and their derivatives $\partial(\sigma_h \circ u)/\partial x_i$ converge in $L_{p,\text{loc}}$ to $(\tau \circ u)\partial u/\partial x_i$. This gives us that $u^+ \in W^1_{p,\text{loc}}(U)$, and

$$\frac{\partial u^+}{\partial x_i}(x) = \tau[u(x)]\frac{\partial u}{\partial x_i}(x).$$

The lemma is proved.

COROLLARY 1. *If* $u \in W_p^1(U)$, *then* $u^- = \max\{-u, 0\}$ *is in* $W_p^1(U)$. *For any two functions* $u, v \in W_p^1(U)$ *the functions*

$$\max\{u, v\} = \tfrac{1}{2}(u + v + |u - v|),$$
$$\min\{u, v\} = \tfrac{1}{2}(u + v - |u - v|)$$

belong to $W_p^1(U)$.

COROLLARY 2. *Suppose that* $u \in W_p^1(U)$, *and let*

$$E_k = \{x \in U | u(x) = k\},$$

where $k \in \mathbf{R}$. *Then the partial derivatives* $\partial u / \partial x_i$ *of* $u(x)$ *vanish almost everywhere on* E_k.

PROOF. Let

$$v(x) = \max\{u(x), k\} = u(x) + [k - u(x)]^+ = k + [u(x) - k]^+.$$

Then for almost all $x \in U$ the partial derivatives $(\partial v / \partial x_i)(x)$ are defined. Using the formula $v(x) = u(x) + [k - u(x)]^+$, we have on the basis of the lemma that

$$\frac{\partial v}{\partial x_i}(x) = \frac{\partial u}{\partial x_i}(x) - \tau[k - u(x)]\frac{\partial u}{\partial x_i}(x).$$

For $x \in E_k$ the quantity $k - u(x)$ is 0, and hence $\tau[k - u(x)] = 0$. This implies that

$$\frac{\partial v}{\partial x_i}(x) = \frac{\partial u}{\partial x_i}(x)$$

for almost all $x \in E$. Using the formula $v(x) = k + [u(x) - k]^+$, we get that for almost all $x \in U$

$$\frac{\partial v}{\partial x_i}(x) = \tau[u(x) - k]\frac{\partial u}{\partial x_i}(x).$$

The function $\tau[u(x) - k]$ is zero on E_k, and hence for almost all $x \in E_k$

$$\frac{\partial v}{\partial x_i}(x) = 0, \quad \text{i.e.,} \quad \frac{\partial u}{\partial x_i}(x) = \frac{\partial v}{\partial x_i}(x) = 0,$$

almost everywhere on E, which is what was required to prove.

LEMMA 5.4. *Let* (u_m), $m = 1, 2, \ldots$, *be a sequence of functions of class* $W_p^1(U)$ *that converges in* $W_p^1(U)$ *to a function* $u_0 \in W_p^1(U)$. *Then* $u_m^+ \to u^+$ *in* $W_p^1(U)$ *as* $m \to \infty$.

PROOF. By the condition of the lemma,

$$\int_U |u_m - u_0|^p \, dx \to 0, \qquad \int_U \left| \frac{\partial u_m}{\partial x_i} - \frac{\partial u}{\partial x_i} \right|^p \, dx \to 0$$

as $m \to \infty$ for all $i = 1, \ldots, m$. Since

$$|u_m^+(x) - u_0^+(x)| \le |u_m(x) - u_0(x)|$$

for any x, this implies, first of all, that

$$\int_U |u_m^+ - u_0^+|^p \, dx \to 0 \quad \text{as } m \to \infty.$$

Further, for each m

$$\left| \frac{\partial u_m^+}{\partial x_i} - \frac{\partial u_0^+}{\partial x_i} \right| \le \tau(u_m) \left| \frac{\partial u_m}{\partial x_i} - \frac{\partial u_0}{\partial x_i} \right|$$
$$+ |\tau(u_m) - \tau(u_0)| \left| \frac{\partial u_0}{\partial x_i} \right|.$$

From this we conclude that

$$\left\| \frac{\partial u_m^+}{\partial x_i} - \frac{\partial u_0^+}{\partial x_i} \right\|_{p,U} \le \left\| \frac{\partial u_m}{\partial x_i} - \frac{\partial u_0}{\partial x_i} \right\|_{p,U} \tag{5.22}$$
$$+ \left(\int_U |\tau[u_m(x)] - \tau[u_0(x)]|^p \left| \frac{\partial u_0}{\partial x_i}(x) \right|^p dx \right)^{1/p}.$$

To prove the lemma it suffices to establish that the integral on the right-hand side of the last inequality tends to zero as $m \to \infty$. The sequence (u_m) has a subsequence for which this integral tends to its limit superior and then we choose a subsequence (u_{m_k}) of this subsequence such that $u_{m_k}(x) \to u_0(x)$ almost everywhere. Let E_0 be the set of all $x \in U$ such that $u_0(x) = 0$. Then, by Corollary 2 to Lemma 5.3, $\partial u_0(x)/\partial x_i = 0$ almost everywhere on E_0, and thus

$$|\tau[u_m(x)] - \tau[u_0(x)]| \frac{\partial u_0}{\partial x_i}(x) = 0$$

almost everywhere on E_0. Suppose that $x \notin E_0$ and that $u_{m_k}(x) \to u_0(x)$ for this x. Since $x \notin E_0$, it follows that $u_0(x) \ne 0$, and hence $\tau[u_{m_k}(x)] \to \tau[u_0(x)]$ as $k \to \infty$. We get that $|\tau \circ u_{m_k} - \tau \circ u_0|(\partial u/\partial x_i)$ tends to zero almost everywhere in U as $k \to \infty$, which, by Lebesgue's theorem on passing to the limit, implies that

$$\int_U \left(|\tau \circ u_{m_k} - \tau \circ u_0| \left| \frac{\partial u}{\partial x_i} \right| \right)^p dx \to 0$$

as $k \to \infty$. It follows from what has been proved that the integral on the right-hand side of (5.22) tends to zero as $m \to \infty$. The lemma is proved.

COROLLARY 1. *Let* (u_m), $m = 1, 2, \ldots$, *be a sequence of functions of class* $W^1_{p,\text{loc}}(U)$ *that converges in* $W^1_{p,\text{loc}}(U)$ *to the function* u_0. *Then* $u^+_m \to u^+_0$ *in* $W^1_{p,\text{loc}}(U)$ *as* $m \to \infty$.

COROLLARY 2. *Let* (u_m), $m = 1, 2, \ldots$, *be a sequence of functions of class* $W^1_p(U)$ *which converges in* $W^1_p(U)$ *to a function* $u_0 \in W^1_p(U)$ *as* $m \to \infty$. *Then* $u^-_m \to u^-_0$ *and* $|u_m| \to |u_0|$ *in* $W^1_p(U)$ *as* $m \to \infty$. *Further, if* (u_m) *and* (v_m) *are two sequences of functions of class* $W^1_p(U)$ *which converge in* $W^1_p(U)$ *to the respective functions* u *and* v *in* $W^1_p(U)$, *then*

$$\max\{u_m, v_m\} \to \max\{u, v\},$$
$$\min\{u_m, v_m\} \to \min\{u, v\}$$

in $W^1_p(U)$ *as* $m \to \infty$.

COROLLARY 3. *Let* (u_m), $m = 1, 2, \ldots$, *be a sequence of functions of class* $W^1_{p,\text{loc}}(U)$ *which converges in* $W^1_{p,\text{loc}}(U)$ *to a function* $u_0 \in W^1_{p,\text{loc}}(U)$ *as* $m \to \infty$. *Then* $u^+_m \to u^+_0$, $u^-_m \to u^-_0$ *and* $|u_m| \to |u_0|$ *in* $W^1_{p,\text{loc}}(U)$ *as* $m \to \infty$. *If* u_m *and* v_m *are two sequences of functions converging in* $W^1_{p,\text{loc}}(U)$ *to the respective functions* u *and* v *of class* $W^1_{p,\text{loc}}(U)$, *then* $\max\{u_m, v_m\} \to \max\{u, v\}$ *and* $\min\{u_m, v_m\} \to \min\{u, v\}$ *in* $W^1_{p,\text{loc}}(U)$ *as* $m \to \infty$.

LEMMA 5.5. *Suppose that the functions* $u, v \in W^1_p(U)$ *are such that* $0 \le u(x) \le v(x)$ *for almost all* $x \in U$. *If* $v \in W^1_p(U/A)$, *where* $A \subset \overline{U}$ *and* A *is closed, then* u *is also in* $W^1_p(U/A)$.

PROOF. Let (v_m), $m = 1, 2, \ldots$, be a sequence of functions of class $\tilde{W}^1_p(U/A)$ which converges to v in $W^1_p(U)$. Let $u_m(x) = \min\{u(x), v_m(x)\}$. For each m the function u_m is in $W^1_p(U)$ in view of Corollary 1 to Lemma 5.3. For every m there is an open set $V_m \supset A$ such that $v_m(x) = 0$ almost everywhere on $V_m \cap U$. This implies that $u_m(x) = 0$ almost everywhere on $V_m \cap U$, i.e., $u_m \in \tilde{W}^1_p(U/A)$ for each m. We have that $u_m \to \min\{u, v\} = u$ in $W^1_p(U)$ as $m \to \infty$, which gives us that $u \in W^1_p(U/A)$, and the lemma is proved.

Let u be an arbitrary function of class $W^1_p(U)$. Then we say that $u \in \overset{+}{W}{}^1_p(U/A)$, where $A \subset \overline{U}$ is closed, if its negative part u^- vanishes on A, i.e., $u^- \in W^1_p(U/A)$. In this case we also say that u is *nonnegative on* A and write $u \ge 0|A$. If u and v are functions in $W^1_p(U)$, then we say that $u \ge v|A$ if $u - v \ge 0|A$. In other words, the relation $u \ge v|A$ means that $(u - v)^- \in W^1_p(U/A)$. The set $\overset{+}{W}{}^1_p(U/\partial U)$ will simply be denoted by $\overset{+}{W}{}^1_p(U)$ below. The definition implies that, in particular, if $u \in W^1(U)$

and there exists an open set $V \supset A$ such that $u^-(x) = 0$ almost everywhere on $V \cap U$, then $u \in \overset{+}{W}{}^1_p(U)$.

LEMMA 5.6. *For every $A \subset \overline{U}$ the set $\overset{+}{W}{}^1_p(U/A)$ is a closed convex cone in the Banach space $W^1_p(U)$.*

PROOF. We are required to prove the following three facts:

a) $\overset{+}{W}{}^1_p(U/A)$ is a closed subset of $W^1_p(U)$.

b) $\lambda u \in \overset{+}{W}{}^1_p(U/A)$ for any $u \in \overset{+}{W}{}^1_p(U/A)$ and any $\lambda \geq 0$, $\lambda \in \mathbf{R}$.

c) The sum of any two functions u and v in $\overset{+}{W}{}^1_p(U/A)$ is in $\overset{+}{W}{}^1_p(U/A)$.

We first prove a). By Corollary 2 to Lemma 5.3, the mapping $\theta: u \to u^-$ of the space $W^1_p(U)$ into itself is continuous. The set $\overset{+}{W}{}^1_p(U/A)$ is the complete inverse image of $W^1_p(U/A)$ with respect to θ, and since $W^1_p(U/A)$ is closed, $\overset{+}{W}{}^1_p(U/A)$ is also closed in $W^1_p(U)$. Let $u \in \overset{+}{W}{}^1_p(U/A)$. For every $\lambda \geq 0$ the relation $(\lambda u)^- = \lambda u^-$ holds; hence if $u^- \in W^1_p(U/A)$, then also $(\lambda u)^- \in W^1_p(U/A)$, i.e., if $u \in \overset{+}{W}{}^1_p(U/A)$, then also $\lambda u \in \overset{+}{W}{}^1_p(U/A)$.

We now prove c). Suppose that $u, v \in \overset{+}{W}{}^1_p(U/A)$, and let $w = u + v$. Since $u = u^+ - u^-$ and $v = v^+ - v^-$, it follows that $w = (u^+ + v^+) - (u^- + v^-)$. From this, $0 \leq w^-(x) \leq u^-(x) + v^-(x)$. The function $u^- + v^-$ obviously belongs to $W^1_p(U/A)$. By Lemma 5.5, this implies that $w^- \in W^1_p(U/A)$, and thus $w \in \overset{+}{W}{}^1_p(U/A)$, which is what was required to prove.

COROLLARY. *Let $A \subset \overline{U}$, and let u, v, and w be functions in $W^1_p(U)$. Then the following statements hold:*

1. *If $u \geq v|A$ and $v \geq u|A$, then $u = v|A$.*
2. *If $u \geq v|A$ and $v \geq w|A$, then $u \geq w|A$.*
3. *If $u \geq v|A$, then $\lambda u \geq \lambda v|A$ for all $\lambda \geq 0$.*
4. *If $u \geq v|A$, then for every function $w \in W^1_p(U)$*

$$u + w \geq v + w|A.$$

The proof is left to the reader.

LEMMA 5.7. *Let U be a bounded open set in $\mathbf{R}^n$, and let $A \subset \overline{U}$ and $B \subset \overline{U}$ be closed sets. If $u \in W^1_p(U/A)$ and $u \in W^1_p(U/B)$, then $u \in W^1_p(U/(A \cup B))$.*

PROOF. Let (v_m) and (w_m) be sequences of functions in $W^1_p(U)$ convergent to u in $W^1_p(U)$ and such that each v_m vanishes in a neighborhood of A, while each w_m vanishes in a neighborhood of B. Let

$$\xi_m = \min\{v_m^+, w_m^+\}, \quad \eta_m = \min\{v_m^-, w_m^-\}.$$

Then clearly each of the functions ξ_m and η_m vanishes in a neighborhood of $A \cup B$. We have that $\xi_m \to u^+$ and $\eta_m \to u^-$ in $W_p^1(U)$ as $m \to \infty$. This implies that u^+ and u^- are in $W_p^1(U/(A \cup B))$, and hence $u = u^+ - u^- \in W_p^1(U/(A \cup B))$, which is what was required to prove.

Lemma 5.3 can be extended to the case of functions of the form $F \circ u$, where F is an arbitrary piecewise smooth function. A function $F: \mathbf{R} \to \mathbf{R}$ is said to be *piecewise smooth* if F is continuous and there exists a finite set E such that the derivative $F'(u)$ is defined and continuous for every point $u \notin E$, and the limits $F'(u - 0)$ and $F'(u + 0)$ exists at each point $u \in E$ and are finite. Obviously, if $u \in E$, then the derivatives

$$F_L'(u) = \lim_{h \to -0} \frac{F(u + h) - F(u)}{h} = F'(u - 0),$$

$$F_R'(u) = \lim_{h \to +0} \frac{F(u + h) - F(u)}{h} = F'(u + 0)$$

exist at u.

LEMMA 5.8. *Let $F: \mathbf{R} \to \mathbf{R}$ be a piecewise smooth function. Assume that there exists a constant $k < \infty$ such that $|F'(u)| \le k$ at each point u where the derivative $F'(u)$ is defined. Then for every function $u \in W_{p,\mathrm{loc}}^1(U)$ (where U is an open set in $\mathbf{R}$) the function $v = F \circ u$ belongs to $W_{p,\mathrm{loc}}^1(U)$, and for almost all $x \in U$*

$$\frac{\partial v}{\partial x_i}(x) = F_L'[u(x)] \frac{\partial u}{\partial x_i}(x). \tag{5.23}$$

REMARK. Equality (5.23) remains true if the left-hand derivative of F in it is replaced by the right-hand derivative.

PROOF. Let $F_\nu = \alpha_{h_\nu} * F$, where α is an arbitrary averaging kernel on $\mathbf{R}$, and $h_\nu \to 0$ as $\nu \to \infty$. Then the sequence of functions F_ν converges to F uniformly in $\mathbf{R}$ as $\nu \to \infty$, $|F_\nu'(u)| \le k$ for all ν, and $F_\nu'(u) \to F'(u)$ at each point $u \notin E$. Let $v_\nu = F_\nu \circ u$. By Lemma 5.3, the functions v_ν are all in $W_{p,\mathrm{loc}}^1(U)$. Further,

$$\frac{\partial v_\nu}{\partial x_i}(x) = F_\nu'[u(x)] \frac{\partial u}{\partial x_i}(x)$$

almost everywhere in U. We have that $v_\nu \to v$ uniformly as $\nu \to \infty$ and the sequence v_ν is clearly locally bounded in $W_{p,\mathrm{loc}}^1(U)$. On the basis of Theorem 1.1 in Chapter I, this implies that $v \in W_{p,\mathrm{loc}}^1(U)$. Let $E = \{u_1, \ldots, u_m\}$ be the finite set of points where F is not differentiable, and let $B_i = \{x \in U | u(x) = u_i\}$ and $B = \bigcup_1^m B_i$. If $x \notin B$, and $u(x)$ and $(\partial u/\partial x_i)(x)$ are defined, then

$$\frac{\partial v_\nu}{\partial x_i}(x) \to F'[u(x)]\frac{\partial u}{\partial x_i}(x)$$

as $\nu \to \infty$ for this x. If $|B| = 0$, then this proves that the right-hand side of (5.23) is the limit of $\frac{\partial v_\nu}{\partial x_i}(x)$ as $\nu \to \infty$. Assume that $|B| > 0$. Then, by Corollary 2 to Lemma 5.4, $\frac{\partial u}{\partial x_i}(x) = 0$ almost everywhere on B. Hence,

$$\frac{\partial v_\nu}{\partial x_i}(x) \to 0 = F'_L[u(x)]\frac{\partial u}{\partial x_i}(x)$$

for all $x \in B$. Thus, the derivatives $\partial v_\nu/\partial x_i$ converge almost everywhere in U to the function

$$x \to F'_L[u(x)]\frac{\partial u}{\partial x_i}(x)$$

as $\nu \to \infty$. This establishes that

$$\frac{\partial v}{\partial x_i}(x) = F'_L[u(x)]\frac{\partial u}{\partial x_i}(x)$$

almost everywhere in U, and the lemma is proved.

§5.4. The Dirichlet problem, extremal functions, and generalized solutions of the Euler equation for functionals of the calculus of variations. Let $U \subset \mathbf{R}^n$ be an arbitrary open set, and let $F(x,q)$ ($x \in U, q \in \mathbf{R}^n$) be a function satisfying all or some of the conditions A–E (see §5.1).

THEOREM 5.2 (Semicontinuity of the functional I_F). *Assume that F satisfies conditions* A, B, *and* C. *Let* (u_ν), $\nu = 1, 2, \ldots$, *be a sequence of functions of class $W_p^1(U)$ converging in $L_{1,\mathrm{loc}}(U)$ to a function $u_0 \in W_p^1(U)$ as $\nu \to \infty$. Then for every measurable set $A \subset U$*

$$I_F(u_0, A) \le \varliminf_{\nu \to \infty} I_F(u_\nu, A).$$

Proofs of this theorem and of certain more general results will be given in Chapter III, §3.3.

Using Theorem 5.2 and some results in §5.3, we can now prove a theorem on the existence of a solution of the Dirichlet problem. For $v \in W_p^1(U)$ denote by $W_p(v, U)$ the collection of all functions $f \in W_p^1(U)$ such that $f = v | \partial U$. Formally, $W_p(v, U) = v + \overset{\circ}{W}{}_p^1(U)$.

THEOREM 5.3. *Suppose that F satisfies conditions* A, B, *and* C *in* §5.1. *Then for every function $v \in W_p^1(U)$ there exists a function $v_0 \in W_p(v, U)$*

at which the functional $u \to I_F(u, U)$ *takes its smallest value on the set* $W_p(v, U)$, *i.e.,* $I_F(v_0, U) \le I_F(u, U)$ *for any function* $u \in W_p(v, U)$.

PROOF. Suppose that F and U satisfy all the conditions of the theorem. Let

$$\gamma = \inf_{u \in W_p(v,U)} I_F(u, U).$$

Since $I_F(u, U) > 0$ for all $u \in W_p(v, U)$ and $I_F(v, U) < \infty$, it obviously follows that $\gamma \ge 0$ and $\gamma < \infty$. Let $u_\nu \in W_p(v, U)$ be such that $I_F(u_\nu, U) < \gamma + (1/\nu)$, and let $w_\nu(x) = u_\nu(x) - v(x)$. For each ν the function w_ν is in the class $\overset{\circ}{W}{}^1_p(U)$. By condition C in §5.1,

$$\int_U |u'_\nu(x)|^p \, dx \le (1/a_1) \int_U F[x, u'_\nu(x)] \, dx$$
$$= (1/a_1) I_F(u_\nu, U) < (1/a_1)(\gamma + (1/\nu)) \le (1/a_1)(\gamma + 1).$$

From this,

$$\|w'_\nu\|_{p,U} \le \|v'\|_{p,U} + [(\gamma + 1)/a_1]^{1/p} = M < \infty, \tag{5.24}$$

where M is a constant. Since w_ν is in $\overset{\circ}{W}{}^1_p(U)$ and U is bounded, it follows from (5.24), in view of (5.19), that the sequence (w_ν) is bounded in $W^1_p(U)$. This allows us to conclude that (w_ν) has a subsequence that converges weakly in $W^1_p(U)$ to some function w_0. In order not to complicate the notation, we assume that (w_ν) is this subsequence. Since $\overset{\circ}{W}{}^1_p(U)$ is convex and closed in $W^1_p(U)$, it is also weakly closed there, which implies that w_0 is in $\overset{\circ}{W}{}^1_p(U)$. Let $u_0 = w_0 + v$. The function $u_\nu = w_\nu + v$ converges weakly in $W^1_p(U)$ to u_0 as $\nu \to \infty$, and hence $u_\nu \to u_0$ in $L_{p,\mathrm{loc}}(U)$ in any case. By Theorem 5.2, this gives us that

$$I_F(u_0, U) \le \varliminf_{\nu \to \infty} I_F(u_\nu, F) \le \gamma.$$

On the other hand, $I_F(u_0, U) \ge \gamma$ by the definition of γ. Consequently, $I_F(u_0, U) = \gamma$. This proves the theorem.

A function $u \in W^1_p(U)$ is called an *extremal* of the functional I_F on an open set U if u gives I_F the smallest possible value on the set $W_p(u, U)$, i.e., $I_F(v, U) \ge I_F(u, U)$ for every function $v \in W^1_p(U)$ such that $v = u|\partial U$.

We now investigate the question of a connection between the extremal functions of I_F and stationary functions, i.e., generalized solutions of the equation $\mathrm{div}[F_q(x, u')] = 0$. This investigation is based on the following statement.

LEMMA 5.9. *Let V be a bounded open set in $\mathbf{R}^n$, and let u and ψ be functions of class $W_p^1(V)$. Assume that the function $F: (x, q) \to F(x, q)$ satisfies conditions A–D in §5.1. Then the function $f(t) = I_F(u + t\psi, V)$ is convex and differentiable in $\mathbf{R}$. Further, for every $t \in \mathbf{R}$*

$$f'(t) = \int_V \langle \psi'(x), F_q[x, u'(x) + t\psi'(x)] \rangle \, dx. \tag{5.25}$$

PROOF. We first prove that f is convex. Let $t_1, t_2 \in \mathbf{R}$ and numbers $\lambda > 0$ and $\mu > 0$ with $\lambda + \mu = 1$ be given arbitrarily, and let $t_0 = \lambda t_1 + \mu t_2$. Then, by condition B on F in §5.1,

$$F(x, u' + t_0\psi') = F[x, \lambda(u' + t_1\psi') + \mu(u' + t_2\psi')]$$
$$\leq \lambda F(x, u' + t_2\psi') + \mu F(x, u' + t_2\psi').$$

Integrating this inequality termwise, we get that $f(t_0) \leq \lambda f(t_1) + \mu f(t_2)$, and it is proved that f is convex.

The validity of (5.25) follows easily from known theorems of analysis on differentiation of integrals depending on a parameter, and from condition D for F.

COROLLARY 1. *If u is an extremal function for the functional I_F in a bounded open set $V \subset \mathbf{R}^n$, where F satisfies conditions A–D in §5.1, then u is a stationary function for I_F on V.*

PROOF. Indeed, let u be an extremal function for I_F on $V \subset \mathbf{R}^n$. Let $\eta \in \overset{\circ}{W}_p^1(V)$ be arbitrary, and define $f(t) = I_F(u + t\eta, V)$. The function f is convex and differentiable on $\mathbf{R}$. For each $t \in \mathbf{R}$ we have that $u + t\eta = u|\partial V$. Since u is extremal,

$$I_F(u + t\eta, V) \geq I_F(u, V)$$

for all t. This means that f attains a minimum at $t = 0$, and hence $f'(0) = 0$, i.e.,

$$\int_V \langle \eta'(x), F_q[x, u'(x)] \rangle \, dx = 0.$$

Since $\eta \in \overset{\circ}{W}_p^1(V)$ is arbitrary, the corollary is proved.

COROLLARY 2. *Let $u \in W_p^1(V)$ be an extremal function for the functional I_F on V, where V is a bounded open set in $\mathbf{R}^n$. If v is a stationary function for I_F such that $v = u|\partial V$, then v coincides identically with u on V.*

PROOF. Let $v \in W_p^1(V)$ be a stationary function coinciding with u on the boundary of V. For all $x \in V$ we have that for $0 < t < 1$

$$F(x, v'(x) + t[u'(x) - v'(x)]) \leq (1 - t)F[x, v'(x)] \\ + tF[x, u'(x)]. \tag{5.26}$$

Since F is essentially convex (condition B in §5.1), equality at a point $x \in V$ for $0 < t < 1$ can hold if and only if $v'(x) = u'(x)$.

Let $f(t) = I_F[v + t(u - v), V]$. Integrating (5.26) termwise, we get that

$$f(t) \leq (1 - t)f(0) + tf(1) = f(0) + t[f(1) - f(0)] \qquad (5.27)$$

for $t \in [0, 1]$. From this, $[f(t) - f(0)]/t \leq f(1) - f(0)$. Passing to the limit as $t \to 0$, we get that $f'(0) \leq f(1) - f(0)$. We have that

$$f'(0) = \int_V \langle \eta'(x), F_q[x, v'(x)] \rangle \, dx,$$

where $\eta = u - v$. Since v is a stationary function, it follows that $f'(0) = 0$, and thus

$$f(1) - f(0) \geq 0. \qquad (5.28)$$

Since u is an extremal function,

$$f(t) = I_F[v + t(u - v), V] \geq f(1) = I_F(u, V) \qquad (5.29)$$

for all $t \in [0, 1]$, and, in particular,

$$f(0) \geq f(1). \qquad (5.30)$$

Comparing (5.28) and (5.30), we get that $f(1) = f(0)$. By (5.27), this implies that $f(t) \leq f(1)$ for $t \in [0, 1]$, and hence $f(t) = f(1) = f(0) + t[f(1) - f(0)]$ for all $t \in [0, 1]$ on the basis of (5.29). Thus, we get equality when we integrate (5.26) termwise. This implies that equality holds in (5.26) for almost all $x \in V$. Hence, $v'(x) = u'(x)$ for almost all $x \in V$, and thus $v - u$ is constant on V. Since $v - u$ vanishes on the boundary of V, $v - u \equiv 0$, i.e., $v \equiv u$, which is what was to be proved.

COROLLARY 3. *Let V be an open set lying strictly inside U, and let $v \in W_p^1(U)$. Then the function $u \in W_p(v, V)$ for which $I_F(u, V)$ attains a minimum on $W_p(v, V)$ is unique.*

§5.5. The maximum principle for extremals of functionals of the calculus of variations.

If a harmonic function is nonnegative at all points of the boundary of a set U, then it is also nonnegative everywhere interior to U by the well-known maximum principle for harmonic functions. If u_1 and u_2 are two harmonic functions such that $u_1(x) \geq u_2(x)$ on the boundary of U, then $u_1(x) \geq u_2(x)$ also everywhere in U. Our immediate goal is to extend this result to functions which are extremals of more general functionals of the calculus of variations.

Below, F denotes a function of the variables (x, q), $x \in U$ (an open subset of $\mathbf{R}^n$) and $q \in \mathbf{R}^n$, satisfying conditions A–D in §5.1.

THEOREM 5.4 [134]. *Suppose that U is a bounded open set in $\mathbf{R}^n$, and u and v are two stationary functions for the functional I_F on U, with $u, v \in W_p^1(U)$. If $u \geq v|\partial U$, then $u(x) \geq v(x)$ almost everywhere in U.*

PROOF. Assume, on the contrary, that $u(x) < v(x)$ on a subset A of U with positive measure. Let $\eta = (u - v)^-$. By assumption, $u - v \in \overset{+}{W}{}_p^1(U)$. This means by definition that $\eta \in \overset{\circ}{W}{}_p^1(U)$. On the basis of Corollary 2 to Lemma 5.2, $\eta'(x) = 0$ for almost all $x \notin A$, since it is obvious that $\eta(x) = 0$ for $x \notin A$. On A we have that $\eta + (u - v) = 0$, so the same Corollary 2 to Lemma 5.2 gives us that

$$\eta'(x) + [u'(x) - v'(x)] = 0$$

for almost all $x \in A$, i.e., $\eta'(x) = v'(x) - u'(x)$ almost everywhere on A. We have that

$$\int_U F(x, u' + \eta') \, dx > \int_U F(x, u') \, dx, \tag{5.31}$$

since $u + \eta = u|\partial U$, and an extremal function with given boundary values is unique. Further,

$$\int_U F(x, u' + \eta') \, dx$$
$$= \int_{U \backslash A} F(x, u' + \eta') \, dx + \int_A F(x, u' + \eta') \, dx \tag{5.32}$$
$$= \int_{U \backslash A} F(x, u') \, dx + \int_A F(x, v') \, dx.$$

Combining (5.31) and (5.32), we conclude that

$$\int_A F(x, v') \, dx > \int_A F(x, u') \, dx. \tag{5.33}$$

Let us now consider the function $v - \eta$. We have that

$$\int_U F(x, v' - \eta') \, dx = \int_{U \backslash A} F(x, v') \, dx + \int_A F(x, u') \, dx$$
$$< \int_{U \backslash A} F(x, v') \, dx + \int_A F(x, v') \, dx = \int_U F(x, v') \, dx.$$

However, this contradicts the fact that v is extremal for I_F and $v - \eta = v|\partial U$. The contradiction obviously proves the theorem.

COROLLARY. *Assume that F also satisfies condition E in §5.1, i.e., $F(x, q) \geq F(x, 0)$ for any $x \in U$ and $q \in \mathbf{R}^n$. Let $u \in W_p^1(U)$ be a stationary function for I_F on U. Suppose that there exists a number $L \in \mathbf{R}$ such that*

$u(x) \leq L|\partial U(u(x) \geq L|\partial U)$. *Then* $u(x) \leq L$ *(respectively,* $u(x) \geq L$*) almost everywhere in* U.

For a proof it suffices to note that the function $v(x) \equiv L$ is a solution of the equation $\operatorname{div} F_q[x, v'(x)] = 0$, by the condition E in §5.1.

§**5.6. Harnack's inequality and its corollaries.** One of the main properties of extremals of functionals of the calculus of variations (used in an essential way in what follows) is connected with the so-called Harnack inequality.

Let $F(x, q)$ be an arbitrary function satisfying conditions A–E in §5.1. Further, let p, a_1, a_2, a_3, and a_4 denote the constants in these conditions.

THEOREM 5.5 (Harnack's inequality for a ball). *There exists a number* $C = C(p, a_1, a_2, a_3, a_4) \geq 1$ *such that if* u *is a nonnegative generalized solution of the equation* $\operatorname{div} F_q[x, u'(x)] = 0$ *in the ball* $B(x_0, 2r)$, *then*

$$\operatorname*{ess\,sup}_{x \in \overline{B}(x_0, r)} u(x) \leq C \operatorname*{ess\,inf}_{x \in \overline{B}(x_0, r)} u(x).$$

The constant C *does not depend on the radius of* B.

This theorem is a special case of a general theorem of Serrin on solutions of nonlinear elliptic equations [156]. The proof of Serrin's theorem follows an idea of Moser [104]. A proof of Theorem 5.5 based on this idea is also presented in Chapter III. We mention some corollaries to the theorem.

THEOREM 5.6. *Let* $U \subset \mathbf{R}^n$ *be an open set, and* $u(x)$ *a generalized solution in* U *of the equation* $\operatorname{div} F_q[x, u'(x)] = 0$, *where* F *satisfies all the conditions* A–E *in* §5.1. *Then* u *is continuous in* U *and satisfies a Hölder condition with exponent* σ *on compact subsets of* U, *where* $0 < \sigma \leq 1$ *and depends only on the constants* p, a_1, a_2, a_3, *and* a_4 *in conditions* A–E *and the number* n.

It is not hard to show that a result analogous to Theorem 5.5 remains true if instead of $B(x_0, 2r)$ one takes an arbitrary bounded domain U in $\mathbf{R}^n$, and instead of $\overline{B}(x_0, r)$ an arbitrary compact subset of U. We confine ourselves to the following particular result.

LEMMA 5.10. *There exists a constant* $C_\lambda = C(\lambda, p, a_1, a_2, a_3, a_4) \geq 1$, *where* $0 < \lambda < 1$, *such that if* u *is a nonnegative solution of the equation* $\operatorname{div} F_q[x, u'(x)] = 0$ *in the domain* $U = B(x_0, r) \backslash \{x_0\}$, *then*

$$\max_{x \in S(x_0, \lambda r)} u(x) \leq C_\lambda \min_{x \in S(x_0, \lambda r)} u(x).$$

THEOREM 5.7 [156] (Liouville's theorem). *Let $u(x)$ be a generalized solution of the equation*

$$\operatorname{div} F_q[x, u'(x)] = 0, \tag{5.33}$$

defined on the whole of $\mathbf{R}^n$. If $u(x)$ is bounded on $\mathbf{R}^n$, then it is identically constant.

PROOF. Let

$$M(r) = \max_{|x| \le r} u(x), \qquad m(r) = \min_{|x| \le r} u(x).$$

For every $r > 0$ we have that $m(r) \le u(x) \le M(r)$ on the ball $B_r = \overline{B}(0, r)$.

Consider the functions

$$v_1(x) = u(x) - m(2r), \qquad v_2(x) = M(2r) - u(x).$$

The functions v_1 and v_2 on B_{2r} are nonnegative, v_1 satisfies the equation $\operatorname{div} F_q[x, v'(x)] = 0$, and v_2 satisfies $\operatorname{div} F_q[x, -v'(x)] = 0$. The function $\overline{F}(x, q) = F(x, -q)$ satisfies conditions A–E in §5.1 with the same constants as F. Therefore, the Harnack constant for $\overline{F}$ has the same value as for F. We have that

$$\max_{x \in B_r} v_1(x) = M(r) - m(2r), \qquad \min_{x \in B_r} v_1(x) = m(r) - m(2r),$$

$$\max_{x \in B_r} v_2(x) = M(2r) - m(r), \qquad \min_{x \in B_r} v_2(x) = M(2r) - M(r).$$

From this,

$$\begin{aligned} M(r) - m(2r) &\le C[m(r) - m(2r)], \\ M(2r) - m(r) &\le C[M(2r) - M(r)], \end{aligned} \tag{5.34}$$

where C is the constant in Theorem 5.5. Adding the inequalities in (5.34) termwise, we get after simple transformations

$$M(r) - m(r) \le [(C - 1)/(C + 1)][M(2r) - m(2r)].$$

From this, induction gives us that

$$M(r) - m(r) \le [(C - 1)/(C + 1)]^\nu [M(2^\nu r) - m(2^\nu r)] \tag{5.35}$$

for all $\nu = 1, 2, \ldots$. If the function $u(x)$ is defined for all $x \in \mathbf{R}^n$ and $|u(x)| \le K = \text{const} < \infty$, then $M(2^\nu r) - m(2^\nu r) \le 2K$ for all ν. Passing to the limit as $\nu \to \infty$ in (5.35), we get that $M(r) - m(r) = 0$ for any $r > 0$, i.e., $u(x) \equiv \text{const}$ in $\mathbf{R}^n$, which is what was to be proved.

§5.7. The concept of the flow of a stationary function in a capacitor. Let $K = (A, B)$ be an arbitrary capacitor in $\mathbf{R}^n$. The symbol $U(K)$ will denote the set $\mathbf{R}^n \setminus (A \cup B)$—the field of the capacitor K. We also introduce the notation $\partial_A U(K) = \overline{U(K)} \cap A$ and $\partial_B U(K) = \overline{U(K)} \cap B$.

Assume that the field U of the capacitor $K = (A, B)$ is a bounded subset of $\mathbf{R}^n$. This last condition will be assumed throughout the subsection unless a statement to the contrary is made. Let $W_p^1(K) = W_p^1(A, B)$ be the collection of all functions $u \in W_p^1[U(K)]$ such that $u(x) = 0$ on $\partial_A u(K)$ and $u(x) = 1$ on $\partial_B U(K)$.

Suppose that a stationary function u for the functional I_F is defined in the field U of the capacitor (A, B), where, as usual, F is assumed to satisfy all the conditions A–E in §5.1. Let $\eta \in W_p^1(A, B)$ be an arbitrary function, and consider the integral

$$I(\eta) = \int_U \langle \eta'(x), F_q[x, u'(x)] \rangle \, dx. \tag{5.36}$$

This integral does not depend on the choice of the function $\eta \in W_p^1(A, B)$. Indeed, let η_1 and η_2 be two arbitrary functions in $W_p^1(A, B)$. Then

$$I(\eta_1) - I(\eta_2) = \int_U \langle \eta_1'(x) - \eta_2'(x), F_q[x, u'(x)] \rangle \, dx.$$

We have that $\partial U = \partial_A U \cup \partial_B U$. The function $\eta_0 = \eta_1 - \eta_2$ is equal to zero on each of the sets $\partial_A U$ and $\partial_B U$. By Lemma 5.7, this implies that $\eta_0 = 0|\partial U$, and hence

$$\int_U \langle \eta_0'(x), F_q[x, u'(x)] \rangle \, dx = 0,$$

which leads to the conclusion that $I(\eta_1) = I(\eta_2)$, as required.

The quantity $I(\eta)$, $\eta \in W_p^1(A, B)$, will be called the *flow* of the extremal function u for the functional I_F in the capacitor (A, B), and denoted by $\Omega(A, B, u, F)$. Note that if $\eta \in W_p(A, B)$, then clearly $1 - \eta \in W_p(B, A)$, which gives us that

$$\Omega(A, B, u, F) = -\Omega(B, A, u, F)$$

for every generalized solution of the equation $\operatorname{div} F_q[x, u'(x)] = 0$ in $U(K) = \mathbf{R}^n \backslash (A \cup B)$.

THEOREM 5.8 [134]. *Suppose that $K = (A, B)$ is a capacitor in $\mathbf{R}^n$ whose field $U(K) = U$ is a bounded subset of $\mathbf{R}^n$, and u_1 and u_2 are stationary functions in $U(K)$ for the functional I_F, where F satisfies conditions A–E in §5.1. If $u_1 = u_2$ on the set $\partial_A U$ and $u_1 \geq u_2$ on $\partial_B U$, then*

$$\Omega(A, B, u_1, F) \geq \Omega(A, B, u_2, F). \tag{5.37}$$

If $u_1 \geq u_2$ on $\partial_A U$ and $u_1 = u_2$ on $\partial_B U$, then

$$\Omega(A, B, u_1, F) \leq \Omega(A, B, u_2, F). \tag{5.38}$$

PROOF. Suppose that $u_1 = u_2$ on $\partial_A U$ and $u_1 \geq u_2$ on $\partial_B U$. Assume that there exists a number $\delta > 0$ such that $u_1 - u_2 - \delta \geq 0$ on $\partial_B U$. Let $\eta = \min\{(u_1 - u_2/\delta, 1\}$. We have that

$$\eta = 1 - \left(\frac{u_1 - u_2}{\delta} - 1\right)^-.$$

The function $((u_1 - u_2)/\delta - 1)^-$ is equal to zero on $\partial_B U$, and hence $\eta = 1$ on $\partial_B U$. Further, $u_1 - u_2 = 0$ on $\partial_A U$. This implies that $\eta = 0$ on $\partial_A U$. We thus get that η is equal to 0 on $\partial_A U$ and 1 on $\partial_B U$, i.e., $\eta \in W_p(A, B)$.

Let

$$f(t) = \int_U F[x, u_2'(x) + t\eta'(x)]\, dx.$$

According to Lemma 5.9, f is convex and differentiable for all $t \in \mathbf{R}$. Further,

$$f'(t) = \int_U \langle \eta', F_q(x, u_2' + t\eta')\rangle\, dx.$$

In particular,

$$f'(0) = \int_U \langle \eta', F_q(x, u_2')\rangle\, dx = \Omega(A, B, u_2, F),$$

$$f'(\delta) = \int_U \langle \eta', F_q(x, u_2' + \delta\eta')\rangle\, dx.$$

We now mention that for almost all $x \in U$

$$\langle \eta'(x), F_q[x, u_2'(x) + \delta\eta'(x)]\rangle = \langle \eta'(x), F_q[x, u_1'(x)]\rangle. \tag{5.39}$$

Indeed, let $P \subset U$ be the collection of all $x \in U$ such that

$$[u_1(x) - u_2(x)]/\delta \geq 1.$$

For almost all $x \in B$ we have that $\eta'(x) = 0$, and hence (5.39) holds for almost all $x \in B$. But if $x \in U \backslash P$, then $[u_1(x) - u_2(x)]/\delta < 1$. By continuity of u_1 and u_2 on U, the last inequality holds also in some neighborhood of x. In this neighborhood $\delta\eta'(x) = u_1'(x) - u_2'(x)$, and hence

$$F_q[x, u_2'(x) + \delta\eta'(x)] = F_q[x, u_1'(x)]$$

in a neighborhood of the given point x. Thus, the proof of (5.39) is complete. It follows from the foregoing that

$$f'(\delta) = \int_U \langle \eta'(x), F_q[x, u_1'(x)]\rangle\, dx = \Omega(A, B, u_1, F).$$

Since f is convex, the derivative f' is an increasing function, and thus $f'(0) \leq f'(\delta)$, i.e.,

$$\Omega(A, B, u_1, F) \geq \Omega(A, B, u_2, F).$$

We now get rid of the condition that $u_1 - u_2 > \delta$ on $\partial_B U$. Let v_τ, where $\tau > 0$, be a stationary function for I_F equal to u_1 on $\partial_A U$ and to $u_1 + \tau$ on $\partial_B U$. By what has been proved, the function of the variable $\tau \to \Omega(A, B, v_\tau, F)$ is nondecreasing on $(0, \infty)$. For each τ

$$\Omega(A, B, v_\tau, F) \geq \Omega(A, B, u_2, F)$$

by what has been proved. To conclude the proof it suffices to establish that

$$\Omega(A, B, v_\tau, F) \to \Omega(A, B, u_1, F) \quad \text{as } \tau \to 0.$$

The function $u_1 + \tau$ is also a stationary function for I_F in U. Further, $u_1 \leq v_\tau \leq u_1 + \tau$ on the boundary of U, which, by Theorem 5.4, implies that $u_1(x) \leq v_\tau(x) \leq u_1(x) + \tau$ for all $x \in U$. Since F is convex with respect to the second argument for any q and q_0, we get

$$F(x, q) \geq F(x, q_0) + \langle q - q_0, F_q(x, q_0)\rangle.$$

From this,

$$\int_U F[x, v_\tau'(x)]\, dx - \int_U F[x, u_1'(x)]\, dx$$
$$\geq \int_U \langle v_\tau'(x) - u_1'(x), F_q[x, u_1'(x)]\rangle\, dx = \tau\Omega(A, B, u_1, F),$$
$$\int_U F[x, u_1'(x)]\, dx - \int_U F[x, v_\tau'(x)]\, dx$$
$$\geq \int_U \langle u_1'(x) - v_\tau'(x), F_q[x, v_\tau'(x)]\rangle\, dx$$
$$\geq -\tau\Omega(A, B, v_\tau, F) \geq -\tau\Omega(A, B, v_1, F).$$

These inequalities obviously give us that

$$\int_U F[x, v_\tau'(x)]\, dx \to \int_U F[x, v_1'(x)]\, dx$$

as $\tau \to 0$. We have that $v_\tau \to u_1$ in U as $\tau \to 0$. On the basis of Theorem 3.6 in Chapter III, it follows from this that

$$\Omega(A, B, v_\tau, F) \to \Omega(A, B, u_1, F) \quad \text{as } \tau \to 0.$$

The proof of the first inequality in the theorem is thus complete.

The case when $u_1 \geq u_2$ on A and $u_1 = u_2$ on B reduces to that already considered if we note that

$$\Omega(A, B, u, F) = -\Omega(B, A, u, F).$$

By what has been proved, if $u_1 \geq u_2$ on A and $u_1 = u_2$ on B, then

$$\Omega(B, A, u_2, F) \geq \Omega(B, A, u_2, F),$$

which implies that

$$\Omega(A, B, u_1, F) \leq \Omega(A, B, u_2, F),$$

as was to be proved.

§5.8. The set of singular points of stationary functions for functionals of the calculus of variations.

THEOREM 5.9 [128]. *Let U be an open set in $\mathbf{R}^n$, $A \subset U$ a set closed with respect to U, and $F(x, q)$ a function satisfying all the conditions in §5.1. Assume that $U\backslash A$ is nonempty and that there exists a function v which is a stationary function for I_F on $U\backslash A$ and is such that $\lim_{x \to x_0} v(x) = \infty$ for every point x_0 in A that is a limit point for $U\backslash A$. Then the p-capacity of A is equal to zero.*

PROOF. 1) Let $x_0 \in A$ be a point of A which is a limit point for $U\backslash A$. Let B_1 be an open ball about x_0 lying strictly inside U, and $B_2 \subset B_1$ a ball concentric to it such that the open ring domain $K = B_1\backslash\overline{B}_2$ contains points of $U\backslash A$. Let S_1 and S_2 be the boundary spheres of B_1 and B_2, and let S_0 be a sphere concentric to S_1 and S_2, lying in the ring K, and containing points of $U\backslash A$. The function v is continuous at each point $x \in B_1$ not in A, and $v(x) \to \infty$ as $x \to y$ if $y \in A$, which implies that v is bounded below on B_1. By assumption, v is a stationary function for I_F. This gives us that $v + C$ is also a stationary function for I_F for any constant C, and in view of this it can be assumed that $v(x) \geq 0$ for all $x \in B_1$.

We extend the definition of v by setting $v(x) = \infty$ for $x \in A$. Let the number $T > 0$ be arbitrary, and denote by P_T the set of all $x \in B_1$ such that $v(x) \geq T$; P_T is obviously closed with respect to B_1. Let $Q_T = \overline{B}_2 \cap P_T$, $H_T = B_1\backslash Q_T$ and $G_T = B_1\backslash P_T$. The sets H_T and G_T are open. Let $v^*(x) = \min\{v(x), T\}$. The function v^* belongs to $W^1_{p,\text{loc}}(U)$. Indeed, let $x \in U$. If $x \notin A$, then there exists a neighborhood V of x lying strictly inside $U\backslash A$. The restriction of v to this neighborhood is a function of class $W^1_p(V)$. On the basis of Lemma 5.8, this implies that the restriction of v^* to V is also a function in $W^1_p(V)$. Let $x \in A$. Then there is a neighborhood V of x in which $v(x) > T$, and hence $v^*(x) \equiv T$ on V. Thus, every point $x \in U$ has a neighborhood V such that the restriction of v^* to V is a function of class W^1_p. From this, $v^* \in W^1_{p,\text{loc}}(U)$. It is not hard to see also that v^* is continuous on U.

We now construct a stationary function v_T for I_F equal to 0 on the sphere S_1 and equal to T on the part of the boundary of H_T lying in B_2.

To do this we take a function $u_0 \in C_0^\infty(\mathbf{R}^n)$ such that $0 \le u_0(x) \le T$ for all x, with $u_0(x) = 0$ in a neighborhood of S_1, and $u_0(x) = T$ in a neighborhood of $\overline{B}_2$. Then v_T is a stationary function for I_F equal to u_0 on the boundary of H_T.

The function $f(x)$ identically equal to T is stationary for I_F on U. Obviously, $T - u_0(x) \ge 0$ everywhere in $\mathbf{R}^n$, and thus $T - v_T(x) \ge 0$ on the boundary of H_T. This gives us that $v_T(x) \le T$ everywhere on H_T. In exactly the same way we get that $v_T(x) \ge 0$ on H_T.

Let us now prove that $v^*(x) \ge v_T(x)$ everywhere on G_T. Note that $v^*(x) = v(x)$ on G_T by definition; hence v^* is a stationary function for I_F on G_T. It suffices to establish that $v^* \ge v_T$ on the boundary of G_T. We construct a sequence (w_m), $m = 1, 2, \ldots$, of functions of class $C_0^\infty(\mathbf{R}^n)$ such that $w_m(x) = 0$ in a neighborhood of S_1, $w_m(x) = u_0(x) = T$ in a neighborhood of P_T, and $\|w_m - v_T\|_{1,p,H_T} \to 0$ as $m \to \infty$. Such a sequence exists, since $v_T - u_0 \in \overset{\circ}{W}{}_p^1(H_T)$, $u_0 \in C^\infty(U) \cap W_p^1(H_T)$, and $C_0^\infty(H_T)$ is dense in $\overset{\circ}{W}{}_p^1(H_T)$. Let $\overline{w}_m = \min\{w_m, T\}$. By Corollary 1 to Lemma 5.4, $\overline{w}_m \in W_p^1(H_T)$, and $\overline{w}_m \to \overline{v}_T = \min\{v_T, T\}$ in $W_p^1(H_T)$ as $m \to \infty$. Since $v_T \le T$, we get that $\overline{w}_m \to v_T$ as $m \to \infty$. Let $\varepsilon > 0$ be arbitrary. Then the function $(\overline{w}_m(x) - v^*(x) - \varepsilon)^+ = h_m(x)$ is equal to zero in a neighborhood of the boundary of G_T. Averaging h_m, we get a function of class C^∞ equal to zero in a neighborhood of the boundary of G_T. This implies that $h_m \in \overset{\circ}{W}{}_p^1(G_T)$. Passing to the limit as $m \to \infty$, we find that the function $h = (v_T - v^* - \varepsilon)^+$ belongs to $\overset{\circ}{W}{}_p^1(G_T)$. This means that $v^* + \varepsilon \ge v_T$ on the boundary of G_T, and hence $v^*(x) + \varepsilon \ge v_T(x)$ everywhere on G_T. Since $\varepsilon > 0$ is arbitrary, this proves that $v^*(x) \ge v_T(x)$ everywhere on G_T.

Let x_1 be a point of S_0 in G_T. Such points exist by assumption. We have that $v_T(x_1) \le v(x_1) = v^*(x_1)$. The function v_T is nonnegative and is an extremal function for I_F on K. Using Harnack's inequality for the domain K and the compact set S_0, we get that $v_T(x) \le L = \mathrm{const}$ for all $x \in S_0$, and the constant L does not depend on T. This finishes the first step of the proof. In this step we constructed an extremal function v_T for I_F on the set $H_T \supset K$. Here $v_T(x) = 0$ on the sphere S_1 which is the outer boundary of H_T, $v_T(x) = T$ on the part of the boundary of H_T lying in the ball $\overline{B}_2$, and $v_T(x) \le L = \mathrm{const}$ on S_0.

Denote by $\Omega(T)$ the quantity $\Omega(S_1, Q_T, v_T, F)$—the flow of the extremal v_T with respect to the capacitor (S_1, Q_T). The function $\eta = (1/T)v_T(x)$ is in the class $W_p(S_1, Q_T)$ and, hence,

$$\Omega(T) = \int_{H_T} \langle \eta'(x), F_q[x, v_T'(x)] \rangle \, dx$$
$$= (1/T) \int_{H_T} \langle v_T'(x), F_q[x, v_T'(x)] \rangle \, dx.$$

From this, by condition D (see §5.1),

$$\int_{H_T} |v_T'(x)|^p \, dx \le kT\Omega(T),$$

where $k < \infty$ is a constant. Considering that $v_T = T\eta$, where $\eta \in W_p(S_1, Q_T)$, we get that

$$C_p(S_1, Q_T) \le \frac{k\Omega(T)}{T^{p-1}}. \tag{5.40}$$

We now show that $\Omega(T)$ is bounded above by a constant independent of T. Suppose that μ is an extremal for I_F in the ring K_0 between the spheres S_1 and S_0, is equal to 0 on S_1, and is equal to $L + 1$ on S_0. Then $\mu = v_T$ on S_1 and $\mu \ge v_T$ on S_0, and hence, by Theorem 5.8,

$$\Omega_0 = \Omega(S_1, S_0, \mu, F) \ge \Omega(S_1, S_0, v_T, F).$$

However, it is not hard to see that

$$\Omega(S_1, S_0, v_T, F) = \Omega(S_1, Q_T, v_T, F) = \Omega(T);$$

thus, $\Omega(T) \le \Omega_0$ for every T.

By what has been proved, it follows from (5.40) that

$$C_p(S_1, Q_T) \le k\Omega_0/T^{p-1}.$$

Since $A \cap \overline{B}_2 \subset Q_T$, this implies that

$$C_p(S_1, A \cap \overline{B}_2) \le k\Omega_0/T^{p-1},$$

and, since $T > 0$ is arbitrary here and $p > 1$, $C_p(S_1, A \cap \overline{B}_2) = 0$.

We have obtained that each point of A that is a boundary point in U has a neighborhood $V = B_2$ such that the p-capacity of $A \cap \overline{V}$ is equal to zero. Let us now prove that all the points A are boundary points in U. Assume that this is not so. Then A has interior points. Let x_0 be a boundary point of A^0 in U. Then x_0 is also a boundary point of A. Hence, by what has been proved, there exists a neighborhood V of x_0 such that $\overline{V} \cap A$ is a set of zero p-capacity, and hence $\overline{V} \cap A$ is a set of measure zero. The set $V \cap A$ has interior points, since x_0 is a boundary point of A^0, and thus the open set $V \cap A^0$ is nonempty. However, this contradicts the fact that $V \cap A$ is a set of measure zero. The contradiction proves that A does not have interior points.

Every point of A therefore has a neighborhood V such that $V \cap A$ is a set of capacity zero. Consequently, $C_p(A) = 0$, and the theorem is proved.

§5.9. Liouville's theorem on conformal mappings in space. A mapping $f: U \to \mathbf{R}^n$ of an open domain $U \subset \mathbf{R}^n$ into $\mathbf{R}^n$ will be called a *generalized conformal mapping* if f is a mapping with bounded distortion that is not identically constant, and its distortion coefficient $K(f)$ is equal to 1.

The purpose of this subsection is to prove that every generalized conformal transformation of $\mathbf{R}^n$ is a Möbius transformation.

If f is a generalized conformal mapping of a domain $U \subset \mathbf{R}^n$, then the linear mapping $f'(x)$ is a general orthogonal mapping for almost all $x \in U$.

THEOREM 5.10. *Every generalized conformal mapping of an open domain $U \subset \mathbf{R}^n$ is a Möbius transformation.*

PROOF. Let U be an open domain in $\mathbf{R}^n$, and $f: U \to \mathbf{R}^n$ a generalized conformal mapping. The matrix-valued function $\theta_f(x)$ is defined almost everywhere in U, where $\theta_f(x) = I$ if $f'(x)$ is a singular mapping, and

$$\theta_f(x) = |\mathscr{J}(x, f)|^{2/n} [f'(x)]^{-1} [f'(x)^*]^{-1}$$

otherwise. Since $f'(x)$ is a general orthogonal transformation, it follows that $[f'(x)]^{-1}[f'(x)^*]^{-1} = \lambda^2 I$, where

$$\lambda = |\det[f'(x)]^{-1}|^{1/n} = |\mathscr{J}(x, f)|^{-1/n}.$$

This implies that $\theta_f(x) = I$ in this case. Thus, if f is a generalized conformal mapping, then $\theta_f(x) = I$ almost everywhere, and equation (5.16) for it takes the form

$$\operatorname{div}[|u'(x)|^{n-2} u'(x)] = 0. \tag{5.41}$$

On the basis of Theorem 5.1 we thus establish that if the function $u = v$ is a solution of (5.41), then the function $u = v \circ f$ also satisfies this equation. In particular, each of the components $f_1, \ldots, f_n$ of the vector-valued function f is a solution of (5.41).

We now use a result of Ural'tseva [77], according to which every solution of (5.41) is a function of class $C^{1,\alpha}$ for some $\alpha > 0$. Further, if G is the set of $x \in U$ for which $|u'(x)| > 0$ (G is open, because $u'(x)$ is continuous), then $u(x)$ belongs to C^∞ on G. If $\mathscr{J}(x, f) = 0$ at a point $x \in U$, then all the derivatives of f are also equal to zero at this point. This implies that if $\mathscr{J}(x, f) = 0$ for all $x \in U$, then the mapping f is identically constant, which is excluded by the definition of a generalized conformal mapping.

Let G be the collection of all $x \in U$ such that $|\mathscr{J}(x, f)| > 0$, and let G_0 be an arbitrary connected component of G. Let $u(x) = f_i(x)$, $i = 1, \ldots, n$, be one of the components of the vector-valued function f. The function u satisfies (5.41), and $|u'(x)| = |\mathscr{J}(x, f)|^{1/n}$ for all $x \in G_0$. From this,

$|u'(x)| > 0$ for all $x \in G_0$. Consequently $u(x)$ belongs to $C^\infty(G_0)$, and hence $f \in C^\infty(G_0)$. The set G_0 is open. We prove that it is closed with respect to U. Let $x_0 \in U$ be an arbitrary limit point of G_0 in U.

Let $\mu(x) = |f'(x)|$. By Ural'tseva's theorem, the derivatives of a solution of (5.41) are continuous in U. This gives us that the function $\mu(x)$ is continuous in U, and hence $\mu(x_0)$ is the limit of $\mu(x)$ as $x \to x_0$ in G_0. Since $|f'(x)| > 0$ for all x for every Möbius transformation, this implies that $\mu(x_0) > 0$. Consequently, f is a Möbius transformation in some neighborhood of x_0. If two Möbius transformations coincide on some open set, then they clearly coincide everywhere. This enables us to conclude that $x_0 \in G_0$. Thus, G_0 contains all its limit points in U; that is, G_0 is closed relative to U. Since U is connected, we have that $G_0 = U$, and the theorem is proved.

REMARK. The theorem was first proved by the author [120] and by Gehring [34] for the case when f is a homeomorphism. In the formulation given here the theorem was proved by the author [126] using a different method.

§5.10. The property of quasi-invariance of conformal capacity.

THEOREM 5.11. *Let U be an open domain in $\mathbf{R}^n$, $f: U \to \mathbf{R}^n$ a homeomorphism with bounded distortion, and $V = f(U)$. Suppose that A and B are arbitrary disjoint closed sets contained in $\overline{U}$, $A_1 = f(A)$, and $B_1 = f(B)$. Then*

$$C_n(A, B, U) \leq K(f)C_n(A_1, B_1, V).$$

PROOF. Let $v \in \tilde{W}(A_1, B_1, V)$ be arbitrary, and let $u = v \circ f$. By Lemma 4.7, u is in $W_{n,\mathrm{loc}}^1(U)$, and $u'(x) = [f'(x)]^* v'[f(x)]$ for almost all $x \in U$. Hence

$$|u'(x)|^n \leq \|f'(x)\|^n |v'[f(x)]|^n \leq K|\mathcal{J}(x, f)||v'[f(x)]|^n.$$

The function u vanishes in a neighborhood of A and is equal to 1 in a neighborhood of B, which easily gives us that $u \in W_n(A, B, U)$. We have

$$\int_U |u'(x)|^n \, dx \leq K(f) \int_U |v'[f(x)]|^n |\mathcal{J}(x, f)| \, dx$$
$$= K(f) \int_V |v'(y)|^n \, dy,$$

and so

$$C_n(A, B, U) \le K(f) \int_V |v'(y)|^n \, dy.$$

Since $v \in \tilde{W}(A_1, B_1, V)$ was arbitrary, this proves the desired inequality.

§6. Topological properties of mappings with bounded distortion

§6.1. **Continuous mappings with nonnegative Jacobian.** Let $U \subset \mathbf{R}^n$ be an open set in $\mathbf{R}^n$, and $f: U \to \mathbf{R}^n$ a continuous mapping of class $W^1_{p,\mathrm{loc}}(U)$. Then we say that f is a *mapping of monotone type* if the Jacobian of f has constant sign in U, i.e., if either $\mathscr{J}(x, f) \ge 0$ almost everywhere in U or $\mathscr{J}(x, f) \le 0$ almost everywhere in U. For a mapping $f: U \to \mathbf{R}^n$ of monotone type let $\sigma(f) = 1$ if $\mathscr{J}(x, f) \ge 0$ almost everywhere in U, and $\sigma(f) = -1$ if $\mathscr{J}(x, f) \le 0$ almost everywhere in U.

Every mapping with bounded distortion is a mapping of monotone type. Denote by Θ the mapping $(y_1, \ldots, y_n) \to (-y_1, y_2, \ldots, y_n)$ of the space $\mathbf{R}^n$. Let $f: U \to \mathbf{R}^n$ be a mapping of monotone type. Then $\tilde{f} = \theta \circ f$ is also a mapping of monotone type. Further, $\sigma(\tilde{f}) = -\sigma(f)$. Let $G \subset U$ be a compact domain, and a an (f, G)-admissible point. Then the point $\theta(a)$ is $(\tilde{f}, G)$-admissible, and $\mu[\theta(a), \tilde{f}, G] = \mu[a, f, G]$. This assertion follows immediately from the definition in §2 of the degree of a mapping.

LEMMA 6.1. *Suppose that U is an open set in $\mathbf{R}^n$ and $f: U \to \mathbf{R}^n$ is a continuous mapping of class $W^1_{n,\mathrm{loc}}$ such that $\mathscr{J}(x, f) \ge 0$ for almost all $x \in U$. Then $\mu(y, f, G)$ is nonnegative for every compact domain $G \subset U$ and any (f, G)-admissible point $y \in \mathbf{R}^n$.*

PROOF. We take an arbitrary compact domain $G \subset U$ and construct a sequence $(f_\nu: V \to \mathbf{R}^n)$, $\nu = 1, 2, \ldots$, of C^∞-mappings, defined on an open set V containing G and lying strictly inside U, that converges uniformly to f on V and is such that $\|f_\nu - f\|_{1,n,V} \to 0$ as $\nu \to \infty$. Such a sequence (f_ν) exists according to Theorem 1.7 in Chapter I. Then as $\nu \to \infty$

$$\int_V |\mathscr{J}(x, f_\nu) - \mathscr{J}(x, f)| \, dx \to 0.$$

Denote by P_ν the collection of all $x \in V$ such that $\mathscr{J}(x, f_\nu) < 0$, and by Q_ν the set of $x \in V$ at which $\mathscr{J}(x, f_\nu) = 0$. According to Theorem 2.2, $f_\nu(Q_\nu)$ is a set of measure zero. The set P_ν is open, and since $\mathscr{J}(x, f_\nu) \ne 0$ for all $x \in P_\nu$, the set $f_\nu(P_\nu)$ is also open. Theorem 2.2 now allows us to conclude that

$$|f_\nu(P_\nu)| \le \int_{P_\nu} |\mathscr{J}(x, f_\nu)| \, dx$$

for all ν. As $\nu \to \infty$ the functions $\mathscr{J}(x, f_\nu)$ converge in $L_1(V)$ to the function $\mathscr{J}(x, f)$, which gives us that as $\nu \to \infty$

$$\int_{P_\nu} |\mathscr{J}(x, f_\nu)| \, dx = \int_V |\mathscr{J}(x, f_\nu)|^- \, dx \to \int_V [\mathscr{J}(x, f)]^- \, dx = 0.$$

This leads to the conclusion that $|f_\nu(P_\nu)| \to 0$ as $\nu \to \infty$. Take an arbitrary point $y_0 \in \mathbf{R}^n \backslash f(\partial G)$, and let $\delta = \rho(y_0, f(\partial G)) > 0$. Since $f_\nu \to f$ uniformly as $\nu \to \infty$, there is a ν_1 such that the ball $B(y_0, \delta/2)$ does not intersect the set $f_\nu(\partial G)$ for $\nu > \nu_1$. Let $\nu_0 \geq \nu_1$ be such that for $\nu \geq \nu_0$ the measure of the set $f_\nu(P_\nu)$ is less than the volume of $B(y_0, \delta/2)$. Then $B(y_0, \delta/2)$ contains a point y_1 such that $y_1 \notin f_\nu(P_\nu \cup Q_\nu)$ for $\nu \geq \nu_0$. The set $f_\nu^{-1}(y_1)$ is then not contained in $P_\nu \cup Q_\nu$, which implies that f_ν is regular with respect to the point y, and the Jacobian of f_ν is nonnegative at each point $x \in f_\nu^{-1}(y)$. This enables us to conclude that $\mu(y_1, f_\nu, G) \geq 0$ for all $\nu \geq \nu_0$. The ball $B(y_0, \delta/2)$ is contained in a connected component of $\mathbf{R}^n \backslash f_\nu(\partial G)$, and hence $\mu(y_0, f_\nu, G) = \mu(y_1, f_\nu, G)$ for $\nu \geq \nu_0$; in particular, $\mu(y_0, f_\nu, G) \geq 0$. Passing to the limit as $\nu \to \infty$, we get that $\mu(y_0, f, G) \geq 0$, which is what was required to prove.

LEMMA 6.2. *Suppose that U is an open set in $\mathbf{R}^n$ and $f: U \to \mathbf{R}^n$ is a continuous mapping of class $W_{p,\mathrm{loc}}^1(U)$ such that $\mathscr{J}(x, f) \geq 0$ for almost all $x \in U$. Let G and $G_1, \ldots, G_m$ be compact domains contained in U and such that no two of $G_1, \ldots, G_m$ have common interior points, and $G \supset \bigcup_1^m G_k$. Then*

$$\mu(y, f, G) \geq \sum_{k=1}^m \mu(y, f, G_k) \tag{6.1}$$

for every point $y \in \mathbf{R}^n$ with $y \notin f(\partial G)$ and $y \notin f(\partial G_k)$ for all $k = 1, \ldots, m$.

PROOF. Consider the set $V = G_0 \backslash \bigcup_1^m G_k$. If V is empty, then (6.1) holds with equality, in view of the additivity property of the degree of a mapping (Proposition II, §2.1). Assume that $V \neq \varnothing$. Let $\{V_{m+1}, V_{m+2}, \ldots\}$ be the collection of all connected components of V, and let $\overline{V}_k = G_k$ for $k \geq m + 1$. Each of the sets G_k is a compact domain. By assumption, $f^{-1}(y)$ is contained in the union of the sequence of open sets $G_1^0, G_2^0, \ldots$, and since $f^{-1}(y)$ is compact, it can be covered by finitely many of these sets. Let $f^{-1}(y)$ be contained in the union of the first l sets G_k, $l \geq m$. Then

$$\mu(y, f, G) = \sum_{k=1}^l \mu(y, f, G_k) \geq \sum_{k=1}^m \mu(y, f, G_k),$$

by the additivity property of the degree of a mapping, since each of the terms on the right-hand side is nonnegative by Lemma 6.1. The lemma is proved.

THEOREM 6.1. *Let* $f: U \to \mathbf{R}^n$ *(where* U *is an open set in* $\mathbf{R}^n$*) be a mapping with bounded distortion. Take a compact domain* $G \subset U$ *and a point* $y \in \mathbf{R}^n$. *If* y *is* (f, G)-*admissible, i.e.,* y *is not in the image of the boundary of* G *under* f, *and* y *belongs to* $f(G)$, *then* $|\mu(y, f, G)| \geq 1$ *and* $\operatorname{sgn} \mu(y, f, G) = \sigma(f)$.

PROOF. Assume first that the Jacobian of f is nonnegative in U, suppose that the compact domain G and the point $y_0 \in \mathbf{R}^n$ satisfy the condition of the theorem. Let $S = f^{-1}(y_0) \cap G$. The set S is closed with respect to G and hence compact, and S does not contain boundary points of G, since $y_0 \notin f(\partial G)$ by assumption. Hence, $S \subset G^0$. Let x_0 be an arbitrary boundary point of S.

We set $d = \rho(y_0, f(\partial G))$ and find a $\delta > 0$ such that if $|x - x_0| < \delta$, then $|f(x) - y_0| = |f(x) - f(x_0)| < d$. Choose an arbitrary r such that $0 < r < \delta$. It is clear that $\overline{B}(x_0, r) \subset G$. Denote by $D(r)$ the collection of all points $x \in B(x_0, r)$ such that $\mathscr{J}(x, f) > 0$. The set $D(r)$ is measurable, and its measure is nonzero. Indeed, assume that $|D(r)| = 0$. Then $\mathscr{J}(x, f) = 0$ for almost all $x \in B(x_0, r)$, and hence each of the partial derivatives $(\partial f/\partial x_i)(x)$ is equal to zero almost everywhere in $B(x_0, r)$, by the inequality defining mappings with bounded distortion. This implies that f is constant on $B(x_0, r)$ and, in particular, $f(x) = f(x_0) = y$ for all $x \in B(x_0, r)$. This enables us to conclude that $B(x_0, r) \subset S$, i.e., x_0 is an interior point of S, which contradicts the choice of x_0.

According to Theorem 1.2, the mapping $f'(x)$ constructed formally from the partial derivatives of f for almost all $x \in U$ is the differential of f. Consequently, there is a point $x_1 \in D(r)$ such that $f'(x_1)$ is the differential of f at x_1. We have that

$$f(x) = f(x_1) + f'(x_1)(x - x_1) + \alpha(x)|x - x_1|,$$

where $\alpha(x) \to 0$ as $x \to x_1$. Since $\det f'(x_1) > 0$, $f'(x_1)$ is a nonsingular mapping, and thus there is a number $\lambda > 0$ such that $|f'(x_1)(h)| \geq \lambda|h|$ for any $h \in \mathbf{R}^n$. Since $\alpha(x) \to 0$ as $x \to x_1$, there is a $\rho_0 > 0$ such that $|\alpha(x)| < \lambda/2$ for $|x - x_1| < \rho_0$. For $\rho < \rho_0$ the mapping f on the sphere $S(x_1, \rho)$ is homotopic in $\mathbf{R}^n \backslash \{f(x_1)\}$ to the affine mapping $x \to f(x_1) + f'(x_1)(x - x_0)$, which implies that for $0 < \rho < \rho_0$ the degree of f at the point $y_1 = f(x_1)$ with respect to $\overline{B}(x_1, \rho)$ is equal to 1. For sufficiently small ρ the ball $\overline{B}(x_1, \rho)$ is contained in G, and hence

$$\mu(y_1, f, G) \geq \mu[y_1, f, \overline{B}(x, \rho)] = 1$$

by Lemma 6.2, i.e., $\mu(y_1, f, G) \geq 1$. We have that

$$|y_1 - y_0| = |f(x_1) - f(x_0)| < d,$$

since $|x - x_0| < \delta$. Consequently, the interval $[y_0, y_1]$ does not contain points of $f(\partial G)$ and thus lies in a single connected component of $\mathbf{R}^n \setminus f(\partial G)$. This implies that

$$\mu(y_0, f, G) = \mu(y_1, f, G) \geq 1,$$

and the theorem is proved for the case when $\mathscr{J}(x, f) \geq 0$ almost everywhere in U.

The case when $\sigma(f) = -1$, i.e., $\mathscr{J}(x, f) \leq 0$ almost everywhere in U, can obviously be reduced to the case already considered by virtue of the remarks made at the beginning of this section. The theorem is proved.

COROLLARY 1. *Let* $f: U \to \mathbf{R}^n$ *be a mapping with bounded distortion. Then for every compact domain* $G \subset U$ *the exterior component of* $\mathbf{R}^n \setminus f(\partial G)$ *does not contain points* $y \in f(G)$.

REMARK. Continuous mappings having the property in Corollary 1 are called *monotone* mappings.

COROLLARY 2. *Let* $f: U \to \mathbf{R}^n$ *be a mapping with bounded distortion. Then for any closed ball* $\overline{B}(a, r) \subset U$ *the function* $|f(x) - f(a)|$ *attains its greatest value on the boundary of* $\overline{B}(a, r)$.

Indeed, let

$$l = \max_{x \in S(a, r)} |f(x) - f(a)|,$$

$$l_1 = \max_{x \in \overline{B}(a, r)} |f(x) - f(a)|.$$

Assume that $l_1 > l$. Let x_1 be a point of $\overline{B}(a, r)$ where $|f(x) - f(a)| = l_1$. Obviously, $|x_1 - a| < r$. The set $f(\partial G)$, where $G = \overline{B}(a, r)$, is contained in $\overline{B}[f(a), l]$. The point $f(x_1)$ lies outside this ball, and hence belongs to the exterior component of $\mathbf{R}^n \setminus f(\partial G)$, which contradicts Corollary 1.

§6.2. Satisfaction of condition N for mappings with bounded distortion.

The goal of this subsection is to prove that every mapping with bounded distortion has property N. The proof uses only a few properties of mappings with bounded distortion. In this connection we establish here an assertion more general than indicated in the heading.

We first prove a statement enabling us to get information about the behavior of a mapping of class W_n^1 on almost all planar sections of its domain.

LEMMA 6.3. *Let U be an open subset of $\mathbf{R}^{n-1}$, and let $f: U \to \mathbf{R}^n$ be a continuous mapping. If f belongs to the class $W^1_{n,\text{loc}}(U)$, then $f(U)$ is a set of measure zero in $\mathbf{R}^n$.*

PROOF. Let f satisfy the conditions of the lemma. For an arbitrary set $A \subset \mathbf{R}^n$ let $d(A) = \operatorname{diam} A$. By the Sobolev imbedding theorem (estimate (2.15) in Chapter I), the following inequality holds for every $(n-1)$-dimensional cube $Q = Q(a, r) \subset \mathbf{R}^{n-1}$ and any mapping $g: Q(a, r) \to \mathbf{R}^n$ of class $W^1_n(Q)$:

$$d[g(Q)] \le Mr^{1/n} \left(\int_Q |g'(x)|^n \, dx \right)^{1/n},$$

where M is a constant.

Let $Q(a, r)$ be an arbitrary $(n-1)$-dimensional cube contained in U. We partition it by planes parallel to the coordinate planes into k^{n-1} equal cubes $Q_1, \dots, Q_m$, where $m = k^{n-1}$. For each $i = 1, \dots, m$

$$d[f(Q_i)] \le M \left(\frac{r}{k} \right)^{1/n} \left(\int_{Q_i} |f'(x)|^n \, dx \right)^{1/n}.$$

Raising both sides to the nth power and summing over i, we get

$$\sum_{i=1}^m (d[f(Q_i)])^n \le \frac{M^n r}{k} \int_Q |f'(x)|^n \, dx. \tag{6.2}$$

The right-hand side of this inequality tends to zero as $k \to \infty$. Each set $f(Q_i)$ is contained in a ball of radius $d[f(Q_i)]$. It thus follows from (6.2) that for any $\varepsilon > 0$ there exists a finite system of balls that covers $f(Q)$ and has sum of volumes less than ε. This implies that $f(Q)$ is a set of measure zero.

Since U can be represented as the union of a sequence of cubes lying strictly inside U, what has been proved implies that $f(U)$ is the union of a countable family of sets of measure zero, and hence $|f(U)| = 0$. The lemma is proved.

We introduce some auxiliary notation. Denote by p_i the mapping of $\mathbf{R}^n$ that assigns to a point $x = (x_1, \dots, x_i, \dots, x_n) \in \mathbf{R}^n$ its ith coordinate x_i. If U is an open subset of $\mathbf{R}^n$, then $p_i(U)$ is clearly an open subset of $\mathbf{R}$. Take a number $h \in \mathbf{R}$ and an integer i with $1 \le i \le n$, and denote by $q_{i,h}$ the mapping of $\mathbf{R}^{n-1}$ into $\mathbf{R}^n$ defined by

$$q_{i,h}(t_1, t_2, \dots, t_{n-1}) = (t_1, \dots, t_{i-1}, h, t_i, \dots, t_{n-1}) \in \mathbf{R}^n.$$

If $U \subset \mathbf{R}^n$ is an open set and $h \in p_i(U)$, then the collection of $t \in \mathbf{R}^{n-1}$ with $q_{i,h}(t) \in U$ forms an open subset of $\mathbf{R}^{n-1}$. This set will be denoted by $P_{i,h}(U)$.

LEMMA 6.4. *Let U be an open subset of $\mathbf{R}^n$, and let $f: U \to \mathbf{R}^n$ be a mapping of class $W^1_{p,\mathrm{loc}}(U)$, where $p \geq 1$. For an integer i between 1 and n let $f_i(t, h) = f[q_{i,h}(t)]$. Then the function $t \mapsto f_i(t, h)$ belongs to the class $W^1_{p,\mathrm{loc}}[P_{i,h}(U)]$ for each i with $1 \leq i \leq n$ and for almost all $h \in p_i(U)$. Furthermore, the derivatives $(\partial f_i/\partial t_k)(t, h)$ are the restrictions of the corresponding derivatives of f. Namely, let $\partial_j f(x) = (\partial f/\partial x_j)(x)$. Then*

$$\frac{\partial f_i}{\partial t_k}(t, h) = (\partial_k f)[q_{i,h}(t)] \quad \text{for } k < i,$$

and

$$\frac{\partial f_i}{\partial t_k}(t) = \partial_{k+1} f[q_{i,h}(t)] \quad \text{for } k \geq i.$$

PROOF. Let Q be an arbitrary cube lying strictly inside U. We construct a sequence (f_ν), $\nu = 1, 2, \ldots$, of functions of class C^∞ defined in Q such that $\|f_\nu - f\|_{1,p,Q} \to 0$ as $\nu \to \infty$. It will be assumed that $\|f_\nu - f\|_{1,p,Q} < \frac{1}{2^\nu}$ for each $\nu = 1, 2, \ldots$. Obviously, this can always be achieved by passing to a subsequence. Define $f_\nu(x) - f(x) = r_\nu(x)$. Let $x = (x_1, \ldots, x_m)$. We regard x as the pair (t, x_i), where t is the point in $\mathbf{R}^{n-1}$ obtained if the ith coordinate of x is crossed out, so that $x = q_{i,x_i}(t)$. Let

$$w(x) = \sum_{\nu=1}^{\infty} \left(|r_\nu(x)|^p + \sum_{k=1}^{n} \left| \frac{\partial r_\nu}{\partial x_k}(x) \right|^p \right). \tag{6.3}$$

The integral of each term on the right-hand side does not exceed $1/2^{\nu p}$; thus, the numerical series obtained by integrating (6.3) converges, and this implies that w is an integrable function. Let (F_ν), $\nu = 1, 2, \ldots$, be a sequence of Lebesgue-integrable functions. If $\sum_{\nu=1}^{\infty} \|F_\nu\|_{L_1}$ converges, then $F_\nu(x) \to 0$ almost everywhere. In view of this observation, $r_\nu(x) \to 0$ and $\frac{\partial r_\nu}{\partial x_i}(x) \to 0$ almost everywhere in Q. Let $A \subset Q$ be a set of measure zero on which the quantity

$$|r_\nu(x)|^p + \sum_{k=1}^{n} \left| \frac{\partial r_\nu}{\partial x_k}(x) \right|^p$$

does not tend to zero as $\nu \to \infty$. The set $p_i(Q)$ is some interval (a_i, b_i). Let $E_i'(Q)$ be the collection of h such that the plane $x_i = h$ intersects A in a set of positive $(n - 1)$-dimensional Lebesgue measure. For such h the set $q_{i,h}^{-1}(A)$ is a set of measure zero in $\mathbf{R}^{n-1}$. (The mapping $q_{i,h}$ maps $q_{i,h}^{-1}(A)$ onto the indicated section of A.) Further, let $E_i''(Q)$ be the set of $h \in (a_i, b_i)$ for which the function $t \mapsto w_i(t, h)$ is not integrable on the $(n - 1)$-dimensional cube $Q_i = P_{i,h}(Q)$. By Fubini's theorem, $E_i''(Q)$ is a set of measure zero. Let $E_i(Q) = E_i'(Q) \cup E_i''(Q)$.

Take an arbitrary $h \in (a_i, b_i)$ not in $E_i(Q)$. Then

$$\int_{Q_i} w_i(t, h)\, dt = \sum_{\nu=1}^{\infty} \left(\int_{Q_i} \left(|r_\nu(t, h)|^p + \sum_{j=1}^{n} \left| \frac{\partial r_\nu}{\partial x_j}(t, h) \right|^p \right) \right) dt < \infty.$$

For this h

$$\int_{Q_i} |r_\nu(t, h)|^p\, dt \to 0, \qquad \int_Q \left| \frac{\partial r_\nu}{\partial x_i}(t, h) \right| dt \to 0,$$

and $r_\nu(t, h) \to 0$ and $\frac{\partial r_\nu}{\partial x_j}(t, h) \to 0$ for almost all $t \in Q_i$. Considering the expressions for r_ν and its derivatives, we get that if $h \notin E_i(Q)$, then the functions $f_{\nu,i}(\cdot, h)$ converge in $L_p(Q_i)$ to $f_i(\cdot, h)$, and their derivatives $\frac{\partial f_{\nu,i}}{\partial t_k}(\cdot, h)$ converge in $L_p(Q)$ to the functions $(\partial f / \partial x_j)(\cdot, h)$, where $j = k$ for $k < i$, and $j = k + 1$ for $k \geq i$. This enables us to conclude that $f_i(\cdot, h)$ belongs to the class $W_p^1(Q_i)$, and the derivative $\frac{\partial f_i}{\partial t_k}(\cdot, h)$ coincides with the corresponding generalized derivative of f_i.

To finish the proof we represent U as the union of a sequence (Q_ν), $\nu = 1, 2, \dots,$ of open cubes. The set $E_i(Q_\nu)$ of zero measure in $\mathbf{R}$ is defined for each ν. Then the set $E_i = \bigcup_\nu E_i(Q_\nu)$ is clearly a set of measure zero. Let $h \notin E_i$. Then on $P_{i,h}(Q_\nu)$ the function $f_i(\cdot, h)$ belongs to the class W_p^1. Since $P_i(U)$ is the union of the sets $P_{i,h}(Q_\nu)$, which are open in $\mathbf{R}^{n-1}$, this implies that f_i is a function of class $W_{p,\mathrm{loc}}^1(P_i(U))$. The lemma is proved.

Let $U \subset \mathbf{R}^n$ be an open set, and $f : U \to \mathbf{R}^n$ a continuous mapping. We say that f is *stable* if it satisfies the following condition S. Let G be an arbitrary compact domain contained in U. Then for every point $y \in f(G) \backslash f(\partial G)$ there is an $\varepsilon > 0$ such that, for every continuous mapping $\varphi : G \to \mathbf{R}^n$ with $|f(x) - \varphi(x)| < \varepsilon$, for all $x \in G$ the set $\varphi(G^0)$ contains y.

We present a certain sufficient condition ensuring the stability of a mapping. Let $f : U \to \mathbf{R}^n$ be a continuous mapping. Assume that for every compact domain $G \subset U$ and any point $y \in f(G)$ not in $f(\partial G)$ the quantity $\mu(y, f, G)$ is nonzero. Then f is a stable mapping. Indeed, assume that f satisfies the given condition. Take an arbitrary compact domain $G \subset U$ and a point $y \in f(G) \backslash f(\partial G)$. Let $\varepsilon = \rho(y, f(\partial G))$. Then $\varepsilon > 0$. Let $\varphi : G \to \mathbf{R}^n$ be a continuous mapping such that $|f(x) - \varphi(x)| < \varepsilon$ for all $x \in G$. The mapping φ is homotopic to f as a mapping of the pair $(G, \partial G)$ into the pair $(\mathbf{R}^n, \mathbf{R}^n \backslash \{0\})$. The required homotopy is given by the mapping

$$\varphi_t(x) = (1 - t)f(x) + t\varphi(x), \qquad t \in [0, 1].$$

This implies that $y \notin \varphi(\partial G)$ and $\mu(y, \varphi, G) = \mu(y, f, G) \neq 0$, and allows us to conclude that $y \in f(G^0)$, which is what was required to prove.

In particular, every homeomorphism $f: U \to \mathbf{R}^n$ is stable. By Theorem 6.1, every mapping with bounded distortion is also stable.

THEOREM 6.2. *Let $U \subset \mathbf{R}^n$ be an open set, and $f: U \to \mathbf{R}^n$ a continuous mapping of the class $W^1_{n,\mathrm{loc}}(U)$. If f is stable, then it has property N.*

PROOF. Let U be a domain in $\mathbf{R}^n$, and $f: U \to \mathbf{R}^n$ a continuous stable mapping in the class $W^1_{n,\mathrm{loc}}(U)$. The plane $L_i(h) = \{x \mid x_i = h\}$ is said to be *normal* if the image under f of the section of U by this plane is a set of measure zero in $\mathbf{R}^n$. It will follow from this theorem that $L_i(h)$ is normal for any h. However, this has not yet been proved, and we must assume the existence of nonnormal planes. Nevertheless, it follows from Lemmas 6.4 and 6.3 that the plane $L_i(h)$ is normal for almost all $h \in p_i(U)$. Indeed, by Lemma 6.4, the function $f_i = f \circ q_{i,h}$ belongs to $W^1_{n,\mathrm{loc}}$ on the $(n-1)$-dimensional domain $P_{i,h}(U)$ for almost all $h \in p_i(U)$. By Lemma 6.3, $f_i[P_{i,h}(U)] = f[L_i(h) \cap U]$ is a set of measure zero for every h such that this holds, and hence the plane $L_i(h)$ is normal.

We say that the closed cube $Q = \overline{Q}(a,r)$ is *normal* if the planes of its faces are all normal. Let $Q \subset U$ be an arbitrary normal cube. Then $f(\partial Q)$ is a set of measure zero. We construct a sequence (f_m), $m = 1, 2, \ldots$, of mappings of class C^∞ such that $f_m \to f$ uniformly in Q, and $\|f_m - f\|_{1,n,Q} \to 0$ as $m \to \infty$. Then

$$\int_Q \mathscr{I}(x, f_m)\, dx \to \int_Q \mathscr{I}(x, f)\, dx, \quad \int_Q |\mathscr{I}(x, f_m)|\, dx \to \int_Q |\mathscr{I}(x, f)|\, dx$$

as $m \to \infty$.

Let χ be the indicator function of the set $f(Q)$, and χ_m the indicator function of the set $f_m(Q)$ in $\mathbf{R}^n$. We prove that $\chi_m(y) \to \chi(y)$ for almost all $y \in \mathbf{R}^n$. Indeed, if $y \notin f(Q)$, then $\chi(y) = 0$, since $f_m \to f$ uniformly, there is an m_0 such that $|f_m(x) - f(x)|$ is less than the positive number $\rho(y, f(\partial Q))$ for all $x \in Q$ when $m \geq m_0$. For such m we have that $y \notin f_m(Q)$; hence $\chi_m(y) = 0 = \chi(y)$. Assume that $y \in f(Q)$ and $y \notin f(\partial Q)$. In view of the stability property of f, there is an $\varepsilon > 0$ such that if $\varphi: Q \to \mathbf{R}^n$ is continuous and $|\varphi(x) - f(x)| < \varepsilon$ for all $x \in Q$, then $y \in \varphi(Q^0)$. Let m_0 be such that $|f_m(x) - f(x)| < \varepsilon$ for all $x \in Q$ when $m \geq m_0$. For all such m the point y belongs to $f_m(Q)$; hence $\chi_m(y) = 1 = \chi(y)$. Thus, it has been established that $\chi_m(y) \to \chi(y)$ for every $y \notin f(\partial Q)$. Since $f(\partial Q)$ is a set of measure zero (the cube Q is normal), it is thereby proved that $\chi_m(y) \to \chi(y)$ for almost all $y \in \mathbf{R}^n$. Each of the mappings f_m belongs to C^∞, and hence has property N. Thus,

$$\int_Q |\mathscr{I}(x, f_m)|\, dx = \int_{\mathbf{R}^n} N(y, kf_m, Q)\, dy \geq f_m(Q). \tag{6.4}$$

(Here $N(y, f_m, Q)$ is the number of elements in $f^{-1}(y) \cap Q$; $N(y, f_m, Q) \geq \chi_m(y)$ for all $y \in \mathbf{R}^n$.) Since $\chi_m(y) \to \chi(y)$ almost everywhere, it follows that

$$|f(Q)| = \int_{\mathbf{R}^n} \chi(y) \, dy \leq \lim_{m \to \infty} \int_{\mathbf{R}} \chi_m(y) \, dy = \lim_{m \to \infty} |f_m(Q)|.$$

The left-hand side of (6.4) tends to the limit $\int_Q |\mathscr{F}(x, f)| \, dx$. Passing in (6.4) to the limit, we thus get that

$$|f(Q)| \leq \int_Q |\mathscr{F}(x, f)| \, dx.$$

Inequality (6.5) has been established so far under the assumption that the cube Q is normal. Assume now that $Q = \overline{Q}(a, r)$ is not normal. Since the cube $Q_h = \overline{Q}(a, h)$ is normal for almost all h such that $U \supset Q_h$, there is a decreasing sequence (r_m), $m = 1, 2, \ldots$, of values such that $r_m \to r$ as $m \to \infty$ and each of the cubes Q_{r_m} is normal. For each m we have that $f(Q) \subset f(Q_{r_m})$, and hence

$$|f(Q)| \leq |f(Q_{r_m})| \leq \int_{Q_{r_m}} |\mathscr{F}(x, f) \, dx.$$

Passing to the limit as $m \to \infty$, we get that (6.5) holds also in the case when Q is not assumed to be normal.

Let E be an arbitrary set of measure zero contained in U, and let $\varepsilon > 0$ be arbitrary. Since the function $x \mapsto |\mathscr{F}(x, f)|$ is locally integrable in U, for the given $\varepsilon > 0$ there is an open set $V \supset E$ with $V \subset U$ such that

$$\int_V |\mathscr{F}(x, f)| \, dx < \varepsilon.$$

The set V is representable as the union of a sequence (Q_ν), $\nu = 1, 2, \ldots$, of disjoint half-open cubes, each of which lies strictly inside V. For each m

$$|f(Q_\nu)| \leq \int_{Q_\nu} |\mathscr{F}(x, f)| \, dx.$$

From this,

$$|f(V)| \leq \sum_{\nu=1}^{\infty} |f(Q_\nu)| \leq \sum_{\nu=1}^{\infty} \int_{Q_\nu} |\mathscr{F}(x, f)| \, dx$$
$$= \int_V |\mathscr{F}(x, f)| \, dx < \varepsilon,$$

and hence $|f(E)| < \varepsilon$. Since $\varepsilon > 0$ is arbitrary, this proves the theorem.

COROLLARY 1 [123]. *Every homeomorphism* $f: U \to \mathbf{R}^n$ *of the class* $W^1_{n,\mathrm{loc}}(U)$ (U *an open subset of* $\mathbf{R}^n$) *has property* N.

COROLLARY 2 [125]. *Every mapping with bounded distortion has property* N.

REMARK 1. Let $f: U \to \mathbf{R}^n$ be a homeomorphism, where U is an open subset of $\mathbf{R}^n$. Assume that f belongs to $W^1_{p,\mathrm{loc}}(U)$, where $1 \le p < n$. Then it can happen that f does not have property N. An example for the case $n = 2$ was constructed by Ponomarev [144]. The construction presented in [144] carries over in an obvious way to the case of arbitrary n.

REMARK 2. The conclusion of Theorem 6.2 is false in general without some additional conditions of a topological nature on f. We give an example showing this in the case $n = 2$. Let G be a domain on the plane with boundary a simple closed curve for which the Lebesgue measure of any arc is nonzero. Let $f(z)$ implement a conformal mapping of the half-disk $\{z \,|\, |z| \le 1 \text{ and } \mathrm{Im}\, z \ge 0\}$ onto G. The mapping f is continuous and belongs to W^1_2. We extend f to the disk $|z| \le 1$ by setting $f(z) = f(\bar{z})$ if $\mathrm{Im}\, z \le 0$. The function f is continuous in the closed disk $|z| \le 1$ and belongs to the class W^1_2 on each of the half-disks $|z| \le 1$, $\mathrm{Im}\, z \ge 0$, and $\mathrm{Im}\, z \le 0$. This implies that $f \in W^1_2$. The mapping f does not have property N: the segment $\mathrm{Im}\, z = 0$, $-1 < \mathrm{Re}\, z < 1$, is carried by this mapping into some arc of the curve L. The Lebesgue measure of this arc is nonzero, and hence f does not have property N: the two-dimensional Lebesgue measure of the image under f of some set of measure zero (a segment in this case) is nonzero.

The author does not know of any examples of this kind for $n > 2$.

§6.3. Topological properties of mappings with bounded distortion.

We first define some concepts relating to the topology of mappings. Let X and Y be arbitrary topological spaces, and let $f: X \to Y$ be continuous. The mapping f is said to be *open* if $f(U)$ is open in Y for every open set U in X. If $f(A)$ is closed for every closed set $A \subset X$, then f is said to be a *closed* mapping. The mapping f is said to be *zero-dimensional* if for every $y \in Y$ the set $f^{-1}(y)$ does not have connected components containing more than one point. If for every $y \in Y$ all the points in $f^{-1}(y)$ are isolated, then f is called an *isolated* mapping. One says that f is a *local homeomorphism* if every $x \in X$ has a neighborhood U such that the restriction of f to U is a homeomorphism of U into Y.

A point $a \in X$ is called a *branch point* of a continuous mapping $f: X \to Y$ if f is not a homeomorphism in any neighborhood of a; the collection of all branch points of f is denoted by B_f.

LEMMA 6.5. *Suppose that U is a domain in $\mathbf{R}^n$ and $f: U \to \mathbf{R}^n$ is a mapping with bounded distortion. If f is not identically constant in U, then the full inverse image of a point under f is a set of zero n-capacity.*

PROOF. Let $y \in \mathbf{R}^n$ be an arbitrary point. The set $A = f^{-1}(y)$ is closed with respect to U, and $V = U \backslash A$ is an open set. The function $u(x) = \ln(1/|x - y|)$ belongs to the class C^∞ and is a solution of the equation $\mathrm{div}[|u'(x)|^{n-2}u'(x)] = 0$ in $\mathbf{R}^n \backslash \{y\}$. Hence, by Theorem 5.1, the function

$$v(x) = u[f(x)] = \ln[1/(|f(x) - y|)]$$

is a solution in the open set V of the equation

$$\mathrm{div}[(\theta(x)u'(x), u'(x))^{(n-2)/2}\theta(x)u'(x)] = 0,$$

where the matrix $\theta(x)$ is positive-definite, and there exist constants $\lambda_1, \lambda_2 > 0$ such that the eigenvalues $\theta(x)$ lie between λ_1 and λ_2 for all x. It is obvious that $v(x) \to \infty$ as x tends in V to an arbitrary boundary point of A in U. We thus see that all the conditions of Theorem 5.9 are satisfied here, and hence $C_n(A) = 0$. The lemma is proved.

LEMMA 6.6. *Let $A \subset \mathbf{R}^n$. If the conformal capacity of A is equal to zero, then for every point $a \in \mathbf{R}^n$ the sphere $S(a,r)$ is disjoint from A for almost all $r > 0$.*

PROOF. By Corollary 2 to Theorem 3.4, the α-dimensional Hausdorff measure of A is equal to zero for every $\alpha > 0$. In particular, the linear Hausdorff measure of A is equal to zero. By Lemma 3.5, this implies that the 1-content of A is equal to zero, i.e., for every $\varepsilon > 0$ there is a sequence $(B(x_k, r_k))$, $k = 1, 2, \ldots$, of balls which covers A and is such that the sum of the radii is less than ε. If the sphere $S(a, r)$ contains points of A, then it intersects at least one of the balls $B(x_k, r_k)$. The set of r such that $S(a, r)$ intersects at least one of the balls $B(x_k, r_k)$ is a union of intervals of lengths $2r_1, 2r_2, \ldots$, and it has measure less than 2ε. Thus, the set of x such that $S(a, r) \cap A$ is nonempty has measure less than 2ε, and the lemma is thereby proved, because $\varepsilon > 0$ is arbitrary.

THEOREM 6.3 [125]. *Every mapping with bounded distortion that is not identically constant is an isolated mapping.*

PROOF. Let $f: U \to \mathbf{R}^n$ be a mapping with bounded distortion which is not identically constant in U. It will be assumed that $\sigma(f) = 1$. The general case can obviously be reduced to this one. Let $x_0 \in U$ be arbitrary, let $y_0 = f(x_0)$, and let $A = f^{-1}(y_0)$. The set A is closed relative to U and is a set of zero capacity. By Corollary 2 to Theorem 3.4, this implies that the linear Hausdorff measure of A is equal to zero.

Let $r > 0$ be such that $B(x_0, r) = B$ lies strictly inside U, and $S(x_0, r) \cap A = \varnothing$. Then $y_0 \notin f(\partial B)$, and hence the number $\mu_0 = \mu(y_0, f, B)$ is defined. We prove that the set $A \cap B$ consists of at most u_0 elements. Indeed, let $x_1, \ldots, x_k$ be an arbitrary finite system of points in $A \cap B$. Since A is a set of zero conformal capacity, Lemma 6.6 gives us that there are ball neighborhoods $B_1, \ldots, B_k$ of the respective points $x_1, \ldots, x_k$ such that the boundary of each set B_i does not intersect A, $B \supset B_i$ for $i = 1, \ldots, k$, and $B_1, \ldots, B_k$ are disjoint. We have that $f(x_i) = y_0$ and $y_0 \notin f(\partial B_i)$ and hence $\mu(y_0, f, B_i)$ is defined. By Lemma 6.2,

$$\mu_0 = \mu(y_0, f, \overline{B}) \geq \sum_{i=1}^{k} \mu(y_0, f, \overline{B}_i).$$

Since $y_0 \in f(\overline{B}_i)$ for each $i = 1, \ldots, k$, Theorem 6.1 gives us that $\mu(y_0, f, B_i) \geq 1$ for all i. From this we have that $\mu_0 \geq k$, and hence $A \cap B$ consists of at most μ_0 elements. Thus, $A \cap B$ is a finite set. We get that every point $x_0 \in U$ has a neighborhood containing finitely many points of $f^{-1}[f(x_0)]$. The theorem is proved.

THEOREM 6.4 [125]. *Every mapping f with bounded distortion from an open domain U to the space $\mathbf{R}^n$ which is not identically constant is an open mapping.*

PROOF. It is required to prove that $f(V)$ is open for every open set $V \subset U$. Indeed, take any point $x \in V$ and let $y = f(x)$. The set $f^{-1}(y)$ has zero capacity, and hence by Lemma 6.6 there is a ball B_0 about x such that $y \notin f(\partial B_0)$ and $B_0 \subset V$. Then $\mu(y, f, B_0)$ is defined, and $|\mu(y, f, B_0)| \geq 1$ in view of Theorem 6.1, since $y \in f(B_0)$. The last inequality will hold at all points of some neighborhood of y. Hence, all the points in this neighborhood belong to $f(B_0)$, i.e., y is an interior point of $f(B_0)$, hence also of V. Thus, $f(x)$ is an interior point of $f(V)$ for any $x \in V$, i.e., $f(V)$ is an open set, which is what was required to prove.

THEOREM 6.5 [129]. *Suppose that U is an open domain in $\mathbf{R}^n$ and $f: U \to \mathbf{R}^n$ is a mapping with bounded distortion. Then the function $y \to \mu(y, f, G)$ is bounded for every compact domain $G \subset U$.*

PROOF. We assume that $\mathscr{J}(x, f) \geq 0$ almost everywhere in U. The general case can clearly be reduced to this one. The assertion is true if f is identically constant, so it will be assumed that this is not the case. Let $x_0 \in U$. There is a $\delta_1(x_0) > 0$ such that $f(x) \neq f(x_0)$ for $0 < |x - x_0| \leq \delta_1(x_0)$. Let $B_0 = B[x_0, \delta_1(x_0)]$ and $S_0 = \partial B_0$. Then $y_0 = f(x_0) \notin f(S_0)$.

Let $\delta_2(x_0) > 0$, $\delta_2(x_0) < \delta_1(x_0)$, be such that for $r < \delta_2(x_0)$ the image of $B(x_0, r)$ under f is contained in a connected component of $\mathbf{R}^n \backslash f(S_0)$. Suppose that $y \in f[B(x_0, r)]$ and $y \notin f[S(x_0, r)]$ where $r < \delta_2(x_0)$. Then

$$\mu[y, f, B(x_0, r_0)] \le \mu(y, f, B_0) = \mu(y_0, f, B_0).$$

This shows that the function $y \to \mu[y, f, B(x_0, r)]$ is bounded for $r < \delta_2(x_0)$.

Using the theorem of Borel, we construct a finite system of balls

$$B(x_1, \rho_1), \ldots, B(x_k, \rho_k)$$

covering the compact domain G, where $\rho_i = \delta_2(x_i)/2$. Let $y \in \mathbf{R}^n$ be an arbitrary point. For each $i = 1, \ldots, k$ let $r_i > 0$ be such that $\rho_i < r_i < 2\rho_i$ and the sphere $S(x_i, r_i)$ does not intersect the set $f^{-1}(y)$. Let $B_i = B(x_i, r_i)$. It is not hard to see that

$$\mu(y, f, G) \le \sum_{i=1}^{k} \mu(y, f, B_i).$$

On the other hand, $\mu(y, f, B_i) \le \mu_i$, where μ_i depends only on the point x_i and is equal to the index of f at x_i. This gives us that

$$\mu(y, f, G) < \mu_0 = \sum_{i=1}^{k} \mu_i = \text{const}.$$

The theorem is proved.

We make some remarks about isolated continuous mappings. Suppose that U is an open domain in $\mathbf{R}^n$, and f is an isolated continuous mapping. Let $a \in U$ be an arbitrary point. The point a is an isolated point of the set $f^{-1}[f(a)]$, and thus there is a $\delta > 0$ such that the closed ball $\overline{B}_\delta = \overline{B}(a, \delta)$ does not contain points of $f^{-1}[f(a)]$ other than a and is contained in U. Let G be an arbitrary compact domain contained in $\overline{B}_\delta$ and such that $a \in G^\circ$. We have that $f^{-1}([f(a)]) \cap \overline{B}_\delta = \{a\} \subset G^\circ$, and hence $\mu[f(a), f, G] = \mu[f(a), f, \overline{B}_\delta]$ in view of Proposition II in §2.1. Thus, $\mu[f(a), f, G]$ has the same value for all compact domains G contained in $\overline{B}_\delta$ and containing a as an interior point. This common value of $\mu[f(a), f, G]$ is called the *index* of f at the point a and denoted by $j(a, f)$. This quantity is a kind of analogue of the concept of the multiplicity of a root of an equation.

Suppose that U is an open set in $\mathbf{R}^n$, $f: U \to \mathbf{R}^n$ is an isolated mapping, and $G \subset U$ is a compact domain. Assume that the point $b \in \mathbf{R}^n$ is (f, G)-admissible. Then it is not hard to see that the set $f^{-1}(b) \cap G$ is finite. Let

$a_1, \ldots, a_m$ be its elements. Then

$$\mu(b, f, G) = \sum_{k=1}^{m} j(a_k, f). \tag{6.6}$$

Indeed, let $G_1, \ldots, G_m$ be the closed balls of radius r about the points $a_1, \ldots, a_m$. If r is sufficiently small, then these balls are contained in G° and are disjoint, and thus, by the additivity property of the degree of a mapping (Proposition II in this chapter, §2.1)

$$\mu(b, f, G) = \sum_{k=1}^{m} \mu(b, f, G_k).$$

On the other hand, it is obvious that $\mu(b, f, G_k) = j(a_k, f)$, and equality (6.6) is thereby proved.

Let $f: U \to \mathbf{R}^n$ be a mapping with bounded distortion. Then it follows from Theorem 6.1 that $|j(a, f)| \geq 1$ for every point $a \in U$. Further, $\operatorname{sgn} j(a, f) = \sigma(f)$.

COROLLARY TO THEOREM 6.5. *Let U be an open domain in $\mathbf{R}^n$, and $f: U \to \mathbf{R}^n$ a mapping with bounded distortion. As before, for $A \subset U$ and $y \in \mathbf{R}^n$ let $N(y, f, A)$ be the number of elements in the set $f^{-1}(y) \cap A$ ($N(y, f, A) = \infty$ if this set is infinite). Then the function $y \mapsto N(y, f, A)$ is bounded for every compact set $A \subset U$.*

PROOF. Let $A \subset U$ be compact. By the Borel theorem, there is a finite set of balls $B_i = B(a_i, r_k)$, $i = 1, \ldots, m$, such that $A \subset \bigcup_1^k B_i$, and for each i the closed ball $G_i = \overline{B}(a_i, 2r_k)$ is contained in U. The function $y \mapsto \mu(y, f, G_i)$ is bounded, by the theorem: $|\mu(y, f, G_i)| \leq \mu_i = \text{const} < \infty$ for all $y \notin f(\partial G_i)$. Take an arbitrary $y \in f(A)$, and let $i_1, \ldots, i_l$ be the indices of the balls B_i containing points of the set $f^{-1}(y) \cap A$. For each i there is a δ_i such that $r_i < \delta_i < 2r_i$ and the sphere $S(a_i, \delta_i)$ does not contain points of $f^{-1}(y)$. Let $H_{i_j} = \overline{B}(a_{i_j}, \delta_{i_j})$. Then the number $\mu(y, f, H_{i_j})$ is defined. By (6.6),

$$N(y, f, H_{i_j}) \leq |\mu(y, f, H_{i_j})|.$$

For every $y \notin f(\partial G_i \cup \partial H_i)$ we have

$$|\mu(y, f, H_i)| \leq |\mu(y, f, G_i)|.$$

Now (6.6) allows us to conclude that $|\mu(y, f, G)|$ increases as G becomes larger, and this implies that $|\mu(y, f, H_{i_j})| \leq \mu_{i_j}$. Obviously,

$$N(y, f, A) \leq \sum_{j=1}^{l} N(y, f, H_{i_j}) \leq \sum_{i=1}^{m} \mu_i,$$

and the corollary is proved.

THEOREM 6.6 [129]. *Suppose that U is an open set in $\mathbf{R}^n$ and $f: U \to \mathbf{R}^n$ is a mapping with bounded distortion which is not identically constant. Let the point $a \in U$ be such that $|j(a, f)| = 1$. Then f is a homeomorphism in some neighborhood of a.*

PROOF. We assume that $\sigma(f) = 1$; the case $\sigma(f) = -1$ can be reduced to this case. To prove the theorem it suffices to establish that there is a neighborhood V of a in which f is one-to-one. Suppose that there is no such neighborhood. Let $\delta_0 > 0$ be such that $x \in U$ and $f(x) \neq f(a)$ for $0 < |x - a| \leq \delta_0$. By our assumption, for every positive integer m there are points x_m' and x_m'' such that $|x_m' - a| < 1/m$, $|x_m'' - a| < 1/m$, and $f(x_m') = f(x_m'') = y_m$. The point y_m converges to $b = f(a)$ as $m \to \infty$. This implies that for sufficiently large m

$$\mu[y_m, f, \overline{B}(a, \delta_0)] = \mu[b, f, \overline{B}(a, \delta_0)] = j(a, f) = 1.$$

On the other hand, using (6.6), we get by the nonnegativity of the index of f that for all $m \geq m_0$

$$\mu[y_m, f, \overline{B}(a, \delta_0)] \geq j(x_m', f) + j(x_m'', f) \geq 2.$$

This is a contradiction, and the theorem is proved.

COROLLARY. *Let $f: U \to \mathbf{R}^n$ be a mapping with bounded distortion. If $|j(x, f)| > 1$ at a point $x \in U$, then f is not a homeomorphism in any neighborhood of x.*

The existence of branch points (and even of whole branch curves) for mappings with bounded distortion is the peculiarity which distinguishes mappings with bounded distortion from arbitrary quasiconformal mappings. The set of branch points is obviously closed with respect to the domain of the mapping. The dimension of B_f does not exceed $n - 2$, as a consequence of the following theorem of Chernavskiĭ [26], [27].

THEOREM 6.7. *Suppose that $U \subset \mathbf{R}^n$ is an open domain and $f: U \to \mathbf{R}^n$ is an arbitrary isolated open continuous mapping. Then the set of branch points of f is at most $(n - 2)$-dimensional.*

A proof of Theorem 6.7 can also be found in Väisälä's paper [172].

§6.4. **A theorem on removable singularities.** Many investigations have dealt with the problem of removable singularities, i.e., the question of when a quasiconformal mapping or a mapping with bounded distortion defined on some set can be extended to a larger set with preservation of its properties (see [33], [100], and [158]). Questions relating especially to the case of arbitrary mappings with bounded distortion were considered in [114]. For lack of space we cannot present here the results obtained in these references despite their importance, and we confine ourselves to the following theorem, which is used in §10.

THEOREM 6.8 (Väisälä [173]). *Suppose that U is an open domain in $\mathbf{R}^n$, and $E \subset U$ is a set closed relative to U with $(n-1)$-dimensional Hausdorff measure zero. Then every quasiconformal mapping $f: U \backslash E \to \mathbf{R}^n$ admits a unique continuous extension $g: U \to \mathbf{R}^n$. Furthermore, g is a quasiconformal mapping of U, $K(g) = K(f)$ and $K_0(g) = K_0(f)$.*

PROOF. Since E does not have interior points, the required extension is unique if it exists. To prove the theorem it thus suffices to establish that the limit $\lim_{x \to x_0} f(x)$ exists for every $x_0 \in E$. Since the last assertion is local, it can be assumed that $U \neq \overline{\mathbf{R}}^n$, $f(U) \neq \overline{\mathbf{R}}^n$, and f is bounded; this can be made true by performing an inversion transformation in addition.

We introduce a function $h: U \to \mathbf{R}^n$ by setting $h(x) = f(x)$ for $x \notin E$ and $h(x) = 0$ for $x \in E$. Then h is a mapping of class $W^1_{n,\text{loc}}(U)$ in view of Theorem 1.5 in Chapter I. For almost all $x \in U \backslash E$, and thus for almost all $x \in U$, we have that $h'(x) = f'(x)$, and thus $|h'(x)|^n \leq K(f)|\mathscr{J}(x, h)|$ almost everywhere in U. On the basis of Theorem 1.1 in this chapter it follows that there exists a continuous mapping $h^*: U \to \mathbf{R}^n$ such that $h^*(x) = h(x)$ almost everywhere in U. Obviously, $h^*(x) = f(x)$ for almost all $x \in U$. The mapping $g = h^*$ is the desired extension of f to U. Obviously, g is a mapping with bounded distortion. Since $g'(x) = f'(x)$ almost everywhere in U, we see that $K(g) = K(f)$ and $K_0(g) = K_0(f)$.

We show that g is one-to-one. Assume, on the contrary, that there are two points $x_1, x_2 \in U$ such that $g(x_1) = g(x_2) = y$, $x_1 \neq x_2$. Let V_1 and V_2 be disjoint neighborhoods of x_1 and x_2, and let $W_i = g(V_i)$, $i = 1, 2$. The sets W_1 and W_2 are open, and $y \in W_i$ for $i = 1, 2$. There is a $\delta > 0$ such that $B(y, \delta) \subset W_1 \cap W_2$. Since E is a set of measure zero, $g(E)$ is a set of measure zero, and thus there is a point $y' \in B(y, \delta)$ such that $y' \notin g(E)$. Suppose that the points $x_1' \in V_1$ and $x_2' \in V_2$ are such that $g(x_1') = g(x_2') \neq y'$. Since the neighborhoods V_1 and V_2 are disjoint, $x_1' \neq x_2'$. Further, $x_1', x_2' \notin E$, which leads us to conclude that $g(x_i') = f(x_i')$. This contradicts the fact that f is a homeomorphism, and the theorem is proved.

§6.5. On the method of moduli. We present some facts from which the reader can get an impression of the method of moduli in the theory of spatial mappings. We do not provide proofs for the main facts connected with this method, but refer the reader to the corresponding literature. In the planar case the method of moduli was developed in a well-known paper of Ahlfors and Beurling [10]. Applications of the method to the theory of quasiconformal mappings and mappings with bounded distortion are considered in [113], [91], and [94]–[96].

A *curve* in $\mathbf{R}^n$ is defined here to be any continuous mapping $\gamma: [0, 1] \to \mathbf{R}^n$ not constant on any interval $[\alpha, \beta] \subset [0, 1]$. Let $\gamma: [0, 1] \to \mathbf{R}^n$ be a given curve. The curve γ determines a certain metric δ in $[0, 1]$, where $\delta(t_1, t_2)$ is the diameter of the set $\gamma([t_1, t_2])$ for $t_1, t_2 \in [0, 1]$. The one-dimensional Hausdorff measure on $[0, 1]$ corresponding to this metric is denoted by s_γ and called the *arc length* of the curve γ. A curve γ is said to be *rectifiable* if $s_\gamma([0, 1]) < \infty$.

Let ρ be an arbitrary Borel-measurable function defined in a domain U of $\mathbf{R}^n$, and let $\gamma: [0, 1] \to U$ be a curve in this domain. Then the integral $\int_0^1 \rho[\gamma(t)] \, ds_\gamma(t)$ (if it makes sense) is called the *integral of the function ρ with respect to arclength along the curve γ*, and is denoted by $\int_\gamma \rho(x) \, ds_x$.

Let Γ be an arbitrary set of curves in $\mathbf{R}^n$. A Borel-measurable function $\rho \geq 0$ on $\mathbf{R}^n$ is said to be *admissible with respect to the family* Γ, or, briefly, Γ *admissible*, if $\int_\gamma \rho(x) \, ds_x \geq 1$ for every curve $\gamma \in \Gamma$. The infimum of the integrals $\int_{\mathbf{R}^n} [\rho(x)]^n \, dx$ on the set of all Γ-admissible functions ρ is called the *modulus* of the family Γ and denoted by $M(\Gamma)$.

Let Γ be an arbitrary family of curves lying in an open subset U of $\mathbf{R}^n$, and let $f: U \to \mathbf{R}^n$ be a continuous mapping. The *image* of the family Γ under f is defined to be the collection $f(\Gamma)$ of all curves $f \circ \gamma: [0, 1] \to \mathbf{R}^n$, where $\gamma: [0, 1] \to \mathbf{R}^n$ is an arbitrary curve in Γ. If Γ is an arbitrary family of curves in the set $V = f(U)$, then let $f^{-1}(\Gamma)$ be the collection of all curves $\gamma: [0, 1] \to U$ such that $f \circ \gamma \in \Gamma$. It is natural to call the family $f^{-1}(\Gamma)$ the *inverse image* of Γ under f.

Application of the method of moduli to the theory of mappings with bounded distortion is based on the following two theorems.

THEOREM 6.9 [113]. *Suppose that U is an open set in $\mathbf{R}^n$ and $f: U \to \mathbf{R}^n$ is a mapping with bounded distortion. Then for every family Γ of curves in U*

$$M(\Gamma) \geq (1/K(f)) M[f(\Gamma)].$$

THEOREM 6.10 [113]. *Suppose that $f: U \to \mathbf{R}^n$ is a mapping with bounded distortion, and $G \subset U$ is a compact domain contained in U and such that $f(\partial G)$ coincides with the boundary of the set $H = f(G)$. Let Γ be a family of curves in H, and $m \geq 1$ an integer such that $|\mu(y, f, G)| = m$ for all $y \in H^0$. Then*

$$M(\Gamma) \leq (K_0(f)/m) M[f^{-1}(\Gamma)].$$

We mention that Väisälä [174] has obtained certain further strengthenings and refinements of these theorems.

It follows from Theorems 6.9 and 6.10 that if f is a homeomorphism with bounded distortion, then for every family of curves in U

$$(1/K(f))M[f(\Gamma)] \leq M(\Gamma) \leq K_0(f)M[f(\Gamma)].$$

This result was established by Shabat [157], [158], and Väisälä [174].

An essential role in applications of the method of moduli is played by estimates for the moduli of certain families of curves. The required estimates can be obtained from estimates of the capacities of certain capacitors in view of the following theorem of Ziemer [181].

THEOREM 6.11. *Suppose that* $K = (A_0, A_1)$ *is a capacitor in* $\mathbf{R}^n$, *where* A_1 *is compact, and let* Γ *be the collection of all curves* $\gamma: [0, 1] \to \mathbf{R}^n$ *such that* $\gamma(0) \in A_0$ *and* $\gamma(1) \in A_1$, *and* $\gamma(t)$ *belongs to the field* U *of the capacitor* (A_0, A_1) *for* $0 < t < 1$. *Then* $M(\Gamma) = C_n(A_0, A_1)$.

We remark that it is possible to use the method of moduli only after establishing that every mapping with bounded distortion is an isolated open mapping. A proof of this fact without the use of the apparatus of differential equations is not known.

§6.6. Bi-Lipschitz mappings. We first make some remarks about mappings of class $W_{n,\mathrm{loc}}^1(U)$ defined on open of $\mathbf{R}^n$.

LEMMA 6.7. *Let* U *be an open subset of* $\mathbf{R}^n$, *and* $f: U \to \mathbf{R}^n$ *a local homeomorphism of class* $W_{n,\mathrm{loc}}^1(U)$. *Then* f *has property* N, f *is differentiable in* U *almost everywhere, and the Jacobian of* f *has constant sign on every connected component of* U. *Further, for every open* $V \subset U$ *the measure of the set of* $x \in V$ *with* $\mathscr{J}(x, f) > 0$ *is nonzero.*

PROOF. Let $f: U \to \mathbf{R}^n$ satisfy all the conditions of the lemma. Then it is easy to see that the conditions of Theorem 6.2 hold for f, and this gives us that f has property N.

Fix an arbitrary point $x_0 \in U$ and a number $\delta > 0$ such that the ball $B(x_0, \delta)$ is strictly inside U and f is a homeomorphism on $\overline{B}(x_0, \delta)$. We use a result in §4.3 of Chapter III. On $B(x_0, \delta)$ the mapping f satisfies the condition $E(T)$ indicated there, with constant $T = 1$. Further, f belongs to the class $W_n^1(B(x_0, \delta))$ on $B(x_0, \delta)$, and is hence differentiable almost everywhere in $B(x_0, \delta)$ according to Theorem 4.3 in Chapter III. Since $x_0 \in U$ was arbitrary, this proves that f is differentiable on U almost everywhere.

Let $\overline{B}(x_0, \delta) = G$, $H = f(G)$, and $W = f[B(x_0, \delta)] = H^0$. Take an arbitrary ball $\mathscr{D} = B(x, r) \subset G$ and let $E = f(\mathscr{D})$. We use Theorem 2.2 in this chapter, taking the indicator χ_E of E as $u(y)$; that is, $u(y) = 1$ for $y \in E$ and $u(y) = 0$ for $y \notin E$. Since $f|_G$ is a homeomorphism, $N_G(y, f) = 1$ for all $y \in W$, and the function $y \mapsto \mu(y, f, G)$ is constant on W. Further, $|\mu(y, f, G)| = 1$, and hence $\mu(y, f, G) = \mu_0$, where $\mu_0 = \pm 1$. In this case $u[f(x)] = 1$ for $x \in \mathscr{D}$, and $u[f(x)] = 0$ for $x \notin \mathscr{D}$. Using Theorem 2.2, we get that

$$\int_{\mathscr{D}} |\mathscr{J}(x, f)| \, dx = \int_E N_G(y, f) \, dy = |E|,$$

$$\int_{\mathscr{D}} \mathscr{J}(x, f) \, dx = \int_E \mu(y, f, G) \, dy = \mu_0 |E|.$$

Since f is a homeomorphism, $|E| > 0$, and hence the set of $x \in \mathscr{D}$ with $\mathscr{J}(x, f) \neq 0$ has nonzero measure. Moreover,

$$\int_{\mathscr{D}} \mathscr{J}(x, f) \, dx = \mu_0 \int_{\mathscr{D}} |\mathscr{J}(x, f)| \, dx.$$

Since $\mathscr{D}$ is an arbitrary ball in $B(x_0, \delta)$, this implies that

$$\mathscr{J}(x, f) = \mu_0 |\mathscr{J}(x, f)|$$

for almost all $x \in B(x_0, \delta)$. In particular, either $\mathscr{J}(x, f) \geq 0$ almost everywhere in $B(x_0, \delta)$, or $\mathscr{J}(x, f) \leq 0$ almost everywhere in $B(x_0, \delta)$.

In particular, the measure of the set of $x \in B(x_0, \delta)$ with $\mathscr{J}(x, f) \neq 0$ is nonzero. Since $x_0 \in U$ is arbitrary, this gives us that the measure of the set $\{x \in V \mid \mathscr{J}(x, f) \neq 0\}$ is nonzero for any open set $V \subset U$.

Let U_+ be the set of $x \in U$ such that $\mathscr{J}(t, f) \geq 0$ for almost all $t \in B(x, \delta)$ for some number $\delta > 0$, and let U_- be the set of $x \in U$ such that $\mathscr{J}(t, f) \leq 0$ almost everywhere in $B(x, \delta)$ for some $\delta > 0$. In view of what has been proved, $U_+ \cup U_- = U$, $U_+ \cap U_- = \varnothing$, and each of the sets U_+ and U_- is open. This implies that every connected component of U is contained in one of the sets U_+ or U_-, and the proof of the lemma is complete.

Let U_- be an open subset of $\mathbf{R}^n$. A mapping $f: U \to \mathbf{R}^n$ is said to be *bi-Lipschitz* if there exists a constant L, $1 \leq L < \infty$, such that for every $x_0 \in U$ there is a number $\delta > 0$ for which the ball $B(x_0, \delta)$ is in U and

$$\frac{|x' - x''|}{L} \leq |f(x') - f(x'')| \leq L|x' - x''|$$

for any $x', x'' \in B(x_0, \delta)$. The infimum of the numbers L satisfying this condition will be denoted by $L(f)$.

If $f: U \to R^n$ is a bi-Lipschitz mapping, then it is a local homeomorphism. By Theorem 2.7 in Chapter I, f belongs to the class $W_{\infty, \text{loc}}(U)$, and, in particular, $f \in W^1_{n, \text{loc}}(U)$. This gives us that f is differentiable almost everywhere. Further, by Lemma 6.7, the Jacobian of f has constant sign for $x \in U$.

If the mapping $f: U \to \mathbf{R}^n$ is bi-Lipschitz and U contains the segment joining x_1 and x_2, then

$$|f(x_2) - f(x_1)| \leq L(f)|x_2 - x_1|. \tag{2}$$

Indeed, assume that $[x_1, x_2]$ is contained in U, and let $L > L(f)$ be arbitrary. By definition, every point $x \in U$ has a neighborhood $B(x, \delta)$ such that (1) holds for any x_1 and x_2 in this neighborhood. Covering $[x_1, x_2]$ by finitely many such neighborhoods, we find a finite sequence of points $a_0 = x_1, a_2, \ldots, a_m = x_2$ arranged in order on $[x_1, x_2]$ such that for each $i = 1, \ldots, m$

$$|f(a_i) - f(a_{i-1})| \leq L|a_i - a_{i-1}|.$$

Summing these inequalities, we get that

$$|f(x_2) - f(x_1)| \leq \sum_{i=1}^{m} |f(a_i) - f(a_{i-1})|$$

$$\leq L \sum_{i=1}^{m} |a_i - a_{i-1}| = L|x_2 - x_1|.$$

Since $L > L(f)$ is arbitrary, this yields (2).

Let $f: U \to \mathbf{R}^n$ be a homeomorphism. In this case if f is bi-Lipschitz, then so is f^{-1}. Further, $L(f) = L(f^{-1})$.

We show that if $f: U \to \mathbf{R}^n$ is a bi-Lipschitz mapping, then for every $x_0 \in U$ there is a $\delta > 0$ such that

$$\frac{1}{L(f)}|x_2 - x_1| \leq |f(x_2) - f(x_1)| \leq L(f)|x_2 - x_1| \tag{3}$$

for any $x_1, x_2 \in B(x_0, \delta)$; that is, the constant L in (1) can be taken equal to $L(f)$. Let $B(x_0, \rho)$ be an arbitrary ball about x_0 on which f is a homeomorphism, and let $V = f[B(x_0, \rho)]$ and $y = f(x_0)$. There is an $\varepsilon > 0$ such that $B(y_0, \varepsilon) \subset V$; let δ, $0 < \delta \leq \rho$, be such that $|f(x) - y_0| < \varepsilon$ for $|x - x_0| < \delta$. By what was proved above,

$$|f(x_2) - f(x_1)| \leq L(f)|x_2 - x_1|$$

for any $x_1, x_2 \in B(x_c, \rho)$. Further, in precisely the same way we conclude that

$$|f^{-1}(y_2) - f^{-1}(y_1)| \leq L(f^{-1})|y_2 - y_1| = L(f)|y_2 - y_1|$$

for any $y_1, y_2 \in B(y_0, \varepsilon)$. Setting $y_1 = f(x_1)$ and $y_2 = f(x_2)$ here, where $x_1, x_2 \in B(x_0, \delta)$, we get that

$$|x_2 - x_1| \le L(f)|f(x_2) - f(x_1)|,$$

and hence both the inequalities in (3) hold for any $x_1, x_2 \in B(x_0, \delta)$.

Let us show that every bi-Lipschitz mapping has bounded distortion. With this goal we first establish that at any point $x \in U$ where f is differentiable the linear mapping $F = f'(x)$ is also bi-Lipschitz, and $L(F) \le L(f)$. For an arbitrary $t \in \mathbf{R}$, $t \ne 0$, and $X \in \mathbf{R}^n$ let

$$F_t(X) = \frac{f(x + tX) - f(x)}{t}.$$

For any $X_1, X_2 \in \mathbf{R}^n$ we have that

$$|F_t(X_1) - F_t(X_2)| = \left| \frac{f(x + tX_1) - f(x + tX_2)}{t} \right|.$$

Let $\delta > 0$ be such that the inequalities in (3) hold for any x_1 and x_2 in $B(x, \delta)$. There is a $t_0 > 0$ such that if $0 < t < t_0$, then $|tX_1| < \delta$ and $|tX_2| < \delta$. For $0 < t < t_0$

$$\frac{1}{L(f)}|X_2 - X_1| \le |F_t(X_2) - F_t(X_1)| \le L(f)|X_2 - X_1|.$$

If f is differentiable at x, then $F_t(X) \to F(X)$ as $t \to 0$, where $F = f'(x)$, and we thus get that for any $X_1, X_2 \in \mathbf{R}^n$

$$\frac{1}{L(f)}|X_2 - X_1| \le |F(X_2) - F(X_1)| \le L(f)|X_2 - X_1|.$$

This means that F is bi-Lipschitz, and $L(F) \le L(f)$.

In particular, it follows from what has been proved that at each point $x \in U$ the principal dilation coefficients of the linear mapping $f'(x)$ lie in $[(1/L(f)), L(f)]$, which implies that $f'(x)$ is nonsingular and $K[f'(x)] \le [L(f)]^{2n-2}$ for almost all $x \in U$. This establishes that f has bounded distortion. Further,

$$K(f) \le [L(f)]^{2n-2}.$$

It is not hard to see that in this case we also have that

$$K_0(f) \le [L(f)]^{2n-2}.$$

§7. Local structure of mappings with bounded distortion

§7.1. Preliminary remarks. Let $f \colon U \to \mathbf{R}^n$ be a mapping with bounded distortion, where U is an open set in $\mathbf{R}^n$. Let $x_0 \in U$ be an arbitrary point, and let $y_0 = f(x_0)$ and $S = f^{-1}(y_0)$. All the points in S are isolated, by what was proved in §6. The function

$$v(y) = \ln(1/|y - y_0|)$$

belongs to the class C^∞ in $\mathbf{R}^n \backslash \{y_0\}$ and is a solution of the equation

$$\operatorname{div}[|v'(y)|^{n-2}v'(y)] = 0.$$

By Theorem 5.1, this implies that the function $u = v \circ f$ is a generalized solution on $U \backslash S$ of the equation

$$\operatorname{div}[\langle \theta(x)u'(x), u'(x) \rangle^{(n-2)/2}\theta(x)u'(x)] = 0,$$

where the matrix $\theta(x)$ is determined from the condition that

$$\theta(x) = |\mathscr{J}(x, f)|^{2/n}[f'(x)]^{-1}([f'(x)]^*)^{-1}$$

if $\mathscr{J}(x, f) \neq 0$, and $\theta(x) = I$ otherwise. Further, if η is the differential form

$$\eta(y) = \sum_{k=1}^{n}(-1)^{k-1}|v'(y)|^{n-2}\frac{\partial v}{\partial y_k}(y)\, dy_1$$
$$\ldots dy_{k-1}\, dy_{k+1} \ldots dy_n,$$

and the form ζ is defined by $\zeta(x) = (f^*\eta)(x)$, then $d\zeta(x) = 0$, and for every function $\varphi \in C^1$

$$d\varphi \wedge \zeta = \operatorname{sgn}\mathscr{J}(x, f)\langle \theta u', u' \rangle^{(n-2)/2}\langle \theta u', \varphi' \rangle\, dx_1, dx_2 \ldots dx_n. \qquad (7.1)$$

We have that

$$\frac{\partial v}{\partial y_i}(y) = -\frac{y_i - y_{0i}}{|y - y_0|^2}.$$

From this,

$$\eta(y) = (-1/|y - y_0|^n)\sum_{k=1}^{n}(-1)^{k-1}(y_k - y_{0k})\, dy_1$$
$$\ldots dy_{k-1}\, dy_{k+1} \ldots dy_n,$$

and hence $-\eta$ is the same exterior form used for computing the degree of a mapping.

Let $t_0 > 0$ be such that the ball $\overline{B}_0 = \overline{B}(x_0, t_0)$ is contained in U and does not contain other points of S. Then for any x with $0 < |x - x_0| \leq t_0$

the point $f(x)$ is different from $y_0 = f(x_0)$. For every compact domain $G \subset \overline{B}_0$ with x_0 in its interior we have that

$$\mu(y_0, f, G) = \mu(y_0, f, \overline{B}_0) = j(x_0, f).$$

Let φ be an arbitrary compactly supported C^1-function with support in $\overline{B}_0$ that is equal to 1 in a neighborhood of x_0. According to Lemma 2.4 in this chapter,

$$\int_{B_0} d\varphi(x) \wedge \zeta(x) = \omega_n \mu[y_0, f, B_0] = \omega_n j(x_0, f).$$

Considering (7.1), we conclude from this that

$$\operatorname{sgn} \mathscr{J}(x, f) \int_{B_0} \langle \theta u', u' \rangle^{(n-2)/2} \langle \theta u', \varphi' \rangle \, dx = \omega_n j(x_0, f).$$

Here $u(x) = -\ln|f(x) - y_0|$, and $\theta(x)$ is the matrix-valued function determined from f in §5.2 of this chapter. Observe now that $j(x_0, f) > 0$ if f preserves orientation, i.e., $\mathscr{J}(x, f) \geq 0$ almost everywhere, and $j(x_0, f) < 0$ otherwise, i.e., if $\mathscr{J}(x, f) < 0$. Because of this, the formula just obtained can be rewritten as

$$\int_{B_0} \langle \theta u', u' \rangle^{(n-2)/2} \langle \theta u', \varphi' \rangle \, dx = \omega_n |j(x_0, f)|. \tag{7.2}$$

The function u is a stationary function for the functional $\int F[x, u'(x)] \, dx$ on $U \backslash S$, where $F(x, q) = \langle \theta(x)q, q \rangle^{n/2}$. In $\mathbf{R}^n$ we consider the capacitor $K = (A, B)$, where $A = \mathbf{R}^n \backslash B(x_0, t_0)$ and $B = \{x_0\}$, and we denote by $\Omega(x_0, f)$ the flow function of u in this capacitor. By definition,

$$\Omega(x_0, f) = \int_{\mathbf{R}^n} \langle F_q[x, u'(x)], \varphi'(x) \rangle \, dx$$

$$= n \int_{B_0} \langle \theta u', u' \rangle^{(n-2)/2} \langle \theta u', \varphi' \rangle \, dx,$$

and thus, by (7.2),

$$\Omega(x_0, f) = n \omega_n j(x_0, f). \tag{7.3}$$

Let $f: U \to \mathbf{R}^n$ (where U is an open set in $\mathbf{R}^n$) be a mapping with bounded distortion that is not identically constant in U. Let $x_0 \in U$ be an arbitrary point. We introduce some quantities characterizing the behavior of f in a neighborhood of this point. By results in the preceding section, there exists a number $\delta > 0$ such that $f(x) \neq f(x_0)$ for $0 < |x - x_0| \leq \delta_0$. The supremum of such numbers δ is denoted by $\delta_f(x_0)$. (The index f in this notation will be omitted in what follows whenever this does not lead to confusion.)

Let $0 < \tau < \delta(x_0)$. Define

$$L_f(x_0, \tau) = \max_{|x - x_0| = \tau} |f(x) - f(x_0)|,$$

$$l_f(x_0, \tau) = \min_{|x - x_0| = \tau} |f(x) - f(x_0)|.$$

Obviously, $L_f(x_0, \tau)$ and $l_f(x_0, \tau)$ are continuous functions of τ; moreover, $0 < l_f(x_0, \tau) \leq L_f(x_0, \tau)$, and $L_f(x_0, \tau) \to 0$ as $\tau \to 0$. The quantity $L_f(x_0, \tau)$ is a nondecreasing function of τ. Indeed, if this were not so, then there would be numbers τ_1 and τ_2 such that $\tau_1 < \tau_2$ and

$$L_f(x_0, \tau_1) > L_f(x_0, \tau_2). \tag{7.4}$$

The image of the sphere $S(x_0, \tau_2)$ is contained in the ball $B(y_0, r)$, where $y_0 = f(x_0)$ and $r = L_f(x_0, \tau_1)$. It follows from (7.4) that there is an $x \in B(x_0, \tau_2)$ such that the point $y = f(x)$ lies outside $B(y_0, r)$. It is not hard to see that $\mu(y, f, \overline{B}(x_0, \tau_1)) = 0$. However, this contradicts Theorem 6.1. Hence for each $\tau \in (0, \delta(x_0))$ there exists a number $\tau_1 = a(\tau) \leq \tau$ such that $L_f(x_0, \tau_1) = l_f(x_0, \tau)$. (If such a number τ_1 is not unique, we choose the largest of them as $a(\tau)$.)

§7.2. Some estimates of a solution of an elliptic equation having one singular point. Let U be an arbitrary bounded domain in $\mathbf{R}^n$. Assume that a function $F(x, q)$ which is a normal kernel in the sense of §5.1 is defined for almost all $x \in U$, i.e., $F(x, q)$ satisfies all the conditions A–E in §5.1, and the exponent p in conditions C and D is equal to n.

Choose any point $a \in U$, and let $U^* = U \backslash \{a\}$. For $\delta > 0$ let $U^*_\delta = U \backslash \overline{B}(a, \delta)$. We consider the generalized solutions of the equation

$$\operatorname{div} F_q[x, u'(x)] = 0, \tag{7.5}$$

in U^* which satisfy the condition that $u(x) \to \infty$ as $x \to a$. By a theorem of Moser proved in Chapter III, every such solution $u(x)$ is continuous in U^*. If $u(x)$ is a solution of (7.5) in U^*, then the conditions $u(x) = 0$ on ∂U, $u(x) \geq 0$ on ∂U, and $u(x) \leq 0$ on ∂U will mean that u is in $W^1_n(U^*_\delta)$ for all sufficiently small δ and vanishes (is nonnegative or nonpositive, respectively) on the component ∂U of the boundary of U^*_δ in the sense of the definitions in §5.3 of the present chapter.

Let $u(x)$ be a generalized solution of (7.5) in $U^* = U \backslash \{a\}$. Let $A_0 = \mathbf{R}^n \backslash U$ and $A_1 = \{x_0\}$. Denote by Ω_0 the flow of the solution $u(x)$ of (7.5) with respect to the capacitor (A_0, A_1).

Our goal is to get estimates for the solution $u(x)$ of (7.5). Let us first recall some facts established in §3. Denote by $A_T(l)$, where $l > 0$, the capacitor (A_0, A_1) with A_0 the interval $-1 \leq x_n \leq 0$ of the Ox_n-axis and A_1 the ray $x_n \geq l$ of this axis. Define

$$C_n[A_T(l)] = \omega_n / [\ln \Phi(l)]^{n-1}. \tag{7.6}$$

As shown in §3, there exists a constant $\lambda_n > 1$ such that for all $l > 0$

$$l + 1 \leq \Phi(l) \leq \lambda_n(l + 1). \tag{7.7}$$

The capacitor $A_T(l)$ has the following extremal property. Assume that (E_0, E_1) is a given capacitor in $\mathbf{R}^n$ such that E_0 contains two distinct points x_0 and y_0 in a single connected component of E_0, and E_1 has an unbounded connected component containing a point z_0 with $l = |z_0 - x_0|/|y_0 - x_0|$. Then

$$C_n(E_0, E_1) \geq C_n[A_T(l)] = \frac{\omega_n}{[\ln \Phi(l)]^{n-1}}. \tag{7.8}$$

LEMMA 7.1. *Suppose that U is a bounded open domain in $\mathbf{R}^n$, $a \in U$, $U^* = U \backslash \{a\}$, and $u(x)$ is a function defined and continuous on $\overline{U} \backslash \{a\} = U^* \cup \partial U$ which is a generalized solution of equation* (7.5) *in U^*. Assume that $u(x) \to \infty$ as $x \to a$, and $u(x) \leq 0$ for $x \in \partial U$. Let $h \geq 0$. Denote by A_h the collection of all points $x \in \overline{U} \backslash \{a\}$ such that $u(x) \leq h$, and let $M_h = U \backslash A_h$. Then M_h is a connected open set, the set A_h is closed, and if the boundary of U is connected, then A_h is also connected.*

PROOF. Since $u(x)$ is continuous in $\overline{U} \backslash \{a\}$ and $u(x) \to \infty$ as $x \to a$, the set A_h is closed. This implies that M_h is open. Assume that M_h is not connected. Then it has at least two different connected components: M' and M''. Since $u(x) \to \infty$ as $x \to a$, there is a neighborhood V of a such that $u(x) > h$ for all $x \in V$, $x \neq a$. Obviously, $V \subset M_h$, and thus $a \in M_h$. It will be assumed that $a \in M''$. At each boundary point of M' we have that $u(x) \leq h$. We show that $u(x) \leq h$ on the boundary of M' also in the generalized sense defined in §5.3 above. Indeed, let $v_m(x) = (h + 1/m - u(x))^-$, where $m = 1, 2, \ldots$ is an integer. Obviously, v_m vanishes in a neighborhood of the boundary of M'. As $m \to \infty$ we have that $v_m \to v = (h - u)^-$ in $W_n^1(M')$, and thus $(h - u)^- \in W_n^1(M'/\partial M')$, i.e., $h - u \in \overset{+}{W}{}_n^1(M')$, which is what was to be proved. From this and from the particular case of the maximum principle for equations of the form (7.5) (the corollary to Theorem 5.4) we get that $h - u(x) \geq 0$ in M'. This contradicts the fact that $u(x) > h$ for all $x \in M'$ by assumption. The contradiction proves the connectedness of M_h.

Assume now that the boundary of U is connected. We prove that A_h is connected in this case. Obviously, $\partial U \subset A_h$. The point a does not belong to A_h, and hence there is a $\delta > 0$ such that the closed ball $\overline{B}(a, \delta)$

does not contain points of A. Let $U_\delta^* = U \backslash \overline{B}(a, \delta)$. Assume, contrary to what is to be proved, that A_h is disconnected. Since $u(x)$ is continuous in $\overline{U} \backslash \{a\}$, A_h is obviously closed, and it follows from our assumption that there exist nonempty closed sets A' and A'' such that $A' \cup A'' = A$ and $A' \cap A'' = \varnothing$. One of the sets A' or A'' contains the connected set ∂U. We assume that $A'' \supset \partial U$. Since A' and A'' do not have common points, there exist open sets G' and G'' such that $A' \subset G'$, $A'' \subset G''$, and the closures of G' and G'' are disjoint. Note that $A' \subset \overline{U} \backslash \overline{B}(a, \delta)$, and since $A' \cap \partial U = \varnothing$, it follows that $A' \subset U \backslash \overline{B}(a, \delta) = U_\delta^*$. In view of this it can be assumed that $G' \subset U_\delta^*$, since this inclusion can always be achieved by replacing G' by the intersection $G' \cap U_\delta^*$ if necessary. The boundary of G' lies in the closure of U_δ^*, and since $\overline{G}'$ is disjoint from ∂U (because $\partial U \subset A'' \subset G''$), all the points of $\partial G'$ are interior points of U, and $a \notin \partial G'$. Since $\partial G' \cap A_h = \varnothing$, $u(x)$ is continuous and $u(x) > h$ at each point $x \in \partial G'$. By continuity, there is a number $\gamma > 0$ such that $u(x) \geq h + \gamma$ at each point $x \in \partial G'$. Again by continuity, $u(x) \geq h + \gamma$ on $\partial G'$ also in the sense of the generalized definition in §5.3. Indeed, for each $m = 1, 2, \ldots$ the function $v_m(x) = (u(x) - h - \gamma + 1/m)^-$ vanishes in a neighborhood of $\partial G'$, and $v_m(x) \to v(x) = (u(x) - h - \gamma)^-$ in $W_n^1(G')$ as $m \to \infty$; consequently, $v \in \overset{\circ}{W}{}_n^1(G')$, and this means that $u(x) \geq h + \gamma$ on $\partial G'$ in the sense of the definition in §5.3. By the minimum principle for a solution of an equation of the form (2.1) (the corollary to Theorem 5.4), this implies that $u(x) \geq h + \gamma$ for all $x \in G'$, which contradicts the fact that $u(x) \leq h$ for all $x \in A' \subset G'$ by assumption. The contradiction proves that A_h is connected, and the lemma is proved.

THEOREM 7.1. *Suppose that U is a bounded open domain in $\mathbf{R}^n$, a is a point in U, $U^* = U \backslash \{a\}$, and $u(x)$ is a function defined and continuous on $\overline{U} \backslash \{a\}$ which is a generalized solution of equation (7.5) in U^*. Assume that $u(x) \leq 0$ on ∂U and $u(x) \to \infty$ as $x \to a$. Let r_0 be the distance from a to the unbounded connected component of $\mathbf{R}^n \backslash \overline{U}$, and let Ω_0 be the flow of the solution of (7.5) in the capacitor $(\mathbf{R}^n \backslash U, \{a\})$. Then $\Omega_0 \geq 0$, and for all $x \in U \backslash \{a\}$*

$$u(x) \leq (\Omega_0 / a_4 \omega_n)^{1/n-1} \ln \Phi_n(r_0 / |x - a|), \qquad (7.9)$$

where Φ_n is the function defined by (7.6), and a_4 is the constant of inequality (5.6) in §5.1.

PROOF. Suppose that $u(x)$ satisfies all the conditions of the theorem. Take an arbitrary point $x_0 \in U$, $x_0 \neq a$, and let $h = u(x_0)$. Assume first that $h > 0$. Let A be the set of all $x \in U$ such that $u(x) \leq h$, and let $M = U \backslash A$. By Lemma 7.2, M is connected. This implies that $\overline{M}$ is also connected. Obviously, $a \in \overline{M}$ and $x_0 \in \overline{M}$. Since $u(x)$ is continuous on $\overline{U} \setminus \{a\}$ and $u(x) \leq 0$ on ∂U, $\overline{M}$ does not intersect ∂U. Let $V = U \setminus \overline{M}$. The boundary of V consists of two parts. One of them is ∂U, and the other (the intersection $\overline{V} \cap \overline{M}$) is denoted by $\partial_M V$. Let $u_0(x)$ be a generalized solution of (7.5) in V satisfying the following boundary conditions: $u_0(x) = 0$ on ∂U and $u_0(x) = h = u(x)$ on $\partial_M v$. Since $u(x)$ is continuous on $\overline{U} \setminus \{a\}$, $u(x) \leq 0$ on ∂U also in the generalized sense of §5.3. Indeed, the function $v_m = [1/m - u(x)]^-$ vanishes in a neighborhood of ∂U by the continuity of $u(x)$. As $m \to \infty$ we have that $v_m \to (-u)^- = u^+$ in $W_n^1(V)$. From this, $u^+ = 0$ on ∂U in the sense of §5.3, i.e., $u(x) \leq 0$ on ∂U. By Theorem 5.8, this implies that

$$\Omega(\mathbf{R}^n \setminus U, \overline{M}, u_0, F) \leq \Omega(\mathbf{R}^n \setminus U, \overline{M}, u, F) = \Omega_0.$$

Let $\zeta(x) = u_0(x)/h$. Then $\zeta \in W_n^1(V)$, $\zeta(x) = 0$ on ∂U, $\zeta(x) = 1$ on $\partial_M V$, and, hence,

$$\Omega(\mathbf{R}^n \setminus U, \overline{M}, u_0, F) = \int_{\mathbf{R}^n} \langle \zeta'(x), F_q[x, u_0'(x)] \rangle \, dx$$
$$= \frac{1}{h} \int_{\mathbf{R}^n} \langle u_0'(x), F_q[x, u_0'(x)] \rangle \, dx.$$

We have that $\langle q, F_q(x, q) \rangle \geq q_4 |q|^n$ (inequality (5.6) in §5.1). Therefore,

$$\Omega_0 \geq \Omega(\mathbf{R}^n \setminus U, \overline{M}, u_0, F) \geq \frac{a_4}{h} \int_{\mathbf{R}^n} |u_0'(x)|^n \, dx$$
$$= a_4 h^{n-1} \int_V |\zeta'(x)|^n \, dx. \tag{7.10}$$

Since the function $u(x)$ takes positive values in U^*, what has been proved gives us that $\Omega_0 > 0$. The last integral on the right-hand side of (7.10) can be estimated as follows:

$$\int_{\mathbf{R}^n} |\zeta'(x)|^n \, dx \geq C_n(\mathbf{R}^n \setminus U, \overline{M}). \tag{7.11}$$

The set $\overline{M}$ contains the points a and x_0. The set $\mathbf{R}^n \setminus U$ has an unbounded connected component, and the distance from a to this component is equal to r_0. Using (7.8), we get that

$$C_n(\mathbf{R}^n \setminus U, \overline{M}) \geq \omega_n / [\ln \Phi(l)]^{n-1},$$

where $l = r_0 / |x_0 - a|$. On the basis of (7.10) and (7.11), this gives us that $a_4 \omega_n h^{n-1} / [\ln \Phi(l)]^{n-1} \leq \Omega_0$, and hence

$$u(x_0) = h \leq (\omega_0 / a_4 \omega_n)^{1/(n-1)} \ln \Phi(l).$$

Since this inequality obviously holds at all points x where $u(x) \leq 0$, the theorem is proved.

§7.3. A measure of the distortion of a small sphere under a mapping with bounded distortion.

THEOREM 7.2. *Suppose that U is an open set in $\mathbf{R}^n$, and $f: U \to \mathbf{R}^n$ is a mapping with bounded distortion that is not identically constant in U. Then for every point $a \in U$ there is a number $r_0 > 0$ such that for every positive $r \leq r_0$*

$$L_f(a, r)/l_f(a, r) \leq \exp \beta_n \alpha_1(a, f), \qquad (7.12)$$

and for every $x \in B(a, r)$

$$|f(x) - f(a)| \geq L_f(a, r)|\gamma_n|x - a|/r|^{\alpha_1(a, f)}, \qquad (7.13)$$

where β_n and γ_n are constants depending only on n,

$$\alpha_1(a, f) = [K(f)|j(a, f)|]^{1/(n-1)},$$

and the number r_0 is determined as follows. First let $r_1 > 0$ be such that $x \in U$ and $f(x) \neq f(a)$ for $0 < |x - a| \leq r_1$. Then r_0 is such that $L_f(a, r_0) = l_f(a, r_1)$.

PROOF. Suppose that f satisfies all the conditions of the theorem. We take an arbitrary point $a \in U$ and from it determine numbers r_1 and r_0 as indicated in the statement of the theorem. Let r be such that $0 < r \leq r_0$. We set $L = L_f(a, r)$ and let $u(x) = \ln(L/|f(x) - f(a)|)$. The function $u(x)$ is a generalized solution of the equation div $F_q[x, u'(x)] = 0$ in $B(a, r_1)\backslash\{a\}$, where $F(x, q) = \langle 0(x)q, q\rangle^{n/2}$, $\theta(x)$ being the matrix-valued function defined from $f(x)$ as indicated in §5.2. The flow of the function $u(x)$ in the capacitor $(\mathbf{R}^n\backslash B(a, r_1), \{a\})$ is equal to $\Omega_0 = n\omega_n|j(a, f)|$. The constant a_4 for this normal kernel F is equal to $n/K(f)$. Let A be the collection of all points $x \in B(a, r_1)$ such that $u(x) \leq 0$. Since $u(x) \leq 0$ for all $x \in S(a, r_1)$ in view of the fact that $l_f(a, r_1) = L_f(a, r_0) \geq L_f(a, r) = L$, it follows that A contains the sphere $S(a, r_1)$. Let $V = B(a, r_1)\backslash A$. On the basis of Lemma 7.2, $a \in V$, and the set V is connected. Further, by the same Lemma 7.2, A is also connected. Obviously, $\mathbf{R}^n\backslash V = A \cup (\mathbf{R}^n\backslash B(a, r_1))$, and thus $\mathbf{R}^n\backslash V$ is connected. For each $x \in B(a, r)$ the point $f(x)$ is an interior point of $f[B(a, r)]$. From this it is obvious that $|f(x) - f(a)| < L_f(a, r) = L$ for all $x \in B(a, r)$, and hence $u(x) > 0$ in $B(a, r)$, i.e., $B(a, r) \subset V$. The sphere $S(a, r)$ contains points of $\mathbf{R}^n\backslash V$, and this implies that the distance from a to $\mathbf{R}^n\backslash V$ is equal to r.

Suppose that $x \in S(a, r)$ is such that $|f(x) - f(a)| = l_f(a, r)$. If $l_f(a, r) = L_f(a, r)$, then, since always $\Phi(l) \geq 1$ in view of (7.7), the required inequality holds in this case. Assume that $l_f(a, r) < L_f(a, r)$. Then

$u(x) > 0$, and thus $x \in V$. Using the estimate in Theorem 7.1, we get from this that

$$\ln \frac{L_f(a, r)}{l_f(a, r)} = u(x) \le [K(f)|j(a, f)|]^{1/n-1} \ln \Phi_n(l).$$

which clearly implies (7.12).

Let us now apply the estimate in Theorem 7.1 to an arbitrary point $x \in B(a, r)$ different from a. We get that

$$u(x) = \ln \frac{L}{|f(x) - f(a)|} \le [K(f)|j(a, f)|]^{1/(n-1)}$$
$$\times \ln \Phi_n(r/|x - a|).$$

From this,

$$\frac{|f(x) - f(a)|}{L} \ge [\Phi_n(r/|x - a|)]^{-\alpha_1(a,f)},$$

where $\alpha_1(a, f) = [K(f)|j(a, f)|]^{1/(n-1)}$. Let us use (7.7). We have that

$$[1/\Phi_n(l)]^{\alpha_1(a,f)} \ge [1/\lambda_n l]^{\alpha_1(a,f)}.$$

Setting $l = r/|x - a|$ here and considering that $|x - a| < r$, we have that

$$|f(x) - f(a)| \ge \gamma_n^{\alpha_1(a,f)}(|x - a|/r)^{\alpha_1(a,f)},$$

where $\gamma_n = 1/2\lambda_n$ is a constant. The proof of the theorem is complete.

§7.4. Behavior of a mapping with bounded distortion near an arbitrary point of the domain.

THEOREM 7.3. *Suppose that U is an open set in $\mathbf{R}^n$, and $f: U \to \mathbf{R}^n$ is a mapping with bounded distortion that is not identically constant in U. Then for every point $a \in U$ there is an $r_0 > 0$ such that if $0 < r \le r_0$, then*

$$|f(x) - f(a)| \le L_f(a, r)M_1(|x - a|/r)^{\alpha(a,f)}$$

for $|x - a| \le r$, where M_1 is the constant on the right-hand side of (7.12)

$$\alpha(a, f) = [|j(a, f)|/K_0(f)]^{1/(n-1)},$$

and r_0 can be determined as in Theorem 7.2.

PROOF. Take an arbitrary point $a \in U$ and determine from it numbers r_0 and r_1 as indicated in Theorem 7.2. If $0 < |x - a| \le r_1$, then $x \in U$, $f(x) \ne f(a)$, and $l_f(a, r_1) = L_f(a, r_0)$. Let

$$u(x) = \ln[1/|f(x) - f(a)|].$$

The function $u(x)$ is a generalized solution of the equation

$$\operatorname{div} F_q[x, u'(x)] = 0 \tag{7.14}$$

in $G = B(a, r_1) \setminus \{a\}$, where $F(x, q) = \langle \theta(x)q, q \rangle^{n/2}$, and $u(x) \to \infty$ as $x \to a$. For $r \in (0, r_0)$ let

$$\Lambda(r) = \ln[1/l_f(a, r)], \qquad \lambda(r) = \ln[1/L_f(a, r)].$$

The inequalities $\lambda(r) \leq u(x) \leq \Lambda(r)$ hold at each point of $S(a, r)$, $u(x) = \lambda(r)$ for at least one point $x \in S(a, r)$, and $u(x) = \Lambda(r)$ for at least one point $x \in S(a, r)$. Fix an arbitrary $r \leq r_0$, $r > 0$, and let $u_0(x) = u(x) - \lambda(r)$. The function u_0 is also a solution of (7.14); u_0 is continuous on $\overline{B}(a, r_1) \setminus \{a\}$, and $u_0(x) \geq 0$ at each point $x \in S(a, r)$. Since $u_0(x)$ is continuous, $u_0(x) \geq 0$ on $S(a, r)$ also in the generalized sense defined in §5.3. Take an arbitrary point $x_0 \in B(a, r)$, $x_0 \neq a$. Let $|x_0 - a| = t$. We have that $0 < t < r$. Let v_1 and v_2 be solutions of (7.14) in the ring domain $H = B(a, r) \setminus \overline{B}(a, t)$ which satisfy the following boundary conditions:

$$v_1(x) = 0 \quad \text{on } S(a, r), \qquad v_1(x) = u_0(x) \quad \text{on } S(a, t),$$
$$v_2(x) = v_1(x) \quad \text{on } S(a, r), \qquad v_2(x) = \Lambda(t) - \lambda(r) \quad \text{on } S(a, t).$$

Let $A_0 = \mathbf{R}^n \setminus B(a, r)$ and $A_1 = \overline{B}(a, t)$. We have that

$$v_1 \geq u_0 \quad \text{on } S(a, r) = \partial_{A_0} H, \qquad v_1 = v_2 \quad \text{on } S(a, t) = \partial_{A_1} H,$$
$$v_2 = v_1 \quad \text{on } \partial_{A_0} H, \qquad v_2 \geq v_1 \quad \text{on } \partial_{A_1} H.$$

On the basis of Theorem 5.8 of this chapter, this implies that

$$\Omega_0 = \Omega(A_0, A_1, u_0, F) \leq \Omega(A_0, A_1, v_1, F) \leq \Omega(A_0, A_1, v_2, F),$$

i.e.,

$$n\omega_n |j(a, f)| \leq \Omega(A_0, A_1, v_2, F). \tag{7.15}$$

Let $\Lambda(t) - \lambda(t) = h$. The function Λ is decreasing, and hence $\Lambda(t) \geq \Lambda(r) \geq \lambda(r)$; thus, $h \geq 0$. The equality $h = 0$ is impossible here, since v_2 would otherwise be identically zero, and then $\Omega(A_0, A_1, v_2, F)$ would be equal to zero, which is impossible in view of (7.15). Thus, $h > 0$. Let $\zeta(x) = v_2(x)/h$. Then $\zeta(x) = 0$ on $S(a, r)$, $\zeta(x) = 1$ on $S(a, t)$, and, hence,

$$\begin{aligned}
\Omega(A_0, A_1, v_2, F) &= \int_H \langle \zeta'(x), F_q[x, v_2'(x)] \rangle \, dx \\
&= \frac{n}{h} \int_H \langle \theta(x)v_2'(x), v_2'(x) \rangle^{n/2} \, dx.
\end{aligned} \tag{7.16}$$

Let $w(x)$ be an arbitrary function of class $W_n^1(H)$ such that $w(x) = 0$ on $S(a, r)$ and $w(x) = 1$ on $S(a, t)$. Since v_2 minimizes the integral

$$\int_H \langle \theta(x)u'(x), u'(x) \rangle^{n/2} \, dx$$

on the set of functions equal to 0 on $S(a,r)$ and equal to h on $S(a,t)$, it follows that

$$\int_H \langle \theta(x)v_2'(x), v_2'(x) \rangle^{n/2}\, dx \leq h^n \int_H \langle \theta(x)w'(x), w'(x) \rangle^{n/2}\, dx$$

$$\leq K_0(f)h^n \int_H |w'(x)|^n\, dx,$$

which, by (7.15) and (7.16), gives us that

$$n\omega_n|j(a,f)| \leq nK_0(f)h^{n-1} \int_H |w'(x)|^n\, dx.$$

Since w is an arbitrary function equal to 0 on $S(a,r)$ and 1 on $S(a,t)$, this implies that

$$n\omega_n|j(a,f)| \leq nK_0(f)h^{n-1}C_p(A_0, A_1).$$

As established in §3 of this chapter,

$$C_n(A_0, A_1) = \omega_n/(\ln r/t)^{n-1},$$

and as a result we arrive at the inequality

$$|j(a,r)| \leq K_0(f)\{[\Lambda(t) - \lambda(r)]/(\ln r/t)\}^{n-1},$$

whence $\Lambda(t) - \lambda(r) \geq [|j(a,f)|/K_0(f)]^{1/(n-1)}\ln(r/t)$, i.e.,

$$\ln \frac{L_f(a,r)}{l_f(a,t)} \geq \alpha(a,f)\ln(r/t).$$

From this,

$$\frac{l_f(a,t)}{L_f(a,r)} \leq (t/r)^{\alpha(a,f)}. \tag{7.17}$$

On the basis of Theorem 7.2,

$$\frac{L_f(a,t)}{l_f(a,r)} \leq M_1. \tag{7.18}$$

We get from (7.17) and (7.18) that

$$\frac{|f(x) - f(a)|}{L_f(a,r)} \leq \frac{L_f(a,t)}{l_f(a,t)} \leq M_1 \left(\frac{|x-a|}{r} \right)^{\alpha(a,f)}.$$

The theorem is proved.

The results in this section are a refinement of results in the author's paper [134]. We remark that estimates analogous to those established here were also obtained by Poletskiĭ [113] and Martio [91] by another method.

§8. Characterization of mappings with bounded distortion by the property of quasiconformality

§8.1. The concept of a mapping which is quasiconformal at a point and in a domain. The purpose of this section is to prove that the properties of mappings with bounded distortion established in the preceding sections can be generalized in a particular sense. Further, a geometric characterization of mappings with bounded distortion will also be obtained.

Suppose that $U \subset \mathbf{R}^n$ is an open set and $f: U \to \mathbf{R}^n$ is an isolated open mapping. Then for every point $x \in U$ the integer $j(x, f)$ (the index of f) is defined. We say that f *is of constant sign* if either $j(x, f) > 0$ for all $x \in U$, or $j(x, f) < 0$ for all $x \in U$. In the first case f will be called *positive*, and in the second *negative*. An isolated open mapping of constant sign will be called a *mapping of type T*, or, briefly, a *T-mapping*. By results in §6, every mapping with bounded distortion is a T-mapping.

Let x_0 be an arbitrary point in $\mathbf{R}^n$. Assume that some closed neighborhood $G_t(x_0)$ of x_0 is defined for every $t \in (0, 1]$. We say that the set $\{G_t(x_0) | t \in (0, 1]\}$ of neighborhoods forms a *normal system* if there exists a continuous function $v: \mathbf{R}^n \to \mathbf{R}$ such that $v(x_0) = 0$, $v(x) > 0$ for $x \neq x_0$, $G_t(x_0) = \{x \in \mathbf{R}^n | v(x) \leq t\}$, and the set $\Gamma_t(x_0) = \{x \in \mathbf{R}^n | v(x) = t\}$ is the boundary of $G_t(x_0)$ for each $t \in (0, 1]$. The function v is called the *generating function* for the normal system $\{G_t(x_0) | t \in (0, 1]\}$. Let

$$r_G(x_0, t) = \inf_{x \in \Gamma_t(x_0)} |x - x_0|, \tag{8.1}$$

$$R_G(x_0, t) = \sup_{x \in \Gamma_t(x_0)} |x - x_0|. \tag{8.2}$$

The limit inferior

$$\alpha = \varliminf_{t \to 0} [r_G(x, t)] / R_G(x, t) \tag{8.3}$$

is called the *regularity parameter* of the family $\{G_t(x_0) | 0 < t \leq 1\}$. The system $\{G_t(x_0) | 0 < t \leq 1\}$ of neighborhoods is said to be *regular* if $\alpha > 0$.

Suppose that $f: U \to \mathbf{R}^n$ is a mapping of type T, $x_0 \in U$ and $\{G_t(x_0) | 0 < t \leq 1\}$ is a normal system of neighborhoods of x_0. Let

$$r_G(x_0, t, f) = \inf_{x \in \Gamma_t(x_0)} |f(x) - f(x_0)|, \tag{8.4}$$

$$R_G(x_0, t, f) = \sup_{x \in \Gamma_t(x_0)} |f(x) - f(x_0)|. \tag{8.5}$$

The mapping f is said to be quasiconformal at x_0 if there exists a normal regular system $\{G_t(x_0) | 0 < t \leq 1\}$ of neighborhoods of x_0 such that

$$\beta = \varliminf_{t \to 0} \frac{r_G(x_0, t, f)}{R_G(x_0, t, f)} > 0. \tag{8.6}$$

The quantity $q(x_0, f) = 1/\alpha\beta$, where α is the regularity parameter of the system $G_t(x_0)$, is called the *coefficient of quasiconformality of f at x_0.*

A mapping $f: U \to \mathbf{R}^n$ of type T is said to be *quasiconformal* in a domain $U \subset \mathbf{R}^n$ if f is quasiconformal at each point $x \in U$, and there exists a constant Q, $1 \le Q < \infty$, such that $q(x, f) \le Q$ for all $x \in U$. The smallest constant Q satisfying this condition is called the *quasiconformality coefficient of f in U* and denoted by $q(U, f)$.

Let $f: U \to \mathbf{R}^n$ be a mapping with bounded distortion. For every domain V lying strictly inside U there exists a constant $k_0 < \infty$ such that $|j(x, f)| \le k_0$ for all $x \in V$. By Theorem 7.3, this implies that f is quasiconformal on V. In the notation used below, the index G indicating the given normal system of neighborhoods will be omitted whenever no confusion is possible.

§8.2. Differentiability almost everywhere of quasiconformal T-mappings.
Let $U \subset \mathbf{R}^n$ be an arbitrary open domain, and let $f: U \to \mathbf{R}^n$ be a T-mapping. We assume that f is a positive mapping. For an arbitrary set $A \subset U$ and any $y \in \mathbf{R}^n$ let $N(y, f, A)$ be the number of elements in $f^{-1}(y) \cap A$. The function $y \to N(y, f, A)$ is the multiplicity function of f considered in §2.4. Lemma 2.4 in §2.4 allows us to conclude that this function is the limit of a sequence of Borel-measurable functions, and thus it is measurable.

LEMMA 8.1. *Suppose that $f: U \to \mathbf{R}^n$ is a positive T-mapping. If a set A lies strictly inside U, then the function $y \to N(y, f, A)$ is bounded.*

PROOF. Suppose that A lies strictly inside U. Let $y \in \mathbf{R}^n$ be an arbitrary point. By the condition of the lemma, all the points of $f^{-1}(y)$ are isolated. This implies that for every point $x \in \overline{A}$ there is a ball about x lying strictly inside U whose boundary does not contain points of $f^{-1}(y)$. The set $\overline{A}$ is compact. Covering $\overline{A}$ by finitely many such balls, we get an open set G such that $\overline{A} \subset G$, $\overline{G}$ is compact and contained in U, and ∂G does not contain points of $f^{-1}(y)$. We have that

$$N(y, f, A) \le N(y, f, G) \le \sum_{x \in f^{-1}(y) \cap G} j(x, f) = \mu(y, f, G).$$

The function $y \to \mu(y, f, G)$ is constant on each connected component of $\mathbf{R}^n \backslash f(\partial G)$. This implies that each point $y \in \mathbf{R}^n$ has a neighborhood in which $N(y, f, A)$ is bounded. By the compactness of $f(\overline{A})$, this gives us that $N(y, f, A)$ is bounded on $f(\overline{A})$. For $y \notin f(\overline{A})$ we have that $N(y, f, A) = 0$. It is thereby established that $N(y, f, A)$ is bounded, and the lemma is proved.

For a T-mapping $f\colon U \to \mathbf{R}^n$ we let

$$l_f(A) = \int_{\mathbf{R}^n} N(y, f, A)\, dy,$$

where A is a Borel set contained in U. Obviously, l_f is a measure on the σ-algebra of Borel sets contained in U. For every compact set $M \subset U$ there exists a constant $k = k(M)$ such that $l_f(A) \leq k|f(A)|$ for any Borel set $A \subset M$.

Fix a domain $U \subset \mathbf{R}^n$ and a T-mapping $f\colon U \to \mathbf{R}^n$ that is quasiconformal in U. Let $Q > q(U, f)$ be an arbitrary number. For each point $x \in U$ let $\{G_t(x)|0 < t \leq 1\}$ be a fixed normal regular system of neighborhoods such that $\alpha(x)\beta(x) > 1/Q$, where $\alpha(x)$ is the regularity parameter of the system $G_t(x)$ and

$$\beta(x) = \lim_{t \to 0} r(x, t, f)/R(x, t, f).$$

Here $r(x, t)$, $R(x, t)$, $r(x, t, f)$, and $R(x, t, f)$ are defined by (8.1)–(8.5). Obviously, $\alpha(x), \beta(x) > 1/Q$ for all $x \in U$. It will be assumed (without loss of generality) that

$$r(x, t)/R(x, t) > 1/Q, \qquad r(x, t, f)/R(x, t, f) > 1/Q$$

for any $x \in U$ and any $t \in (0, 1]$.

LEMMA 8.2. *Let* $f\colon U \to \mathbf{R}^n$ *be a quasiconformal T-mapping. For* $x_0 \in U$ *let*

$$k(x_0) = \varlimsup_{x \to x_0} |f(x) - f(x_0)|/|x - x_0|.$$

Then $k(x_0) < \infty$ *for almost all* $x_0 \in U$, *and* $\int_A [k(x)]^n\, dx \leq Q^{2n} l_f(A)$ *for every Borel set* $A \subset U$.

PROOF. By the classical Lebesgue theorem on differentiating set functions, the limit

$$\lim_{t \to 0} \frac{l_f[G_t(x)]}{|G_t(x)|} = \gamma(x)$$

exists and is finite for almost all $x \in U$. Further, for every Borel set $A \subset U$

$$\int_A \gamma(x)\, dx \leq l_f(A). \tag{8.7}$$

We show that if $\gamma(x) < \infty$ at a point $x \in U$, then also $k(x) < \infty$; moreover, $k(x) \leq Q^2[\gamma(x)]^{1/n}$. Fix an arbitrary point $x_0 \in U$ such that $\gamma(x_0) < \infty$.

Let (x_ν), $\nu = 1, 2, \ldots$, be a sequence of points of U such that $x_\nu \to x_0$ as $\nu \to \infty$, $x_\nu \neq x_0$ for all ν, and

$$|f(x_\nu) - f(x_0)|/|x_\nu - x_0| \to k(x_0)$$

as $\nu \to \infty$. Let v be the generating function for the system of neighborhoods $G_t = G_t(x_0)$. Let $t_\nu = v(x_\nu)$ and $y_\nu = f(x_\nu)$. Obviously, x_ν is a boundary point of G_t, which implies that

$$|x_\nu - x_0| \geq r(x_0, t_\nu).$$

Further, since $x_\nu \in \partial G_{t_\nu}$, it clearly follows that

$$|y_\nu - y_0| \leq R(x_0, t_\nu, f).$$

From this,

$$R(x_0, t_\nu) \leq Qr(x_0, t_\nu) \leq Q|x_\nu - x_0| \tag{8.8}$$

and

$$r(x_0, t_\nu, f) \geq (1/Q)R(x_0, t_\nu, f) \geq (1/Q)|y_\nu - y_0|. \tag{8.9}$$

We have that

$$|f(G_{t_\nu})| \leq l_f(G_{t_\nu}), \tag{8.10}$$

whence

$$l_f(G_{t_\nu})/|G_{t_\nu}| \geq |f(G_{t_\nu})|/|G_{t_\nu}|.$$

The set G_{t_ν} is conained in the ball about x_0 with radius $R(x_0, t_\nu)$, hence also in the ball of radius $Q|x_\nu - x_0|$ about x_0, by (8.8). From this,

$$|G_{t_\nu}| \leq \sigma_n Q^n |x_\nu - x_0|^n. \tag{8.11}$$

The set $f(G_{t_\nu})$ clearly contains the ball of radius $r(x_0, t_\nu, f)$, hence also the ball of radius $|y_\nu - y_0|/Q$, by (8.9). From this,

$$|f(G_{t_\nu})| \geq (\sigma_n/Q^n)|y_\nu - y_0|^n. \tag{8.12}$$

Inequalities (8.10)–(8.12) allow us to conclude that

$$\frac{l_f(G_{t_\nu})}{G_{t_\nu}|} \geq \frac{1}{Q^{2n}} \frac{|y_\nu - y_0|^n}{|x_\nu - x_0|^n}.$$

Passing to the limit as $\nu \to \infty$, we get that

$$\gamma(x_0) \geq \lim_{\nu \to \infty} \frac{1}{Q^{2n}} \left(\frac{|y_\nu - y_0|}{|x_\nu - x_0|} \right)^n = [k(x_0)]^n/Q^{2n}.$$

Inequality (8.7) allows us to conclude that

$$\int_A [k(x_0)]^n \, dx \leq Q^{2n} l_f(A)$$

for every Borel set $A \subset U$. The lemma is proved.

COROLLARY. *If $f : U \to \mathbf{R}^n$ is a quasiconformal T-mapping, then f is differentiable almost everywhere in the domain U.*

Indeed, we have the following classical result due to Stepanov.

THEOREM 8.1 [163]. *Let U be an open set in $\mathbf{R}^n$, and $f: U \to \mathbf{R}^n$ a continuous function. If for almost all $x \in U$*

$$\varlimsup_{y \to x} \frac{|f(y) - f(x)|}{|y - x|} < \infty,$$

then f is differentiable almost everywhere in $\mathbf{R}^n$.

By Lemma 8.2, every quasiconformal T-mapping obviously satisfies the conditions of Stepanov's theorem, and this leads to the corollary.

§8.3. **The condition of absolute continuity for a real function of a single variable.** We make some remarks about continuous functions of a single real variable.

Let $f: [a, b] \to \mathbf{R}$ be a given function. The function f is said to *satisfy condition N* if the image $f(E)$ of every closed set $E \subset [a, b]$ with measure zero is a set of measure zero.

Let $f: [a, b] \to \mathbf{R}$ be a continuous function. For $x \in [a, b]$

$$Df(x) = \varlimsup_{h \to 0} \left| \frac{f(x + h) - f(x)}{h} \right|.$$

The function Df is measurable. Indeed, let Δ_n be the set of all rational numbers $h \neq 0$ such that $|h| < 1/n$, and let

$$D_n f(x) = \sup_{0 < |h| < 1/n} \left| \frac{f(x + h) - f(x)}{h} \right|.$$

Then $D_n f(x) \to Df(x)$ for all x. By continuity,

$$D_n f(x) = \sup_{h \in \Delta_n} \left| \frac{f(x + h) - f(x)}{h} \right|.$$

From this it is clear that $D_n f$ is measurable, being the upper envelope of a countable set of measurable functions, and hence $Df = \lim_{n \to \infty} D_n f$ is measurable.

LEMMA 8.3. *Let $f: [a, b] \to \mathbf{R}$ be a continuous function. If f has property N and $\int_a^b Df(x)\, dx < \infty$, then f is absolutely continuous.*

PROOF. Suppose that $f: [a, b] \to \mathbf{R}$ satisfies all the conditions of the lemma. Let $\varphi(x) = \int_a^x Df(t)\, dt$, and let A be the collection of all points $x \in [a, b]$ at which $Df(x) < \infty$, φ is differentiable, and $\varphi'(x) = Df(x)$. Then $[a, b] \backslash A$ is clearly a set of measure zero.

Take an arbitrary number $\varepsilon > 0$. Let $[p, q] \subset [a, b]$. We say that $[p, q]$ is an *interval of type H* if for all $x \in [p, q]$

$$\varphi(x) - \varphi(p) + \varepsilon(x - p) \geq f(x) - f(p)$$
$$\geq -[\varphi(x) - \varphi(p) + \varepsilon(x - p)]. \tag{8.13}$$

If $[p, q]$ is an interval of type H, then the measure of the set $f([p, q])$ is clearly at most $2[\varphi(q) - \varphi(p)] + 2\varepsilon(q - p)$. Define $\psi_1(x) = \varphi(x) - f(x)$ and $\psi_2(x) = \varphi(x) + f(x)$. Suppose that $x \in A$ and $x < b$. Then

$$\lim_{x' \to x+0} \frac{\psi_1(x') - \psi_1(x)}{x' - x} \geq 0, \qquad \lim_{x' \to x+0} \frac{\psi_2(x') - \psi_2(x)}{x' - x} \geq 0,$$

which implies that there exists a $\delta(x) > 0$ such that if $x < x' < x + \delta(x)$, then

$$\frac{\psi_1(x') - \psi_1(x)}{x' - x} > -\varepsilon, \qquad \frac{\psi_2(x') - \psi_2(x)}{x' - x} > -\varepsilon.$$

This leads to the conclusion that if $x < x' < x + \delta(x)$, then

$$\varphi(x') - \varphi(x) - [f(x') - f(x)] = \psi_1(x') - \psi_1(x) > -\varepsilon(x' - x),$$
$$\varphi(x') - \varphi(x) + [f(x') - f(x)] = \psi_2(x') - \psi_2(x) > -\varepsilon(x' - x).$$

It obviously follows from these inequalities that every interval $[x, x']$ with $x < x' < x + \delta(x)$ is an interval of type H.

Take arbitrary points $x_1, x_2 \in [a, b]$ such that $x_1 < x_2$. We prove that

$$|f([x_1, x_2])| \leq 2 \int_{x_1}^{x_2} Df(t)\, dt. \tag{8.14}$$

Let $A_0 = A \cap [x_1, x_2]$. On the basis of the Vitali covering theorem, there exists an at most countable set $\{[p_m, q_m],\ m = 1, 2, \ldots\}$ of disjoint intervals of type H contained in $[x_1, x_2]$ such that if $U_1 = \bigcup [p_m, q_m]$, then A_0/U_1 is a set of measure zero. Denote by U the union of the open intervals (p_m, q_m). Obviously, $|U_1| = |U| = x_2 - x_1$, and thus $E_0 = [x_1, x_2] \setminus U$ is a closed set of measure zero. We have that

$$[x_1, x_2] = \left(\bigcup_m (p_m, q_m) \right) \cup E_0,$$
$$f([x_1, x_2]) = \bigcup_m f([p_m q_m]) \cup f(E_0).$$

Since E_0 is a closed set of measure zero, and f has the property N by assumption, it follows that $|f(E_0)| = 0$, which gives us that

$$|f([x_1, x_2])| \leq \sum_m |f([p_m, q_m])|. \tag{8.15}$$

Since each of the intervals $[p_m, q_m]$ is an interval of type H,

$$|f([p_m, q_m])| \leq 2[\varphi(q_m) - \varphi(p_m)] + 2\varepsilon(q_m - p_m)$$
$$= 2 \int_{p_m}^{q_m} (Df)(t)\, dt + 2\varepsilon(q_m - p_m).$$

Summing these inequalities termwise, we get by (8.15) that

$$|f([x_1, x_2])| \leq 2 \int_U Df(t)\, dt + 2\varepsilon |U|$$

$$\leq 2 \int_{x_1}^{x_2} Df(t)\, dt + 2\varepsilon(x_2 - x_1).$$

Since $\varepsilon > 0$ is arbitrary, this proves (8.14).

Since $|f(x_2) - f(x_1)| \leq f([x_1, x_2])$, it follows from what was proved that for any $x_1, x_2 \in [a, b]$, $x_1 \leq x_2$,

$$|f(x_2) - f(x_1)| \leq 2 \int_{x_1}^{x_2} Df(t)\, dt.$$

By well-known properties of the Lebesgue integral, the last inequality implies that f is absolutely continuous, and the lemma is proved.

§8.4. The analytic nature of quasiconformal T-mappings.

We now establish that *every Q-quasiconformal mapping of type T belongs to the class $W_{n,\mathrm{loc}}^1(U)$.*

The proof of this is split into several steps and is based on an idea in Men'shov's classical paper [97].

Let U be an open domain in $\mathbf{R}^n$, and $f: U \to \mathbf{R}^n$ a quasiconformal mapping of type T. Take an arbitrary integer m such that $q(U, f) < m$. For each point $x \in U$ fix a regular normal system $\{G_t(x), 0 < t \leq 1\}$ of neighborhoods such that if $\alpha(x)$ is the regularity parameter of this family and

$$\beta(x) = \lim_{t \to 0} r(x, t, f)/R(x, t, f),$$

then $\alpha(x)\beta(x) \geq 1/q(U, f) > 1/m$. We assume that

$$\frac{r(x, t)}{R(x, t)} > \frac{1}{m} \quad \text{and} \quad \frac{r(x, t, f)}{R(x, t, f)} > \frac{1}{m}$$

for all $t \in (0, 1]$. Obviously, there is no loss of generality in this.

Let $a = (a_1, \ldots, a_n)$ be an arbitrary point in U, and let $h > 0$ be such that the cube $\overline{Q} = \overline{Q}(a, h)$ is in U. Denote by C_k $(1 \leq k \leq n)$ the section of $\overline{Q}$ by the plane $P_k(a) = \{x \in \mathbf{R}^n | x_k = a_k\}$. If $x \in C_k$, then let $p(x)$ be the interval $|x_k - a_k| \leq h/2$ of the line passing through x parallel to the kth coordinate axis of $\mathbf{R}^n$. If A is a subset of C_k, then let $p(A)$ be the union of all the intervals $p(x)$ with $x \in A$. For an arbitrary set E lying in some l-dimensional plane P, $|E|_l$ denotes the l-dimensional Lebesgue outer measure of E.

LEMMA 8.4. *The function f has property N on the interval $p(x)$ for almost all points $x \in C_k$ in the sense of $(n-1)$-dimensional Lebesgue measure in the plane $P_k(a)$.*

PROOF. We define in C_k a measure μ by setting $\mu(E) = l_f[p(E)]$ for a Borel set $E \subset C_k$. The set $p(E)$ is obviously also a Borel set, so $\mu(E)$ is defined.

Let $K(x,r)$ be the $(n-1)$-dimensional ball in $P_k(a)$ about x with radius r. Define

$$D(x) = \varlimsup_{r \to 0} \mu[K(x,r)]/|K(x,r)|_{n-1}.$$

By the classical results of Lebesgue on differentiation of set functions, D is finite for almost all $x \in C_k$.

Suppose that the point $x \in C_k$ is such that f does not have property N on $p(x)$. We show that $D(x) = \infty$ in this case. By the preceding remark, this will prove the lemma.

According to the definition of property N, there is a closed set $E \subset p(x)$ such that $|E_1| = 0$ and $m_1[f(E_1)] > 0$. (Here m_1, as earlier, denotes the one-dimensional Hausdorff measure on $\mathbf{R}^n$.)

Let E_s be the collection of all $x \in E$ such that $r(x,1) > 1/s$, where $s > 0$ is an integer. Obviously,

$$E = \bigcup_{s=1}^{\infty} E_s, \qquad f(E) = \bigcup_{s=1}^{\infty} f(E_s),$$

which implies that

$$m_1[f(E)] \le \sum_{s=1}^{\infty} m_1[f(E_s)],$$

and $m_1[f(E_s)] > 0$ for at least one s, because $m_1[f(E)] > 0$. Fix an arbitrary such value of s, and let $\gamma = \gamma_1[f(E_s)]$ be the 1-content of $f(E_s)$. By the choice of s we have that $\gamma > 0$, and for every system of balls covering $f(E_s)$ the sum of the radii is at least $\gamma/2$.

Let $\varepsilon > 0$ be arbitrary, and let $N > 0$ be an integer such that $h/2N(m+1) < 1/s$. We divide the interval $p(x)$ into $2N(m+1)$ equal intervals. Let V_n be the union of all the intervals in the partition which contain points of E. Since $|E|_1 = 0$ and E is closed, it follows that $|V_n|_1 \to 0$ as $N \to \infty$. Let N_0 be such that $|V_N| < \varepsilon$ for $N \ge N_0$. Fix an arbitrary $N \ge N_0$. From the intervals forming V_N choose all those which contain points of E_s, and take any point of E_s in each of them. Let $\xi_1, \ldots, \xi_l$ be all the points obtained in this way. Let $\rho = h/2N(m+1)$. For each $i = 1, \ldots, l$ there is a $t_i > 0$ with $t_i < 1$ such that $r(\xi_i, t_i) = \rho$.

The required number t_i can be found as follows. Let v_i be the generating function of the system $\{G_t(\xi_i)|0 < t \le 1\}$ of neighborhoods. The largest value of v_i in $\overline{B}(\xi_i, \rho)$ is the desired value t_i. Let $G_{t_i}(\xi_i) = G_i$. The sets $G_1, \ldots, G_l$ cover the set E_s, and the sets $H_i = f(G_i)$ cover $f(E_s)$. The ball of radius $mr(\xi_i, t_i, f)$ about $\eta_i = f(\xi_i)$ contains H_i. This gives us that $\sum_i r(\xi_i, t_i, f) > \gamma/2m$.

Let us divide the intervals in the partition of $p(x)$ constructed above into $2m + 2$ groups by putting into a single group the intervals obtainable one from another by parallel translation by a multiple of h/N. Corresponding to this, the points $\xi_1, \ldots, \xi_l$ are distributed into $2m+2$ classes by putting the points belonging to the intervals of a single group into a single class. For at least one of these classes the sum of the values $r(\xi_i, t_i, f)$ is at least $\gamma/4(m + 2)m = \theta$. We assume that $\xi_1, \ldots, \xi_q$ are all the points ξ_i in this class. By construction

$$|\xi_i - \xi_j| \ge h/N - 2h/(2m + 2)N = mh/(m + 1)N$$

for any $i, j \le q$, $i \ne j$. Denote by B_i the ball about ξ_i of radius $mr(\xi_i, t_i) = m\rho = mh/N(m + 1)$. The balls B_i with $i = 1, \ldots, q$ are disjoint, and $B_i \supset G_i$, which implies that the neighborhoods G_i are disjoint. The number q does not exceed the number of intervals forming the set V_N; therefore, the sum of the lengths of these intervals, which equals $q\rho$, does not exceed $|V_n| < \varepsilon$. Consider the ball $K(x, m\rho)$. Let $Z_\rho = p[K(x, \rho)]$ be the cylinder constructed on this ball. The set $f(G_i)$ contains the n-dimensional ball about η_i with radius $r(\xi_i, t_i, f)$. We have that $G_i \subset Z_\rho$, and since the G_i are disjoint, $l_f(Z_\rho)$ is not less than the sum of the volumes of the balls about η_i with radii $r(\xi_i, t_i, f)$, i.e.,

$$l_f(Z_\rho) \ge \sigma_n \sum_{i=1}^{q} [r(\xi_i, t_i, f)]^n.$$

By the elementary inequality $u_1^n + \cdots + u_q^n \ge q^{1-n}(u_1 + \cdots + u_q)^n$, where the u_i are nonnegative, we get that

$$l_f(Z_\rho) \ge \sigma_n q^{1-n} \left[\sum_{i=1}^{q} r(\xi_i, t_i, f) \right]^n \ge \sigma_n (\gamma/2m)^n q^{1-n}.$$

On the other hand, we have that $q\rho < \varepsilon$, which implies that $1/q < \rho/\varepsilon$, and hence $q^{1-n} > \rho^{n-1}/\varepsilon^{n-1}$. From this, $l_f(Z_\rho) \ge \sigma_n(\gamma/2m)^n \rho^{n-1}/\varepsilon^{n-1}$ and, consequently, $l_f(Z_\rho)/K(x, m\rho) \ge C/\varepsilon^{n-1}$, where $C > 0$ does not depend on N. We have that $\rho = h/2N(m + 1)$. Since $N \ge N_0$ here is arbitrary, this implies that $\overline{D}(x) \ge C/\varepsilon^{n-1}$, and thus the arbitrariness of $\varepsilon > 0$ gives us that $\overline{D}(x) = \infty$, as was to be proved.

LEMMA 8.5. *Let* $f: U \to \mathbf{R}^n$ *be a mapping of type T, where U is an open domain in* $\mathbf{R}^n$. *If f is quasiconformal in U, then f belongs to the class* $W^1_{n,\text{loc}}(U)$.

PROOF. Let $Q = Q(x_0, h)$ be an arbitrary cube such that $Q \subset U$, and let C_k be the section of Q by the plane $x_k = x_{0k}$. By Lemma 8.4, the vector-valued function f has property N on $p(x)$ for almost all $x \in C_k$. Let

$$Df(x) = \overline{\lim_{x' \to x}} \frac{|f(x') - f(x)|}{|x' - x|}.$$

The function Df is integrable to the power n on Q in view of Lemma 8.2. This gives us that Df is integrable over $p(x)$ for almost all $x \in C_k$. By Lemma 7.4, f is absolutely continuous on $p(x)$ for every $x \in C_k$ such that f has property N on $p(x)$ and Df is integrable on $p(x)$. We have that

$$\left| \frac{\partial f_j}{\partial x_k}(x) \right| = \lim_{h \to 0} (1/h)|f_j(x + he_k) - f_j(x)|$$

$$\leq \overline{\lim_{h \to 0}} \frac{|f(x + h) - f(x)|}{|h|} = Df(x).$$

The function Df is integrable to the power n. Thus, we get that for each $k = 1, 2, \ldots$ the function f is absolutely continuous on almost all lines parallel to the kth coordinate axis, and its derivative $\partial f / \partial x_k$ is integrable to the power n on Q. By the test in §1 of Chapter I for a function to belong to W^1_n, this implies that f is in $W^1_n(Q)$. Since the cube $Q \subset U$ was taken arbitrarily, this establishes that $f \in W^1_{n,\text{loc}}(U)$, and the lemma is proved.

§8.5. Main result.

LEMMA 8.6. *Suppose that U is an open domain in* $\mathbf{R}^n$, *and* $f: U \to \mathbf{R}^n$ *is a continuous mapping of type T. Assume that f is differentiable at a point a, with* $df_a \neq 0$, *and that there exists a normal system* $\{G_t(a), 0 < t \leq 1\}$ *of neighborhoods of a with regularity coefficient* $\alpha > 0$ *such that*

$$\lim_{t \to 0} \frac{r(a, t, f)}{R(a, t, f)} = \beta > 0.$$

Then

$$\mathscr{J}(a, f) = \det df_a \neq 0, \qquad \text{sgn}\, \mathscr{J}(a, f) = j(a, f),$$

where $j(a, f)$ *is the index of f at a, and the quasiconformality coefficient of the linear mapping* $L = df_a$ *does not exceed* $1/\alpha\beta$.

PROOF. If $\mathscr{J}(a, f) \neq 0$, then $\text{sgn}\, \mathscr{J}(a, f) = j(a, f)$ in view of the properties of the index indicated in §2. Suppose that the linear mapping $L = df_{x_0}$ is nonzero, $u_1, \ldots, u_n$ is an orthonormal system of principal direction vectors of L, and $v_1, \ldots, v_n$ are orthogonal unit vectors such that $Lu_i = \mu_i v_i$, where $\mu_1 \geq \cdots \geq \mu_n$ are the principal dilations of L.

To simplify the notation we assume that $a = 0$ and $b = f(a) = 0$. Clearly this can always be arranged by suitably choosing the coordinate system. Let $r(t) = r(0, t)$, $R(t) = R(0, t)$, $r_f(t) = r(0, t, f)$, $R_f(t) = R(0, t, f)$, $A_t = \partial G_t$, and $B_t = f(A_t)$. Since f is differentiable by assumption at the point $a = 0$, we have that $f(x) = L(x) + \varepsilon(x)|x|$, where $\varepsilon(x) \to 0$ as $x \to 0$.

Suppose that $a(t) > 0$ and $b(t) > 0$ are such that $a(t)u_1 \in A_t$ and $b(t)u_n \in A_t$. Then

$$r_f(t) \leq |f[b(t)u_n]|, \qquad R_f(t) \geq |f[a(t)u_1]|.$$

From this,

$$
\begin{aligned}
\frac{r_f(t)}{R_f(t)} &\leq \frac{|f[b(t)u_n]|}{|f[a(t)u_1]|} \leq \frac{|f[b(t)u_n]|}{b(t)} \cdot \frac{a(t)}{|f[a(t)u_1]|} \cdot \frac{b(t)}{a(t)} \\
&\leq \frac{|f[b(t)u_n]|}{b(t)} \cdot \frac{a(t)}{|f[a(t)u_1]|} \cdot \frac{R(t)}{r(t)}.
\end{aligned}
\tag{8.16}
$$

We have that

$$f[b(t)u_n] = \mu_n b(t)u_n + \varepsilon_2(t)b(t),$$
$$f[a(t)u_1] = \mu_1 a(t)u_1 + \varepsilon_1(t)a(t),$$

where $\varepsilon_1(t), \varepsilon_2(t) \to 0$ as $t \to 0$. Hence,

$$\frac{|f[b(t)u_n]|}{b(t)} \to \mu_n, \qquad \frac{|f[a(t)u_1]|}{a(t)} \to \mu_1$$

as $t \to 0$. Further,

$$\varlimsup_{t \to 0} R(t)/r(t) = 1/\alpha,$$

and, passing in (8.16) to the limit as $t \to 0$, we get that $\beta \leq \mu_n/\mu_1 \cdot 1/\alpha$. This gives us that $\mu_n > 0$, and thus L is nonsingular, i.e., $\det L = \mathscr{J}(a, f) \neq 0$. Further, $\mu_1/\mu_n \leq 1/\alpha\beta$, and the lemma is proved.

THEOREM 8.2 [134]. *Every mapping $f: U \to \mathbf{R}^n$ of type T which is quasiconformal in a domain $U \subset \mathbf{R}^n$ is a mapping with bounded distortion.*

PROOF. Suppose that $f: U \to \mathbf{R}^n$ satisfies all the conditions of the theorem. Then, by Lemma 8.5, f belongs to $W^1_{n,\mathrm{loc}}(U)$. By the corollary to Lemma 8.2, f is differentiable almost everywhere in U. Lemma 8.7 now allows us to conclude that the Jacobian of f has constant sign in U. Finally, by the same Lemma 8.7, the distortion coefficient of the linear mapping $f'(x)$ does not exceed $q(U, f) \leq \infty$. Thus, f satisfies all the conditions in the definition of a mapping with bounded distortion, and the theorem is proved.

COROLLARY. *Let U be an open subset of $\mathbf{R}^n$, and $f: U \to \mathbf{R}^n$ an isolated open mapping. Assume that f is quasiconformal in every domain $\mathscr{D}$ lying strictly inside U, and there exists a constant $Q \geq 1$, $Q < \infty$, such that $q(x, f) \leq Q$ at each point $x \in U \backslash B_f$. Then f is a mapping with bounded distortion.*

PROOF. It follows from Theorem 8.2 that f is a mapping with bounded distortion on every domain $\mathscr{D}$ lying strictly inside U. For almost all $x \in B_f$ we have that $\mathscr{J}(x, f) = 0$, and hence $f'(x) = 0$. The inequality $q(x, f) \leq Q$ holds at each point $x \notin B_f$. This implies that $K[f'(x)] \leq Q^n$ at each point $x \notin B_f$ such that f is differentiable and $\mathscr{J}(x, f) \neq 0$. Consequently, for almost all $x \in U$

$$\|f'(x)\|^n \leq Q^n \mathscr{J}(x, f),$$

and hence f is a mapping with bounded distortion.

§8.6. Homeomorphic quasiconformal mappings.
As an application of the above results, we prove the following assertion.

THEOREM 8.3. *Let U be an open set in $\mathbf{R}^n$, and $f: U \to \mathbf{R}^n$ a homeomorphism. If f is quasiconformal, then the inverse mapping $g = f^{-1}$ is also quasiconformal. Moreover, $q(V, g) = q(U, f)$, where $V = f(U)$.*

PROOF. Let $\varepsilon > 0$ be arbitrary. For every point $x \in U$ we construct a normal system $\{G_t(x) | 0 < t \leq 1\}$ of neighborhoods such that

$$\lim_{t \to 0} \frac{r_G(x, t)}{R_G(x, t)} = \alpha, \qquad \lim_{t \to 0} \frac{r_G(x, t, f)}{R_G(x, t, f)} = \beta,$$

with

$$\alpha\beta > k = 1/[q(U, f) + \varepsilon].$$

Let $H_t(y) = f[G_t(x)]$, where $y = f(x)$. The sets $H_t(x)$ clearly form a normal family of neighborhoods of the point $y = f(x)$. Further, if v is the generating function of the family of neighborhoods $G_t(x)$, then $w(x) = v[g(x)]$ is the generating function for the family of neighborhoods $H_t(x)$. We have that

$$r_G(x, t, f) = \inf_{y' \in \partial H_t} |y' - y| = r_H(y, t),$$

$$R_G(x, t, f) = \sup_{y' \in \partial H_t} |y' - y| = R_H(y, t),$$

$$r_G(x, t) = \inf_{y' \in \partial H_t} |g(y') - g(y)| = r_H(y, t, g),$$

$$R_G(x, t) = \sup_{y' \in \partial H_t} |g(y') - g(y)| = R_H(y, t, g).$$

It is clear from this that g is quasiconformal at the point y, and the quasiconformality coefficient for it at y does not exceed $1/\alpha\beta < q(U, f) + \varepsilon$. Since $y \in V$ and $\varepsilon > 0$ are arbitrary, the proof of the theorem is complete.

We prove the following assertion as an application of the last theorem.

THEOREM 8.4. *Suppose that $U \subset \mathbf{R}^n$ is an open set, and $f: U \to \mathbf{R}^n$ is a homeomorphic mapping with bounded distortion. Then $\mathscr{J}(x, f) \neq 0$ for almost all $x \in U$.*

PROOF. Suppose that $f: U \to \mathbf{R}^n$ satisfies the condition of the theorem. Let $V = f(U)$ and $g = f^{-1}$. The mapping g is quasiconformal by the theorem. Let E_1 be the set of all points $x \in U$ at which f is not differentiable. Then $|E_1| = 0$. Let E_2 denote the collection of all points $x \in U$ at which f is differential and $\mathscr{J}(x, f) = 0$. Since f has property N, $|f(E_1)| = 0$. By Theorem 2.2,

$$|f(E_2)| = \int_{E_2} |\mathscr{J}(x, f)| \, dx = 0.$$

Consequently, $A = f(E_1 \cup E_2)$ is a set of measure zero. Since g is a quasiconformal mapping and thus has property N, $|g(A)| = 0$. We have that $g(A) = E_1 \cup E_2$, and this proves the theorem.

§9. Sequences of mappings with bounded distortion

§9.1. A theorem on local boundedness of sequences of mappings with bounded distortion.

THEOREM 9.1. *Suppose that U is an open domain in $\mathbf{R}^n$, and $(f_m: U \to \mathbf{R}^n)$, $m = 1, 2, \ldots$, is a sequence of mappings with bounded distortion such that the sequence $(K(f_m))$ of their distortion coefficients is bounded. If (f_m) is locally bounded in $L_n(U)$, then it is locally bounded also in the sense of $W_n^1(U)$.*

PROOF. Let $f_m = (f_{m1}, \ldots, f_{mn})$. We take an arbitrary value of i such that $1 \leq i \leq n$ and let $u_m = f_{mi}$. For each m the function u_m is a generalized solution in U of the differential equation

$$\operatorname{div}(\langle \theta_m(x)u'(x), u'(x) \rangle^{(n/2)-1} \theta_m(x)u'(x)) = 0,$$

where $\theta_m(x) = \theta(x, f_m)$ is the matrix defined from f_m as indicated in §5. This means that for every $\varphi \in W_n^1(U)$ with compact support in U

$$\int_U \langle \theta_m u'_m, u'_m \rangle^{n/2-1} \langle \theta_m u'_m, \varphi' \rangle \, dx = 0. \tag{9.1}$$

The matrix θ_m is symmetric. Since the sequence $K(f_m)$ is bounded by an assumption of the theorem, there is a constant α, $0 < \alpha < 1$, such that for every m and for almost all $x \in U$

$$\alpha^2 |\xi|^2 \leq \langle \theta_m(x)\xi, \xi \rangle \leq 1/\alpha^2 |\xi|^2 \tag{9.2}$$

for any vector $\xi \in \mathbf{R}^n$. The constant α here depends only on the supremum K of the sequence $(K(f_m))$, $m = 1, 2, \ldots$.

Let $A \subset U$ be an arbitrary compact set, and let $\delta > 0$ be such that the closed δ-neighborhood $\overline{U}_\delta(A)$ of A is contained in U. We set $G = \overline{U}_\delta(A)$ and construct a function $\zeta \in C_0^\infty(U)$ such that $\zeta(x) = 1$ for $x \in A$, $\zeta(x) = 0$ for $x \notin G$, and $0 \leq \zeta(x) \leq 1$ for all x. Let $\varphi = u_m \zeta^n$ in (9.1). We transform the integrand. We have that $\varphi' = u'_m \zeta^n + n u_m \zeta^{n-1} \zeta'$. Further,

$$\langle \theta_m u'_m, u'_m \rangle^{n/2-1} \langle \theta_m u'_m, \varphi' \rangle = \langle \theta_m u'_m, u'_m \rangle^{n/2} \zeta^n$$
$$+ n \langle \theta_m u'_m, u'_m \rangle^{n/2-1} \langle \theta_m u'_m, \zeta' \rangle u_m \zeta^{n-1}.$$

The inequalities (9.2), along with the inequalities

$$|\langle \theta_m \xi, \eta \rangle| \leq |\theta_m \xi||\eta| \leq 1/\alpha |\xi||\eta|,$$

give us that

$$\langle \theta_m u'_m, u'_m \rangle^{(n-2)/2} \langle \theta_m u'_m, \varphi' \rangle \geq \alpha^n |u'_m|^n \zeta^n - \frac{n}{\alpha^{n-1}} |u'_m|^{n-1} \zeta^{n-1} |u_m||\zeta'|.$$

Let us use Young's inequality:

$$ab \leq (a^n/n) + (n-1)b^{(n-1)/n}/n$$

for any $a \geq 0$ and $b \geq 0$. Setting $a = |u_m||\zeta'|/t$ and $b = |u'_m|\zeta^{n-1}t$, where $t > 0$ will be chosen later, we get that

$$\langle \theta_m u'_m, u'_m \rangle^{(n-2)/2} \langle \theta_m u'_m, \varphi' \rangle$$
$$\geq \left(\alpha^n - \frac{n-1}{\alpha^n} t(n-2)/n \right) |u'_m|^n \zeta^n - u_m^n |\zeta'|^n/t^n.$$

The number t is now chosen from the condition

$$\alpha^n - (n-1)/\alpha^n t^{(n-1)/n} = \alpha^n/2.$$

As a result

$$\langle \theta_m u'_m, u'_m \rangle^{(n-2)/2} \langle \theta_m u'_m, \varphi' \rangle \geq \alpha^n/2 \cdot |u'_m|^n \zeta^n - u_m^n |\zeta'|^n/t^n. \tag{9.3}$$

Integrating inequality (9.3) termwise and considering (9.1), we get that

$$0 \geq \frac{\alpha^n}{2} \int_U |u'_m(x)|^n [\zeta(x)]^n \, dx - \frac{1}{t^n} \int_U |u_m|^n |\zeta'|^n \, dx,$$

whence

$$\int_U |u'_m(x)|^n [\zeta(x)]^n \, dx \leq C \int_U |u_m(x)|^n |\zeta'(x)|^n \, dx,$$

where $C = 2/(\alpha t)^n$ is a constant, and C depends only on the constant K (the supremum of the sequence $(K(f_m))$, $m = 1, 2, \ldots$). We have that $\zeta(x) = 1$ for $x \in A$ and $\zeta'(x) = 0$ for $x \notin G$. From this,

$$\int_A |u'_m(x)|^n \, dx \leq C_1 \int_G |u_m(x)|^n \, dx. \tag{9.4}$$

Since $G \subset U$ and G is compact, and since the sequence (f_m), $m = 1, 2, \ldots$, is locally bounded in $L_n(U)$ by hypothesis, (9.4) gives us that the sequence $(\int_A |u'_m(x)|^n \, dx)$, $m = 1, 2, \ldots$, is bounded. Since the compact set $A \subset U$ was taken arbitrarily, the theorem is proved.

§9.2. A theorem on the limit of a sequence of mappings with bounded distortion.

THEOREM 9.2. *Suppose that* $(f_m: U \to \mathbf{R}^n)$, $m = 1, 2, \ldots$, *is an arbitrary sequence of mappings with bounded distortion that converges locally in* $L_n(U)$ *to a mapping* $f_0: U \to \mathbf{R}^n$. *Assume that the sequence* $(K(f_m))$, $m = 1, 2, \ldots$, *of distortion coefficients is bounded. Then the limit mapping* f_0 *is a mapping with bounded distortion, and*

$$K(f_0) \leq \varliminf_{m \to \infty} K(f_m).$$

PROOF. It obviously follows from the conditions of the theorem that the sequence (f_m), $m = 1, 2, \ldots$, is bounded in $L_{n,\text{loc}}(U)$. Hence, it is locally bounded also in $W_n^1(U)$, by Theorem 9.1. According to the corollary to Lemma 4.9 of this chapter, the conditions of the theorem imply that the functions $\mathscr{J}_m: x \to \mathscr{J}(x, f_m)$ converge weakly in $L_{1,\text{loc}}(U)$ to the function $\mathscr{J}_0: x \to \mathscr{J}(x, f_0)$ as $m \to \infty$, i.e., for every continuous function φ with compact support in U

$$\int_U \varphi(x)\mathscr{J}(x, f_m) \, dx \to \int_U \varphi(x)\mathscr{J}(x, f_0) \, dx. \tag{9.5}$$

For each of the mappings f_m the Jacobian has constant sign in U. We say that f_m is a positive mapping if $\mathscr{J}(x, f_m) \geq 0$ almost everywhere in U, and a negative mapping if $\mathscr{J}(x, f_m) \leq 0$ almost everywhere in U. The relation (9.5) allows us to conclude that if there exist positive mappings with arbitrarily large indices, then $\int_U \varphi(x)\mathscr{J}(x, f_0) \, dx \geq 0$ for every nonnegative function $\varphi \in C_0(U)$, and hence $\mathscr{J}(x, f_0) \geq 0$ almost everywhere in U. Similarly, if there exist negative mappings with arbitrarily large indices, then $\mathscr{J}(x, f_0) \leq 0$ almost everywhere in U. It follows from the foregoing that if there exist both positive and negative mappings with arbitrarily large indices, then $\mathscr{J}(x, f_0) = 0$ almost everywhere in U, and the mapping

f_0 is identically constant. In this case all the assertions of the theorem are obviously satisfied.

We assume in what follows that the limit mapping f_0 is not equivalent (in the sense of the theory of the integral) to a constant mapping. Then there exists an index m_0 such that the mappings f_m with $m \geq m_0$ are either all positive or all negative. For simplicity it will be assumed that the f_m are positive for all m. Then $\mathscr{J}(x, f_0) \geq 0$ almost everywhere in U.

Let $K_0 = \varliminf_{m \to \infty} K(f_m)$. Take an arbitrary number $\varepsilon > 0$ and choose from the sequence (f_m), $m = 1, 2, \ldots$, a subsequence (f_{m_k}), $m_1 < m_2 < \ldots$, such that $K(f_{m_k}) \leq K_0 + \varepsilon$ for all k. For each k and for almost all x,

$$|f'_{m_k}(x)|^n \leq (K_0 + \varepsilon)\mathscr{J}(x, f_{m_k}).$$

Let φ be an arbitrary nonnegative function in $C_0(U)$. Multiplying both sides of the last inequality by $\varphi(x)$ and integrating with respect to x, we get that

$$\int_U \varphi(x)|f'_{m_k}(x)|^n\, dx \leq (K_0 + \varepsilon)\int_U \varphi(x)\mathscr{J}(x, f_{m_k})\, dx. \qquad (9.6)$$

As $k \to \infty$

$$\int_U \varphi(x)\mathscr{J}(x, f_{m_k})\, dx \to \int_U \varphi(x)\mathscr{J}(x, f_0)\, dx.$$

Further, by the theorem proved in §3 of Chapter III on semicontinuity for functionals of the calculus of variations,

$$\int_U \varphi(x)|f'_0(x)|^n\, dx \leq \varliminf_{k \to \infty} \int_U \varphi(x)|f'_{m_k}(x)|^n\, dx.$$

On the basis of the foregoing, it follows from (9.6) that

$$\int_U \varphi(x)|f'_0(x)|^n\, dx \leq (K_0 + \varepsilon)\int_U \varphi(x)\mathscr{J}(x, f_0)\, dx,$$

and $|f'_0(x)|^n \leq (K_0 + \varepsilon)\mathscr{J}(x, f_0)$ almost everywhere in U, because here φ is an arbitrary nonnegative function of class $C_0(U)$. The mapping f_0 belongs to the class $W^1_{n,\mathrm{loc}}(U)$ in view of Theorem 9.1. It is thereby established that f_0 is a mapping with bounded distortion. Further, we get that $K(f_0) \leq K_0 + \varepsilon$. Since $\varepsilon > 0$ is arbitrary, it follows that $K(f_0) \leq K_0 = \varliminf_{m \to \infty} K(f_m)$, and this completes the proof of the theorem.

COROLLARY. *Let* (f_m), $m = 1, 2, \ldots$, *be an arbitrary sequence of quasiconformal mappings of an open domain* $U \subset \mathbf{R}^n$ *that converges in* $L_{n,\mathrm{loc}}(U)$ *to some mapping* f_0 *which is not identically constant in* U. *Then the limit mapping* f_0 *is also quasiconformal.*

PROOF. Let G be an arbitrary compact domain contained in U. Then for every point $y \notin f_0(\partial G)$ the degree $\mu(y, f_m, G)$ of f_m with respect to G

is defined for sufficiently large m. Further, $\mu(y, f_m, G) = \mu(y, f_0, G)$ for all sufficiently large m. Since each of the mappings f_m is quasiconformal, $|\mu(y, f_m, G)| \leq 1$ for all m. Consequently, $|\mu(y, f_0, G)| \leq 1$ for every point $y \notin f_0(\partial G)$. The set $f_0^{-1}(y) \cap G$ is finite, and $\mu(y, f_0, G)$ is equal to the sum of the indices of f at the points of $f_0^{-1}(y)$. By Theorem 6.1 in §6, all these indices have the same sign, and each of them is a nonzero integer. It follows from the inequality $|\mu(y_0, f, G)| \leq 1$ that $f^{-1}(y_0) \cap G$ consists of at most one point. Since $y \notin f_0(\partial G)$ was taken arbitrarily, this permits us to conclude that f_0 is one-to-one on G^0. The compact domain $G \subset U$ was taken arbitrarily, so this implies that f_0 is one-to-one in U. (Since U is connected, for any two points $x_1, x_2 \in U$ there is clearly a compact domain $G \subset U$ such that x_1 and x_2 belong to the interior of G.) Consequently, f is a homeomorphic mapping of U, and this proves the corollary.

Theorems 9.1 and 9.2 were established by the author in [130].

§9.3. A sufficient condition for relative compactness of a family of mappings with bounded distortion.

THEOREM 9.3. *Let U be an open subset of $\mathbf{R}^n$, and Φ a family of mappings with bounded distortion from U to $\mathbf{R}^n$. Assume the following conditions: for every closed ball $G = \overline{B}(x_0, r) \subset U$ there exist numbers $K(G) < \infty$ and $L(G) < \infty$ such that for any $f \in \Phi$ the distortion coefficient of $f'(x)$ does not exceed $K(G)$ almost everywhere in G, and*

$$\int_G |f(x)|^n \, dx \leq L(G).$$

Then the family Φ is uniformly equicontinuous on every compact set $A \subset U$.

PROOF. Take an arbitrary point $x_0 \in U$ and let $r > 0$ be such that $G = \overline{B}(x_0, r) \subset U$. We prove that there exists a constant $M(G) < \infty$ such that

$$\|f\|_{W_n^1[B(x_0, r/2)]}^n = \int_{B(x_0, r/2)} |f'(x)|^n \, dx + \int_{B(x_0, r/2)} |f(x)|^n \, dx \leq M(G).$$

Indeed, if not, then there exists a sequence (f_m), $m = 1, 2, \ldots$, of functions in Φ such that $\|f_m\|_{W_n^1[B(x_0, r/2)]} \to \infty$ as $m \to \infty$. By a condition of the theorem, this sequence is locally bounded in $L_n(B(x_0, r))$, and $K(f_m) \leq K(G) = \text{const} < \infty$. Hence, by Theorem 9.1, this sequence is locally bounded in $W_n^1[B(x_0, r)]$, and we clearly get a contradiction. This proves the existence of the required constant $M(G)$.

The family Φ is thus bounded in $W_n^1[B(x_0, r/2)]$. The estimates in §1 allow us to conclude that it is uniformly equicontinuous in $B(x_0, r/4)$. We get that every point $x_0 \in U$ has a neighborhood $B(x_0, \delta)$ in which Φ is uniformly equicontinuous. Let $A \subset U$ be a compact set. Covering A by finitely many neighborhoods of the indicated kind, we get that Φ is uniformly equicontinuous on A, and the theorem is proved.

§10. The set of branch points of a mapping with bounded distortion and locally homeomorphic mappings

§10.1. The measure of the set of branch points.

The main result in this subsection is the following statement.

THEOREM 10.1 [137]. *The intersection of the set B_f of branch points of a mapping with bounded distortion with any $(n-1)$-dimensional plane is a set whose $(n-1)$-dimensional Lebesgue measure is equal to zero.*

The proof is based on certain auxiliary constructions, to which we now proceed. Let $f: U \to \mathbf{R}^n$ be an arbitrary mapping with bounded distortion, where U is an open set in $\mathbf{R}^n$. As usual, B_f denotes the set of branch points of f. A disk in $\mathbf{R}^n$ is defined to be any $(n-1)$-dimensional open ball in $\mathbf{R}^n$. The disk about a with radius $r > 0$ and lying in an $(n-1)$-dimensional plane P will be denoted by $K(a, r, P)$. In cases when knowledge of the plane of a disk is not necessary we use the simpler notation $K(a, r)$. A disk $K(a, r)$ is said to be *clean* if it does not contain branch points of f.

Let $K(a_0, r, P) \subset U$ be an arbitrary disk in $\mathbf{R}^n$. We consider the collection of all clean disks $K(b, \rho)$ contained in $K(a_0, r, P)$. Since B_f is closed with respect to U and is at most $(n-2)$-dimensional, it does not contain any disk, which implies that the indicated set of disks is nonempty. The supremum of the ratios ρ/r on the set of all these disks $K(b, \rho)$ is denoted by $\alpha(a, r, P)$ and will be called the *measure of cleanness* of the disk $K(a, r, P)$. The infimum of $\alpha(a, r, P)$ on the collection of all numbers $r < 0$ with $r < \rho(a, \partial U)$ and all planes P passing through a is denoted by $\alpha(a)$.

LEMMA 10.1. *$\alpha(a)$ is positive for every point $a \in U$.*

PROOF. Assume, on the contrary, that $\alpha(a) = 0$ for some point $a \in U$. Then there exist a sequence (r_ν), $\nu = 1, 2, \ldots$, of values and a sequence (P_ν), $\nu = 1, 2, \ldots$, of planes passing through a such that $\alpha_\nu = \alpha(a, r_\nu, P_\nu) \to 0$ as $\nu \to \infty$. It can obviously be assumed without loss of generality that the numbers r_ν tend to some limit r_0 ($0 \le r_0 \le \infty$) as $\nu \to \infty$, and that the planes P_ν converge to some plane P_0 passing through a. Two cases are possible: 1) $r_0 > 0$, and 2) $r_0 = 0$.

Let us take the first case. We have the disk $K(a, r_0, P_0)$. In the case $r_0 = \infty$ it clearly coincides with P_0. Let x_0 be an arbitrary point of $K(a, r_0, P_0)$, and let x_ν be the orthogonal projection of it on P_ν. Then $x_\nu \to x_0$ as

$\nu \to \infty$, $|x_\nu - a_0| \to |x_0 - a_0| < r_0$, and hence $|x_\nu - a_0| < r_\nu$ for sufficiently large ν, i.e., $x_\nu \in x_\nu \in K(a_0, r_\nu, P_\nu)$. By assumption, $\alpha_\nu \to 0$ as $\nu \to \infty$. Let $K(b_\nu, \rho_\nu) \subset K_\nu$ be a disk that contains x_ν and such that $\alpha_\nu r_\nu < \rho_\nu < \alpha_\nu r_\nu + (1/\nu)$. By the definition of the number $\alpha_\nu = \alpha(a, r_\nu, P_\nu)$ the ball $K(b_\nu, \rho_\nu)$ contains points of the set B_f. Let y_ν be one of these. Obviously, $|x_\nu - y_\nu| \le 2(\alpha_\nu r_\nu + 1/\nu)$, and hence $|x_\nu - y_\nu| \to 0$ as $\nu \to \infty$. This implies that $y_\nu \to x_0$ as $\nu \to \infty$. Thus, x_0 is the limit of a sequence of branch points of f. Since B_f is closed relative to U, this gives us that $x_0 \in B_f$. Because $x_0 \in K(a, r_0, P_0)$ was taken arbitrarily, we get that $K(a_0, r_0, P_0) \subset B_f$, which is impossible, since B_f is at most $(n - 2)$-dimensional. Thus, the case $r_0 > 0$ leads to a contradiction.

Let us now consider the case $r_0 = 0$. There is a $\delta_1 > 0$ such that $\delta_1 < \rho(a, \partial u)$ and $f(x) \ne f(a)$ for $0 < |x - a| \le \delta_1$. Let

$$L_1 = \min_{|x-a|=\delta_1} |f(x) - f(a)|.$$

There is a $\delta_2 > 0$ such that $0 < \delta_2 \le \delta_1$ and

$$l = \max_{|x-a|\le\delta_2} |f(x) - f(a)| \le L_1.$$

Assume that $r_\nu \le \delta_2$ for all ν, and let

$$\Lambda_\nu = L_f(a, r_\nu) = \max_{|x-a|=r_\nu} |f(x) - f(a)|,$$

$$\lambda_\nu = l_f(a, r_\nu) = \min_{|x-a|=r_\nu} |f(x) - f(a)|.$$

We now define a sequence of mappings F_ν, $\nu = 1, 2, \ldots$, of the unit ball $B_1 = B(0, 1)$ by setting $F_\nu(X) = (1/\Lambda_\nu)[f(a + r_\nu X) - f(a)]$ for $X \in B_1$ and $\nu = 1, 2, \ldots$. Obviously, F_ν is a mapping with bounded distortion. Further, $K(F_\nu) \le K(f)$ and $|F_\nu(X)| \le 1$ for all ν. For each point $y \in \mathbf{R}^n$ we have that

$$\mu(y, F_\nu, \overline{B}(0, 1)) = \mu(y_\nu, f, \overline{B}(a, r_\nu)),$$

where $y_\nu = f(a) + \lambda_\nu y$. In view of the condition that the index is bounded (Theorem 6.5) there exists a constant $\mu_0 < \infty$ such that $|\mu[y, f, \overline{B}(0, r_\nu)]| \le \mu_0$ for all ν. The sequence of mappings F_ν, $\nu = 1, 2, \ldots$, satisfies all the conditions of Theorem 9.2, and hence the sequence (F_ν) of mappings is equi-uniformly continuous on every ball $B(0, t)$, where $0 < t < 1$. From the sequence (F_ν) we choose a subsequence converging to some mapping $F_0 \colon B_1 \to \mathbf{R}^n$ uniformly on every ball $B(0, t)$. In order not to complicate the notation we assume that F_ν is the required subsequence.

By Theorems 7.2 and 7.3, for each ν

$$\Lambda_\nu M_2(|x - a|/r_\nu)^{\alpha_1} \le |f(x) - f(a)| \le \Lambda_\nu M_1(|x - a|/r_\nu)^{\alpha_2}$$

if $0 \le |x - a| \le r_\nu$, where M_1 and M_2 are constants, and

$$\alpha_1 = K(f)|j(a, f)|^{1/(n-1)}, \qquad \alpha_2 = |j(a, f)|^{1/(n-1)}/K_0(f).$$

Setting $x = a + r_\nu X$ here, we get that for each ν

$$M_2|X|^{\alpha_1} \le |F_\nu(X)| \le M_1|X|^{\alpha_2}$$

for all $X \in B(0, 1)$. Passing to the limit as $\nu \to \infty$, we conclude that also

$$M_2|X|^{\alpha_1} \le |F_0(X)| \le M_1|X|^{\alpha_2}$$

in $B(0, 1)$. In particular, this implies that F_0 is not identically constant in $B(0, 1) = B_1$. By Theorem 9.1, F_0 is a mapping with bounded distortion.

Denote by Q_ν the $(n - 1)$-dimensional plane passing through the point O and parallel to P_ν. The planes Q_ν converge to a plane Q_0 parallel to P_0 as $\nu \to \infty$. Let K_ν be the disk obtained by intersecting B_1 with the plane Q_ν, $\nu = 0, 1, \ldots$. The mapping $X \to a + r_\nu X$ obviously establishes a one-to-one correspondence between the set of branch points of F_ν and the set $B_f \cap B(a, r_\nu)$. This implies that the measure of cleanness of the disk $K(a, r_\nu, P_\nu)$ is equal to the measure of cleanness of the disk K_ν with respect to the mapping F_ν, and $\alpha(0, 1, Q_\nu) = \alpha_\nu \to 0$ as $\nu \to \infty$.

Take an arbitrary point x_0 of K_0, and let $\varepsilon > 0$ be such that $|x_0| < 1 - \varepsilon$ and $F_0(x) \ne F_0(x_0)$ for any x with $0 < |x - x_0| < \varepsilon$. Let $y_0 = F_0(x_0)$. Then $y_0 \notin F_0(S(x_0, \varepsilon))$, and thus the quantity $\mu(y_0, F_0, B(x, \varepsilon))$ is defined. Let x_ν be the point of K_ν closest to x_0. Obviously, $x_\nu \to x_0$ as $\nu \to \infty$. It is also clear that $|x_\nu| \le |x_0| < 1 - \varepsilon$. By assumption, $\alpha_\nu \to 0$ as $\nu \to \infty$. In the plane Q_ν we construct a disk $K(b, \rho)$ contained in K_ν and such that $\alpha_\nu < \rho < \alpha_\nu + 1/\nu$. By the definition of the number α_ν, this disk contains branch points of F_ν. Let ξ_ν be one of these. Then it is clear that $|\xi_\nu - x_\nu| < 2(\alpha_\nu + 1/\nu)$, and hence $\xi_\nu \to x_0$ as $\nu \to \infty$. Let $y_\nu = F_\nu(\xi_\nu)$. Then $F_\nu \to F_0$ uniformly on every compact subset of $B(0, 1)$ and, in particular, on $\overline{B}(x_0, \varepsilon)$ as $\nu \to \infty$. This gives us that $y_\nu \to y_0 = F_0(x_0)$ as $\nu \to \infty$. For sufficiently large ν

$$\mu(y_\nu, F_\nu, \overline{B}(x_0, \varepsilon)) = \mu(y_0, F_0, \overline{B}(x_0, \varepsilon)).$$

For sufficiently large ν (namely, for ν such that $y_\nu \in B(x_0, \varepsilon)$ and $y_\nu \notin F_\nu[S(x_0, \varepsilon)]$) the inequality

$$|\mu[y_\nu, F_\nu, \overline{B}(x_0, \varepsilon)]| \ge |j(\xi_\nu, F_\nu)| \ge 2$$

holds since ξ_ν is a branch point of F_ν. We conclude from this that $|\mu(y_0, F_0, \overline{B}(x_0, \varepsilon))| \ge 2$ also holds. The number $\varepsilon > 0$ with $\varepsilon < 1 - |x_0|$ was arbitrary, so it follows that $|j(x_0, F_0)| \ge 2$, i.e., x_0 is a branch point of F_0. Since $x_0 \in K_0$ was taken arbitrarily, we get that the set of branch points of some mapping with bounded distortion contains a disk, which is impossible. Thus the case $r_0 = 0$ leads to a contradiction, and the lemma is proved.

PROOF OF THEOREM 10.1. Let P be an arbitrary $(n - 1)$-dimensional plane, and let $E = P \cap B_f$. For a set $A \subset P$ the symbol $|A|$ here denotes its $(n - 1)$-dimensional Lebesgue measure. Take an arbitrary point $x \in E$ and consider the ratio

$$\lambda(x, r) = |E \cap K(x, r, P)|/|K(x, r, P)|.$$

For every $\varepsilon > 0$ each disk $K(x, r)$ contains a disk of radius at least $[\alpha(x) - \varepsilon]r$ which does not contain points of B_f. This implies that $\lambda(x, r) \leq 1 - [\alpha(x)]^{n-1}$ for each r. Let $\bar{\lambda}(x) = \overline{\lim}_{r \to 0}(x, r)$. The number $\bar{\lambda}(x)$ is the so-called *upper density* of E at the point x. We get that $\bar{\lambda}(x) \leq 1 - [\alpha(x)]^{n-1} < 1$. Thus, the set E has upper density less than 1 at each of its points x. By classical theorems (see, for example, [149]) on differentiation of set functions, this implies that E is a set of measure zero in P. The theorem is proved.

COROLLARY 1. *The set of branch points of any mapping with bounded distortion is a set of measure zero.*

COROLLARY 2. *Let $f: U \to \mathbf{R}^n$, where $U \subset \mathbf{R}^n$ is an open set. Then the differential of f at a point $x \in U$ is a nonsingular linear mapping for almost all $x \in H$.*

PROOF. The set B_f is closed. Suppose that $x \notin B_f$. Then there exists a neighborhood V of x such that f is a homeomorphism on V. By Theorem 8.3, $f'(x)$ is nonsingular almost everywhere on V. Covering $U \backslash B_f$ by countably many neighborhoods on each of which f is a homeomorphism, we get the required result.

§10.2. **Some lemmas on local homeomorphisms.** Let $U \subset \mathbf{R}^n$ be an open set, and $f: U \to \mathbf{R}^n$ an open mapping. The mapping f is called proper if $f^{-1}(A)$ is compact for every compact set $A \subset f(U)$.

LEMMA 10.2. *Let U be a domain in $\mathbf{R}^n$, $f: U \to \mathbf{R}^n$ a continuous open mapping, G a domain lying strictly inside $f(U)$, and V a connected component of $f^{-1}(G)$. If V lies strictly inside U, then $f(V) = G$, G does not contain images of boundary points of V, and the restriction of f to V is a proper mapping.*

PROOF. Assume, on the contrary, that there is a point $y_0 \in G$ such that $y_0 \in f(\partial V)$. Suppose that the point $x_0 \in \partial V$ is such that $f(x_0) = y_0$. Since

f is continuous, there is an $\varepsilon > 0$ such that $B(x_0, \varepsilon) \subset U$ and $f[B(x_0, \varepsilon)] \subset G$. Since x_0 is a boundary point of V, $B(x_0, \varepsilon) \cap V$ is nonempty. Hence, $V' = B(x_0, \varepsilon) \cap V$ is connected. Obviously, $f(V') \subset G$. At the same time, $V \subset V'$ and $V \neq V'$, since $x_0 \in V'$ but $x_0 \notin V$. We thus get a contradiction to the fact that V is a connected component of $f^{-1}(G)$.

Let us now prove that $f(V) = G$. The set $f(V)$ is open relative to G. Since G is connected, for the proof it suffices to establish that $f(V)$ is closed relative to G. Let $y_0 \in G$ be a limit point of $f(V)$. Then there is a sequence (y_ν), $\nu = 1, 2, \ldots$, of points in $f(V)$ such that $y_\nu \to y$ as $\nu \to \infty$. Let $y_\nu = f(x_\nu)$, where $x_\nu \in V$. It will be assumed that x_ν tends to some point $x_0 \in \overline{V}$ as $\nu \to \infty$. Then $y_0 = f(x_0)$. Since G does not contain images of boundary points of V, it now follows that $x_0 \in V$, and $y_0 \in f(V)$. Thus, $f(V)$ contains all its limit points in G, and hence $f(V)$ is closed relative to G.

We now prove that $f|_V$ is a proper mapping. Let $Q \subset G$ be compact. The set $P = f^{-1}(Q) \cap V$ is closed relative to V. We prove that it is closed in $\mathbf{R}^n$. Assume that this is not so. Then there is a point $x_0 \in \overline{P}$ such that $x_0 \notin V$. Then $x_0 \in \partial V$, and hence $y_0 = f(x_0) \in \partial G$. By the continuity of f, $y_0 \in Q$, and we get a contradiction to the fact that $Q \subset G$. The lemma is proved.

LEMMA 10.3. *Let G be an open domain in $\mathbf{R}^n$, and $f \colon G \to \mathbf{R}^n$ a continuous local homeomorphism. If f is one-to-one on some compact set $K \subset G$, then f is a homeomorphism of some neighborhood $U \subset G$ of K.*

PROOF. Let $U_\delta(K)$ be the δ-neighborhood of K, where $\delta > 0$. We prove that f is one-to-one on $U_\delta(K)$ for some $\delta > 0$. Assume that this is not so. Then for every m there is a pair of points $x'_m, x''_m \in G$ such that $\rho(x'_m, K) < 1/m$, $\rho(x''_m, K) < 1/m$, $x'_m \neq x''_m$, and $f(x'_m) = f(x''_m)$. Without loss of generality it can be assumed that the sequences (x'_m) and (x''_m) converge to points $x', x'' \in K$ as $m \to \infty$. By the continuity of f,

$$f(x') = \lim_{m \to \infty} f(x'_m) = \lim_{m \to \infty} f(x''_m) = f(x'').$$

Since f is one-to-one on K, the equality $f(x') = f(x'')$ gives us that $x' = x''$, and thus the sequences (x'_m) and (x''_m) converge to the same point $x_0 \in K$. However, this contradicts the fact that f is a local homeomorphism.

Since every continuous one-to-one mapping of a domain in $\mathbf{R}^n$ is a homeomorphism, this proves the lemma.

We say that a domain $G \subset \mathbf{R}^n$ can be exhausted simply connectedly if there exists a sequence (G_m), $m = 1, 2, \ldots$, of simply connected domains such that $\overline{G}_m \subset G_{m+1}$ for each m and $\bigcup G_m = G$.

Let X be an arbitrary topological space. A *path* in X is defined to be any continuous mapping $\sigma: [0, 1] \to X$. Here the point $x = \sigma(0)$ is called the *start* of the path σ, $y = \sigma(1)$ is called the *end*, and the path σ is said to *join* the points x and y. A path σ is said to be *closed* if $\sigma(0) = \sigma(1)$.

A topological space X is said to be *arcwise connected* if for any two points x and y in X there is a path joining x and y.

An arcwise connected topological space X is said to be *simply connected* if for any closed path $\sigma: [0, 1] \to X$ there exists a continuous mapping $\zeta: [0, 1] \times [0, 1] \to X$ such that $\zeta(t, 0) = \sigma(t)$, $\zeta(0, u) = \zeta(1, u) = \sigma(0) = \sigma(1)$ for all u, and $\zeta(t, 1) = \sigma(0)$ for all $t \in [0, 1]$. The mapping ζ is called a *homotopy* of the path σ into the point $\sigma(0)$.

Let X and Y be connected topological spaces, and $f: X \to Y$ a continuous mapping. Such an f is called a *covering map* if the following conditions hold:

a) f maps X onto Y.

b) Every point $y \in Y$ has a connected neighborhood V such that every connected component of $f^{-1}(V)$ is mapped homeomorphically onto V by f.

We have the following statement [55].

THEOREM 10.2. *Let $f: X \to Y$ be a continuous mapping. Assume that X is connected and Y simply connected. If f is a covering map, then f is a homeomorphism.*

LEMMA 10.4. *A proper local homeomorphism φ of a bounded domain $G \subset \mathbf{R}^n$ onto a domain G' that can be simply connectedly exhausted is a homeomorphism.*

PROOF. Let (G'_m), $m = 1, 2, \ldots$, be a simply connected exhaustion of G'. We construct a sequence (G_m), $m = 1, 2, \ldots$, of subsets of G by induction. Let G_1 be a connected component of $\varphi^{-1}(G'_1)$. If G_m has already been constructed, then G_{m+1} is the connected component of $\varphi^{-1}(G'_{m+1})$ which contains G_m. Let φ_m be the restriction of φ to G_m. On the basis of Lemma 10.2, φ_m maps G_m onto G'_m and is a proper mapping for each m.

Take an arbitrary point $y \in G'_m$. The set $\varphi_m^{-1}(y)$ is compact, since φ is a proper mapping. Because φ_m is a local homeomorphism, all the points of $\varphi_m^{-1}(y)$ are isolated. This implies that $\varphi_m^{-1}(y)$ is a finite set. Let $x_1, \ldots, x_k$ be all the elements of $\varphi_m^{-1}(y)$. Denote by U_i a neighborhood of x_i such that $\overline{U}_i \subset G_m$ and φ_m is a homeomorphism on U_i, and let $V = \bigcap_{i=1}^k \varphi_m(U_k)$. Finally, let $U'_i = \varphi_m^{-1}(V) \cap U_i$. Then φ_m maps U'_i onto V for each i, and $\varphi_m^{-1}(V)$ is the union of the sets U'_i. We thus get that every point $y \in G'_m$ has a neighborhood V such that each connected component of $\varphi_m^{-1}(V)$ lies strictly inside G_m and is mapped homeomorphically onto V. This means that φ_m is a covering map. On the basis of Theorem 10.2, this implies that φ_m is a homeomorphism of G_m onto G'_m.

Let $\tilde{G} = \bigcup_1^\infty G_m$. Then $\tilde{G} \subset G$, and φ is clearly a homeomorphism on $\tilde{G}$. We prove that φ is a homeomorphism of the whole domain G. To do this it suffices to establish that $G = \tilde{G}$. Obviously, $\tilde{G}$ is open. We show that $\tilde{G}$ is closed with respect to G. Indeed, let x_0 be a limit point of $\tilde{G}$, and let $y_0 = \varphi(x_0)$. There is an m_1 such that $y_0 \in G'_{m_1}$. Let U be a connected neighborhood of x_0 such that $\varphi(U) \subset G'_m$. There is an m_2 such that U contains points of G_{m_2}; let $m > m_1$ and $m > m_2$. The set $G_{m_2} \cup U$ is connected, and $\varphi(G_{m_2} \cup U) \subset G'_m$. This implies that $G_{m_2} \cup U \subset G'_m$, and thus $x_0 \in G_m$; hence $x_0 \in \tilde{G}$. Therefore, $\tilde{G}$ contains all its limit points, and is thus closed relative to G. Since G is connected, what has been proved implies that $\tilde{G} = G$. The lemma is proved.

§10.3. The measure of the image of the set of branch points for mappings with bounded distortion.

Let a and b be arbitrary points in $\mathbf{R}^n$, with $a \neq b$. Denote by $P_a(b)$ the ray in $\mathbf{R}^n$ formed by the points $x = a + t(b-a)$, where $t \geq 1$. Let E be an arbitrary set in $\mathbf{R}^n$, and a a point not in E. The union of the rays $P_a(x)$ with $x \in E$ is denoted by $P_a(E)$.

LEMMA 10.5. *Let E be a set in $\mathbf{R}^n$ such that $m_\alpha(E) = 0$ for some $\alpha > 0$, and let a be a point not in E. Then $m_{\alpha+1}[P_a(E)] = 0$.*

PROOF. Take an arbitrary integer $k \geq 1$, and let T_k be the ring consisting of all points $x \in \mathbf{R}^n$ with $1/k \leq |x - a| \leq k$. Let $E_k = E \cap T_k$ and $P_k = P_a(E_k) \cap T_k$. Obviously, $m_\alpha(E_k) = 0$ for each k. Further, $P_a(E) = \bigcup_1^\infty P_k$, and the lemma will be proved if we establish that $m_{\alpha+1}(P_k) = 0$ for each k, because the Hausdorff measure is countably additive. Let $\varepsilon > 0$ be arbitrary. By the definition of a Hausdorff measure, there is a sequence of balls $B_i = B(x_i, r_i)$ covering E_k such that $r_i < 1/4k$ for each i and $\sum_i r_i^\alpha < \varepsilon$. We can obviously assume that each B_i contains points of E_k, since the balls for which this is not true can be excluded from the sequence (B_i), $i = 1, 2, \ldots$, without detriment to the rest of its properties. Under this condition, $\rho(a, B_i) > 1/2k$ for all i, and hence $a \notin B_i$ for all i. Let $H_i = P_a(B_i) \cap T_k$. The sets $H_1, H_2, \ldots$ obviously cover P_k. Let $\nu_i = (x_i - a)/|x_i - a|$. Denote by $B_{i,m}$ the ball about $a + (2m - 1)r_i\nu_i$ with radius $r_{i,m} = (2k^2 + 1)r_i$, where $m \geq 1$ is an integer. The balls $B_{i,m}$ cover H_i for $m = 1, \ldots, m_i$, where m_i is the largest integer m such that $(2m - 1)r_i \leq k$. Consequently, we get a set $\{B_{i,m}, \ i = 1, 2, \ldots, 1 \leq m \leq m_i\}$ of balls that covers P_k. We have that

$$\sum_i \sum_{m=1}^{m_i} r_{i,m}^{\alpha+1} \leq \sum_i (2k^2 + 1)^{\alpha+1} \frac{k+1}{2r_i} r_i^{\alpha+1} = M \sum_i r_i^{\alpha} < M\varepsilon,$$

where M is a constant. Since $\varepsilon > 0$ is arbitrary, this implies that $m_{\alpha+1}(P_k)$ = 0, and the lemma is proved.

THEOREM 10.3 [44]. *Let U be a domain in $\mathbf{R}^n$, and $f\colon U \to \mathbf{R}^n$ a mapping with bounded distortion. If its set B_f of branch points is nonempty, then $(n-2)$-dimensional Hausdorff measure of $f(B_f)$ is positive.*

PROOF. Let $f\colon U \to \mathbf{R}^n$ satisfy the condition of the theorem. Assume, contrary to what is to be proved, that $m_{n-2}[f(B_f)] = 0$. Take an arbitrary point $x_0 \in B_f$, and let $y_0 = f(x_0)$. Let $\delta > 0$ be such that $\overline{B}(x_0, \delta) \subset U$ and $f(x) \neq y_0$ for $0 < |x - x_0| \leq \delta$. Such a δ exists since f is an isolated mapping. Let $\gamma > 0$ be the smallest value taken by $|f(x) - y_0|$ on the sphere $S(x_0, \delta)$. Let $G = B(y_0, \gamma/2)$. Obviously, $G \subset f[B(x_0, \delta)]$. Let V be the connected component of $f^{-1}(G)$ containing x_0. Since V is connected, and $|f(x) - y_0| < \gamma/2$ for all $x \in V$, it follows that $V \subset B(x_0, \delta)$, because otherwise V would contain points of $S(x_0, \delta)$, and $|f(x) - y_0| > \gamma$ for $x \in S(x_0, \delta)$. Further, it is also clear that $\overline{V} \subset B(x_0, \delta)$, since otherwise the supremum of $|f(x) - y_0|$ on V would be at least γ. On the basis of Lemma 10.2 we get that $f(V) \subset G$ and $f(\partial V) \subset \partial G$.

We show that $f(B_f \cap V)$ is a set closed relative to G. Indeed, let $z \in G$ be a limit point of $f(B_f \cap V)$. Then there is a sequence (t_ν), $\nu = 1, 2, \ldots,$ of points in $B_f \cap V$ such that $f(t_\nu) \to z$. Without loss of generality it can be assumed that $t_\nu \to t_0 \in \overline{V}$ as $\nu \to \infty$. Since B_f is closed relative to U and $\overline{V} \subset U$, it follows that $t_0 \in B_f$. Obviously, $f(t_0) = z$. Since G does not contain boundary points of V, we have that $t_0 \in V$, hence $z \in f(B_f \cap V)$, which is what was to be proved.

Let $f(B_f \cap V) = A$. By assumption, the $(n-2)$-dimensional Hausdorff measure of A is equal to zero. Hence, there is a point $b \in G$ such that $b \notin A$. We construct the set $P_b(A)$ and let $H = G \backslash P_b(A)$. Then the $(n-1)$-dimensional Lebesgue measure of $P_b(A)$ is equal to zero on the basis of Lemma 10.5. Since A is closed relative to G, $G \cap P_b(A)$ is also closed, and hence H is open. The set H is simply connected. Indeed, since $b \in A$, every closed curve in H is shrunk to the point b without leaving H by the deformation defined by the formula $f(t, x) = b + t(x - b)$.

Let W be one of the connected components of $f^{-1}(H)$ contained in V. Then $\overline{W} \subset B(x_0, \delta)$, and Lemma 10.2 gives us that $f(W) = H$, $f(\partial W) \subset \partial H$, and the restriction of f to W is a proper mapping. On the basis of

Lemma 10.4, for every connected component of $f^{-1}(H)$ lying in V the restriction of f to this component is a homeomorphism.

Let $W_1, \ldots, W_\nu, \ldots$ be all the connected components of $f^{-1}(H) \cap V$, and let f_ν^{-1} be the mapping inverse to the restriction of f to W_ν. Then f_ν^{-1} is quasiconformal on the set $H = G \backslash P_b(A)$, and since $m_{n-1}[P_b(A)] = 0$, it follows by Theorem 6.8 that f_ν^{-1} extends to a homeomorphism of G into V. Let $W_0 = f_V^{-1}(G)$. Then $W_0 \subset V$. Obviously, W_0 is both open and closed relative to V, and since V is connected, $W_0 = V$. Consequently, f is a homeomorphism of V into $\mathbf{R}^n$, and this contradicts the fact that V contains branch points of f. The theorem is proved.

§10.4. A local homeomorphism theorem.

The main result in this section is the proof of a theorem of V. M. Gol'dshteĭn which says that a mapping with bounded distortion whose distortion coefficient is close to 1 is a local homeomorphism.

LEMMA 10.6. *Suppose that $U \subset \mathbf{R}^n$ is an open set, $f: U \to \mathbf{R}^n$ is a mapping with bounded distortion, $k \geq 1$ is a constant, and B_k is the set of all points $x \in U$ such that $|j(x, f)| > k$. Then B_k is closed relative to U.*

PROOF. It will be assumed that $j(x, f) > 0$ for all $x \in U$. The general case can obviously be reduced to this one. Let $x_0 \in U$ be an arbitrary limit point of B_k. Take an $r > 0$ such that if $G = \overline{B}(x_0, r)$, then $y_0 = f(x_0)$ is an (f, G)-admissible point, and $\mu(y_0, f, G) = j(x, f)$. Let $(x_\nu), \nu = 1, 2, \ldots,$ be an arbitrary sequence of points in B_k convergent to x_0, and let $y_\nu = f(x_\nu)$. Then $y_\nu \to y_0$ as $\nu \to \infty$. Hence, there is a ν_0 such that if $\nu \geq \nu_0$, then y_ν is also (f, G)-admissible and $\mu(y_\nu, f, G) = \mu(y_0, f, G)$. We have that $\mu(y_\nu, f, G) \geq j(x_\nu, f) > k$, and hence $j(x_0, f) = \mu(y_0, f, G) > k$, i.e., $x_0 \in B_k$, which is what was to be proved.

LEMMA 10.7. *Suppose that $U \subset \mathbf{R}^n$ is an open set, $f: U \to \mathbf{R}^n$ is a mapping with bounded distortion, $A \subset U$ is a compact set, and $j_0 = \max_{x \in A} |j(x, f)|$. For $x \in U$ let $\tau(x)$ be the distance to the element of $f^{-1}[f(x)]$ closest to x and different from x. Let B be the collection of all points $x \in A$ such that $|j(x, f)| = j_0$. Then there is a number $\delta > 0$ such that $\tau(x) \geq \delta$ for all $x \in B$.*

PROOF. Assume, on the contrary, that the required $\delta > 0$ does not exist. Then there is a sequence of points $(x_\nu), \nu = 1, 2, \ldots,$ in B such that $\tau(x_\nu) \to 0$ as $\nu \to \infty$. Without loss of generality it can be assumed that the sequence $(x_\nu), \nu = 1, 2, \ldots,$ converges to some $x_0 \in A$ as $\nu \to \infty$.

Let $k = j_0 - 1$, and let $B_k = \{x \in U | |j(x, f)| > k\}$. By the theorem, B_k is closed relative to U, which implies that $B = A \cap B_k$ is compact, and hence $x_0 \in B$. There is an $r > 0$ such that $G = \overline{B}(x_0, r) \subset U$, the point $y_0 = f(x_0)$ is (f, G)-admissible, and $\mu(y_0, f, G) = j(x_0, f)$. Let $y_\nu = f(x_\nu)$. Then $y_\nu \to y_0$ as $\nu \to \infty$, and hence there is a ν_1 such that y_ν is (f, G)-admissible for $\nu \geq \nu_1$, and $\mu(y_\nu, f, G) = \mu(y_0, f, G)$. Let x'_ν be the closest point to x_ν such that $f(x'_\nu) = f(x_\nu)$ and $x'_\nu \neq x_\nu$. Then $|x'_\nu - x_\nu| \to 0$ as $\nu \to \infty$ in view of our assumption, and thus there exists a $\nu_2 \geq \nu_1$ such that x'_ν and x_ν belong to $B(x_0, r)$ for $\nu \geq \nu_2$. For $\nu \geq \nu_2$ we have

$$|\mu(y_\nu, f, G)| \geq |j(x_\nu, f)| + |j(x'_\nu, f)| \geq j_0 + 1.$$

However, this contradicts the fact that

$$|\mu(y_\nu, f, G)| = |\mu(y_0, f, G)| = j_0$$

for $\nu \geq \nu_1$. Accordingly, the assumption that the lemma is false leads to a contradiction, which is what was to be proved.

We next use Theorems 7.2 and 7.3. The conditions of the latter theorem contain constants r_1 and r_0 whose choice depends on the given point. Recall how r_1 and r_0 were chosen. Let $f: U \to \mathbf{R}^n$ be a mapping with bounded distortion, where U is an open set in $\mathbf{R}^n$, and let $a \in U$. Then r_1 is such that $0 < r_1 < \rho(a, \partial U)$, and $f(x) \neq f(a)$ for $0 < |x - a| \leq r_1$. As r_0 we can take any number $r > 0$ such that

$$L_f(a, r) = \max_{|x-a|=r} |f(x) - f(a)| < l_f(a, r_1)$$

$$= \min_{|x-a|=r_1} |f(x) - f(a)|.$$

LEMMA 10.8. *Let $f: U \to \mathbf{R}^n$ be a mapping with bounded distortion, where U is an open set in $\mathbf{R}^n$, let $A \subset U$ be a compact set, and let $\delta = \text{dist}(A, \partial U) > 0$. Assume that there exists a constant $r_1 = r_1(A) > 0$ such that $f(x)$ is different from $f(a)$ for every $a \in A$ and any $x \in U$ with $0 < |x - a| \leq r_1$. Then there is a number $r_0 = r_0(A) > 0$ such that the estimates in Theorems 7.2 and 7.3 hold if $a \in A$, $0 < r \leq r_0$, and $|x - a| \leq r$, $x \in U$.*

PROOF. Take an arbitrary $r' < \min\{\delta, r_0\}$ and denote by $\rho(x)$ the distance from $f(x)$ to the image under f of the sphere $S(x, r')$, $\rho(x) = l_f(x, r')$. It is not hard to see that ρ is continuous on A, and $\rho(x) > 0$ for all $x \in A$. Hence, there is a number $\rho_0 > 0$ such that $\rho(x) \geq \rho_0$ for all $x \in A$.

Let

$$L_t(x) = L_f(x, t) = \max_{|y-x|=t} |f(y) - f(x)|,$$

where $t > 0$. The function L_t is continuous on A, and $L_t(x) > 0$ for all $x \in A$. Further, $L_t(x)$ is a nondecreasing function of t for each fixed x. As

$t \to 0$ we have that $L_t(x) \to 0$ for all $x \in A$. Since A is compact, $L_t(x) \to 0$ uniformly on A as $t \to 0$, by the classical theorem of Dini. Hence, there is an $r_0 > 0$ such that $L_t(x) \le \rho_0$ for $t \le r_0$. Since $\rho_0 \le \rho(x) = l_f(x, r')$, r_0 is clearly the desired number.

THEOREM 10.4 [44]. *Suppose that U is a domain in $\mathbf{R}^n$, $f: U \to \mathbf{R}^n$ is a mapping with bounded distortion, and $N \ge 2$ is an integer. Let B_N be the collection of all points $x \in U$ such that $|j(x, f)| \ge N$. Then for every $\gamma > (K_0(f)/N)^{1/(n-1)}$ the Hausdorff measure $m_\gamma[f(B_N)]$ of the set $f(B_N)$ is zero.*

PROOF. Denote by B_N^1 the collection of all points $x \in U$ such that $|j(x, f)| = N$, and let B_N^2 be the collection of all points $x \in U$ such that $|j(x, f)| \ge N+1$. According to Lemma 10.6, B_N^2 is closed relative to U, and thus $U_1 = U \backslash B_N^2$ is an open set. Obviously, $B_N^1 \subset U_1$. Take an arbitrary open set V strictly inside U_1. For all $x \in U_1$ we have that $|j(x, f)| \le N$; hence B_N^1 is the collection of all $x \in U_1$ such that $|j(x, f)| > N - 1$. On the basis of Lemma 10.6, B_N^1 is closed relative to U_1.

Let $A = \overline{V} \cap B_N^1$. The set A is compact, and $|j(x, f)| = N$ for every $x \in A$. By Lemmas 10.7 and 10.8, there exist numbers $r_1 > 0$ and $r_0 > 0$ such that $L_f(a, r_0) \le l_f(a, r_1)$ and $f(x) \ne f(a)$ for $0 < |x - a| \le r_1$ if $a \in A$.

By Theorem 7.3, for every $a \in A$ and for $|x - a| \le r_0$

$$|f(x) - f(a)| \le L_f(a, r_0)M_1(|x - a|/r_0)^{\alpha_2},$$

where

$$\alpha_2 = [|j(a, f)|/K_0(f)]^{1/(n-1)} = [N/K_0(f)]^{1/(n-1)}$$

and M_1, is a constant. Since A is compact,

$$|f(x) - f(a)| \le M(|x - a|/r_0)^{\alpha_2} \tag{10.1}$$

for all $a \in A$, where $M < \infty$ is a constant.

According to the corollary to Theorem 10.1, B_N^1 is a set of measure zero. Let $\varepsilon > 0$ be arbitrary. Then there is a sequence of balls B_ν, $\nu = 1, 2, \dots$, covering A such that the radius r_ν of B_ν does not exceed $r_0/2$, and $\sum_\nu r_\nu^n < \varepsilon$. Moreover, it will be assumed that B_ν contains points of A.

We consider the set $f(B_\nu)$. Let $a \in B_\nu \cap A$. For all $x \in B_\nu$ we have that $|x - a| \le 2r_\nu \le r_0$, and hence, by (10.1),

$$|f(x) - f(a)| \le M(|x - a|/r_0)^{\alpha_2} \le (2^{\alpha_2}M/r_0)r_\nu^{\alpha_2}.$$

Let $(2^{\alpha_2}M/r_0)r_\nu^{\alpha_2} = \rho_\nu$. The set $f(B_\nu)$ is thus contained in some ball B_ν' of radius ρ_ν. This gives us a sequence of balls B_ν' which covers $f(A)$. We have that

$$\sum_{\nu} \rho_{\nu}^{n/\alpha_2} = 2^n (M/r_0)^{n/\alpha_2} \sum_{\nu} r_{\nu}^n < L\varepsilon,$$

where $L = 2^n (M/r_0)^{n\alpha_2} < \infty$. Since $\varepsilon > 0$ is arbitrary, this implies that the γ-content of the set $f(A) = f(B_n^1 \cap A)$ is zero, where $\gamma = n/\alpha_2$. Consequently, $m_\nu[f(B_N^1 \cap \overline{V})] = 0$.

Covering U_1 by countably many balls lying strictly inside U_1, we get from what has been proved that $m_\nu[f(B_N^1)] = 0$ for $\gamma = nK_0(f)/N^{1/(n-1)}$. For $M \geq N$ let B_M^1 be the collection of all points $x \in U$ such that $|j(x, f)| = M$. On the basis of what has been proved, $m_\gamma[f(B_M^1)] = 0$ for $\gamma = \gamma_M = nK_0(f)/M^{1/(n-1)}$. From this, $m_\gamma[f(B_M^1)] = 0$ also for $\gamma = \gamma_N$. We have that $f(B_N) = \bigcup_{M=N}^\infty f(B_M^1)$. By the countably additivity of a Hausdorff measure, this implies that $m_\gamma[f(B_N)] = 0$ for any $\gamma \geq nK_0(f)/N^{1/(n-1)}$. The theorem is proved.

COROLLARY. *Let $f: U \to \mathbf{R}^n$ be a mapping with bounded distortion such that B_f is nonempty. Then there exists at least one point $x \in B_f$ at which*

$$|j(x, f)| \leq [n/(n-2)]^{n-1}[K_0(f)]^{n-1}. \tag{10.2}$$

Indeed, let N be the smallest integer such that $nK_0(f)/N^{1/(n-1)} < n-2$. Assume that $|j(x, f)| \geq N$ for all $x \in B_f$. Then it follows from the theorem that $m_{n-2}[F(B_f)] = 0$, which is impossible in view of Theorem 10.3.

THEOREM 10.5 [44]. *There exists a $K_0 > 1$ such that every mapping $f: U \to \mathbf{R}^n$ with bounded distortion $K(f) \leq K_0$ is a local homeomorphism.*

PROOF. Assume, on the contrary, that no such constant K_0 exists. Then for each integer $\nu \geq 1$ there is a mapping with bounded distortion $f_\nu: U_\nu \to \mathbf{R}^n$ such that $K(f_\nu) \leq 1 + 1/\nu$ and f is not a local homeomorphism; hence, the set of branch points of f_ν is nonempty for each ν.

By the corollary to Theorem 10.4, there exists a constant $N < \infty$ such that for each ν there is an $x_\nu \in B_\nu$ with $|j(x_\nu, f_\nu)| \leq N$. On the basis of Theorems 7.2 and 7.3, for each ν there exists a number $r_\nu > 0$ such that if $|x - x_\nu| < r_\nu$, then

$$M_1 \Lambda_\nu \left(\frac{|x - x_\nu|}{r_\nu} \right)^{\alpha_{1,\nu}} \leq |f_\nu(x) - f_\nu(x_\nu)|$$
$$\leq M_2 \Lambda_\nu \left(\frac{|x - x_\nu|}{r_\nu} \right)^{\alpha_{2,\nu}}, \tag{10.3}$$

where

$$\Lambda_\nu = L_{f_\nu}(x_\nu, r_\nu) = \max_{|x-x_\nu|=r_\nu} |f(x) - f(x_\nu)|,$$

$$\alpha_{1,\nu} = K(f_\nu)|f(x_\nu, f_\nu)|^{1/(n-1)} \le 2N_0^{1/(n-1)} = \alpha_1,$$

$$\alpha_{2,\nu} = |j(x_\nu, f_\nu)|^{1/(n-1)}/K_0(f_\nu) \ge \alpha_2 > 0.$$

From this,

$$M_1\Lambda_\nu \left(\frac{|x-x_\nu|}{r_\nu}\right)^{\alpha_1} \le |f_\nu(x) - f_\nu(x_\nu)|$$

$$\le M_2\Lambda_\nu \left(\frac{|x-x_\nu|}{r_\nu}\right)^{\alpha_2}.$$

Let $B_1 = B(0,1)$ be the unit ball in $\mathbf{R}^n$. For $X \in B_\nu$ let

$$F_\nu(X) = (1/\Lambda_\nu)[f_\nu(X_\nu + r_\nu X) - f_\nu(x_\nu)].$$

Obviously, F_ν is a mapping with bounded distortion. Further, $K(F_\nu) \le K(f_\nu) \le 1 + (1/\nu)$, so that $K(F_\nu) \to 1$ as $\nu \to \infty$, and the point 0 is a branch point of F_ν. By (10.3),

$$M_1|X|^{\alpha_1} \le F_\nu(X) \le M_2|X|^{\alpha_2}$$

for $\nu = 1, 2, \ldots$. In particular, it follows from this that the sequence (F_ν) of mappings is bounded in $L_n(B_1)$. Hence, by Theorem 9.1, this sequence is uniformly continuous on every interior ball $B(0,t)$, where $0 < t < 1$. Hence, (F_ν) has a subsequence converging in $L_{n,\text{loc}}(B_1)$ to some mapping F_0. In order not to complicate the notation, we assume that (F_ν), $\nu = 1, 2, \ldots$, is this subsequence. By Theorem 9.2, $K(F_0) \le \underline{\lim}_{\nu \to \infty} K(F_\nu) = 1$, and hence $K(F_0) = 1$. It clearly follows from (4.3) that $M_1|X|^{\alpha_1} \le |F_0(X)| \le M_2|X|^{\alpha_2}$, and thus F_0 is an identically constant mapping. On the basis of Theorem 5.10, F_0 is a Möbius mapping, and hence a local homeomorphism. On the other hand, $|j(0, F_\nu)| \ge 2$. From this, $|\mu(0, F_\nu), \overline{B}(0,t)| \ge 2$ for every ball $\overline{B}(0,t)$, $0 < t < 1$, which gives us, after passage to the limit as $\nu \to \infty$, that $|\mu(0, F_0), \overline{B}(0,t)| \ge 2$ for all $t \in (0, 1)$, and hence $|j(0, F_0)| \ge 2$. This is a contradiction, and the theorem is proved.

Theorems 10.4 and 10.5 were also obtained independently by Martio, Rickman, and Väisälä [96]. On the whole, their proof coincides with the proof of Gol'dshteĭn given here. The precise value of the constant K_0 in Theorem 10.5 is unknown. In this connection we mention the following result of Poletskiĭ [113]. Suppose, as in Theorem 10.4, that B_N^1 denotes the collection of all points $x \in B_f$ such that $|j(x, f)| = N$. If $n = 3$, and the set B_N^1 is a rectifiable curve, then $K(f) \ge N$. The estimate is sharp; equality holds when f is a twist about an axis.

§**10.5. A theorem on the radius of injectivity.** The purpose of this subsection is to prove Theorem 10.6, due to Martio, Rickman, and Väisälä [96]. The proof given here (written out at the author's request by V. M. Gol'dshteĭn) coincides on the whole with that in [96]. The only difference is that we do not use the concept of the modulus of a family of curves.

Theorem 10.6 implies the well-known theorem of Zorich [182] (Corollary 2 to Theorem 10.6). The proof of Theorem 10.6 follows a scheme close to that used in [182]. In particular, Lemmas 10.3 and 10.9, which play an important role in the proof of Theorem 10.6, are due to Zorich. We note that the theorem in [182] was formulated earlier by Lavrent'ev [80], but it was first proved only in [182].

LEMMA 10.9. *Suppose that* $f: U \to \mathbf{R}^n$ *is a local homeomorphism of a domain* U *in* $\mathbf{R}^n$, *and* A *and* B *are subsets of* U *such that* f *is a homeomorphism on each of them. If* $A \cap B$ *is nonempty and* $f(A) \cap f(B)$ *is connected, then* f *is one-to-one on* $A \cup B$. *If each of the sets* $f(A) \backslash f(B)$ *and* $f(B) \backslash f(A)$ *is open relative to* $f(A) \cup f(B)$, *then* f *is a homeomorphism on* $A \cup B$.

PROOF. The restrictions of f to A and B are denoted by f_1 and f_2, respectively. Let $f(A) \cap f(B) = Q$, and let $C_1 = f_1^{-1}(Q)$ and $C_2 = f_2^{-1}(Q)$. To prove that f is one-to-one on $A \cup B$ it suffices to establish that $C_1 = C_2$. We first prove that $C_1 \cap C_2 = A \cap B$. It will follow from this, in particular, that $C_1 \cap C_2$ is nonempty. Since $C_1 \subset A$ and $C_2 \subset B$, we have that $C_1 \cap C_2 \subset A \cap B$. Take an arbitrary point $x \in A \cap B$. Then $f(x) \in Q$. Since $x \in A$, it follows that $x \in f_1^{-1}(Q) = C_1$, and $x \in f_2^{-1}(Q) = C_2$ because $x \in B$, i.e., $x \in C_1 \cap C_2$. This leads to the conclusion that $A \cap B \subset C_1 \cap C_2$, i.e., $A \cap B = C_1 \cap C_2$.

We now prove that $f(C_1 \cap C_2)$ is open relative to Q. Take an arbitrary point $x \in C_1 \cap C_2$, and let $y = f(x)$. Since f is a local homeomorphism, there exists a neighborhood V_0 of x such that f maps V_0 homeomorphically onto some neighborhood W_0 of y. Since f is homeomorphic on each of the sets A and B, there exists a neighborhood $W_1 \subset W_0$ of y in $\mathbf{R}^n$ such that $f_1^{-1}(W_1 \cap A) \subset V_0$ and $f_2^{-1}(W_1 \cap B) \subset V_0$. Let $V_1 = f^{-1}(W_1) \cap V_0$. Since f is a homeomorphism on V_0, it follows that $f(V_1 \cap A) = W_1 \cap f(A)$ and $f(V_1 \cap B) = W_1 \cap f(B)$, and hence $f(A \cap B \cap V_1) = Q \cap W_1$. Since $A \cap B = C_1 \subset C_2$, we have that $f(C_1 \cap C_2 \cap V_1) = Q \cap W_1$. This proves that $f(C_1 \cap C_2)$ is open relative to Q.

We now prove that $f(C_1 \cap C_2)$ is closed relative to Q. Suppose that (x_ν), $\nu = 1, 2, \ldots$, is a sequence of points in $C_1 \cap C_2$ such that the sequence $(y_\nu = f(x_\nu))$, $\nu = 1, 2, \ldots$, converges to some point $y_0 \in Q$. Let $\xi_1 = f_1^{-1}(y_0) \in C_1$ and $\xi_2 = f_2^{-1}(y_0) \in C_2$. Since f is a homeomorphism on each of C_1 and C_2, the sequence (x_ν) must converge to each of the points ξ_1 and ξ_2, i.e., $\xi_1 = \xi_2 = \xi_0 \in C_1 \cap C_2$, and hence $y_0 \in f(C_1 \cap C_2)$, which establishes that $f(C_1 \cap C_2)$ is closed. Since Q is connected by assumption, what has been proved implies that $f(C_1 \cap C_2) = Q$, and thus $Q_1 = Q_2$. It is proved that f is one-to-one on $A \cup B$.

To establish that f is a homeomorphism on $A \cup B$ it suffices to verify that the inverse f^{-1} of the restriction of f to $A \cup B$ is a homeomorphism on the set $P = f(A) \cup f(B)$. If $y \in Q = f(A) \cap f(B)$, then the continuity of f^{-1} at y follows from the fact that f^{-1} is continuous on each of $f(A)$ and $f(B)$. Suppose that $y \notin Q$. Then either $y \in f(A) \backslash f(B)$ or $y \in f(B) \backslash f(A)$. In both cases there is a neighborhood W of y such that $W \cap [f(A) \cup f(B)]$ coincides either with $W \cap f(A)$ or with $W \cap f(B)$. Since f^{-1} is continuous on each of $f(A)$ and $f(B)$, this implies that the restriction of f^{-1} to $W \cap [f(A) \cup f(B)]$ is continuous, and hence f^{-1} is continuous on $f(A) \cup f(B)$. The proof of the lemma is complete.

LEMMA 10.10. *Suppose that* $f: U \to \mathbf{R}^n$ *is a quasiconformal mapping,* $B(x_0, r) \subset U$, $y_0 = f(x_0)$, *and*

$$L_0 = L_f(x_0, r) = \max_{|x - x_0| = r} |f(x) - y_0|, \quad l_0 = l_f(x_0, r) = \min_{|x - x_0| = r} |f(x) - y_0|.$$

If $\overline{B}(y_0, L_0) \subset f(U)$, *then*

$$L_0/l_0 \le \exp[(\gamma_n K(f))^{1/(n-1)}],$$

where $\gamma_n > 0$ *is a constant.*

PROOF. Let

$$G = f^{-1}[\overline{B}(y_0, L_0)], \qquad u(x) = \ln(L/|f(x) - f(a)|).$$

The function u is a generalized solution of the equation

$$\operatorname{div}[\langle \theta_f(x) u'(x), u'(x) \rangle^{(n/2)-1} \theta_f(x) u'(x)] = 0 \qquad (*)$$

in U. Further, $u(x) = 0$ on the boundary of G. The complement $\mathbf{R}^n \backslash G$ in $\mathbf{R}^n$ is obviously an unbounded connected set whose boundary coincides with that of G. The boundary of G contains points of the sphere $S(x_0, r)$. The flow Ω_0 of the solution $u(x)$ of $(*)$ with respect to the capacitor (A_0, A_1) with $A_0 = \mathbf{R}^n \backslash G^0$ and $A_1 = \{x_0\}$ is equal to $n\omega_n$. By Theorem 7.1,

$$u(x) \le [K(f)]^{1/(n-1)} \ln \Phi_n(r/|x - a|)$$

for all $x \in G$. Let x_1 be a point of $S(x_0, r)$ such that $|f(x) - f(x_0)| = l_0$. For this point $u(x) = \ln(L/l)$, and we get that

$$\ln(L/l) \le [K(f)]^{1/(n-1)} \ln \Phi_n(1).$$

This proves the lemma.

THEOREM 10.6 [96]. *Let $n \geq 3$. Then there exists a function $\rho_n(K)$ on the interval $1 \leq K < \infty$ such that if f is a local homeomorphism with bounded distortion on the ball $B(0,1) \subset \mathbf{R}^n$ with $K(f) \leq K$, then f is a homeomorphism on the ball $B(0, \rho_n(K))$.*

PROOF. Let $f \colon B(0,1) \to \mathbf{R}^n$ be a local homeomorphism with bounded distortion. Take an arbitrary number τ with $0 < \tau < 1$. Then f is continuous in the closed ball $\overline{B}(0,\tau)$. We assume that $f(0) = 0$; this clearly involves no loss of generality.

For an arbitrary connected set $A \subset \mathbf{R}^n$ such that $0 \in A$ let $\psi(A)$ be the connected component of $f^{-1}(A)$ containing 0. Let $B_r = B(0,r)$ and let $G_r = \psi(B_r)$. Since f is a local homeomorphism, it follows that $G_r \subset B(0,\tau)$ for sufficiently small r, and the restriction of f to G_r is a proper mapping. By Lemma 10.4, the restriction of f to G_r is a homeomorphism for each such r. Let r_0 be the supremum of the values of r such that f is a homeomorphism on G_r. By Lemma 10.4, f is a homeomorphism also on G_{r_0}.

We show that G_{r_0} contains points of $S(0,\tau)$. Assume that this is not so. We prove, first, that f is one-to-one on $\overline{G}_{r_0}$. Indeed, assume, for example, that there exist points $x_1, x_2 \in \overline{G}_{r_0}$ such that $x_1 \neq x_2$, while $f(x_1) = f(x_2) = y_0$. The mapping f is a homeomorphism on G_{r_0}, and hence at least one of x_1 and x_2 (say x_1) must be a boundary point of G_{r_0}; consequently, $y_0 \in S(0,r_0)$. Since f is open, x_2 cannot be an interior point of G_{r_0}, for otherwise the point $f(x_2) = y_0$ would be an interior point of $B(0,r_0)$. Let V_1 and V_2 be disjoint neighborhoods of x_1 and x_2 such that f is a homeomorphism on each of V_1 and V_2. Let $\delta > 0$ be such that $B(y_0,\delta) \subset f(V_1) \cap f(V_2)$. Denote by f_i the restriction of f to V_i, and let $U_i = f_i^{-1}[B(y_0,\delta)]$, $i = 1,2$. Let $D = B(0,r_0) \cap B(y_0,\delta)$ and $f_i^{-1}(D) = Q_i \subset U_i$.

We prove that $Q_i = U_i \cap G_{r_0}$. Indeed, if $x \in U_i \cap G_{r_0}$, then $f(x) \in B(y_0,\delta)$, while $f(x) \in B(0,r_0)$, and hence $f(x) \in D$. This leads to the conclusion that $x \in Q_i$; therefore, $U_i \cap G_{r_0} \subset Q_i$. Assume that Q_i contains points not in $U_i \cap G_{r_0}$. Since x_i is a boundary point of G_{r_0}, $U_i \cap G_{r_0}$ is nonempty. Hence, $Q_i \cup G_{r_0}$ is connected. Since $f(Q_i \cup G_{r_0}) \subset f(Q_i) \cup f(G_{r_0}) = B(0,r_0)$, this gives us that $Q_i \cup G_{r_0} \subset G_{r_0}$, and, in particular, $Q_i \subset G_{r_0}$. From this, $Q_i \subset G_{r_0} \cap U_i$, i.e., $Q_i = G_{r_0} \cap U_i$. The sets Q_1 and Q_2 do not have common elements, and their images coincide with D. We get a contradiction to the fact that f is one-to-one on G_{r_0}. By Lemma 10.2, this implies that f is a homeomorphism on some neighborhood H of $\overline{G}_{r_0}$. Denote by f_0 the restriction of f to H. Then for $r = r_0 + \delta$, where $\delta > 0$ is sufficiently small, f_0^{-1} maps the ball $B(0, r_0 + \delta)$ homeomorphically onto a set lying strictly inside $B(0, \tau)$, and we get a contradiction to the definition of r_0.

Thus we have a set $G_{r_0} \subset B(0, \tau)$ such that f is a homeomorphism on G_{r_0}. Further, G_{r_0} turns out to be sufficiently "long": it contains the point O and points of $S(0, \tau)$. To prove the theorem it is necessary to see that G_{r_0} is also a sufficiently "thick" set.

Let $h > 0$ be the distance from 0 to the boundary of G_{r_0}. We construct the ball $B(0, h)$ and let $r_1 = \min_{|x|=h} |f(x)|$. Next, $\max_{|x|=h} |f(x)|$ is obviously equal to r_0. By Theorem 7.2, $r_0/r_1 \leq \exp[\gamma_n(K(f))^{1/(n-1)}]$. It follows from the definition of r_0 that there is a point $y_0 \in S(0, r_0)$ which is the image of some point $x_0 \in \overline{G}_{r_0}$ belonging to $S(0, \tau)$. For simplicity of the notation it will be assumed that $y_0 = r_0 \mathbf{e}_n$. The general case can obviously be reduced to this one.

Denote by S_t, where $|t| < r_1/2$, the sphere about y_0 passing through the point $x_t = t\mathbf{e}_n$. The symbol $K_{t,u}$, where $|t| < r_1$ and $t < u \leq 2r_0 - t$, denotes the spherical segment consisting of all points $x = (x_1, \ldots, x_{n-1}, x_n)$ in S_t for which $x_n \leq u$. Let f_0 be the restriction of f to G_{r_0}. Denote by $M_{t,u}$ the connected component of $f^{-1}(K_{t,u})$ containing $f_0^{-1}(x_t)$. For $t \in (-r_1, r_1)$ let u_t be the supremum of the numbers $u \in (t, 2r_0 - t)$ such that f is a homeomorphism on $M_{t,u}$. The union of all segments $K_{t,u}$ such that $-r_1 < t < r_1$ and $t < u < u_t$ is denoted by V.

We prove that the set V constructed in this way is open. Take an arbitrary point $\xi \in V$, and let (t_0, u_0) be such that $-r_1 < t_0 < r_1$, $t_0 < u_0 < u_{t_0}$ and ξ belongs to K_{t_0, u_0}. The mapping f is a homeomorphism on the compact set M_{t_0, u_0}, and hence f is a homeomorphism on some neighborhood W of M_{t_0, u_0}, by Lemma 10.3. Let $W' = f(W)$. Obviously, there exists a $\delta > 0$ such that t lies in $(-r_1, r_1)$ and $K_{t,u}$ is contained in W' for $|t - t_0| < \delta$ and $u < u_0 + \delta$. By this, the mapping f is a homeomorphism on $M_{t,u}$. This implies that the union of all the segments $K_{t,u}$ with $|t - t_0| < \delta$ and $u < u_0 + \delta$ is contained in V. The point ξ is an interior point of this union. Hence, it is an interior point of V, and so it is proved that V is open, because $\xi \in V$ was arbitrary.

Let $\tilde{K}_{t,u}$ be the open spherical segment consisting of the points $x = (x_1, \ldots, x_{n-1}, x_n)$ of S_t such that $x_n < u$. Let $\tilde{M}_{t,u_t}$ denote the union of the sets $M_{t,u}$ such that $t < u < u_t$. The function f maps $\tilde{M}_{t,u_t}$ onto $\tilde{K}_{t,u_t}$. The mapping is a homeomorphism, by Lemma 10.4.

We show that $\tilde{M}_{t,u_t}$ does not lie strictly inside $B(0, \tau)$ for any t. Assume, on the contrary, that $\tilde{M}_{t_0, u_{t_0}}$ lies strictly inside $B(0, \tau)$ for some

$t_0 \in (-r_1, r_1)$, and let M_0 be the closure of $\tilde{M}_{t_0, u_{t_0}}$. By the assumption, M_0 is contained in $B(0, \tau)$. The mapping f is one-to-one on M_0. Indeed, assume, on the contrary, that there are points $x_1, x_2 \in M_0$ such that $x_1 \neq x_2$ and $f(x_1) = f(x_2) = y_0$. Let $(x_{1\nu})$ and $(x_{2\nu})$, $\nu = 1, 2, \ldots$, be sequences of points in $\tilde{M}_{t_0, u_{t_0}}$ such that $x_{1\nu} \to x_1$ and $x_{2\nu} \to x_2$ as $\nu \to \infty$. Let $y_{i\nu} = f(x_{i\nu})$. Then $y_{1\nu} \to y_0$ and $y_{2\nu} \to y_0$ as $\nu \to \infty$. Let us join the points $x_{1\nu}$ and $y_{2\nu}$ by an arc γ_ν lying in $\tilde{K}_{t_0, u_{t_0}}$ in such a way that this arc shrinks to the point y_0 as $\nu \to \infty$. (We remark that here we are using the fact that $n \geq 3$.) Let $\tilde{\gamma}_\nu = f_0^{-1}(\gamma_\nu)$, where f_0 is the restriction of f to $\tilde{M}_{t_0, u_{t_0}}$. The ends of the arc $\tilde{\gamma}_\nu$ converge to x_1 and x_2. The upper topological limit of the sequence of arcs $\tilde{\gamma}_\nu$ is a connected set whose points are all mapped into x_0. We get a contradiction to the assumption that f is a local homeomorphism. This proves that f is one-to-one on M_0. By the compactness of M_0, f maps M_0 homeomorphically onto $M_{t_0, u_{t_0}}$. On the basis of Lemma 10.3, this implies that f extends to a homeomorphism of some neighborhood V of M_0. Let $W = f(V)$. If $K_{t_0, u_{t_0}}$ does not coincide with S_{t_0}, then f is a homeomorphism on the set $M_{t_0, u_{t_0}} + \delta$ for some $\delta > 0$, and we arrive at a contradiction to the definition of u_{t_0}.

Therefore, $K_{t_0, u_{t_0}} = S_{t_0}$, and f maps M_0 homeomorphically onto S_{t_0}. The bounded component P of $\mathbf{R}^n \backslash M_0$ is contained in $B(0, \tau)$. Since M_0 passes through interior points of G_{r_0}, P intersects G_{r_0}. The boundary of P is mapped homeomorphically onto the sphere S_{y_0}. This implies that $\mu(y, f, \overline{P}) = 1$ for every $y \in B(y_0, t_0)$, and $\mu(y, f, \overline{P}) = 0$ for $y \notin B(y_0, t_0)$, which allows us to conclude that f maps P onto $B(y_0, t_0)$. The mapping f is a homeomorphism on each of the sets G_{r_0} and P, and the intersection $f(G_{r_0}) \cap f(P) = B_{r_0} \cap B(y_0, t_0)$ is connected. By Lemma 10.9, this implies that f is a homeomorphism on $G_{r_0} \cup P$. In G_{r_0} there is a sequence (x_ν), $\nu = 1, 2, \ldots$, converging to some point $x_0 \in S(x, \tau)$ and such that $y_\nu = f(x_\nu) \to y_0$ as $\nu \to \infty$. For sufficiently large ν the points y_ν are in $B(y_0, t_0)$. This implies that the points x_ν belong to P for sufficiently large ν, and $\lim x_\nu$ is an interior point of $P \subset B(0, \tau)$. We thereby get a contradiction, and hence the given case is also impossible. Thus, for every $t \in (-r_1, r_1)$ the set M_{t, u_t} does not lie strictly inside $B(0, \tau)$, and hence the closure of M_{t, u_t} contains points of $S(0, \tau)$. Let ζ_t be some limit point of M_{t, u_t} on $S(0, \tau)$, and let $z_t = f(\zeta_t)$. The closure of the set formed by the points z_t is denoted by F_1. Finally, let F_0 be the interval consisting of all the points $x = t \mathbf{e}_n$ with $|t| \leq r_1$. Let M be the union of all the sets M_{t, u_t} with $|t| < r_1$. The mapping f is one-to-one from M onto V. The inverse mapping f^{-1} is continuous on V. This implies that M is open and f is a homeomorphism on M. The theorem will be proved if we establish that

M is also a sufficiently "thick" set. The required result will be obtained by certain estimates for the capacity.

Since M_{t,u_t} has limit points on $S(0, \tau)$, the spherical segment $\tilde{K}_{t,u_t}$ cannot lie strictly inside B_{r_0}. This implies that for every $t \in (-r_1/2, r_1/2)$ the vectors $y - y_0$, where $y \in K_{t,u_t}$, form an angle $\theta(t)$ at least $\theta_0 = \arccos(3/4)$ with the vector $-\mathbf{e}_n$. On the basis of Lemma 3.1, this allows us to deduce that

$$C_n(F_0, F_1, V) \geq C_n \ln \frac{r_0 + r_1/2}{r_0 - r_1/2} \geq C_n r_1/r_0,$$

where $C_n > 0$ is a constant.

Let $A_0 = f_M^{-1}(F_0)$ and $A_1 = f_M^{-1}(F_1)$, where f_M denotes the restriction of f to $\overline{M}$. Then, by Theorem 5.11,

$$C_n(F_0, F_1, V) \leq K(f) C_n(A_0, A_1, M)$$
$$\leq K(f) C_n(A_0, A_1, B(0, \tau)).$$

The set A_0 is contained in $S(0, \tau)$, and A_1 lies in $B(0, \delta) \subset G_{r_0}$. From this,

$$C_n(A_0, A_1, B(0, \tau)) \leq C_n(S(0, \tau), \overline{B}(0, \delta), B(0, \tau))$$
$$= C_n(\mathbf{R}^n \backslash B(0, \tau), \overline{B}(0, \delta)) = \omega_n \ln^{1-n}(\tau/\delta);$$

hence

$$\omega_n K(f) \ln^{1-n}(\tau/\delta) \geq C_n r_1/r_0 \geq \lambda(n, K),$$

and passage to the limit as $\tau \to 0$ gives us that $\delta \geq \varphi(K, n) > 0$, which is what was to be proved.

COROLLARY 1. *Let* $f: U \to \mathbf{R}^n$ *be a mapping with bounded distortion. If* f *is a local homeomorphism and* $B(x_0, r) \subset U$, *then* f *is a homeomorphism in the ball* $B[x_0, r\rho_n(K)]$, *where* $\rho_n(K)$ *is the constant in the theorem, and* $K = K(f)$.

This is obtained by applying the theorem to the mapping $\tilde{f}(X) = f(x_0 + rX)$, which is defined in $B(0, 1)$ and is a homeomorphism on $B(0, \rho_n(K))$ by the theorem. Hence, f is a homeomorphism in $B[x_0, r\rho_n(K)]$, which is what was to be proved.

COROLLARY 2 [182]. *Let* $f: \mathbf{R}^n \to \mathbf{R}^n$ *be a local homeomorphism with bounded distortion. Then* f *is a homeomorphism.*

Indeed, by Corollary 1, f is a homeomorphism in the ball $B(0, \rho_n(K)r)$ for any $r > 0$. Therefore, f is a homeomorphism of the whole space $\mathbf{R}^n$, which is what was required.

§11. Extremal properties of mappings with bounded distortion

§11.1. The homomorphism generated on the algebra of exterior forms by a mapping with bounded distortion. Let U and G be open sets in $\mathbf{R}^n$, and $f: U \to G$ a continuous mapping of class $W^1_{r,\mathrm{loc}}(U)$. Then, as shown in §4, a certain exterior form $f^*\omega$ can be defined for every exterior C^1-form ω of degree k $(0 \le k \le r)$ on G. The requirements imposed on the exterior forms in this result can be weakened in the case when f is a mapping with bounded distortion. It is the purpose of this subsection to prove the last fact.

LEMMA 11.1. *Let U be an open set in $\mathbf{R}^n$, $f: U \to \mathbf{R}^n$ a mapping with bounded distortion that is not identically constant in U, and $u(y)$ an integrable function on $\mathbf{R}^n$. Then the function $u[f(x)]\mathcal{J}(x, f) = v(x)$ is locally integrable in U, and for every compact domain G contained in U with boundary a set of measure zero*

$$\int_G v(x)\mathcal{J}(x, f)\,dx = \int_{\mathbf{R}^n} u(y)\mu(y, f, G)\,dy.$$

PROOF. By Theorem 2.2 of this chapter, v is locally integrable in U if u is such that for every chapter domain $G \subset U$ the function $u(\cdot)N(\cdot, f, G)$ is integrable. What was proved in §6 gives us that $N(y, f, G) \le |\mu(y, f, G)|$ for any compact domain G. The function $\mu(\cdot, f, G)$ is bounded and has compact support in $\mathbf{R}^n$, by what was proved in §6. From this, the function $y \to u(y)N(y, f, G)$ is integrable for every function $u \in L_{1,\mathrm{loc}}(\mathbf{R}^n)$, and the lemma is proved.

COROLLARY. *Let $f: U \to \mathbf{R}^n$ be a mapping with bounded distortion, and E an arbitrary set of measure zero in $\mathbf{R}^n$. If f is not identically zero, then $f^{-1}(E)$ is a set of measure zero.*

This proposition can be proved by applying the lemma to the indicator function of E and using the fact that if f is a mapping with bounded distortion, then $\mathcal{J}(x, f) \ne 0$ for almost all $x \in U$.

LEMMA 11.2. *Suppose that U and G are open sets in $\mathbf{R}^n$, and $f: U \to G$ is a mapping with bounded distortion. Let $\omega(y)$ be an exterior form of degree $k \ge 1$ defined on G and belonging to $L_{p,\mathrm{loc}}(G)$, where $p = n/k$. Then $f^*\omega$ is a form of class $L_{p,\mathrm{loc}}(A)$, and for every compact set $A \subset U$ there is a constant $L(A) < \infty$ such that $\|f^*\omega\|_{p,A} \le L(A)\|\omega\|_{p,f(A)}$. If φ is a form of degree $k + 1$ in the class $L_{p_1,\mathrm{loc}}(G)$, where $p_1 = n/(k + 1)$, such that $\varphi = d\omega$, then also $f^*\varphi = d(f^*\omega)$.*

PROOF. Suppose that $\omega \in L_{p,\mathrm{loc}}(G)$ and $\omega(y) = \sum_I a_I(y)\,dy_I$. Then, by definition,

$$(f^*\omega)(x) = \sum_I a_I[f(x)]\,df_I(x),$$

where $df_I(x) = df_{i_1}(x) \wedge \cdots \wedge df_{i_k}(x)$ for $I = (i_1, \ldots, i_k)$. Using inequality (4.2) in §4, we get that

$$|df_I(x)| \le C|df_{i_1}(x)||df_{i_2}(x)| \ldots |df_{i_k}(x)|$$
$$\le C|\mathscr{F}(x,f)|^{k/n}$$

for all x, where C is a constant depending only on k and n. From this,

$$|f^*\omega(x)| \le C|\omega|[f(x)]|\mathscr{F}(x,f)|^{k/n}.$$

Let $B_0 = \overline{B}(x_0, r)$ be an arbitrary closed ball in U. Then

$$\int_{B_0} |f^*\omega(x)|^p\,dx \le C_1^p \int_{B_0} [|\omega|(f(x))]^p |\mathscr{F}(x,f)|\,dx$$
$$\le C_1^p \left| \int_{f(B_0)} [|\omega|(y)]^p \mu(y,f,B_0)\,dy \right| \le [L(B_0)]^p \|\omega\|_{p,f(B_0)},$$

and the first assertion of the lemma is clearly proved.

We prove the second assertion. Let $\varphi \in L_{p_1,\mathrm{loc}}(U)$. Replacing ω by φ in the preceding arguments, we get that $f^*\varphi \in L_{p_1,\mathrm{loc}}(U)$. It remains to show that $f^*\varphi = d(f^*\omega)$. This follows from Lemma 4.7 in the case when φ and ω are forms of class C. We consider the general case. Take an arbitrary compact set $A \subset U$, and let $B = f(A)$. Let ω_h and φ_h be Sobolev averagings of the forms ω and φ. Then $\varphi_h = d\omega_h$ for each h. As $h \to 0$ we have that $\|\omega_h - \omega\|_{p,B} \to 0$ and $\|\varphi_h - \varphi\|_{p_1,B} \to 0$. By inequality $(*)$, this gives us that $f^*\varphi_h \to f^*\varphi$ in $L_{p_1,\mathrm{loc}}(U)$, and $f^*\omega_h \to f^*\omega$ in $L_{p,\mathrm{loc}}(U)$ as $h \to 0$. By Lemma 4.7 in this chapter, $f^*\omega_h = d(f^*\omega_h)$ for each h. This implies that $f^*\varphi = d(f^*\omega)$ in view of Lemma 4.2, and the lemma is proved.

LEMMA 11.3. *Suppose that U and G are open sets in $\mathbf{R}^n$, $f: U \to G$ is a mapping with bounded distortion, and $u: G \to \mathbf{R}$ is a function of class $W_{n,\mathrm{loc}}^1(G)$. Then the function $v(x) = u[f(x)]$ is a function of class $W(U)$. Further,*

$$v'(x) = [f'(x)]^* u'[f(x)].$$

PROOF. We construct a sequence u_ν, $\nu = 1, 2, \ldots,$ of functions of class $C^\infty(G)$ such that $u_\nu \to u$ almost everywhere in G and $du_\nu \to du$ in $L_{n,\mathrm{loc}}(U)$. Then $(f^*u_\nu)(x) \to (f^*u)(x)$ almost everywhere. Each of the functions f^*u_ν is continuous. The forms r^*du_ν belong to $L_{n,\mathrm{loc}}(U)$, and they converge to $f^* du$ in $L_{n,\text{loc}}(U)$ as $\nu \to \infty$. By Lemma 4.7, $f^* du_\nu = d(f^* u_\nu)$ for each ν. The coefficients of the forms $f^* du_\nu$ are thus generalized derivatives of the functions $f^* u_\nu$, and they converge in $L_{n,\text{loc}}(U)$ to the corresponding coefficients of the form $f^* du$ as $\nu \to \infty$. The functions $f^* u_\nu$ converge almost everywhere to $f^* u$. This implies that the coefficients of the form $f^* du$ are generalized derivatives of the function $v = f^* u$. It follows immediately from the definition of $f^* u$ that $v'(x) = [f'(x)]^* u'[f(x)]$, and the lemma is proved.

THEOREM 11.1. *Let U and V be open sets in $\mathbf{R}^n$, and let $f : U \to V$ and $g : V \to \mathbf{R}^n$ be mappings with bounded distortion. Then the composition $h = g \circ f : U \to \mathbf{R}^n$ is also a mapping with bounded distortion. Further,*

$$K(g \circ f) \le K(g)K(f), \qquad K_0(g \circ f) \le K_0(g)K_0(f).$$

PROOF. On the basis of Lemma 11.3 each of the real functions $g_i[f(x)]$ belongs to the class $W^1_{n,\text{loc}}(U)$. Let $E \subset V$ be a set of measure zero such that g is differentiable at each point $y \notin E$, and let $A = f^{-1}(E)$. By the corollary to Lemma 11.1, A is a set of measure zero. Let B be the set of points where f is not differentiable, and let $H = A \cup B$. Then H is a set of measure zero, $g \circ f$ is differentiable at each point $x \in U \backslash H$, and $(g \circ f)'(x) = g'(y)f'(x)$, where $y = f(x)$. From this it follows that $\mathscr{J}(x, g \circ f)$ has constant sign in U, and

$$\begin{aligned}
\|(g \circ f)'(x)\|^n &\le \|g'(y)\|^n \|f'(x)\|^n \\
&\le K(g)K(f)|\mathscr{J}(y, g)||\mathscr{J}(x, f)| \\
&= K(g)K(f)|\mathscr{J}(x, g \circ f)|
\end{aligned}$$

at each point $x \in U \backslash H$. This proves that $g \circ f$ is a mapping with bounded distortion, and $K(g \circ f) \le K(g)K(f)$. The estimate for $K_0(g \circ f)$ can be established similarly by considering the linear mappings $(g \circ f)'(x)$, $g'(y)$, and $f'(x)$. The theorem is proved.

§11.2. **Main theorem.** Suppose that G is an open set in $\mathbf{R}^n$. Assume that for almost all $y \in G$ a symmetric matrix $\sigma(y)$ is defined with elements which are Borel-measurable functions of y, and that there exist constants α and β, $0 < \alpha \le \beta < \infty$, such that for every $\xi \in \mathbf{R}^n$

$$\alpha|\xi|^2 \le \langle \sigma(y)\xi, \xi \rangle \le \beta|\xi|^2. \tag{11.1}$$

Let u be a function of class $W^1_{n,\text{loc}}(G)$. We introduce the exterior form

$$\begin{aligned}
\omega(y, \sigma, u) = \langle \sigma(y)u'(y), u'(y) \rangle^{n/2-1} \sum_{i=1}^{n}(-1)^{i-1} \\
\times (\sigma(y)u'(y))_i \, dy_1 \dots dy_{i-1} \, dy_{i+1} \dots dy_n.
\end{aligned} \tag{11.2}$$

For every function $\eta \in C^\infty(D)$

$$d\eta \wedge \omega(y, \sigma, u) = \langle\sigma(y)u'(y), u'(y)\rangle^{n/2-1}$$
$$\times \langle\sigma(y)u'(y), \eta'(y)\rangle \, dy_1 \, dy_2 \ldots dy_n. \tag{11.3}$$

Let us consider the differential equation

$$\operatorname{div}\{\langle\sigma(y)u'(y), u'(y)\rangle^{(n/2)-1}\sigma(y)u'(y)\} = 0. \tag{11.4}$$

Comparing Lemma 4.8 and the definition of a generalized solution of (11.3), we get that a function $u(y)$ is a generalized solution of (11.4) if and only if the generalized differential of the form $\omega(y, u, \sigma)$ is equal to zero.

Let $f: U \to D$ be an arbitrary mapping with bounded distortion, where U is an open set in $\mathbf{R}^n$. For almost all $x \in U$ the linear mapping $f'(x)$ is defined and is nonsingular. If $f'(x)$ is defined and nonsingular at a point $x \in U$, let $\tau(x)$ be the matrix defined by

$$\tau(x) = [f'(x)]^{-1}\sigma[f(x)]([f'(x)]^*)^{-1}|\mathscr{J}(x, f)|^{2/n}. \tag{11.5}$$

If $f'(x) = 0$, let $\tau(x) = I$. By what was proved in §5.2 of this chapter, the matrix $\tau(x)$ is symmetric and positive-definite, and its eigenvalues lie in the interval $[\alpha[K(f)]^{-2/n}, \beta[K_0(f)]^{2/n}]$.

THEOREM 11.2. *Let U and G be open sets in $\mathbf{R}^n$, and $f: U \to G$ a mapping with bounded distortion. If $u(y)$ is a solution on G of equation (11.4), then $w(x) = u[f(x)]$ is a solution of the equation*

$$\operatorname{div}\{\langle\tau(x)w'(x), w'(x)\rangle^{(n/2)-1}\tau(x)w'(x)\} = 0, \tag{11.6}$$

where the matrix-valued function $\tau(x)$ is determined from $\sigma(y)$ according to (11.5).

PROOF. The function u belongs to $W^1_{n,\mathrm{loc}}(G)$. By Lemma 11.3, this implies that w belongs to $W^1_{n,\mathrm{loc}}(U)$. Further, $w'(x) = [f'(x)]^*u'[f(x)]$ for all $x \in U$.

We introduce the exterior form

$$\zeta(x) = \langle\tau(x)w'(x), w'(x)\rangle^{(n/2)-1}\sum_{i=1}^{n}(-1)^{i-1}(\tau(x)w'(x))_i$$
$$\times dx_1 \ldots dx_{i-1} \, dx_{i+1} \ldots dx_n.$$

The condition that w be a solution of (11.6) is equivalent to the conditions that $d\zeta(x) = 0$. Since $d\omega = 0$, where ω is the form in (11.2), to prove that $d\zeta(x) = 0$ it suffices by Lemma 11.2 to establish that $\zeta(x) = f^*\omega(x)$.

For almost all $x \in U$, $\mathscr{F}(x, f) \neq 0$. We have

$$\omega(y) = \sum_{i=1}^{n} (-1)^{i-1} \omega_i(y) \, dy_1 \ldots dy_{i-1} \, dy_{i+1} \ldots dy_n.$$

Let

$$\varphi(x) = (f^* \omega)(x) = \sum_{i=1}^{n} (-1)^{i-1} \varphi_i(x) \, dx_1 \ldots dx_{i-1} \, dx_{i+1} \ldots dx_n.$$

We consider the vectors $(*\omega)(y) = (\omega_1(y), \ldots, \omega_n(y))$ and $(*\varphi)(x) = (\varphi_1(x), \ldots, \varphi_n(x))$. Obviously,

$$(*\omega)(y) = \langle \sigma(y)v'(y), v'(y) \rangle^{(n/2)-1} \sigma(y)v'(y).$$

By (5.13) in §5,

$$*\varphi(x) = |\mathscr{F}(x, f)|^{2/n} [f'(x)]^{-1} (*\omega)[f(x)](f'(x)^*)^{-1}.$$

After obvious transformations, we get from this that $*\zeta(x) = *\varphi(x)$, and hence $\zeta = \varphi$, which is what was to be proved.

COROLLARY [133]. *Let* $f \colon \mathbf{R}^n \to \mathbf{R}^n$ *be a mapping with bounded distortion of* $\mathbf{R}^n$ *into itself that is not identically constant, and let* $A = \mathbf{R}^n \setminus f(\mathbf{R}^n)$. *Then* A *is a set of zero n-capacity.*

PROOF. The set $f(\mathbf{R}^n)$ is open, and hence A is closed.

Assume that the n-capacity of A is positive. A point $x_0 \in A$ is said to be *essential* if the n-capacity of $A \cap \overline{B}(x_0, r)$ is positive for every $r > 0$. The collection of all nonessential points of A is denoted by E. If $x_0 \in E$, then there is an $r > 0$ such that the n-capacity of $A \cap B(x_0, r)$ is equal to zero. For every point $x \in A \cap B(x_0, r)$ the n-capacity of $A \cap B(x, \delta)$ is equal to zero in this case, where $\delta = r - |x - x_0| > 0$, and hence all the points $x \in A \cap B(x_0, r)$ are nonessential. We thus get that $A \cap B(x_0, r) \subset E$. This proves that E is open relative to A. Let $x_0 \in E$ and $r > 0$ be such that the n-capacity of $A \cap B(x_0, r)$ is equal to zero. Then there exist a point y with rational coordinates and a rational number $\rho > 0$ such that $x_0 \in B(y, \rho) \subset B(x_0, r)$. Obviously, $B(y, \rho) \cap A$ is a set of zero n-capacity. Since $x_0 \in E$ is taken arbitrarily and the set of balls $B(y, \rho)$ of the indicated type is countable, it follows that E is the union of a countable collection of sets of zero n-capacity, and hence has zero n-capacity.

The set $B = A \setminus E$ is closed, and its n-capacity is nonzero. Let a and b be any two boundary points of B, $a \neq b$. Since E clearly does not have interior points, a and b are boundary points also for A. Let ρ be an arbitrary number such that $0 < 2\rho < |b - a|$, and let $B_1 = B \cap \overline{B}(a, \rho)$ and

$B_2 = B \cap \overline{B}(b, \rho)$. Obviously, B_1 and B_2 are sets of positive n-capacity. Let $W_n^1(B_1, B_2)$ be the set of all functions in $W_n^1(\mathbf{R}^n)$ equal to zero on B_1 and equal to 1 on B_2. Since B_1 and B_2 have nonzero capacity, the capacity of the capacitor (B_1, B_2) is nonzero, and hence there is a function $u(x) \in W_n^1(B_1, B_2)$ minimizing the functional $\int_{\mathbf{R}^n} |u'(x)|^n \, dx$. The function $u(x)$ is a solution of the equation

$$\operatorname{div}(|u'(x)|^{n-2} u'(x)) = 0$$

on $\mathbf{R}^n \backslash (B_1 \cup B_2)$, hence also on $f(\mathbf{R}^n)$. The function $u(x)$ is not constant in $f(\mathbf{R}^n)$. Indeed, since a is a boundary point of A, any neighborhood of a contains points of $f(\mathbf{R}^n)$, and hence, by continuity, $u(x)$ takes values arbitrarily close to 0 on $f(\mathbf{R}^n)$. Considering the behavior of $u(x)$ near b, we get in exactly the same way that $u(x)$ takes values arbitrarily close to 1 on $f(\mathbf{R}^n)$. It follows that $u(x)$ is not constant on $f(\mathbf{R}^n)$.

Obviously, $0 \leq u(x) \leq 1$ for all x.

By Theorem 11.2, $v(x) = u[f(x)]$ is a solution on $\mathbf{R}^n$ of the equation

$$\operatorname{div}[\langle \theta(x) v'(x), v'(x) \rangle^{n/2-1} \theta(x) v'(x)] = 0. \tag{11.6}$$

For all $x \in \mathbf{R}^n$ we have that $0 \leq v(x) \leq 1$, and hence it is identically constant in view of the Liouville theorem for generalized solutions of (11.6) (Theorem 5.7 in this chapter). However, this contradicts the fact that $v(x)$ is not constant on $\mathbf{R}^n$ by construction.

Thus, the assumption that the n-capacity of A is nonzero leads to a contradiction, and the corollary is proved.

This result is given here as an example of the application of Theorem 11.2. Picard's theorem for complex functions of a single variable is well known. If $f(z)$ is an analytic function of a single variable defined on the whole complex plane $\mathbf{C}$, then $\mathbf{C} \backslash f(\mathbf{C})$ is finite and consists of at most two elements. It is natural to ask whether there is an analogue of Picard's theorem for mappings with bounded distortion on $\mathbf{R}^n$ for an arbitrary $n \geq 2$. There is indeed such an analogue, as established by Rickman [147]. Namely, he showed that for every $n \geq 2$ there exists an integer $q = q(n, K)$ such that for every mapping $f : \mathbf{R}^n \to \mathbf{R}^n$ with bounded distortion that is not identically constant the set $\mathbf{R}^n \backslash f(\mathbf{R}^n)$ is finite, and the number of elements in it does not exceed q. The proof of Rickman's theorem requires the use of a very refined technique which is an extension of the Nevanlinna theory of the distribution of values of entire functions to the case of mappings with bounded distortion. Certain estimates of solutions of equations of elliptic type are also used. In particular, Theorem 11.2 is used in [147].

§12. Some further results

Below we give a survey of some investigations in the theory of n-space mappings with bounded distortion, and on closely related questions not reflected in the main text. This survey of materials is mainly determined by the personal interests of the author and does not pretend to be complete. There are many excellent publications on the theory of quasiconformal mappings. Here we should mention in the first place the remarkable investigations of Ahlfors, Gehring, and my Finnish colleagues. Space does not permit a discussion of all these publications.

§12.1. Classes of domains in $\mathbf{R}^n$. We describe some classes of domains that arise naturally in the study of mappings with bounded distortion. These classes can be useful in other questions when the requirement of a smooth boundary is too restrictive and at the same time it is known that an arbitrary domain in $\mathbf{R}^n$ can fail to have the needed properties.

1. *Domains of John type* [60]. Let r and R be numbers such that $0 < r \le R < \infty$, and let U be an open domain in $\mathbf{R}^n$. We say that U belongs to the class $\mathscr{J}(r, R)$ if there is a point $a \in U$ such that any point $x \in U$ can be joined to a in U by a rectifiable curve $x(s)$, $0 \le s \le l$ (the parameter s is the arclength), for which $x(0) = x$, $x(l) = a$, the length l of the curve does not exceed R, and for all $s \in [0, l]$

$$\rho[x(s), \partial U] \ge rs/l. \tag{12.1}$$

The point a in this definition is called the *marked point* of U.

We say that U is a domain of class $\mathscr{J}$, or, in other words, a *domain of John type*, if $U \in \mathscr{J}(r, R)$ for some r and R.

It follows from the definition that if $U \in \mathscr{J}(r, R)$, then U is contained in the ball $B(a, R)$. Further, setting $s = l$ in (12.1), we get that $\rho(a, \partial U) \ge r$, and hence $B(a, r)$ is contained in U.

We set $r/l = \gamma$ in (12.1) and let $B(s) = B[x(s), s\gamma]$ and $\Gamma(x) = \bigcup B_s$. By (12.1), $B(s) \subset U$ for each s, and thus $\Gamma(x) \subset U$. In the case when $x(s)$, $0 \le s \le l$, is a segment the set $\Gamma(x)$ is the ordinary circular cone consisting of all possible segments joining x to points of $B(a, r)$. In the general case $\Gamma(x)$ can be regarded as a distinctive kind of curvilinear cone with vertex x. Intuitively, a domain $U \in \mathscr{J}(r, R)$ can be characterized as a domain such that any point x in it can be reached by the vertex of a cone made from some elastic material by bending it in such a way that it does not go out of U.

Domains of the class $\mathscr{J}$ were first used in the theory of space mappings with bounded distortion in the author's paper [140] (if one disregards the work of John himself in which he studied a question that also can be related to the theory of mappings with bounded distortion.)

Domains that are star-like with respect to a ball (see §2.2 in Chapter I) are a special case of the domains of class $\mathscr{J}$.

Every bounded domain whose boundary is a smooth manifold of class C^1 belongs to $\mathscr{J}$. Every domain that is a union of finitely many domains of class $\mathscr{J}$ belongs to $\mathscr{J}$ (see [48]). In particular, every domain belonging to the class S defined in §2.2 of Chapter I is a domain of John type.

The Sobolev imbedding theorems presented in §2.2 of Chapter I are true also in the case when U is a domain of John type (see [48]).

Besov introduced a certain class of domains, which he called domains satisfying the flexible horn condition, that is a generalization of the class $\mathscr{J}$. In particular, domains with the flexible horn condition can be unbounded. Function classes generalizing the classes W_p^l are considered in [17] and [16], and Sobolev-type imbedding theorems are established for them.

We present some results of a technical nature that concern domains of John type. First of all, we give two modifications of the original definition.

LEMMA 12.1 [168]. *An open domain U in $\mathbf{R}^n$ is a domain of class $\mathscr{J}$ if and only if there exist numbers α and R and a point $a \in U$ such that $0 < \alpha \leq 1$, $0 < R < \infty$, for every $x \in U$ there is a rectifiable curve $x(s)$, $0 \leq s \leq l$ (the parameter s is the arclength), with length l at most R such that $x(0) = x$, $x(l) = a$, and*

$$\rho[x(s), \partial U] \geq \alpha s \qquad (12.2)$$

for all $s \in [0, l]$.

(There is also a proof of the lemma in [144].) The fact that a domain of class $\mathscr{J}$ satisfies the condition of the lemma is obvious. The hard part is to prove the converse assertion. (In principle, the factor multiplying s in (12.1) can be arbitrarily large; in (12.2) this factor is a certain fixed constant.)

Let δ be a number such that $0 < \delta \leq 1$. We say that U belongs to the class $\mathscr{C}_B(\delta)$ if it is bounded and there is a point $a \in U$ such that for every point $x \in U$ we can construct an $x: [0, 1] \to U$ with $x(0) = x$ and $x(1) = a$ for which

$$|x(t_2) - x(t_1)| < \frac{1}{\delta} \rho[x(t_2), \partial U]$$

for any t_1 and t_2 with $0 \leq t_1 \leq t_2 \leq 1$. We say that $U \in \mathscr{C}_B$ if $U \in \mathscr{C}_B(\delta)$ for some $\delta \in (0, 1]$.

LEMMA 12.2 [93]. *Every domain U of class $\mathscr{F}(r, R)$ belongs to the class $\mathscr{C}_B(r/R)$. Conversely, if $U \in \mathscr{C}_B(\delta)$ for some $\delta \in (0, 1]$, then $U \in \mathscr{F}(l/\varphi^2, \varphi l)$ depends on n and δ; here $l = \operatorname{diam} U$ and $\varphi \geq 1$.*

We note one more useful property of domains of John type. Let U be a bounded domain in $\mathbf{R}^n$. Then we say that U is a domain of type I_γ, where $0 < \gamma < 1$, if

$$\int_U \frac{dx}{\rho(x, \partial U)^\gamma} < \infty. \tag{12.3}$$

An arbitrary domain in $\mathbf{R}^n$ can fail to have the property I_γ. We give an example. Let $n = 2$, and let V be the set of all points $x = (x_1, x_2) \in \mathbf{R}^n$ lying in $B(0, 1)$ and not belonging to the spiral with equation

$$r = 1 - \frac{\ln \pi}{\ln(\varphi + \pi)}, \qquad 0 \leq \varphi < \infty,$$

in polar coordinates. It is easy to verify that the integral (12.3) is equal to ∞ for the domain $U = V$ for any $\gamma \in (0, 1]$. The product $V \times B_{\mathbf{R}^{n-2}}(0, 2)$ is an example of a domain not having the property I_γ in the case $n > 2$.

The following statement was proved by Trotsenko [168].

LEMMA 12.3 ([144], [168]). *If U is a domain of class $\mathscr{F}(r, R)$ in $\mathbf{R}^n$, then U has the property I_γ, where $\gamma = \gamma_1(r/R)^n$, $0 < \gamma_1 \leq 1$, and γ_1 depends only on n. The estimate*

$$\int_U \frac{dx}{[\rho(x, \partial U)]^\gamma} \leq C \left(\frac{R}{r} \right)^k R^{n-\gamma}$$

is valid, where C and k depend only on n, $k > 0$, and $0 < C < \infty$.

The proof of Lemma 12.3 is based on an estimate of the number of cubes in a Whitney decomposition of the domain that have distance to the boundary of U lying between h and $2h$, where $h > 0$. As shown in [168], the condition "U is a domain of John type" in the statement of the lemma can be replaced by a weaker condition. Namely, let $0 < \alpha \leq 1$. We say that a domain U in $\mathbf{R}^n$ has the property A_α if there is a $\sigma > 0$ such that for any $x \in \partial U$ and any $t \in (0, \sigma]$ there is a point $y \in U$ such that $|x - y| \leq t$ and $\rho(y, \partial U) > \alpha t$. If a domain U in $\mathbf{R}^n$ is bounded and has the property A_α, then U has the property I_γ for some $\gamma \in (0, 1)$.

2. *Uniform domains.* A domain U in $\mathbf{R}^n$ is called a domain of class $U(\alpha, \beta)$, where $0 < \alpha \leq \beta < \infty$, if for any two points $x_1, x_2 \in U$ there exists a domain $G \subset U$ belonging to the class $\mathscr{F}(\alpha|x_1 - x_2|, \beta|x_1 - x_2|)$ such that $x_1, x_2 \in G$. A domain U is said to be *uniform* if U is a domain of class $U(\alpha, \beta)$ for some α and β. The class of uniform domains was

introduced in [93]. The interest in and significance of this class for the theory of quasiconformal mappings has to do with the following result of Martio and Sarvas [93].

THEOREM 12.2. *Let* $f: \mathbf{R}^n \to \mathbf{R}^n$ *be a quasiconformal mapping, and* U *a domain of class* $U(\alpha, \beta)$. *Then* $f(U)$ *is a domain of class* $U(\varepsilon^2, 1/\varepsilon)$, *where* $\varepsilon \in (0, 1)$ *depends only on* n, $K(f)$, α, *and* β.

In particular, it follows from Theorem 12.2 that the image of a ball with respect to a quasiconformal mapping of $\mathbf{R}^n$ onto itself is a homogeneous domain.

Let $n = 2$. A domain $U \subset \mathbf{R}^n$ is called a *quasidisk* if there exists a quasiconformal mapping $f: \overline{\mathbf{R}^2} \to \overline{\mathbf{R}^2}$ such that $U = f(B(0, 1))$. A planar curve Γ is called *quasicircle* if it is the boundary of some quasidisk. Ahlfors obtained a geometric characterization of a quasicircle in 1963 [2].

Let G be a domain in $\mathbf{R}^n$ whose boundary is a simple closed curve L, and let x and y be two arbitrary points of L. These points divide L into two arcs. Denote by $d(x, y)$ the smallest of the diameters of these arcs. One says that G *satisfies the arc condition* if there exists a constant c such that $1 \le c < \infty$ and $d(x, y) \le c|x - y|$ for any $x, y \in L$.

THEOREM 12.3 [2]. *Let* G *be a domain in* $\mathbf{R}^n$. *Then* G *is a quasidisk if and only if its boundary is a simple closed curve and satisfies the arc condition.*

It follows from Theorems 12.2 and 12.3 that if a planar domain G satisfies the arc condition (G is bounded by a simple closed curve), then G is a homogeneous domain. The converse is also true: if a planar domain G bounded by a simple closed curve is homogeneous in the sense of the definition given here, then it satisfies the arc condition, as shown in [93].

We mention further a certain characterization obtained in [92] for homogeneous domains.

Let x_1 and x_2 be two different points in $\overline{\mathbf{R}}^n = \mathbf{R}^n \cup \{\infty\}$, and let Γ be a continuum joining them; that is, a continuum containing x_1 and x_2. Assume that $\delta \in (0, 1]$. Denote by $V_1(\Gamma, \delta)$ the set of all points $y \in \overline{\mathbf{R}}^n$ for which

$$\langle x, y, x_1, x_2 \rangle = \frac{|x - y| \cdot |x_1 - x_2|}{|x - x_1| \cdot |y - x_2|} < \delta$$

for some $x \in \Gamma$. Similarly, let $V_2(\Gamma, \delta)$ be the set of all points $y \in \overline{\mathbf{R}}^n$ with

$$\langle x, y, x_2, x_1 \rangle < \delta$$

for some $x \in \Gamma$. Let $V(\Gamma, \delta) = V_1(\Gamma, \delta) \cap V_2(\Gamma, \delta)$.

To understand the intuitive meaning of $V_1(\Gamma, \delta)$ and $V_2(\Gamma, \delta)$ we consider the case when Γ is the segment with endpoints x_1 and x_2. Let us transfer the point x_1 to ∞ by inversion with respect to the sphere S_0 about x_2 passing through x_1. The segment Γ then does into a ray λ emanating from x_1. The cross ratio is invariant with respect to Möbius transformations. We have that

$$\langle x, y, x_1, \infty \rangle = |x - y|/|x - x_1|,$$

and the condition $\langle x, y, x_1, \infty \rangle < \delta$ means that $|y - x| < \delta |x - x_1|$; that is, that y belongs to $B(x, \delta|x - x_1|)$. The image of $V_1(\Gamma, \delta)$ is the union of these balls and is a right circular cone whose generators form the angle $\varphi = \arcsin \delta$ with its axis λ. This cone is the intersection of the half-spaces with boundary planes passing through x_1 and having outward normals at an angle $\pi/2 + \varphi$ to λ. Inversion with respect to S_0 carries each such half-space into a ball whose boundary passes through x_1 and x_2 and whose radius to x_1 forms the angle $\pi/2 + \varphi$ with the vector $x_2 - x_1$. In this case the set $V_1(\Gamma, \delta)$ is a "spindle", the intersection of all such balls. In this case it is clear that $V_2(\Gamma, \delta) = V_1(\Gamma, \delta)$. In the case of an arbitrary continuum Γ the set $V(\Gamma, \delta)$ is a distinctive kind "curvilinear" spindle joining x_1 and x_2.

Following Martio [92], we say that a domain $U \subset \overline{\mathbf{R}}^n$ has the property $U(\delta)$ ($0 < \delta \leq 1$), and write $U \in U(\delta)$, if for any two points $x_1, x_2 \in U$ there exists a continuum Γ joining x_1 and x_2 such that $V(\Gamma, \delta) \subset U$.

LEMMA 12.4 [92]. *Let U be a domain of class $U(\alpha, \beta)$ in $\overline{\mathbf{R}}^n$. Then there exists a $\delta \in (0, 1]$ depending only on α, β, and n such that $U \in U(\delta)$. Conversely, if $U \in U(\delta)$ for some $\delta \in (0, 1]$, then $U \in U(\alpha, \beta)$, where α and β depend on δ and n.*

The concept of a quasidisk turns out to be useful in many questions of analysis not directly connected with the theory of quasiconformal mappings. In this connection we refer the reader to the monograph [40], where a number of such results are given. In particular, the concept of a homogeneous domain arises naturally in connection with the problem of extension of the classes $W_p^l(U)$, that is, with the problem of conditions under which there exists for every $u \in W_p^l(U)$ a function $\theta u \in W_p^l(\mathbf{R}^n)$ such that $(\theta u)(x) = u(x)$ for all $x \in U$. Further, it is required that the operator $u \mapsto \theta u$ be a continuous mapping from $W_p^l(U)$ into $W_p^l(\mathbf{R}^n)$. Many papers have been devoted to the problem of extending the classes W_p^l. The reader can find more complete information about this in [48]. (In this connection see also [45], [47], [64], [179], and [180].)

In the case $n = 2$ the concept of a uniform domain enables us to give an exhaustive characterization of domains that are images of a ball under a quasiconformal mapping of $\mathbf{R}^n$. For $n \geq 3$ this is not the case. Tukia[*] showed that a cylinder constructed on any locally nonrectifiable unbounded quasicircle (that is, a curve that is the image of a line under a quasiconformal mapping of $\mathbf{R}^2$) separates the space into two homogeneous domains homeomorphic to a ball that cannot be transformed into a ball by a quasiconformal mapping of $\overline{\mathbf{R}}^n$. Another example (due to Trotsenko): Let U_1 and U_2 be the domains in $\overline{\mathbf{R}}^n$, $n \geq 3$, given by $U_1 = \{(x_1, \ldots, x_n) | x_n < \sqrt{|x_1|}\}$ and $U_2 = \{(x_1, \ldots, x_n) | x_n > \sqrt{|x_1|}\}$. The common boundary of U_1 and U_2 is the surface F in $\mathbf{R}^n$ defined by the equation $x_n = \sqrt{|x_1|}$. Assume also that $\infty \in F$. It is easy to verify the U_1 is a homogeneous domain, while U_2 is not (because the angle at the point 0 is zero). This gives us that there does not exist a quasiconformal mapping $f: \overline{\mathbf{R}}^n \to \overline{\mathbf{R}}^n$ such that U_1 is the image of some ball under this mapping. Indeed, if there were such a mapping, then by replacing f by a composition with a Möbius mapping we would get that U_2 is a quasiconformal image of a ball, hence a homogeneous domain in $\mathbf{R}^n$, according to Theorem 12.2.

§12.2. Stability in the Liouville theorem on conformal mappings of a space and related questions.

1. According to Liouville's theorem, proved in §5.9, every mapping f with bounded distortion not identically constant and such that $K(f) = 1$ is a Möbius mapping. In [81] Lavrent'ev posed the question of whether a mapping with $K(f)$ close to 1 is close to being a Möbius mapping. He also obtained the first results in this direction (see below for a history of the question). The study of mappings with distortion coefficient $K(f)$ close to 1 led to proofs that such mappings have many properties not holding in the general case. Some results of this kind will be presented below.

2. First of all note that as $K(f) \to 1$ the differentiability properties of f improve in the sense that the power p to which the derivatives of f are locally integrable increases without bound. Namely, the following statement is valid.

THEOREM 12.4. *There exist constants $\delta_0 > 0$ and $C < \infty$ such that every mapping $f: U \to \mathbf{R}^n$ with bounded distortion and with $K(f) \leq 1 + \delta_0$ belongs to the class $W_{p,\mathrm{loc}}^1(U)$ for any*

$$p \leq C/(K(f) - 1). \tag{12.4}$$

---

[*]Pekka Tukia, *A quasiconformal group not isomorphic to a Möbius group*, Ann. Acad. Sci. Fenn. Ser. A I Math. **6** (1981), 149–160.

For $K > 1$ denote by $p_n(K)$ the supremum of the numbers p such that every mapping $f: U \to \mathbf{R}^n$ with $K(f) \leq K$ belongs to $W^1_{p,\text{loc}}(U)$.

Let $f: U \to \mathbf{R}^n$ be the nonlinear homothety $f(x) = |x|^{\alpha-1}x$ for all $x \in \mathbf{R}^n$, where $0 < \alpha < 1$. As shown above, $K(f) = 1/\alpha$ and $f \in W^1_{p,\text{loc}}(\mathbf{R}^n)$ for every $\rho < n/(1-\alpha) = nK(f)/(K(f)-1)$. This gives an upper bound for $p_n(K)$:

$$p_n(K) \leq nK/(K-1). \tag{12.5}$$

In particular, it follows from (12.4) and (12.5) that

$$p_n(K) = O(1/(K-1)) \tag{12.6}$$

as $K \to 1$.

The estimate in Theorem 12.1 holds only under the assumption that $K(f)$ is less than some value $K_0 = 1 + \delta_0$. For $K(f) \geq 1 + \delta_0$ the theorem does not give any information about the value of $p_n(K)$. The fact that $p(K) > n$ was established earlier for quasiconformal mappings by Gehring [38], and for mappings with bounded distortion by Meyers and Elcrat [98]. The results in [38] and [98] do not imply that $p_n(K) \to \infty$ as $K \to 1$. Comparing Theorem 12.4 and the result in [98], we get that there exists a function $p_n(K)$, $1 \leq K < \infty$, such that $p_n(K) > n$ for all K, $p_n(K) = O(1/(K-1))$ as $K \to 1$, and every mapping f with bounded distortion in $\mathbf{R}^n$ belongs to the class $W^1_{p,\text{loc}}$ for any $p < p_n[K(f)]$.

For the case $n = 2$ Boyarski in [19] and [20] established that $p(K) > 2$. The estimate $p(K) = O(1/(K-1))$ as $K \to 1$ was obtained by Lehto for the case $n = 2$ [84]. The fact that $p_n(K) \to \infty$ as $K \to 1$ was proved also by Iwaniec [58]. The estimate $p_n(K) = O(\ln(1/(K-1)))$ as $K \to 1$ was established for $p_n(K)$ in [58].

The question of the exact value of $p_n(K)$ remains open. There is a conjecture that $p_n(K) = nK/(K-1)$, so that the worst mapping with respect to its differentiability properties is precisely the nonlinear homothety $x \mapsto x|x|^{\alpha-1}$, $\alpha = 1/K$.

3. The next statement is the main result in the problem of stability in the Liouville theorem on conformal mappings.

THEOREM 12.5. *Let U be a domain of class $\mathscr{J}(r, R)$ in $\mathbf{R}^n$, where $n \geq 3$. Then for every $p > n$ there exists a number $\delta_0 = \delta(p)$, depending only on n and p, such that the following assertions hold.*

If $f: U \to \mathbf{R}^n$ is a mapping with bounded distortion for which $K(f) \leq \delta_0(r/R)^2 + 1$, then there exists a Möbius mapping φ such that

$$|\varphi^{-1}[f(x)] - x| \leq C_1 \frac{R^3}{r^2}(K(f) - 1) \tag{12.7}$$

for all $x \in U$, and if $g(x) = \varphi^{-1}[f(x)]$, then

$$\int_U |g'(x) - I|^p \, dx \leq C_2 \frac{R^{n+2p+k}}{r^{2p+k}} [K(f) - 1]^p, \tag{12.8}$$

where I denotes the identity matrix, the numbers C_1 and C_2 depend only on p and n, and k is the constant in Lemma 12.3.

The theorem shows that if $K(f)$ is sufficiently close to 1, then there exists a Möbius mapping φ close to f in the sense that the mapping $g(x) = \varphi^{-1}[f(x)]$ is almost the identity. The estimate (12.8) shows that the derivatives of $g(x)$ are close to the corresponding derivatives of the identity mapping in the sense of the L_p-metric.

THEOREM 12.6 [169]. *Let $U \in U(\alpha, \beta)$ be a domain in $\overline{\mathbf{R}}^n$ with $\infty \in U$. Then there exist constants $\varepsilon_0 > 0$ and $c < \infty$, depending only on α, β, and n, such that any mapping $f: U \to \overline{\mathbf{R}}^n$ with bounded distortion carrying ∞ into ∞ has the following property. If $K_f = 1 + \varepsilon \leq 1 + \varepsilon_0$, then for any ball $B(x_0, r)$ there is a similarity T such that*

$$|Tf(x) - x| \leq cr\varepsilon$$

for all $x \in \overline{B}(x_0, r) \cap U$.

REMARK 1. For a half-space results analogous to Theorem 12.6 were obtained by Belinskiĭ in [13] and [14], and Kopylov in [68] and [72].

In Theorems 12.5 and 12.6 the closeness of a mapping to a Möbius mapping is measured by the extent to which the composition $\varphi \circ f$, φ a Möbius mapping, differs from the identity mapping. It would be more natural to consider the difference $f(x) - \varphi(x)$, where φ is an arbitrary Möbius mapping. Dairbekov [28] established the following statement.

THEOREM 12.7. *Let $U \subset \overline{\mathbf{R}}^n$ be a domain of class $U(\alpha, \beta)$. Then there exist numbers $\varepsilon_0 > 0$ and $C < \infty$ depending only on n, α, and β such that for any quasiconformal mapping $f: U \to \mathbf{R}^n$ with $K(f) < 1 + \varepsilon_0$ there exists a Möbius mapping φ such that*

$$|f(x) - \varphi(x)| \leq C\mathscr{D}[f(U)](K(f) - 1),$$

where $\mathscr{D}[f(U)]$ denotes the diameter of the domain

$$\mathscr{D}[f(U)] = \sup_{x_1 \in U, x_2 \in U} |f(x_2) - f(x_1)|.$$

4. We say a few words about the history of the question. Lavrent′ev [81] established the stability of Möbius mappings in the class $C^{1,\alpha}$. Namely, he proved that if $f: B(0, 1) \to \mathbf{R}^n$ is a quasiconformal mapping of $B(0, 1)$ into itself such that

$$|f'(x_1) - f'(x_2)| \leq H|x_1 - x_2|^\alpha$$

for any $x_1, x_2 \in B(0, 1)$, where $0 < H < \infty$ and $0 < \alpha \leq 1$ are constants, then for every ball $B(0, r)$ with $0 < r < 1$ there exists a Möbius mapping φ such that

$$|f(x) - \varphi(x)| \leq \mu[K(f) - 1, r, H, \alpha]$$

for all $x \in B(0, r)$, where $\mu(\varepsilon, r, H, \alpha)$ depends only on ε, r, H, and α, and $\mu(\varepsilon, r, H, \alpha) \to 0$ as $\varepsilon \to 0$.

It was shown by the author in [121] and [120] that there exists a function $\mu(\varepsilon, r)$ of the variables $\varepsilon > 0$ and $r \in (0, 1)$ such that $\mu(\varepsilon, r) \to 0$ for each r, and for every quasiconformal mapping $f: B(0, 1) \to B(0, 1)$ there exists a Möbius mapping φ such that for all $x \in B(0, r)$

$$|f(x) - \varphi(x)| \leq \mu(K(f) - 1, r). \tag{12.9}$$

The proof of this in [121] was based on the following considerations. We first define a certain measure of the deviation of a mapping f from a Möbius mapping. For an arbitrary Möbius mapping φ let

$$\delta(f, \varphi, r) = \sup_{x \in B(0,r)} |f(x) - \varphi(x)|, \quad \text{and} \quad \delta(f, r) = \inf_\varphi \delta(f, \varphi, r)$$

(the infimum is over the set of all Möbius mappings of $B(0, 1)$). The quantity $\delta(f, r)$ measures the difference between the mapping $f: B(0, 1) \to B(0, 1)$ and a Möbius mapping. Now let

$$\mu_0(\varepsilon, r) = \sup_{K(f) \leq 1 + \varepsilon} \delta(f, r).$$

It is required to prove that $\mu_0(\varepsilon, r) \to 0$ as $\varepsilon \to 0$. Obviously, $\mu_0(\varepsilon, r)$ is a nondecreasing function of ε; hence the limit $\lim_{\varepsilon \to 0} \mu_0(\varepsilon, r) = \eta$ exists and is nonnegative. Assume that $\eta > 0$. Then there is a sequence (f_m), $m = 1, 2, \ldots,$ of quasiconformal mappings of $B(0, 1)$ into itself such that $K(f_m) \to 1$ as $m \to \infty$, and $\delta(f_m, r) > \eta/2$ for each m. Since the sequence of integrals

$$\int_{B(0,1)} |f'_m(x)|^n \, dx, \quad m = 1, 2, \ldots,$$

is bounded, the sequence (f_m) is uniformly equicontinuous on each ball $B(0, r)$ with $0 < r < 1$. In view of this it can be assumed that (f_m) converges to some mapping $f_0: B(0, 1) \to B(0, 1)$, uniformly on every ball $B(0, r)$ (this can obviously always be achieved by passing to a subsequence). With regard to the limit mapping f_0 it is established below that

f_0 is either a Möbius mapping or identically constant. In the case when f_0 is a Möbius mapping we get that, on the one hand, $\delta(f_m, f_0, r) \to 0$ as $m \to \infty$. On the other hand,

$$\delta(f_m, f_0, r) \geq \delta(f_m, r) > \frac{n}{2} > 0$$

for each m, and we thus get a contradiction. If f_0 is identically constant, then for each m there is a Möbius mapping $\varphi_m \colon B(0, 1) \to B(0, 1)$ such that $|\varphi_m(x) - f_0(x)| < \frac{1}{m}$ for all $x \in B(0, 1)$. Obviously,

$$\delta(f_m, \varphi_m, r) \leq \delta(f_m, f_0, r) + \frac{1}{m},$$

and hence $\delta(f_m, \varphi_m, r) \to 0$ as $m \to \infty$. On the other hand,

$$\delta(f_m, \varphi_m, r) \geq \delta(f_m, r) > \frac{n}{2} > 0,$$

and we again arrive at a contradiction. (The author's paper [121] was subjected to the criticism when it appeared that the case $f_0 \equiv \text{const}$ was not considered in it; as is clear from the foregoing, this is trivial.)

The fact that the function $\mu(\varepsilon, r)$ in (12.9) can be replaced by $\mu(\varepsilon)$, where $\mu(\varepsilon) \to 0$ and $\mu(\varepsilon)$ does not depend on r, was established by Belinskiĭ [12] under the assumption that the given mapping f is normalized in a definite way. It was later proved in [14] that $\mu(\varepsilon)$ can be taken equal to $C\varepsilon$, where C is a constant. The final result, contained in Theorem 12.5, was obtained by the author in [140]–[142] (see also the author's book [144]).

5. We can consider different kinds of integral characteristics of the nonconformality of a mapping. If such a characteristic is given, then the question arises as to whether a mapping is close to a Möbius mapping if this characteristic is small for it. (It is assumed that the zero value of the characteristic corresponds to a Möbius mapping.) Here we present a result of this kind.

Let $f \colon U \to \mathbf{R}^n$ be a mapping of class $W_n^1(U)$, $f(x) = (f_1(x), \ldots, f_n(x))$, where U is a domain in $\mathbf{R}^n$. For an arbitrary measurable set $A \subset U$ let

$$\mathscr{D}(f, A) = \int_A \sum_{i=1}^n |f_i'(x)|^n \, dx.$$

For each $x \in U$ such that the linear mapping $f'(x)$ is defined we have that

$$\sum_{i=1}^n |f_i'(x)|^n = \sum_{i=1}^n \left(\sum_{j=1}^n \left(\frac{\partial f_i}{\partial x_j}(x) \right)^2 \right)^{n/2} \geq n \det f'(x).$$

Equality holds here if and only if $f'(x)$ is a general orthogonal transformation and $\det f'(x) > 0$. This implies that if for f

$$\mathscr{D}(f, U) = n \int_U \mathscr{J}(x, f) \, dx,$$

then $\mathscr{J}(x, f) \geq 0$ and $f'(x)$ is a general orthogonal transformation for almost all $x \in U$; consequently, f is a Möbius transformation by Liouville's theorem.

Assume that U is bounded. Let the constants M and μ be such that $0 < \mu \leq M < \infty$, and let V be a set of positive measure strictly inside U. Denote by $W(\mu, M, V, \varepsilon)$ the collection of all mappings $f: U \to \mathbf{R}^n$ in $W_n^1(U)$ such that:

(I) $\|f\|_{W_n^1(U)} \leq M$;

(II) $\mathscr{D}(f, V) \geq \mu$; and

(III) $\int_U \sum_{i=1}^n |f_i'(x)|^n \, dx \leq n(1 + \varepsilon) \int_U \mathscr{J}(x, f) \, dx$.

Let A be a set lying strictly inside U. For $f \in W_n^1(U)$ denote by $\delta(f, A)$ the infimum of $\|f - \varphi\|_{W_n^1(A)}$ when φ runs through the set of all Möbius transformations. Let

$$\mu(\varepsilon, A) = \sup_{f \in W(\mu, M, V, \varepsilon)} \delta(f, A).$$

THEOREM 12.8. *The relation* $\mu(\varepsilon, A) = O(\varepsilon^{1/n})$ *as* $\varepsilon \to 0$ *holds for every measurable set* A *lying strictly inside* U *and such that* $V \subset A$.

6. As consequences of the stability result (Theorem 12.5) we present some theorems on smoothness of quasiconformal mappings and conformal mappings of Riemannian spaces.

Assume that the distortion coefficient for the mapping $f: U \to \mathbf{R}^n$ at the point x tends to 1 as $x \to a \in U$. It turns out that if this convergence is sufficiently fast, then f is differentiable at a.

We give the precise formulations.

Let U be an open subset of $\mathbf{R}^n$, and $f: U \to \mathbf{R}^n$ a mapping with bounded distortion. Then f is differentiable in U almost everywhere. If f is differentiable at a point $x \in U$, then let $K_f(x) = 1$ if $\mathscr{J}(x, f) = 0$, and $K_f(x) = K[f'(x)]$ otherwise.

Let $\omega: [0, \infty) \to \mathbf{R}$ be a given function. We say that ω *satisfies the Dini condition* if it is a nondecreasing function and

$$\int_0^k \frac{\omega(t)}{t} \, dt < \infty$$

for some $k > 0$. (In this case the indicated integral is clearly finite for every $k > 0$.)

THEOREM 12.9. *Let* $f: U \to \mathbf{R}^n$ *be a mapping with bounded distortion, a a point in* U, *and* $\omega: [0, \infty) \to \mathbf{R}^n$ *a function satisfying the Dini condition. If*

$$K_f(x) - 1 \leq \omega(|x - a|), \tag{12.10}$$

then f *is differentiable at* a, *and* $f'(x)$ *is a nonsingular general orthogonal transformation. Further, there exists an* $r_0 > 0$, *depending only on* ω *and* n, *such that for* $0 < r < r_0$ *the estimate*

$$|f(x) - f(a) - f'(a)(x - a)| \leq \frac{\mathscr{D}}{r} C(r, |x - a|)^{|x-a|} \tag{12.11}$$

holds in the ball $B(a, r)$, *where* $\mathscr{D} = \sup_{|x-a|=r/4} |f(x) - f(a)|$, *and* $C(r, h)$ *depends only on* n, ω, r, *and* h; $C(r, h) \to 0$ *as* $h \to 0$. *It is possible to take*

$$C(r, h) = C_0 \left[\frac{h}{r} + \omega(4h) + h \int_h^r \frac{\omega(t)}{t^2} \, dt + \int_0^{4h} \frac{\omega(t)}{t} \, dt \right], \tag{12.12}$$

where $C_0 = \text{const} \geq 0$.

Theorem 12.9 yields a theorem on regularity of conformal mappings of Riemannian spaces. Let $B_1 = B(0, 1)$. Assume that Riemannian metrics are given in B_1 by positive-definite quadratic differential forms:

$$ds_1^2 = g_{kl}(x) \, dx^k \, dx^l, \qquad ds_2^2 = h_{ij}(y) \, dy^i \, dy^j \tag{12.13}$$

with continuous coefficients g_{kl} and h_{ij}. The mapping $f: B_1 \to B_1$ is said to be *conformal* with respect to the Riemannian metrics ds_1^2 and ds_2^2 if $f \in W_{n,\text{loc}}^1$, $\mathscr{I}(x, f)$ has constant sign in B_1, and there exists a function $\lambda(x) \geq 0$ such that for almost all $x \in B_1$

$$h_{ij}[f(x)] \frac{\partial f_i}{\partial x_k}(x) \frac{\partial f_j}{\partial x_l}(x) = [\lambda(x)]^2 g_{kl}(x). \tag{12.14}$$

(Repeated indices are understood to be summed from 1 to n.) The mapping f is said to be *isometric* with respect to the Riemannian metrics ds_1^2 and ds_2^2 in the case when (12.14) holds for f with the function $\lambda(x) \equiv 1$. If f is an isometric mapping, then the derivatives $\partial f_i / \partial x_j$ are bounded on every compact subset of B_1.

THEOREM 12.10. *Assume that Riemannian metrics are given in* B_1 *by the quadratic differential forms* (12.13), *and let* $f: B_1 \to B_1$ *be conformal with respect to these metrics. If for every* $r \in (0, 1)$ *there exists a function* $\omega_r: [0, \infty) \to \mathbf{R}$ *satisfying the Dini condition that is the modulus of continuity for each of the functions* g_{kl} *and* $h_{ij}, k, l, i, j = 1, \ldots, n$, *on the ball* $\overline{B}(0, r)$, *then* f *belongs to the class* C^1, $\det f'(x) \neq 0$ *for all* $x \in U$, *and on every ball* $\overline{B}(0, r)$ *with* $0 < r < 1$ *the derivatives* $(\partial f_i / \partial x_j)(x)$ *have modulus* *of continuity θ_r admitting the estimate $\theta_r(h) \leq C_r(\eta, h)$ for $0 < h < \eta$, where $C_r(\eta, h)$ is obtained by replacing ω by ω_r in* (12.12).

In particular, if g_{kl} and h_{ij} belong to the class C^α, where $0 < \alpha \leq 1$, then f belongs to $C^{1,\alpha}$.

See [144] and [143] for the proofs of Theorems 12.9 and 12.10.

A result analogous to Theorem 12.10 was stated by Calabi and Hartman [21] for mappings. $f: B_1 \to B_1$ isometric with respect to the Riemannian metrics ds_1^2 and ds_2^2. However, an unfortunate mistake crept into their arguments (the authors conclude from the inequalities $0 < x < \pi$ and $|\sin x - 1| < \varepsilon$ that $|x - \pi/2| < C\varepsilon$). If the correct estimate $|x - \pi/2| < C\sqrt{\varepsilon}$ is used, then it turns out that the arguments in [21] are valid only when instead of the Dini condition ω satisfies the stronger condition

$$\int_0^k (\sqrt{\omega(t)}/t)\, dt < \infty.$$

This error does not affect the other results in [21].

If under the conditions of Theorem 12.9 the function ω with $K_f(x) \leq \omega(|x - a|)$ does not satisfy the Dini condition, then the conclusion of the theorem can turn out to be false. An example is given in [144].

The proof of Theorem 12.9 is based on the estimate (12.7) in Theorem 12.5. A sequence of balls $B_m = B(a, \rho_m)$ is constructed with $\rho_m = 2^{-m-3} r$ and $r > 0$ such that $B(a, r) \subset U$. For each m the Möbius mapping φ_m is such that

$$|\varphi_m^{-1}(f(x)) - x| \leq C\omega(4\rho_m)\rho_m$$

in the ball $B_{m-2} = B(a, 4\rho_m)$. The difference $|\varphi_m'(0) - \varphi_{m-1}'(0)|$ is majorized by the quantity $C\omega(\rho_m)$. The Dini condition is equivalent to convergence of $\sum \omega(\rho_m)$. This implies that the sequence $(\varphi_m'(0))$ of linear mappings converges, which allows us to conclude finally that f is differentiable at a.

Theorem 12.10 is obtained by using Theorem 12.9.

If the functions g_{kl} and h_{ij} satisfy stronger regularity conditions than in Theorem 12.10, then f is also regular in a stronger sense than follows from Theorem 12.10. Namely, we have the following statement.

THEOREM 12.11 ([160],[59]). *If the functions g_{kl} and h_{ij} in* (12.13) *belong to the class $C^{r,\alpha}$, where $0 < \alpha < 1$, then every mapping $f: B_1 \to B_1$ conformal with respect to the Riemannian metrics ds_1^2 and ds_2^2 belongs to the class $C^{r+1,\alpha}$.*

Theorem 12.11 was proved in [87] for the case $r = \infty$. In [159] it was established under the additional assumption that the function λ in

(12.14) satisfies a Lipschitz condition. The proof of Theorem 12.11 given by Shefel' [160] is based on some geometric considerations not connected with Theorem 12.6. Iwaniec [59] derived the desired result from estimates of solutions of elliptic equations.

A result analogous to Theorem 12.11 was obtained by Calabi and Hartman [21] for mappings isometric with respect to Riemannian metrics ds_1^2 and ds_2^2. The case of a metric with variable sign is also considered in [21] (that is the forms ds_1^2 and ds_2^2 are not required to be positive definite).

7. The following injectivity theorem of Martio and Sarvas is another application of Theorem 12.5.

THEOREM 12.12 ([93]). *Let U be a domain of class $U(\alpha, \beta)$ in $\mathbf{R}^n$, $n \geq 3$. There exists a constant $c > 0$ depending only on n, α, and β such that every mapping $f: U \to \mathbf{R}^n$ with bounded distortion for which $1 \leq K(f) \leq 1 + c$ is a homeomorphism of U into $\mathbf{R}^n$.*

8. Let $f: B(0, 1) \to \mathbf{R}^n$ be a given mapping, and take an arbitrary Möbius transformation φ. We find the quantity

$$\sup_{x \in B(0,1)} |\varphi[f(x)] - x| = \delta(\varphi, f)$$

and let

$$\delta(f) = \inf_{\varphi \in \mathbf{M}_n} \delta(\varphi, f).$$

Denote by $\mu_n(\varepsilon)$ the supremum of $\delta(f)$ on the set of all mappings $f: B(0, 1) \to \mathbf{R}^n$ with bounded distortion such that $K(f) \leq 1 + \varepsilon$. Theorem 12.5 allows us to conclude that $\mu(\varepsilon) \leq C\varepsilon$ for $0 < \varepsilon \leq \varepsilon_0$, where $C = \text{const}$. The exact value of the constant C, like that of ε_0, is unknown. In this connection we mention the following result of Semenov [153], [170].

THEOREM 12.13.

$$K(n) - \overline{\lim_{\varepsilon \to 0}} \frac{\mu(\varepsilon)}{\varepsilon} \leq 2 + \frac{4}{n-2} - \frac{1}{n-1}$$
$$+ \frac{2n}{3(n-1)B(\frac{1}{2}, \frac{n-1}{2})} \left(\frac{4}{n+1} + \frac{11}{n-2} \right)$$
$$= 2 + O\left(\frac{1}{\sqrt{n}} \right).$$

In particular, for $n = 3, 4, 5,$ and 6,

$$K(3) \leq 11.5, \qquad K(4) \leq \frac{11}{3} + \frac{56}{5\pi},$$
$$K(5) \leq 5.71, \qquad K(6) \leq 2.8 + \frac{242}{34\pi}.$$

§12.3. Stability of isometric and Lorentz transformations.

1. Here we present results from John's important paper [60] on stability (in the sense of the present book) of isometric mappings. First we make some remarks about the definition of quasi-isometric mappings.

Let U be an arbitrary open domain in $\mathbf{R}^n$. Let $f: U \to \mathbf{R}^n$ be a given mapping and $L \geq 1$ a given number. Then we say that f belongs to the class $I_1(L)$ if its Jacobian $\mathcal{J}(x, f)$ is either nonnegative almost everywhere or nonpositive almost everywhere, and for almost all $x \in U$ the principal dilation coefficients of the linear mapping $f'(x)$ either are all equal to zero or lie in the interval (L^{-1}, L).

If the mapping $f: U \to \mathbf{R}^n$ belongs to the class $I_1(L)$, then, as follows from the definition, the function $x \mapsto |f'(x)|$ is bounded, and hence f belongs to the class $W^1_{p,\mathrm{loc}}$ for any $p > 0$. Further, it follows from the definition that for each x outside a set of measure zero either $f'(x) \equiv 0$ or the principal dilation coefficients $\lambda_1 \leq \cdots \leq \lambda_n$ of $f'(x)$ lie in (L^{-1}, L). In the last case

$$K[f'(x)] = \lambda_n^{n-1}/\lambda_1 \ldots \lambda_{n-1} \leq L^{2n-2},$$

and we get that for almost all $x \in U$

$$\|f'(x)\|^n \leq L^{2n-2}|\det f'(x)|.$$

Thus, if $f \in I_1(L)$, then f is a mapping with bounded distortion. Further, $K(f) \leq L^{2n-2}$.

An arbitrary mapping of class $I_1(L)$ can fail to be a local homeomorphism. The mapping f in Example 3 of §4.3 in Chapter I, which is an m-fold twisting about the $(n-2)$-dimensional plane $\mathbf{R}^{n-2} = \{x | x_{n-1} = x_n = 0\}$, belongs to $I_1(L)$ with $L = m$. At the same time, f is not a homeomorphism in any neighborhood of an arbitrary point $x \in \mathbf{R}^{n-2}$.

Theorem 10.5 of the present chapter yields the following proposition.

LEMMA 12.5. *There exists a number $L_0 > 1$ such that if $1 \leq L < L_0$, then every mapping $f: U \to \mathbf{R}^n$ of class $I_1(L)$ is a local homeomorphism.*

PROOF. By Theorem 10.5, there exists a $K_0 > 1$ such that if $f: U \to \mathbf{R}^n$ is a mapping with bounded distortion and $K(f) < K_0$, then f is a local homeomorphism. The number $L_0 = K_0^{1/(2n-2)}$ clearly works.

We say that $f: U \to \mathbf{R}^n$ is a mapping of class $I_2(L)$, where $L \geq 1$, if f is a bi-Lipschitz mapping and $L(f) \leq L$, that is, every point $x \in U$ has a neighborhood G such that for any $y, z \in G$ with $y \neq z$

$$\frac{1}{L} \leq \frac{|f(y) - f(z)|}{|y - z|} \leq L. \tag{12.15}$$

LEMMA 12.6. *For every* $L \geq 1$

$$I_2(L) \subset I_1(L), \tag{12.16}$$

and there exists an $L_0 > 1$ *such that the classes* $I_1(L)$ *and* $I_2(L)$ *coincide for* $1 \leq L < L_0$.

PROOF. The inclusion (12.16) is a consequence of results in §6.6 of this chapter.

Let $L_0 > 1$ be the constant in Theorem 12.14, and let $f \in I_1(L)$, where $1 \leq L < L_0$. Then f is a local homeomorphism by Theorem 12.4. Take an arbitrary point $x_0 \in U$, and let $\delta > 0$ be such that $B(x_0, \delta)$ is contained in U and f is a homeomorphism on $B(x_0, \delta)$. For almost all $x \in B(x_0, \delta)$ we have that $|f'(x)| \leq L$, which implies (see the remark after Theorem 2.7 in Chapter I) that $|f(y) - f(z)| \leq L|y - z|$ for any $y, z \in B(x_0, \delta)$. Let $H = f[B(x_0, \delta)]$ and $y_0 = f(x_0)$. The set H is open. There is an $\varepsilon > 0$ such that $B(y_0, \varepsilon) \subset H$; let $G = f^{-1}[B(y_0, \varepsilon)] \cap B(x_0, \delta)$. Let $B \subset B(y_0, \varepsilon)$ be the set of $y \in B(y_0, \varepsilon)$ at which either f^{-1} is nondifferentiable or $\mathscr{J}(y, f^{-1}) = 0$, and let A be the set of $x \in G$ at which either f is nondifferentiable or $\mathscr{J}(x, f) = 0$. Then $E = A \cup f^{-1}(B)$ is a set of measure zero, and $|f(E)| = 0$. At each point $y \in B(y_0, \varepsilon)$ not in $f(E)$ the mapping f^{-1} is differentiable, and $df^{-1}(y) = (df(x))^{-1}$, where $x = f^{-1}(y)$. This implies that $|df^{-1}(y)| \leq L$, and hence

$$|f^{-1}(u) - f^{-1}(v)| \leq L|u - v|$$

for any $u, v \in B(y_0, \varepsilon)$. Choosing $y, z \in G$ arbitrarily and setting $u = f(y)$ and $v = f(z)$ in the last inequality, we get that

$$\frac{1}{L}|y - z| \leq |f(y) - f(z)|,$$

and hence (12.15) holds for any $y, z \in G$. Since $x_0 \in U$ was arbitrary, $f \in I_2(L)$. In view of (12.16), the lemma is proved.

Let U be an open domain in $\mathbf{R}^n$. A mapping $f \colon U \to \mathbf{R}^n$ is said to be *quasi-isometric* if it belongs to the class $I_2(L)$ for some $L \geq 1$. The smallest number $L \geq 1$ such that $f \in I_2(L)$ (it is not hard to show that such a number exists) is denoted by $L(f)$.

THEOREM 12.14 (JOHN [60]. *Let* U *be a domain of class* $\mathscr{J}(r, R)$ *in* $\mathbf{R}^n$, *and* $f \colon U \to \mathbf{R}^n$ *a mapping of class* $I_2(L)$. *Then there exist a point* $x_0 \in U$ *and an orthogonal transformation* γ *such that for all* $x \in U$

$$|f(x) - f(x_0) - \gamma(x - x_0)| \leq C[L(f) - 1]\frac{R^2}{r},$$

where C *depends only on* n.

The theorem shows that if the difference $L - 1$ is sufficiently small, then the mapping $f \in I_2(L)$ is uniformly close to some isometric mapping of $\mathbf{R}^n$. The next theorem shows that the derivatives of $f \in I_2(L)$ are close to those of an orthogonal transformation if L is close to 1.

THEOREM 12.15 ([60]). *Assume that the domain* $U \subset \mathbf{R}^n$ *is a cube and* $f: U \to \mathbf{R}^n$ *is a mapping of class* $I_2(L)$. *Then there is an orthogonal matrix* γ *such that for every* $p \geq 1$

$$\left(\frac{1}{|U|} \int_U |f'(x) - \gamma|^p \, dx \right)^{1/p} \leq C(L - 1)p,$$

where the constant C *depends only on* n.

Theorem 12.15 extends easily to an arbitrary domain of class $\mathscr{J}(r, R)$.

By Lemma 12.6, Theorems 12.14 and 12.15 remain true if $I_2(L)$ is replaced by $I_1(L)$ in their statements.

The proof of Theorem 12.15 makes essential use of a theorem of John and Nirenberg [63] on functions with bounded mean oscillation. The formulation and proof of the latter theorem will be given below (Theorem 1.1 in Chapter III). A certain analogue of the theorem of John and Nirenberg is used in the proof of Theorem 12.5.

The problem arises of the exact value of the constants in Theorems 12.14 and 12.15. Let U be a domain in $\mathbf{R}^n$, and $f: U \to \mathbf{R}^n$ a quasi-isometric mapping. For an arbitrary isometric transformation $\varphi: \mathbf{R}^n \to \mathbf{R}^n$ let

$$\delta(f, \varphi) = \sup_{x \in U} |\varphi^{-1}[f(x)] - x| \quad \text{and} \quad \delta(f) = \inf_{\varphi} \delta(f, \varphi),$$

where the infimum is over and set of all isometries φ of $\mathbf{R}^n$. The supremum of $\delta(f)$ over the collection of all mappings $f: U \to \mathbf{R}^n$ in $I_2(L)$ is denoted by $\mu_U(L)$.

THEOREM 12.16 [153]. *If* U *is a ball in* $\mathbf{R}^n$, *then*

$$\varlimsup_{L \to 1} \frac{\mu_u(L)}{L - 1} \leq 2n + 1.$$

2. We mention some other properties of quasi-isometric mappings.

LEMMA 12.7 ([60]). *Let* U *be a domain in* $\mathbf{R}^n$, $f: U \to \mathbf{R}^n$ *a mapping of class* $I_2(L)$, *and* x_1 *and* x_2 *two arbitrary points of* U. *If the ellipsoid of revolution*

$$S = \{x \mid |x - x_1| + |x - x_2| \leq L^2 |x_1 - x_2|\}$$

is contained in U, *then*

$$\frac{1}{L}|x_1 - x_2| \le |f(x_1) - f(x_2)| \le L|x_1 - x_2|.$$

COROLLARY ([60]). *Let* $f: U \to \mathbf{R}^n$ *be a mapping of class* $I_2(L)$, *and* $V \subset U$ *a convex set. Let* $h = \frac{1}{2}d(V)\sqrt{L^4 - 1}$. *If the h-neighborhood* $U_h(V)$ *of* V *is contained in* U, *then for any* $x_1, x_2 \in V$

$$\frac{1}{L}|x_1 - x_2| \le |f(x_1) - f(x_2)| \le L|x_1 - x_2|,$$

and hence f *is a homeomorphism on* V.

For a domain $\mathscr{D} \subset \mathbf{R}^n$ let $L(\mathscr{D})$ be the supremum of the numbers $L \ge 1$ such that every mapping $f: \mathscr{D} \to \mathbf{R}^n$ of class $I_2(L)$ is one-to-one. If $\mathscr{D} = \mathbf{R}^n$, then, by a theorem of Zorich (Corollary 2 to Theorem 10.6), every mapping $f: \mathscr{D} \to \mathbf{R}^n$ is a homeomorphism of $\mathbf{R}^n$ onto itself, and hence $L(\mathscr{D}) = \infty$ in this case. It is easy to construct examples of domains with $L(\mathscr{D}) = 1$. Any planar domain whose boundary has a cusp point directed inward (see Figure 2 in Chapter I, §4.3) is such a domain. John [62] showed that if $\mathscr{D}$ is a ball or half-space, then $L(\mathscr{D}) \ge 2^{1/4}$. Let $n = 3$, and let z, r, φ be cylindrical coordinates in $\mathbf{R}^3$, $-\infty < z < \infty$, $r \ge 0$. Let f be the mapping

$$(z, r, \varphi) \mapsto (z/\sqrt{2 + \varepsilon}, r/\sqrt{2 + \varepsilon}, (2 + \varepsilon)\varphi).$$

It is easy to verify that f is quasi-isometric, $L(f) = \sqrt{2 + \varepsilon}$, and f is not one-to-one on the half-space $0 \le \varphi \le \pi$. This shows that for a half-space we have $L(\mathscr{D}) \le \sqrt{2 + \varepsilon}$ for any $\varepsilon > 0$; hence $L(\mathscr{D}) \le \sqrt{2}$ in this case. The construction of the last example extends easily to the case of arbitrary n, and we thus get that $L(\mathscr{D})$ is finite for a half-space.

Following John, we call a domain $\mathscr{D}$ in $\mathbf{R}^n$ *rigid* if $L(\mathscr{D}) > 1$.

By the theorem of Martio and Sarvas (Theorem 12.12), every uniform domain in $\mathbf{R}^n$ is rigid. The problem arises of determining necessary and sufficient conditions for a domain in $\mathbf{R}^n$ to be rigid. This problem was solved by Gehring [39] for the case $n = 2$. He showed that a domain U in $\mathbf{R}^2$ is rigid if and only if every connected component of its boundary with more than one point is a quasicircle.

3. Gurov established an analogue of Theorems 12.14 and 12.15 for Lorentz transformations [50]–[53]. Here we state his results.

Denote by $\mathscr{J}_n$ an $n \times n$ diagonal matrix with the first $n - 1$ diagonal elements equal to 1 and with last diagonal element equal to -1:

$$\mathscr{J}_n = \begin{pmatrix} 1 & & 0 \\ & \ddots & \\ 0 & & 1_{-1} \end{pmatrix}.$$

Let U be an open domain in $\mathbf{R}^n$, and $f(x) = (f_1(x), \ldots, f_n(x))$ a mapping of class $W^1_{1,\text{loc}}(U)$ from U into $\mathbf{R}^n$. We say that f is an ε-quasi-Lorentz mapping if f satisfies the following conditions:

a) The generalized derivative $(\partial f_n / \partial x_n)(x)$ is nonnegative in U.

b) For almost all $x \in U$

$$f'(x)^* \mathscr{I} f'(x) = \mathscr{I}_n + A(x), \tag{12.17}$$

where the matrix $A(x)$ is such that $|A(x)| < \varepsilon$.

Let $A(x) = (\alpha_{ij}(x))$, $i, j = 1, \ldots, n$. Then (12.17) is equivalent to the following system of equalities:

$$f'_{1,i}(x)f'_{1,j}(x) + \cdots + f'_{n-1,i}(x)f'_{n-1,j}(x) - f'_{n,i}(x)f'_{n,j}(x)$$
$$= \sigma_{ij} + \alpha_{ij},$$

where $\sigma_{ij} = 1$ for $i \neq j$, $\sigma_{ii} = 1$ for $i < n$, and $\sigma_{nn} = -1$. (As earlier, we set $f'_{k,i}(x) = (\partial f_k / \partial x_i)(x)$.)

Mappings for which

$$f'(x) \mathscr{I} f'(x) = \mathscr{I}, \qquad \frac{\partial f}{\partial x_n}(x) > 0,$$

are Lorentz mappings. A proof of this assertion when f is sufficiently smooth can be found in any fairly complete guide to relativity theory. Theorem 12.17 below gives us the assertion under the assumptions made here about the smoothness of f.

The collection of all ε-quasi-Lorentz mappings of an open domain $U \subset \mathbf{R}^n$ is denoted by $\text{QL}(\varepsilon, U)$.

THEOREM 12.17 (GUROV [50], [51]). *There exist constants $\delta_n > 0$ and $A_n < \infty$ such that if U is a domain of class $\mathscr{I}(r, R)$ and $f \colon U \to \mathbf{R}^n$ an ε-quasi-Lorentz mapping with $\varepsilon < \delta_n$, then there exists a Lorentz transformation φ such that for all $x \in U$*

$$|\varphi[f(x)] - x| \leq A_n \varepsilon (R^2 / r).$$

Certain additional assumptions about U ensure that the derivatives of an ε-quasi-Lorentz mapping are close to the derivatives of a Lorentz mapping.

Let U be a domain in $\mathbf{R}^n$, and denote by $\partial_h U$ $(h > 0)$ the set of all points $x \in U$ with $\rho(x, \mathbf{R}^n \backslash U) < h$. Define

$$\Omega_\alpha = \sup |\partial_h U| / h^\alpha,$$

where $\alpha > 0$.

THEOREM 12.18 (GUROV [52], [53]). *Let U be a domain of class $\mathscr{F}(r, R)$ such that Ω_α is finite for some $0 < \alpha \leq 1$. Let*

$$0 \leq \varepsilon \leq A_n \frac{r}{R}, \qquad 1 \leq p \leq B_n \frac{r\alpha}{\varepsilon R \ln(R/r)}.$$

If $f \in \mathrm{QL}(\varepsilon, U)$, then the derivatives of f are p-integrable, and there is a Lorentz transformation φ such that

$$\||(\varphi \circ f)' - I\||_{L_p(U)} \leq C_n \varepsilon p \frac{R}{r} \ln \frac{R}{r} \frac{1}{\alpha} \left(\frac{\Omega_\alpha R}{|U|} \right)^{1/p}.$$

Here A_n, B_n, and C_n are constants depending only on n.

The proof of Theorem 12.17 in [53] makes essential use of an analogue of the John-Nirenberg theorem on functions with bounded mean oscillation.

THEOREM 12.19 ([52], [53]). *Suppose that the vector-valued function $u(x) = (u_1(x), \ldots, u_m(x))$ is integrable in the cube $Q_0 = Q(a, r) \subset \mathbf{R}^n$. Assume that there is a constant $\beta \leq (1/2)^{n+3}$ such that for every cube $Q \subset Q_0$ there exists a vector $u_Q \in \mathbf{R}^m$ for which*

$$\int_Q |u(x) - u_Q| \, dx \leq \beta |u_Q| \cdot |Q|.$$

For $\sigma \geq 0$ let $E_u(\sigma)$ denote the Lebesgue set

$$\{x \in Q_0 | |u(x) - u_{Q_0}| > \sigma |u_{Q_0}|\}.$$

If $\sigma > e\beta |u_{Q_0}|$, then

$$|E_u(\sigma)| \leq B_n (1 + \sigma)^{-b_n/\beta} |Q_0|,$$

where B_n and b_n are constants and depend only on n.

COROLLARY. *If the function $u: Q_0 \to \mathbf{R}^n$ satisfies the conditions of the theorem, then it is integrable to any power $p < b_n/\beta$.*

We remark that the class of quasi-Lorentz mappings has received little attention on the whole. The mapping inverse to a quasi-Lorentz mapping can fail to be quasi-Lorentz. The composition of two quasi-Lorentz mappings is not in general quasi-Lorentz. In this connection one might think that the class itself is badly defined. However, attempts to modify the definition in such a way that the mapping inverse to a quasi-Lorentz mapping and the composition of two quasi-Lorentz mappings are also quasi-Lorentz mappings lead to loss of the stability property in Theorems 12.17 and 12.18. A generalization of Theorem 12.19 is used in the proof of Theorem 12.5 on stability in the Liouville theorem.

§12.4. Quasiconformal and quasi-isometric deformations. Semigroups of quasiconformal transformations.

1. In the study of quasiconformal and quasi-isometric mappings it is useful to consider certain special differential operators well known in continuum mechanics.

Let U be an open subset of $\mathbf{R}^n$. For a mapping $v: U \to \mathbf{R}^n$ of class $W^1_{1,\mathrm{loc}}(U)$ let

$$(Q_1 v)(x) = \tfrac{1}{2}[v'(x) + v'(x)^*],$$
$$(Q_2 v)(x) = \tfrac{1}{2}[v'(x) + v'(x)^*] - (n^{-1} \operatorname{div} v(x))I,$$

(I is the $n \times n$ identity matrix). For each $x \in U$ the quantities $(Q_1 v)(x)$ and $(Q_2 v)(x)$ are $n \times n$ matrices whose elements are expressed in terms of the derivatives of the components of the vector-valued function v as follows:

$$(Q_1 v)_{ij} = \frac{1}{2}\left(\frac{\partial v_i}{\partial x_j} + \frac{\partial v_j}{\partial x_i}\right),$$
$$(Q_2 v)_{ij} = \frac{1}{2}\left(\frac{\partial v_i}{\partial x_j} + \frac{\partial v_j}{\partial x_i}\right) - \frac{\delta_{ij}}{n} \operatorname{div} v,$$

where δ_{ij} is the Kronecker symbol.

If $v(x)$ is interpreted as the velocity field of a deformable body (in the case $n = 3$), then the matrix-valued function $(Q_1 v)(x)$ will be identified with the velocity affinor of the deformation of the body, and $(Q_2 v)(x)$ will be the part of this affinor responsible for the interior forces (see [117]). In connection with the theory of n-space mappings the operators Q_1 and Q_2 were first considered by Ahlfors [3] and the author [135].

We present arguments for the likelihood that Q_1 and Q_2 can be useful in the study of quasiconformal and quasi-isometric mappings.

Below, U denotes an open domain in $\mathbf{R}^n$. Let $f: U \to \mathbf{R}^n$ be a quasi-isometric mapping, i.e. a mapping of class $I_2(L)$. Assume that L is sufficiently close to 1, while $f(x)$ is close to the identity mapping $f(x) = x + u(x)$, with not only $u(x)$ but also its derivatives small. Let $L = 1 + \varepsilon$. For almost all $x \in U$

$$\frac{1}{(1+\varepsilon)^2} \le \langle f'(x)\xi, f'(x)\xi \rangle \le (1+\varepsilon)^2, \tag{12.18}$$

where ξ is an arbitrary unit vector. We have that

$$\langle f'(x)\xi, f'(x)\xi \rangle = \langle f'(x)^* f'(x)\xi, \xi \rangle.$$

It follows from (12.18) that the eigenvalues of the symmetric matrix $f'(x)^* f'(x)$ lie in the interval $(1/(1+\varepsilon)^2, (1+\varepsilon)^2)$. We have that

$$(f'(x)^*)f'(x) = I + u'(x) + u'(x)^* + (u'(x))^* u'(x).$$

Throwing out terms of higher than second order with respect to the derivatives of $u(x)$ in the expression for $f'(x)^*f'(x)$, we get that the eigenvalues of the matrix $u'(x) + u'(x)^* = 2(Q_1 u)(x)$ lie in the interval $((1 + \varepsilon)^{-2} - 1, (1 + \varepsilon)^2 - 1)$ (to within quantities of higher than second order of smallness compared with ε).

A vector-valued function $v\colon U \to \mathbf{R}^n$ will be called a *quasi-isometric deformation* if it belongs to the class $W^1_{1,\text{loc}}(U)$ and for almost all $x \in U$ the eigenvalues of the matrix $(Q_1 u)(x)$ lie in $(-M, M)$, where $m = \text{const}$ is finite.

The operator Q_2 arises similarly as a result of linearization of the quasiconformality condition.

Let $f\colon U \to \mathbf{R}^n$ be a mapping with bounded distortion. Assume that $K(f)$ is close to 1, $K(f) = 1 + \varepsilon$, where ε is small, and $f(x)$ is close to the identity mapping, $f(x) = x + u(x)$, where the vector-valued function $u(x)$ is small. It will be assumed that the derivatives of $u(x)$ are also small. Consider the matrix-valued function

$$\theta_f(x) = (f'(x))^{-1}(f'(x)^*)^{-1}|\mathscr{F}(x, f)|^{2/n}$$

for almost all x. We have that $f'(x) = I + u'(x)$. This gives us that, to within quantities of higher than first order of smallness,

$$(f'(x))^{-1} \approx I - u'(x), \qquad (f'(x)^*)^{-1} \approx I - u'(x)^*,$$

$$|\mathscr{F}(x, f)|^{2/n} \approx 1 + \frac{2}{n}\operatorname{tr} u'(x),$$

and

$$\theta_f(x) \approx I - [u'(x) + u'(x)^* - \tfrac{2}{n}\operatorname{tr} u'(x)I] = I - 2(Q_2 u)(x).$$

The eigenvalues of the matrix $\theta_f(x)$ lie in the interval $(K(f)^{-2}, K_0(f)^2)$. We have that $K_0(f) \le K(f)^{n-1}$. This leads to the conclusion that, to within quantities of higher than second order of smallness compared with ε, the eigenvalues of $\theta_f(x)$ lie in $(1 - 2\varepsilon, 1 + 2(n - 1)\varepsilon)$, and hence the eigenvalues of $(Q_2 u)(x)$ are in $(-(n - 1)\varepsilon, \varepsilon)$.

Following Ahlfors [3], we say that a vector-valued function $v\colon U \to \mathbf{R}^n$ in a *quasiconformal deformation* if $v \in W^1_{1,\text{loc}}(U)$ and there exists a constant $M < \infty$ such that for almost all $x \in U$ the eigenvalues of $(Q_2 u)(x)$ lie in $(-M, M)$.

If $v\colon U \to \mathbf{R}^n$ is a quasi-isometric deformation, then f is a quasiconformal deformation. Indeed, in this case each of the functions

$$\frac{1}{2}\left(\frac{\partial v_i}{\partial x_j} + \frac{\partial v_j}{\partial x_i}\right)$$

is bounded in U. In particular, the function $\partial v_i / \partial x_i$ is bounded, and this implies that

$$(Q_2 v)_{ij} = \frac{1}{2}\left(\frac{\partial v_i}{\partial x_j} + \frac{\partial v_j}{\partial x_i}\right) - \frac{\partial_{ij}}{n}\,\mathrm{div}\,v$$

is a bounded function on U.

Above we obtained an interval nonsymmetric with respect to 0 for the eigenvalues of $(Q_2 u)(x)$, namely, the interval $(-(n-1)\varepsilon, \varepsilon)$. To avoid this it is expedient instead of $K(f)$ and $K_0(f)$ to consider the quantity

$$\hat{K}(f) = \max\{K(f), K_0(f)\}.$$

If $K(f)$ and $K_0(f)$ are replaced by $\hat{K}(f)$ in the above arguments, then for $\hat{K}(f) = 1 + \varepsilon$ we get that the eigenvalues of $(Q_2 u)(x)$ lie in the interval $(-\varepsilon, \varepsilon)$.

For any two mappings f and g with bounded distortion on $\mathbf{R}^n$ we have

$$\hat{K}(f \circ g) \le \hat{K}(f)\hat{K}(g).$$

2. Let $\mathbf{I}_n$ be the group of motions of $\mathbf{R}^n$, and let $\mathbf{M}_n$, as above, denote the collection of all Möbius transformations of $\overline{\mathbf{R}}^n$; $\mathbf{I}_n$ and $\mathbf{M}_n$ are Lie groups, and, as such, each is endowed with the structure of a C^∞-differentiable manifold. A vector-valued function $u: \mathbf{R}^n \to \mathbf{R}^n$ is said to be an *infinitesimal motion* if there exists a smooth path $\varphi: [0,1] \to \mathbf{I}_n$ in $\mathbf{I}_n$ such that $\varphi(0) = I$ (here and below, I is the identity mapping) and

$$u(x) = \frac{d}{dt}\varphi(t,x)\Big|_{t=0}$$

for all $x \in \mathbf{R}^n$. A vector-valued function $u: \mathbf{R}^n \to \mathbf{R}^n$ is called an *infinitesimal Möbius transformation* if there exists a smooth path $\psi: [0,1] \to \mathbf{M}_n$ such that $\psi(0) = I$ and

$$u(x) = \frac{d}{dt}\psi(t,x)\Big|_{t=0}$$

for all $x \in \mathbf{R}^n$. Denote by $\Sigma_1(n)$ the collection of all infinitesimal motions of $\mathbf{R}^n$, and by $\Sigma_2(n)$ the set of all infinitesimal Möbius transformations of $\mathbf{R}^n$.

If f is a motion of $\mathbf{R}^n$, then $f'(x)^* f'(x)$ is the identity matrix. Let $f(x) = \varphi(t,x)$, where $\varphi: [0,1] \to \mathbf{I}_n$ is a path in $\mathbf{I}_n$ emanating from the identity of the group $\mathbf{I}_n$. Differentiating the equality $(\varphi'_x(t,x))^* \varphi'_x(t,x) = I$ with respect to t and setting $t = 0$, we get after obvious transformations that the vector-valued function $u(x) = (\partial\varphi/\partial t)(0,x)$ satisfies the equation

$$(Q_1 u)(x) = 0. \tag{12.19}$$

It can be shown that every solution of the system (12.19) is a vector-valued function $u(x)$ of the form

$$u(x) = a + Kx, \qquad (12.20)$$

where $a = \text{const}$ is a vector in $\mathbf{R}^n$, and K is a skew-symmetric matrix. It is easy to verify that every function of the form (12.20) satisfies (12.19) and belongs to $\Sigma_1(n)$. Thus, $\Sigma_1(n)$ coincides with the set of all functions of the form (12.20). This also implies that $\Sigma_1(n)$ coincides with the collection of all solutions of (12.19).

Let $f: \overline{\mathbf{R}}^n \to \overline{\mathbf{R}}^n$ be a Möbius transformation. Then for all $x \in \mathbf{R}^n$ with $f(x) \neq \infty$ the matrix

$$\theta_f(x) = (f'(x))^{-1}(f'(x)^*)^{-1}|\mathscr{F}(x, f)|^{2/n}$$

is the identity, $\theta_f(x) = I$. Consequently, also

$$(\theta_f(x))^{-1} = f'(x)^* f'(x) |\mathscr{F}(x, f)|^{-2/n} = I.$$

Setting $f(x) = \psi(t, x)$, where $\psi: [0, 1] \to \mathbf{M}_n$ is a smooth path in $\mathbf{M}_n$ with $\psi(0, x) \equiv x$, differentiating the resulting equality with respect to t, and setting $t = 0$, we find after obvious transformations that the vector-valued function $u(x) = (\partial\psi/\partial t)(0, x)$ satisfies the system of equations

$$Q_2 u(x) = 0. \qquad (12.21)$$

Every solution of (12.21) has the form

$$u(x) = a + (K + \alpha I)x + 2\langle b, x\rangle x - |x|^2 b, \qquad (12.22)$$

where $\alpha \in \mathbf{R}$, a and b are vectors in $\mathbf{R}^n$, and K is a skew-symmetric matrix. It is easy to verify that every function $u(x)$ representable by (12.22) satisfies (12.21) and belongs to $\Sigma_2(n)$. The set $\Sigma_2(n)$ of functions thus coincides with the set of solutions of (12.21) and, at the same time, with the collection of all functions of the form (12.22).

3. One of the main results in the theory of the operators Q_1 and Q_2 involves estimates of the norm of $u(x)$ in W_p^1 in terms of the L_p-norm of the matrix-valued function $Q_i u(x)$, $i = 1, 2$. The estimates of this kind for the operator Q_1 are called *Korn's inequalities*.

We give the necessary definitions. Let X be a vector space. A linear mapping $P: X \to X$ is called a *projection* if $P(Px) = Px$ for every x, or, what is the same, $P(y) = y$ for every $y \in P(X)$.

Recall that a domain U in $\mathbf{R}^n$ is called a *domain of class S* if U is a union of finitely many domains that are star-like with respect to a ball.

THEOREM 12.20. *Let U be a domain of class S in $\mathbf{R}^n$, and let $p > 1$ be given. Then for every continuous projection $\Pi\colon W_p^1(U) \to \Sigma_1(n)$ there exists a constant $C_1 = C_1(n, p, \Pi)$ such that for every function $u \in W_p^1(U)$*

$$\|u - \Pi u\|_{W_p^1(U)} \le C_1 \|Q_1 u\|_{L_p(U)}.$$

THEOREM 12.21. *Let U be a domain of class S in $\mathbf{R}^n$, and let a number $p > 1$ be given. Then for every continuous projection $\Pi\colon W_p^1(U) \to \Sigma_2(n)$ there exists a constant $C_2 = C_2(n, p, \Pi)$ such that*

$$\|u - \Pi u\|_{W_p^1(U)} \le C_2 \|Q_2 u\|_{L_p(U)}$$

for every $u \in W_p^1(U)$.

It was apparently Friedrichs who first proved Theorem 12.20 rigorously for the case $n = 2$ [32]. Theorem 12.20 was proved in the general case formulated here independently by the author [135], Nečas [108], and Mosolov and Myasnikov [105]. Theorem 12.21 was proved by the author [135].

The proofs of Theorems 12.20 and 12.21 are based on the following considerations. Assume first that U is bounded and star-like with respect to some ball $B(x_0, r) \subset U$. Then

$$
\begin{aligned}
u(x) = &\int_U K_0^{(l)}(x, y) u(y)\, dy \\
&+ \sum_{i,j=1}^n \int_U K_{ij}^{(l)}(x, y)(Q_l u)_{ij}(y)\, dy
\end{aligned}
\tag{12.23}
$$

for any function $u \in W_p^1(U)$, where $l = 1, 2$, and $K_0^{(l)}(x, y)$ is an $n \times n$ matrix whose elements are bounded C^∞-functions on $U \times U$ for any x and y. As a function of x for fixed y, $K_0^{(l)}(x, y)$ belongs to $\Sigma_l(n)$. The function $K_{ij}^{(l)}(x, y)$ takes values in $\mathbf{R}^n$, belongs to the class C^∞ on the set of all $(x, y) \in U \times U$ for which $x \ne y$, and admits the representation

$$K_{ij}^{(l)}(x, y) = H_{ij}^l(x, y) + \omega_{ij}^l\left(\frac{x - y}{|x - y|}\right)|x - y|^{1-n},$$

where the function $H_{ij}^l(x, y)$ is bounded and continuous and has bounded derivatives that are continuous for $x \ne y$, and $\omega_{ij}^l(z)$ is a function of class C^∞ in the domain $\mathbf{R}^n \setminus \{0\}$.

The integral representations (12.23) can be established for domains starlike with respect to a ball. Theorem 12.20 and 12.21 can be obtained for such domains by using a theorem of Calderón. The case when U is an arbitrary domain of type S is easily reduced to this case.

The integral representation (12.23) is nonunique for each of the operators Q_1 and Q_2. The technique used in [135] to get (12.23) is described in general form in [138]. Ahlfors [5] also obtained an integral representation for $Q_2 v$ analogous to (12.23) for the case when the domain is a ball.

THEOREM 12.23 (AHLFORS [6]). *Let* $f: U \to \mathbf{R}^n$ *be a quasiconformal deformation, and* $B(a, r)$ *an arbitrary ball in* U. *Then for any* $x, y \in B(a, r)$

$$|f(x) - f(y)| \le C\|Q_2 f\|_{L_\infty(U)}|x - y| \left(1 + \log \frac{2r}{|x - y|}\right).$$

In particular, it follows from the theorem that if f is a quasiconformal deformation, then f satisfies a Hölder condition with exponent α for all $\alpha \in (0, 1)$.

Since, as mentioned above, a quasi-isometric deformation is automatically quasiconformal, Theorem 12.22 automatically remains valid if Q_2 is replaced by Q_1 and quasi-isometric deformations are used instead of quasiconformal deformations in its formulation.

Of other results touching on the operators $Q_1 u$ and $Q_2 u$ proper, we mention criteria for solvability of the system of equations $Q_1 u = F$ and $Q_2 u = F$. Properly speaking, one criterion for solvability of the system $Q_1 u = F$ is the Saint-Venant condition well known in mechanics. Let $F(x) = (F_{ij}(x))_{i,j=1,\dots,n}$, and $F_{ij}(x) = F_{ji}(x)$ for any i and j. If the functions F_{ij} belong to the class $C^2(U)$, then the system of equations

$$\frac{\partial u_i}{\partial x_j} + \frac{\partial u_j}{\partial x_i} = F_{ij}, \qquad i, j = 1, \dots, n,$$

(this is clearly the system $Q_1 u = F$ in expanded notation) is solvable if and only if

$$\frac{\partial^2 F_{ij}}{\partial x_k \partial x_l} + \frac{\partial^2 F_{kl}}{\partial x_i \partial x_j} - \frac{\partial^2 F_{il}}{\partial x_k \partial x_j} - \frac{\partial^2 F_{kj}}{\partial x_i \partial x_l} = 0 \qquad (12.24)$$

for any $i, j, k, l = 1, \dots, n$. If the derivatives are understood as distributions, then this criterion for solvability of the system $Q_1 u = F$ is valid without any assumptions about the differentiability of the functions F_{ij}. Namely, we have the following assertion.

THEOREM 12.23. *Suppose that* U *is an open domain in* $\mathbf{R}^n$, *and* $F(x) = (F_{ij}(x))$, $i, j = 1, \dots, n$, *is a matrix-valued function whose elements are functions of class* $L_{p,\text{loc}}(U)$ $(1 < p < \infty)$ *with* $F_{ij}(x) = F_{ji}(x)$ *for any* i *and* j. *If* (12.24) *holds for any* i, j, k, l, *and the derivatives there are understood as distributions, then the system of equations* $Q_1 u = F$ *is solvable in* U.

REMARK. Under the conditions of the theorem, (12.24) is equivalent to the following: for every function $\varphi \in C_0^\infty(U)$

$$\int_U \left(F_{ij}\frac{\partial^2 \varphi}{\partial x_k \partial x_l} + F_{kl}\frac{\partial^2 \varphi}{\partial x_i \partial x_j} - F_{il}\frac{\partial^2 \varphi}{\partial x_k \partial x_j} - F_{kj}\frac{\partial^2 \varphi}{\partial x_i \partial x_l}\right) dx = 0.$$

Ahlfors [4] established an analogous criterion for solvability of the system $Q_2 u = F$.

Some further questions about Q_2 are considered in [5], [7], and [8]. In particular, some special differential operators connected in a natural way with Q_2 are introduced there, and their properties are investigated.

4. *Flows of quasiconformal mappings.* The following is a well-known problem in the theory of spatial quasiconformal mappings (see [83]). Let $f : U \to \mathbf{R}^n$ be a quasiconformal mapping. Is it possible for every $\varepsilon > 0$ to construct a finite sequence of quasiconformal mappings $f_1, \ldots, f_m$ such that $K(f_i) \leq 1 + \varepsilon$ for each $i = 1, \ldots, m$ and $f = f_m \circ f_{m-1} \circ \cdots \circ f_1$? This question is of interest because quasiconformal mappings with $K(f)$ close to 1 are considerably more convenient to study than general quasiconformal mappings, due to the stability in the Liouville theorem (Theorem 12.5). The question posed is usually called the *composition problem.*

An analogous composition problem can also be posed for quasi-isometric mappings.

In connection with this question the thought occurs to give special consideration to mappings that can be represented for any $\varepsilon > 0$ as a composition of finitely many quasiconformal mappings with $K(f) < 1 + \varepsilon$. Among such mappings it is natural to single out those that can be included in a quasiconformal flow. We give the necessary definitions.

Below, as above, we let $\hat{K}(f) = \max\{K(f), K_0(f)\}$ for any quasiconformal mapping f.

Let X be an arbitrary set. A mapping $F : \mathbf{R} \times X \to X$ is called a *flow* if $F(0, x) = x$ and $F(t + s, x) = F(t, F(s_2 x))$ for all $x \in X$ and $t, s \in \mathbf{R}$. The concept that arises when the half-line $\mathbf{R}^+ = [0, \infty)$ is used in this definition instead of the set $\mathbf{R}$ of all real numbers is called a *semiflow.* (That is, a semiflow is a mapping $F : \mathbf{R}^+ \times X \to X$ such that $F(0, x) = x$ and $F(t + s, x) = F(t, F(s, x))$ for any $t, s \geq 0$.) In the case when X is a topological space and F a continuous mapping, one says that F is a *topological flow* (respectively, a *topological semiflow*). Instead of $F(t, x)$ we also write $F_t(x)$. For an arbitrary mapping $f : X \to X$ we let $f^\circ = \mathrm{id}$ and $f^k = f \circ f^{k-1}$ for any integer $k > 0$. If f is a bijection, and let $f^k = (f^{|k|})^{-1}$ for an integer $k < 0$. A *fractional (rational) power* of $f : X \to X$ is defined to be any mapping $g : X \to X$ such that $g^l = f^k$ for some integers k and l, where k is not a multiple of l.

We say that a mapping $f : X \to X$ is *included in* a flow $F : \mathbf{R} \times X \to X$ (a *semiflow* $F : \mathbf{R}^+ \times X \to X$) if $f = F_{t_0}$ for some $t_0 \in \mathbf{R}$ (respectively, $t_0 \in \mathbf{R}^+$).

Let $A \subset \mathbf{R}^n$ be a given set. It is assumed that every point $x \in A$ is a limit point for the set of interior points of A. A *quasiconformal flow (semiflow)* on A is defined to be a topological flow (semiflow) in A such that for each t the restriction of F_t to A^0 is a quasiconformal mapping, and there exists a constant $\lambda \geq 0$ such that $\hat{K}(F_t) \leq \exp \lambda |t|$ for all t. A *quasi-isometric flow (semiflow)* on A is defined to be a topological flow (semiflow) F on A such that for each t the restriction of F_t to A^0 is a quasi-isometric mapping, and there exists a constant $\lambda \geq 0$ such that $L(F_t) \leq \exp \lambda |t|$ for all t. Assume that the limit

$$u(x) = \lim_{t \to 0} \frac{F(t, x) - x}{t} = \frac{\partial F}{\partial t}(0, x)$$

exists for all $x \in A$. The function $u(x)$ thus defined is called the *infinitesimal generator* of the flow (semiflow) F.

THEOREM 12.24. *Let F_t be a quasiconformal flow (semiflow) in $B = \overline{B}(0, 1) \subset \mathbf{R}^n$. Then F has an infinitesimal generator $u(x)$ in B. The vector-valued function $u(x)$ is a quasiconformal deformation, and if F is a flow, then $\langle x, u(x) \rangle = 0$ for all $x \in S(0, 1)$; if F is a semiflow, then $\langle x, u(x) \rangle \leq 0$ for all $x \in S(0, 1)$.*

Conversely, let $u(x)$ be a quasiconformal deformation in B. In this case if $\langle x, u(x) \rangle \leq 0$ for all $x \in S(0, 1)$, then there exists a quasiconformal semiflow (F_t) whose infinitesimal generator is $u(x)$. Such a semi-flow is unique. If a quasiconformal deformation $u \colon B \to \mathbf{R}^n$ is such that $\langle x, u(x) \rangle = 0$ for all $x \in S(0, 1)$, then there exists a quasiconformal flow such that $u(x)$ is the infinitesimal generator.

Theorem 12.24 was proved independently by Semenov [150], Ahlfors [6], and Riemann [118].

If $u(x)$ is the infinitesimal generator of the quasiconformal flow F_t, then $\|Q_2 u\|_{L_\infty}$ can be taken as a constant λ such that $\hat{K}(F_t) \leq \exp \lambda t$.

The coefficient $\hat{K}(F_t)$ in the definition of a quasiconformal flow can be replaced by $q(F_t)$, as Semenov did. The concept arising as a result of this change is equivalent to the preceding concept. For a quasiconformal flow F the distortion coefficient $q(F_t)$ admits the estimate $q(F_t) \leq \exp(\mu|t|)$, where

$$\mu = \min_k \operatorname{ess\,sup}_B \left\{ \sum_{i<j} \left| \frac{\partial u_i}{\partial x_j} + \frac{\partial u_j}{\partial x_i} \right| + \sum_{i=1} \left| \frac{\partial u_i}{\partial x_i} - \frac{\partial u_k}{\partial x_k} \right| \right\}.$$

If $f \colon \overline{B} \to \mathbf{R}^n$ can be included in a quasiconformal flow, then it has every fractional power, and the composition problem is solved positively for it in the obvious way. In the general case it is impossible to get a solution of the composition problem by using Theorem 12.24 directly, as shown by the example (constructed by Ivanov [57]) described below. We present the result in [57].

Let $\varphi: [0, 1] \to [0, 1]$ be a continuous bijection such that $\varphi(x) > x$ for all $x \in (0, 1)$. A point $a \in (0, 1)$ is said to be a *relatively attracting point* for the mapping φ if for all $x \in [a, \varphi(a)]$

$$\lim_{m \to \infty} \frac{\varphi^m(x) - \varphi^m(a)}{\varphi^{m+1}(a) - \varphi^m(a)} = 0.$$

Let $\mathscr{K}$ denote the class of bi-Lipschitz mappings φ of $[0, 1]$ onto itself that have relatively attracting points.

Let U be an open set in $\mathbf{R}^n$, let $a \in U$, and let $r > 0$ be such that $\overline{B}(a, r) \subset U$. Take an arbitrary function $\varphi \in K$ and define a mapping $f: U \to U$ by

$$f(x) = a + \frac{r}{|x - a|} \varphi \left(\frac{x - a}{r} \right) (x - a)$$

for $0 < |x - x_0| \le r$ and $f(x) = x$ for $x = a$ and for $|x - a| > r$. Clearly f is a bi-Lipschitz mapping, and hence a quasiconformal mapping of U onto itself. The mapping f preserves orientation. The collection of all mappings f that can be obtained by this construction is denoted by $\mathscr{K}(U)$.

THEOREM 12.25 [57]. *No mapping in the class $\mathscr{K}(U)$ has a bi-Lipschitz fractional power, and for $n \ge 2$ none has a quasiconformal fractional power. Every mapping of class $\mathscr{K}(U)$ can be included in a topological flow on U. Every C^1-neighborhood of the identity transformation on U contains a C^1-diffeomorphism in $\mathscr{K}(U)$. On the other hand, $\mathscr{K}(U)$ does not contain a mapping of class C^2.*

A result analogous to Theorem 12.24 is true also for quasi-isometric mappings, as Semenov has shown [151]. The statement of the corresponding theorem is obtained by replacing the word "quasiconformal" in Theorem 12.24 by the word "quasi-isometric".

Semenov also obtained an analogue of Theorem 12.24 for quasi-Lorentz transformations [151]. Here the quasi-Lorentz property in [151] is understood in a broader sense than in the work of Gurov cited above.

5. Domains U_1 and U_2 in $\mathbf{R}^n$ are said to be *quasiconformally equivalent* if there exists a quasiconformal homeomorphism $f: U_1 \to \mathbf{R}^n$ such that $U_2 = f(U_1)$. Let $\kappa(f)$ denote one of the distortion coefficients of f (that is, $\kappa(f)$ is $K(f)$, or $K_0(f)$, or $q(f)$, or some other distortion coefficient). The infimum of $\kappa(f)$ over the set of all quasiconformal homeomorphisms f

from U_1 onto U_2 is denoted by $\delta_\kappa(U_1, U_2)$. If U_1 and U_2 are not quasiconformally equivalent, then we let $\delta_\kappa(U_1, U_2) = \infty$. The quantity $\delta_\kappa(U_1, U_2)$ is called the *quasiconformality coefficient* of the pair (U_1, U_2) of domains corresponding to the given distortion coefficient κ. The problem arises of determining $\delta_\kappa(U_1, U_2)$, at least for certain concrete pairs of domains U_1 and U_2 in $\mathbf{R}^n$. Only scattered results are known in this direction (see [42], [41], and [166]). Here we consider a closely related problem. Assume that U_1 and U_2 are domains with smooth boundary extremely close to each other. It is required to investigate the asymptotic behavior of $\delta_\kappa(U_1, U_2)$ for various choices of distortion coefficient κ. This leads to a certain extremal problem for the operator Q_2, and it is in this connection that we consider the question here. The results below are due to Vasil'chik [176].

We consider bounded domains in $\mathbf{R}^n$ whose boundaries are $(n-1)$-dimensional manifolds of class C^2. If a domain U satisfies this condition, then we say that U is a domain of class C^2.

Let U be a domain of class C^2 in $\mathbf{R}^n$. A *regular deformation* of U is defined to be a family $(U_t)_{t \in [0,\delta)}$ of domains of class C^2 such that $\overline{U}_t = \Phi(t, \overline{U})$ for every $t \in [0, \delta)$, where $\Phi: [0, \infty] \times \overline{U} \to \mathbf{R}^n$ is a mapping of class C^2 with $\Phi(0, x) = x$ for all $x \in U$, so that $U = U_0$. Denote by $\nu(s)$ the outward unit normal vector at a point $s \in \partial U$. For $s \in \partial U$ let

$$\psi(s) = \frac{\partial \Phi}{\partial t}(0, x), \quad \varphi(s) = \langle \nu(s), \psi(s) \rangle.$$

We call φ the *velocity function* of the boundary of U with respect to the deformation $(U_t)_{t \in [0,\delta)}$ of it.

Let $(U_t)_{t \in [0,\delta)}$ be a given regular deformation of a domain U of class C^2. As $t \to 0$ the function Φ_t converges uniformly on $\overline{U}$ together with the first and second derivatives to the identity mapping and its corresponding derivatives. Therefore, Φ_t is a homeomorphism of U onto U_t for sufficiently small $t \in [0, \delta)$ (it is assumed below that Φ_t is a homeomorphism of U onto U_t for all $t \in [0, \delta)$). Further, $\kappa(\Phi_t) \to 1$ for any distortion coefficient κ, and, moreover, $L(\Phi_t)$ also goes to 1 as $t \to 0$. We have that $\delta_\kappa(U, U_t) \le \kappa(\Phi_t)$, which leads to the conclusion that $\delta_\kappa(U, U_t) \to 1$ as $t \to 0$. The relation $\delta_\kappa(U, U_t) = 1 + t^r R_\kappa + o(t^r)$ is valid as $t \to 0$, where r is either 1 or 2 (r is 1 for the basic distortion coefficients K, K_0, and q, while r is 2 for a certain special distortion coefficient defined below). The determination of the coefficient R_κ in the asymptotic representation of $\delta_\kappa(U, U_t)$ reduces to the solution of a certain special extremal problem.

For an arbitrary symmetric matrix X we set

$$M(x) = \max_{|t|=1}\langle Xt, t\rangle, \qquad m(X) = \min_{|t|=1}\langle Xt, t\rangle,$$

$$\mu(X) = \max_{|t|=1}|\langle Xt, t\rangle| = \max\{|M(X)|, |m(X)|\},$$

$$|X| = \sqrt{\operatorname{tr} X^*X} = \sqrt{\sum_{i,j=1}^{n} x_{ij}^2}.$$

(Obviously, $M(X)$ is the largest and $m(X)$ the smallest eigenvalue of X.)

In addition to the distortion coefficients considered earlier we introduce the quantity

$$H(f) = \operatorname{ess\,sup}_{x\in U} \frac{|f'(x)|^n}{n^{n/2}\mathscr{J}(x, f)}.$$

Let $(U_t)_{t\in[0,\delta)}$ be a regular deformation of a bounded domain U, and $\varphi(s)$, $s \in \partial U$, the velocity function of the deformation. Denote by $W(\varphi)$ the collection of all vector-valued functions $v \in W_1^1(U)$ such that $\langle v(s), \nu(s)\rangle = \varphi(s)$ for all $s \in \partial U$. We consider the functionals

$$I(v) = \operatorname{ess\,sup}_{x\in U} M[(Q_2v)(x)],$$

$$I_0(v) \operatorname{ess\,sup}_{x\in U}\{-m(Q_2v(x))\},$$

$$I_q(v) = \operatorname{ess\,sup}_{x\in U}\{M[Q_2v(x)] - m[Q_2v(x)]\},$$

$$I_F(v) = \operatorname{ess\,sup}_{x\in U}\{|Q_2v(x)|^2\},$$

$$I_L(v) = \operatorname{ess\,sup}_{x\in U} M[(Q_1v)(x)]$$

on the set $W(\varphi)$. The infima of these functionals on the set of functions $v \in W(\varphi)$ will be denoted by $\lambda(U, \varphi)$, $\lambda_0(U, \varphi)\lambda q(U, \varphi)$, $\lambda_F(U, \varphi)$, and $\lambda_L(U, \varphi)$, respectively.

THEOREM 12.26 [176]. *Let $(U_t)_{t\in[0,\delta)}$ be a regular deformation of a bounded domain U of class C^2, and let $\varphi(s)$ be the velocity vector of the deformation on the boundary of U. Then*

$$\lambda(U, \varphi) = \lim_{t\to 0}\frac{\delta_K(u, u_t) - 1}{t}, \qquad \lambda_0(U, \varphi) = \lim_{t\to 0}\frac{\delta_{K_0}(U, U_t) - 1}{t},$$

$$\lambda_q(U, \varphi) = \lim_{t\to 0}\frac{\delta_q(U, U_t) - 1}{t}, \qquad \lambda_F(U, \varphi) = \lim_{t\to 0}\frac{\delta_H(U, U_t) - 1}{t^2},$$

$$\lambda_L(U, \varphi) = \lim_{t\to 0}\frac{\delta_L(U, U_t) - 1}{t}.$$

(*Here $\delta_L(U, U_t)$ is the infimum of $L(f)$ over the set of all quasi-isometric mappings of U onto U_t.*)

§12.5. Mappings with distortion coefficient close to 1. If for a mapping $f: U \to \mathbf{R}^n$ with bounded distortion the quantity $K(f) - 1$ is sufficiently small, then it has properties that do not hold in general for mappings with arbitrary $K(f)$. Some results establishing properties of this kind were indicated above (Theorems 10.5 and 12.5). We shall present other results of the same type.

1. We consider the problem of approximating an arbitrary mapping with bounded distortion by mappings of a simpler nature.

Let U be a domain in $\mathbf{R}^n$. A mapping $\varphi: U \to \mathbf{R}^n$ is said to be *simplicial* if there exists a simplicial complex (infinite in general) such that every compact subset of U is covered by finitely many simplexes of the complex, φ is continuous, and φ is affine on every simplex on the complex.

We say that *the simplicial approximation problem is solvable* for a mapping $f: U \to \mathbf{R}^n$ with bounded distortion if for every $\varepsilon > 0$ there exists a simplicial mapping $\varphi: U \to \mathbf{R}^n$ with bounded distortion such that $|\varphi(x) - f(x)| < \varepsilon$ for all $x \in U$, and the distortion coefficient of φ can be estimated in terms of the distortion coefficient of f:

$$K(\varphi) \leq F_1[K(f)],$$

where $F_1: [1, \infty) \to \mathbf{R}$ is a function bounded on every interval $[1, a]$, $1 < a < \infty$. (The function F_1 can obviously be assumed to be nondecreasing—this can be achieved by replacing $F_1(t)$ by $\sup_{1 \leq u \leq t} F_1(u)$.)

Assume that for every $\varepsilon > 0$ it is possible to construct a mapping $\varphi: U \to \mathbf{R}^n$ of class C^1 such that $\mathscr{J}(x, \varphi) \neq 0$ for all $x \in U$, $|f(x) - \varphi(x)| < \varepsilon$ for all $x \in U$, and $K(\varphi) \leq F_2[K(f)]$, where $F_2: [1, \infty) \to \mathbf{R}$ is bounded on each interval $[1, a]$, $1 < a < \infty$. In this case we say that *the smooth approximation problem is solvable for f*.

The smooth approximation problem is easy to solve in the case $n = 2$ with the help of known existence theorems in the theory of planar quasiconformal mappings. The solution of the simplicial approximation problem is easily obtained from this. There are no analogous existence theorems for the case $n \geq 3$.

The question of solvability of the simplicial approximation problem for an arbitrary mapping with bounded distortion remains open. However, if the distortion coefficient of f is close to 1, then this question admits a positive solution, as does the question of smooth approximation. The corresponding results were obtained by Kopylov in [66] and [67]. We give their formulations.

Suppose that U is a given domain in $\mathbf{R}^n$, and $f: U \to \mathbf{R}^n$ is a given mapping. For $h > 0$ let $\hat{U}_h = \{x \in \mathbf{R}^n | \rho(x, CU) > h\}$. Let h_0 be such that if $0 < h < h_0$, then $\hat{U}_h$ is nonempty. Assume that f belongs to $L_{1,\text{loc}}(U)$. Then for every $x \in \hat{U}_h$ the number

$$f_h(x) = \frac{1}{\sigma_n h^n} \int_{|y| \leq h} f(x + y)\, dy \qquad (12.25)$$

is defined. The function $f_h \colon \hat{U}_h \to \mathbf{R}^n$ is the average of f in the Steklov sense. The function f_h is continuous. If f is continuous, then f_h belongs to C^1.

THEOREM 12.27 (KOPYLOV [67]). *There exists a number $\varepsilon_0 > 0$ such that if $f \colon \mathbf{R}^n \to \mathbf{R}^n$ is a quasiconformal mapping with $K(f) < 1 + \varepsilon_0$, then for every $h > 0$ the mapping $f_h(x)$ defined by (12.25) is quasiconformal, $\mathscr{J}(x, f_h) \neq 0$ for all x, and*

$$K_h(f) \leq F[K(f)],$$

where the function $F \colon [1, 1 + \varepsilon_0) \to \mathbf{R}$ is nondecreasing, and $F(t) \to 1$ as $t \to 1$.

THEOREM 12.28 (KOPYLOV [67]). *Let $f \colon U \to \mathbf{R}^n$ be a quasiconformal mapping. If $K(f) \leq 1 + \varepsilon_1$, where $\varepsilon_1 = \text{const} > 0$, then f_h is quasiconformal on the domain $\hat{U}_\lambda = \{x \mid \rho(x, CU) > \lambda\}$, if h is sufficiently small, and $\mathscr{J}(x, f_h) \neq 0$ for all $x \in \hat{U}_\lambda$. The quasiconformality coefficient of f_h tends to 1 as $K(f) \to 1$. More precisely, there exist a constant $\delta > 0$ and a function $G(t, u)$ of the variables $t \in (0, \infty)$ and $u \in [1, 1 + \varepsilon_1]$ such that*

$$\lim_{t \to 0, u \to 1} G(t, u) = 1,$$

and if $h < \delta\lambda$ and $K(f) < 1 + \varepsilon_1$, then f_h is quasiconformal on $\hat{U}_\lambda$, with

$$K(f_h) \leq G\left(\frac{h}{\lambda}, K(f)\right).$$

Theorems 12.27 and 12.28 remain true if the Steklov average in their formulations is replaced by an averaging in the sense of Sobolev.

2. The class of uniform domains in $\overline{\mathbf{R}}^n$ was defined in part 2 of §12.1. It coincides with the class $U(\delta)$ of domains, where $0 < \delta \leq 1$.

By a ball in $\overline{\mathbf{R}}^n$ we shall mean a set which is a ball $B(x_0, r)$ in $\mathbf{R}^n$, or the complement $\overline{\mathbf{R}}^n \setminus \overline{B}(x_0, r)$ of such a ball, or an open half-space in $\mathbf{R}^n$; a k-dimensional sphere in $\overline{\mathbf{R}}^n$ is either an ordinary k-dimensional sphere in $\mathbf{R}^n$ or a set of the form $P \cup \{\infty\}$, where P is a k-dimensional plane in $\mathbf{R}^n$.

In studying quasiconformal mappings close to conformal mappings it is expedient to investigate the structure of domains of class $U(\delta)$ with coefficient δ close to 1. Here we present results of Trotsenko [169], [170] about the structure of the domains $U(\delta)$ for δ close to 1. In particular, he established various characterizations of domains of this class that are equivalent as $\delta \to 1$.

First of all we note that along with the class $U(\delta)$ it is expedient to consider also a certain other class $\tilde{U}(\delta)$.

Let x_1 and x_2 be two arbitrary points in $\overline{\mathbf{R}}^n$, and Γ an arbitrary continuum in $\overline{\mathbf{R}}^n$ joining them. For $x \in \Gamma$ denote by $B(x, x_1, x_2, \delta)$ $(\delta > 0)$ the set of all points $y \in \overline{\mathbf{R}}^n$ such that

$$\langle x, y, x_1, x_2 \rangle = \frac{|x - y| \cdot |x_1 - x_2|}{|x - x_1| \cdot |y - x_2|} < \delta.$$

The set $B(x, x_1, x_2, \delta)$ is a ball in $\overline{\mathbf{R}}^n$. Let $V_1(\Gamma, \delta)$ be the union of the balls $B(x, x_1, x_2, \delta)$ as x runs through Γ, let $V_2(\Gamma, \delta)$ be the union of all the balls $B(x, x_2, x_1, \delta)$ (the order of x_1 and x_2 has been changed), and let $V(\Gamma, \delta) = V_1(\Gamma, \delta) \cup V_2(\Gamma, \delta)$. We say that a domain U belongs to the class $\tilde{U}(\delta)$ if for any points $x_1, x_2 \in U$ there exists a continuum $\Gamma \subset U$ such that $x_1, x_2 \in \Gamma$ and $V_1(\Gamma, \delta) \subset U$. (For comparison we remark that the definition of the class $U(\delta)$ requires that $V(\Gamma, \delta) \subset U$.)

If $U \in \tilde{U}(1)$, then U also belongs to the class $U(1)$, and U is either a ball in $\overline{\mathbf{R}}^n$ or such that $E = \overline{\mathbf{R}}^n \setminus U$ is a closed subset of an $(n-2)$-dimensional sphere in $\overline{\mathbf{R}}^n$.

Obviously, if $U \in U(\delta)$, then also $U \in \tilde{U}(\delta)$. Trotsenko [170] showed that for every $\delta \in (0, 1]$

$$\tilde{U}(\delta) \subset U\left[1 - \sqrt{(1-\delta)/(1+\delta)}\right].$$

Let $x_1, x_2 \in \overline{\mathbf{R}}^n$ be given points, and let B be a ball in $\overline{\mathbf{R}}^n$. Define

$$(x_1, x_2, B) = \sup_{x \in B} \inf_{y \notin B} \{x_1, x_2, x, y\}.$$

If $x_1 = 0$, $x_2 = \infty$, $B = B(a, r)$, and $x_1 \notin B$, then

$$(x_1, x_2, B) = r/|a| \le 1,$$

as is easy to verify directly. If $x_1 \in B$, then $(x_1, x_2, B) = \infty$, and if $x_2 \in \overline{B}$ and $x_1 \notin B$, then $(x_1, x_2, B) = 1$. In particular, this makes it clear that if $(x_1, x_2, B) < 1$, then x_1 and x_2 do not belong to the closure of B. Let $x_1 = 0$, $x_2 = \infty$ and $B = B(a, r)$, where $|a| > r$. The sphere $S(0, \sqrt{|a|^2 - r^2}$ is orthogonal to the ball $B(a, r)$. Inversion with respect to this sphere transforms $B(a, r)$ into itself, the point 0 into ∞, and ∞ into 0. Since the cross ratio $\langle x_1, x_2, x, y \rangle$ is invariant with respect to Möbius transformations, this implies that in this case $\langle x_1, x_2, B \rangle = \langle x_2, x_1, B \rangle$, and hence $(x_1, x_2, B) = (x_2, x_1, B)$ whenever $(x_1, x_2, B) < 1$.

A family (B_t), $t \in [0, 1)$, of balls in $\overline{\mathbf{R}}^n$ is said to be *continuous* if in some spherical metric introduced in $\overline{\mathbf{R}}^n$ their centers and radii are continuous functions of t. We say that $B_t \to x_0$ as $t \to t_0$ if $x(t) \to x_0$ and $r(t) \to 0$ as $t \to t_0$ ($x(t)$ is the center of B_t and $r(t)$ is its radius).

We introduce some classes of domains in $\overline{\mathbf{R}}^n$.

Let $\varepsilon \geq 0$, $\varepsilon < 1$, be given, and let $U \subset \overline{\mathbf{R}}^n$ be a domain such that $\overline{\mathbf{R}}^n \setminus U$ is nonempty and $\infty \in \partial U$. Then we can make the following definitions:

(I) U belongs to $V_0(\varepsilon)$ if $U \in \tilde{U}(1 - \varepsilon)$.

(II) U belongs to the class $V_1(\varepsilon)$ if for any $x_1, x_2 \in \partial U$ there is a continuous family $(B_t \subset U)$, $t \in (0, 1)$, of balls such that $(x_1, x_2, B_t) \geq 1 - \varepsilon$ for all t, $B_t \to x_1$ as $t \to 0$, and $B_t \to x_2$ as $t \to 1$.

(III) U belongs to the class $V_2(\varepsilon)$ if for any $x \in \partial U \setminus \{\infty\}$ and any $r > 0$ there exists balls $B(y, r(1 - \varepsilon)) \subset U$ and $B(z, r(1 - \varepsilon)) \subset \mathbf{R}^n \setminus U$ such that $|x - y| = |x - z| = r$.

(IV) U belongs to the class $V_3(\varepsilon)$ if for each point $x \in \partial U \setminus \{\infty\}$ and any $r > 0$ there is a hyperplane $P = P(x, r)$ passing through x and such that $\rho(y, P) \leq \varepsilon r$ for every $y \in \partial U \cap B(x, r)$.

We say that U *satisfies condition* $V_i(\varepsilon)$ if it belongs to the class $V_i(\varepsilon)$. Condition V_0 is an interior condition (for interior points of the domain), V_1 is a one-sided boundary condition, and V_2 and V_3 are two-sided boundary conditions. We remark that condition $V_3(\varepsilon)$ means that the boundary of the domain is flat to within ε.

Obviously, if $U \in V_2(\varepsilon)$, then $\overline{\mathbf{R}}^n \setminus \overline{U}$ also belongs to $V_2(\varepsilon)$, and if $U \in V_3(\varepsilon)$, then $\overline{\mathbf{R}}^n \setminus \overline{U} \in V_3(\varepsilon)$.

THEOREM 12.29 (TROTSENKO [169], [170]). *For sufficiently small* $\varepsilon > 0$ *the classes* V_i ($i = 0, 1, 2, 3$) *are equivalent; namely, for any* $i, j = 0, 1, 2, 3$ *there exist an* $\varepsilon_{ij} > 0$ *and a function* $g_{ij}(\varepsilon)$ *such that* $g_{ij}(\varepsilon) \to 0$ *as* $\varepsilon \to 0$, *and for* $0 \leq \varepsilon < \varepsilon_{ij}$

$$V_i(\varepsilon) \subset V_j[g_{ij}(\varepsilon)].$$

Trotsenko obtained the following values of the functions g_{ij} and the constants ε_{ij}:

$$g_{01}(\varepsilon) = \varepsilon, \ \varepsilon_{01} = 1; \quad g_{12}(\varepsilon) = 43\varepsilon, \ \varepsilon_{12} = 10^{-4}; \quad g_{23}(\varepsilon) = \sqrt{8\varepsilon},$$
$$\varepsilon_{23} = 0.1; \quad g_{32}(\varepsilon) = 2\varepsilon, \ \varepsilon_{32} = 1; \quad g_{21}(\varepsilon) = 230\sqrt{\varepsilon}, \ \varepsilon_{21} = 10^{-5};$$
$$g_{20}(\varepsilon) = 690\sqrt{\varepsilon}, \ \varepsilon_{20} = 2 \cdot 10^{-6}.$$

The g_{ij} and ε_{ij} can be estimated for the other values of i and j in terms of these six.

COROLLARY. *There exist a number $\varepsilon_0 > 0$ and a function $\mu(\varepsilon)$ defined on $[0, \varepsilon)$ such that $\mu(\varepsilon) \to 0$ as $\varepsilon \to 0$, and if a domain U in $\mathbf{R}^n$ belongs to the class $U(1 - \varepsilon)$ and the set $U^* = \overline{\mathbf{R}}^n \backslash \overline{U}$ is nonempty, then $U^* \in U(1 - \mu(\varepsilon))$.*

From the values of the functions g_{ij} and the constants ε_{ij} given above it is not hard to get explicit values for the constant ε_0 and the function $\mu(\varepsilon)$ in the corollary.

As $\delta \to 1$ the domains of the class $U(\delta)$ become arbitrarily close to a sphere. This fact is expressed by the next assertion.

THEOREM 12.30 [170]. *Let U be a bounded domain in $\mathbf{R}^n$. Assume that $U = \varphi(V)$, where $\varphi \in \mathbf{M}_n$, and V is a domain of class $V_1(\varepsilon)$. If $\varepsilon < 0.01$, then there exist an $a \in U$ and an $r > 0$ such that*

$$B(a, r) \subset U \subset B\left(a, r\left(1 + 3.3\sqrt{\varepsilon}\right)\right).$$

THEOREM 12.31 ([169], [170]). *There exist an $\varepsilon_0 > 0$ and a function $\delta(\varepsilon)$, $0 \leq \varepsilon < \varepsilon_0$, such that $\delta(\varepsilon) \to 1$ as $\varepsilon \to 0$, and for every quasiconformal mapping $f: \overline{\mathbf{R}}^n \to \overline{\mathbf{R}}^n$ with $K(f) = 1 + \varepsilon < 1 + \varepsilon_0$ the image of a ball is a domain of class $U(\delta(\varepsilon))$.*

The proof is obtained by using the special case of Theorem 12.6 corresponding to $U = \overline{\mathbf{R}}^n$.

Let U be a domain in $\overline{\mathbf{R}}^n$ such that $\infty \in \partial U$. A mapping $f: \overline{\mathbf{R}}^n \to \overline{\mathbf{R}}^n$ is called a *reflection* with respect to the boundary of U if $f(x) = x$ for every point $x \in \partial U$ and f is an involution, i.e. $f(f(x)) = x$ for any $x \in \overline{\mathbf{R}}^n$. Let $f: \overline{\mathbf{R}}^n \to \overline{\mathbf{R}}^n$ be a reflection with respect to the boundary of U. Then we say that f is a $(1 + \sigma)$-*quasi-isometric reflection* if

$$|f(x_1) - f(x_2)| \leq (1 + \sigma)|x_1 - x_2|$$

for any $x_1, x_2 \neq \infty$. Obviously, in this case it is also true that

$$(1 + \sigma)^{-1}|x_1 - x_2| \leq |f(x_1) - f(x_2)|$$

for any $x_1, x_2 \neq \infty$.

THEOREM 12.32 ([169], [170]). *Let U be a domain in the class $V_3(\varepsilon)$, where $0 \leq \varepsilon \leq 5 \cdot 10^{-5} n^{-2}$. Then there exists a $(1 + \sigma)$-quasi-isometric reflection $f: \overline{\mathbf{R}}^n \to \overline{\mathbf{R}}^n$ with respect to the boundary of U, where $\sigma \leq 70n\sqrt{\varepsilon}$.*

THEOREM 12.33 ([169], [170]). *Let U be a domain of class $U(\alpha, \beta)$ and let $\infty \in U$; then there exist an ε and a $\delta > 0$, depending on α, β, and n, such that any mapping $f: U \to \overline{\mathbf{R}}^n$ with bounded distortion for which*

$f(\infty) = \infty$ and $K(f) < 1 + \delta$ *has the following property: if the points* x_0, x_1, z_1, z_2 *belong to* U *and* $|x_0 - x_1| \geq |x_0 - z_i|$, $i = 1, 2$, *then*

$$(1 - c_1\varepsilon)\left(\frac{|z_1 - z_0|}{|x_1 - x_0|}\right)^{1 + c_2\varepsilon} \leq \frac{|f(z_1) - f(z_2)|}{|f(x_1) - f(x_0)|} \leq (1 + c_1\varepsilon)\left(\frac{|z_1 - z_0|}{|x_1 - x_0|}\right)^{1 - c_2\varepsilon}.$$

Here c_1 *and* c_2 *are positive constants that depend only on* n.

3. Now we give some results of Trotsenko relating to the problem of extending mappings with bounded distortion.

THEOREM 12.34 [169]. *Let* $U \subset \overline{\mathbf{R}}^n$ *be a domain in the class* $\hat{U}(\delta)$. *Then there exist constants* $\varepsilon_0 > 0$ *and* $c < \infty$, *depending only on* δ *and* n, *such that for every mapping* $f: U \to \overline{\mathbf{R}}^n$ *with bounded distortion* $K(f) = 1 + \varepsilon \leq 1 + \varepsilon_0$ *there is a quasiconformal mapping* $F: \overline{\mathbf{R}}^n \to \overline{\mathbf{R}}^n$ *with* $F(x) = f(x)$ *for all* $x \in U$ *and* $K(F) \leq 1 + c\varepsilon$.

The following theorems give a complete characterization of the image of a line under a quasiconformal mapping of $\overline{\mathbf{R}}^n$ with distortion coefficient close to 1.

Let $\overline{\mathbf{R}} = \mathbf{R} \cup \{\infty\}$, and let $\gamma: \overline{\mathbf{R}} \to \overline{\mathbf{R}}^n$ be a continuous curve with $\gamma(\infty) = \infty$.

If a and b are points in $\mathbf{R}^n$, then $[a, b]$ denotes the segment with endpoints a and b; that is, the set of points x of the form $x = \lambda a + (1 - \lambda)b$, where $0 \leq \lambda \leq 1$.

We say that a curve γ *satisfies condition* $T_1(\varepsilon)$, where $\varepsilon = \text{const} > 0$, if

$$\rho(\gamma(t), [\gamma(t_1), \gamma(t_2)]) \leq \varepsilon|\gamma(t_1) - \gamma(t_2)|$$

for every $t \in [t_1, t_2]$ for any $t_1, t_2 \in \mathbf{R}$ with $t_1 < t_2$. The curve γ is said to *satisfy condition* $T_2(\varepsilon)$ if

$$(1 + \varepsilon^2)|\gamma(t) - \gamma(v)| \leq |\gamma(t) - \gamma(u)| + |\gamma(u) - \gamma(v)|$$

for any $t, u, v \in \mathbf{R}$ with $t < u < v$.

Conditions $T_1(\varepsilon)$ and $T_2(\varepsilon)$ are equivalent in the sense that there exist functions $\lambda(\varepsilon)$ and $\mu(\varepsilon)$ such that $\lambda(\varepsilon) \to 0$ and $\mu(\varepsilon) \to 0$ as $\varepsilon \to 0$, and if γ satisfies condition $T_1(\varepsilon)$, then γ satisfies $T_2(\lambda(\varepsilon))$, while if γ satisfies $T_2(\varepsilon)$, then it satisfies $T_1(\mu(\varepsilon))$.

THEOREM 12.35 ([169], [170]). *There exist constants* $\varepsilon_0 > 0$ *and* $c_0 < \infty$, *depending only on* $n \geq 3$, *such that for any quasiconformal mapping* $f: \overline{\mathbf{R}}^n \to \overline{\mathbf{R}}^n$ *with* $f(\infty) = \infty$ *and* $K(f) = 1 + \varepsilon \leq 1 + \varepsilon_0$, *the image* $f(\overline{\mathbf{R}})$ *of a line satisfies conditions* $T_1(c_0\varepsilon)$ *and* $T_2(c_0\varepsilon)$.

This theorem can be established by a simple application of Theorem 12.33. It is considerably harder to prove the converse. Namely, we have the following assertion.

THEOREM 12.36 ([169], [170]). *There exist numbers $\varepsilon_0 > 0$ and $c < \infty$, depending only on $n \geq 2$, such that for any curve γ in $\overline{\mathbf{R}}^n$ satisfying condition $T_1(\varepsilon)$, where $\varepsilon < \varepsilon_0$, it is possible to construct a quasiconformal mapping $f: \overline{\mathbf{R}}^n \to \overline{\mathbf{R}}^n$ for which*

$$1) \; f(\overline{\mathbf{R}}) = \gamma(\overline{\mathbf{R}}); \quad 2) \; K(f) \leq 1 + c\varepsilon.$$

§12.6. The general concept of stability classes.

1. Here we describe a certain general notion of stability for classes of mappings by developing on a conceptual level the results in §12.2 on stability in Liouville's theorem on conformal mappings. This concept was worked out by Kopylov in the series of papers [65], [69]–[72], [29], and [30]. These papers contain a new direction in the metric theory of spatial mappings. (For simplicity some of the special questions in the theory discussed here are presented in a less general form than was done by Kopylov.)

Let $n, m \geq 1$ be integers, and let $\mathfrak{G}$ be a set of mappings defined on open subsets of $\mathbf{R}^n$ and taking values in $\mathbf{R}^m$. It will be assumed that the class $\mathfrak{G}$ satisfies some of the conditions below (the numbering differs from that in the original papers of Kopylov).

A *homothety* of $\mathbf{R}^n$ is defined to be a transformation $f: \mathbf{R}^n \to \mathbf{R}^n$ of the form $f(x) = \lambda x + a$, where $\lambda > 0$, $\lambda \in \mathbf{R}$, and $a \in \mathbf{R}^n$.

(K_1). The class $\mathfrak{G}$ is invariant under homothety transformations in the spaces $\mathbf{R}^n$ and $\mathbf{R}^m$; that is, for every $f \in \mathfrak{G}$ and homotheties $\varphi: \mathbf{R}^n \to \mathbf{R}^n$ and $\psi: \mathbf{R}^n \to \mathbf{R}^m$ the mapping $\psi \circ f \circ \varphi$ belongs to $\mathfrak{G}$.

(K_2). Every mapping $f \in \mathfrak{G}$ is continuous, and for any $U \subset \mathbf{R}^n$ every set of functions $G \subset \mathfrak{G}$ such that $|f(x)| \leq M = \text{const} < \infty$ for all $x \in U$ is uniformly equicontinuous on every compact subset of U.

(K_3). The class $\mathfrak{G}$ is closed with respect to locally uniform convergence; that is, if for an $f: U \to \mathbf{R}^m$ there exists a sequence $(f_\nu: U \to \mathbf{R}^m)$, $\nu = 1, 2, \ldots$, of functions in $\mathfrak{G}$ such that $f_\nu \to f$ uniformly on every compact subset of U as $\nu \to \infty$, then $f \in \mathfrak{G}$.

(K_4). If $f: U \to \mathbf{R}^m$ is in $\mathfrak{G}$, then for every open set $V \subset U$ the restriction of f to V belongs to $\mathfrak{G}$. If $f: U \to \mathbf{R}^m$ is such that every point $x \in U$ has a neighborhood $U_x \subset U$ such that the restriction of f to U_x is in $\mathfrak{G}$, then f is in $\mathfrak{G}$.

If the class $\mathfrak{G}$ of functions satisfies all these conditions (K_1)–(K_4), then we call it *K-normal*. In the sense of restrictions on the structure of the class $\mathfrak{G}$, the most stringent of these conditions is (K_1). This is clear, for example, from the fact that if $\mathfrak{G}$ is a K-normal class of mappings, then every $f \in \mathfrak{G}$ satisfies a Hölder condition locally. (This assertion follows from a general result of Kopylov [69], but is easy to establish directly.) At the same time, it is easy to construct classes of mappings satisfying conditions (K_2), (K_3), and (K_4) such that the Hölder condition is not satisfied in general.

Basic in the Kopylov theory is the class $\mathscr{H}_{n,m}$ of holomorphic mappings of the complex space $\mathbf{C}^n$ into the complex space $\mathbf{C}^m$ (here $\mathbf{C}^n$ is identified in the natural way with $\mathbf{R}^{2n}$). This class is obviously K-normal.

Another example of a K-normal class is the collection $\mathbf{M}_n$ of all Möbius mappings $f: U \to \mathbf{R}^n$, where U is an arbitrary open domain in R^n.

The collection of all identically constant mappings is clearly also a K-normal class.

2. In broad outline, the basic content of the Kopylov theory of stability of classes of mappings is as follows. Assume that $\mathfrak{G}$ is a given class of mappings defined on open domains of $\mathbf{R}^n$ and taking values in $\mathbf{R}^m$. Certain other classes are constructed from $\mathfrak{G}$ by a kind of perturbation. For this purpose Kopylov introduces functionals characterizing how far a given mapping $f: U \to \mathbf{R}^m$ (U an open domain in $\mathbf{R}^n$) is from the given class $\mathfrak{G}$. Functionals of two types are introduced. A functional of the first type characterizes the closeness of the given mapping f to the class $\mathfrak{G}$ globally. A functional of the second type characterizes the deviation of f from $\mathfrak{G}$ in an infinitesimal neighborhood of an arbitrary point of the domain. The class $\mathfrak{G}$ is stable if the mapping is close to $\mathfrak{G}$ globally whenever the value of the local closeness functional is small (that is, the value of the functional of the first type is also small for f).

First, there arises the problem of finding classes of mappings stable in the Kopylov sense, and, second, for classes satisfying the stability condition there arises the problem of describing the classes of mappings close to it. Kopylov has obtained far-reaching results in the solution of each of these problems. Some of them are formulated below.

We note at once that the whole theory proves to be substantive only if the class $\mathfrak{G}$ satisfies definite conditions. We assume that it is K-normal.

Accordingly, let $\mathfrak{G}$ be a K-normal class of mappings defined on domains in $\mathbf{R}^n$ and taking values in $\mathbf{R}^m$.

Assume that $f: U \to \mathbf{R}^m$ is a given continuous function, where U is a domain in $\mathbf{R}^n$, and let $\rho \in (0,1]$ be arbitrary. Take a ball $B = B(x, r) \subset U$, and denote by B_ρ the ball $B(x, \rho r)$ concentric with it and by $\Delta(f, B)$ the diameter of the set $f(B)$. We define a certain number $\xi_{\rho, B}(f, \mathfrak{G})$ from B and f. If $\Delta(f, B)$ is either 0 or ∞, then let $\xi_{\rho, B}(f, \mathfrak{G}) = 0$. But if $0 < \Delta(f, B) < \infty$, we first put $f_B^* = (1/\Delta(f, B))f$, and then define $\xi_{\rho, B}(f, \mathfrak{G})$

to be the infimum of the quantity

$$\|f_B^* - g\|_{C(B_\rho)} = \sup_{|t-x|<\rho r} |f_B^*(t) - g(t)|$$

over the set of all mappings $g \in \mathfrak{G}$ defined in B. Since for every vector $h \in \mathbf{R}^m$ the mapping $g \equiv h$ belongs to $\mathfrak{G}$, it is clear that always $\xi_{\rho,B}(f,\mathfrak{G}) \leq 1$ (to see this it suffices to take $g \equiv f_B^*(x)$, where x is the center of B). Denote by $\xi_\rho(f,\mathfrak{G})$ the supremum of $\xi_{\rho,B}(f,\mathfrak{G})$ over the set of all balls $B \subset U$. This the first closeness functional introduced by Kopylov.

Another closeness functional is defined as follows. For every ball $B \subset U$ the quantity $\xi_{\rho,B}(f,\mathfrak{G})$, as a function of ρ, is nondecreasing. Let

$$\xi_B(f,\mathfrak{G}) = \int_0^1 \xi_{\rho,B}(f,\mathfrak{G}) \, d\rho$$

and, further,

$$\xi(f,\mathfrak{G}) = \sup_{B \subset U} \xi_B(f,\mathfrak{G}).$$

For this point all the constructions are natural enough. To some extent the appearance of the integral may seem to have little motivation. The introduction of $\xi_B(f,\mathfrak{G})$, and with it $\xi(f,\mathfrak{G})$, was dictated by the desire to have a characteristic of the deviation of f from $\mathfrak{G}$ independent of the choice of a concrete value of ρ. On the other hand, the functionals $\xi(f,\mathfrak{G})$ and $\xi_\rho(f,\mathfrak{G})$, where $0 < \rho < 1$, are equivalent in the sense that for every $\varepsilon > 0$ there is a $\delta > 0$ such that if $\xi(f,\mathfrak{G}) < \delta$, then $\xi_\rho(f,\mathfrak{G}) < \varepsilon$; conversely, if $\xi_\rho(f,\mathfrak{G}) < \delta$, then $\xi(f,\mathfrak{G}) < \varepsilon$. Intuitively, if one of the functionals $\xi(f,\mathfrak{G})$ and $\xi_\rho(f,\mathfrak{G})$ tends to zero, then so does the other.

The functional $\xi_\rho(\cdot,\mathfrak{G})$ is invariant with respect to homothety transformations; that is, if $\varphi: \mathbf{R}^n \to \mathbf{R}^n$ and $\psi: \mathbf{R}^m \to \mathbf{R}^m$ are homotheties, then

$$\xi_\rho(\psi \circ f \circ \varphi, \mathfrak{G}) = \xi_\rho(f,\mathfrak{G})$$

for every $f: U \to \mathbf{R}^m$. This implies that the functional $\xi(\cdot,\mathfrak{G})$ has the same property. If $f: U \to \mathbf{R}^m$ is a continuous function and $\xi_\rho(f,\mathfrak{G}) = 0$ for some $\rho \in (0,1]$, then f belongs to the class $\mathfrak{G}$.

If $\xi_\rho(f,\mathfrak{G})$ $(0 < \rho < 1)$ is small, then f is close to the class $\mathfrak{G}$ in the sense of the uniform norm. More precisely, the following assertion holds. Let $U \subset \mathbf{R}^n$ be a given domain, $A \subset U$ a compact set, and V an open domain strictly inside U and containing A. Then for every $\varepsilon > 0$ there is a $\delta > 0$ such that if $\xi_\rho(f,\mathfrak{G}) < \delta$ for an $f: U \to \mathbf{R}^m$, then there is a $g: V \to \mathbf{R}^m$ in $\mathfrak{G}$ such that

$$|f(x) - g(x)| \leq \varepsilon$$

for all $x \in A$. The number δ depends on ε, ρ, the ratio $\mathrm{dist}(A, \mathbf{R}^n \backslash V)/d(A)$, and the given class $\mathfrak{G}$.

An analogous assertion remains true if $\xi(f, \mathfrak{G})$ is considered instead of $\xi_\rho(f, \mathfrak{G})$.

The functional $\xi(f, \mathfrak{G})$ is the main global closeness functional considered here.

We now define a certain functional characterizing the closeness of a mapping to a given class $\mathfrak{G}$ locally.

Thus, let $\mathfrak{G}$ be an arbitrary K-normal class of mappings of domains in $\mathbf{R}^n$ into $\mathbf{R}^m$, and let $f: U \to \mathbf{R}^m$ be a given continuous mapping. Then for any point $x \in U$ let

$$\xi(x, f, \mathfrak{G}) = \varlimsup_{r \to \infty} \xi_{B(x,r)}(f, \mathfrak{G})$$

and

$$\Xi_l(f, \mathfrak{G}) = \sup_{x \in U} \xi(x, f, \mathfrak{G})$$

(our notation is somewhat different from Kopylov's).

Obviously,

$$\Xi_l(f, \mathfrak{G}) \leq \xi(f, \mathfrak{G})$$

for every $f: U \to \mathbf{R}^m$.

Denote by $W^1(n, m)$ the set of all mappings $f: U \to \mathbf{R}^m$, where U is an arbitrary domain in $\mathbf{R}^n$, that belong to one of the classes $W_{p,\mathrm{loc}}^1(U)$ for some $p > n$.

THEOREM 12.37. *There exist a number $\varepsilon_0 > 0$ and a function $\alpha: [0, \varepsilon_0) \to [0, \infty)$ such that $\alpha(\varepsilon) \to 0$ as $\varepsilon \to 0$, and if $f: U \to \mathbf{C}^m$ is a mapping in $W^1(2n, 2m)$ on an open domain U in $\mathbf{C}^n$ and $\Xi(f, \mathscr{H}_{n,m}) < \varepsilon$, where $0 \leq \varepsilon \leq \varepsilon_0$, then*

$$\xi(f, \mathscr{H}_{n,m}) < \alpha(\varepsilon).$$

This theorem shows that the class of holomorphic mapping is stable in the sense that smallness for a given mapping f of the functional measuring local closeness to the class $\mathscr{H}_{n,m}$ enables us to conclude that the global functional is also small for f.

The next problem arising here is to give a complete description of the class of mappings $f \in W^1(2n, 2m)$ for which $\Xi(f, \mathscr{H}_{n,m})$ is small.

For every mapping $f: U \to \mathbf{C}^m$ in $W^1(2n, 2m)$, where U is a domain in $\mathbf{C}^n$, the generalized derivatives $(\partial f/\partial z_i)(z)$ and $(\partial f/\partial \bar{z}_i)(z)$, $i = 1, \ldots, n$ are defined. Let $f^1, \ldots, f^m$ be the components of f. A *multidimensional system of Beltrami equations* is defined to be a system of equations of the form

$$\frac{\partial f^i}{\partial \bar{z}_k} = \sum_{\alpha=1}^n \sum_{\rho=1}^m Q_{k\rho}^{i\alpha}(z) \frac{\partial f^\rho}{\partial z_\alpha}. \tag{12.26}$$

The collections of functions $Q_{k\rho}^{i\alpha}(z)$, $i, \rho = 1, \ldots, m$, $k, \alpha = 1, \ldots, n$, will be denoted by the symbol $Q(z)$, and system (12.26) will be written briefly as

$$\frac{\partial f}{\partial \bar{z}} = Q(z) \frac{\partial f}{\partial z}. \tag{12.27}$$

For an arbitrary $n \times m$ matrix $Z = (Z_\alpha^\rho)$, $\rho = 1, \ldots, m$, $\alpha = 1, \ldots, n$, we set

$$|Z| = \sqrt{\sum_{\rho=1}^{m} \sum_{\alpha=1}^{n} |Z_\alpha^\rho|^2}$$

and let $Q(z)Z$ denote the $n \times m$ matrix $W = (W_k^i)$ with

$$W_k^i = \sum_{\alpha=1}^{n} \sum_{\rho=1}^{m} Q_{k\rho}^{i\alpha}(z) Z_\alpha^\rho, \qquad i = 1, 2, \ldots, m, k = 1, 2, \ldots, n.$$

The supremum of $|Q(z)Z|$ over the set of all Z with $|Z| < 1$ is denoted by $\|Q(z)\|$. System (12.27) is elliptic if and only if

$$\|Q\|_\infty = \sup_z \|Q(z)\| < 1.$$

THEOREM 12.38. *Suppose that U is an open domain in $\mathbf{C}^n$ and $f: U \to \mathbf{C}^m$ a mapping of class $W^1(2n, 2m)$. If $\Xi(f, \mathscr{H}_{n,m}) < \varepsilon < 1/16\sqrt{n}$, then f is a solution of some system of the form (12.27) with $\|Q\|_\infty \leq 16\sqrt{n}\varepsilon < 1$.*

Kopylov also considered other closeness functionals, and established theorems analogous to Theorem 12.37 for other classes of mappings. In particular, a certain analogue of Theorem 12.37 holds for Möbius mappings, which is a corollary to the stability theorem given in §12.2. Kopylov (together with Dairbekov) also investigated the questions of stability of solutions of systems of first-order elliptic equations (see [29] and [30]).

Use of the closeness functionals introduced by Kopylov has also led to the determination of a characterization of the boundary values of mappings with bounded distortion on a ball that have distortion coefficient close to 1 (see [65] and [72]).

§12.7. A characterization of quasiconformal mappings as mappings preserving the space W_n^1. Let U and V be open sets in $\mathbf{R}^n$, and $f: U \to V$ a quasiconformal mapping. Then for every function $u \in W_n^1(V)$ the function $u \circ f$ also belongs to the class $W_n^1(U)$. We thus get a linear mapping

$$f^*: u \in W_n^1(V) \mapsto u \circ f \in W_n^1(U).$$

This mapping is also bounded. The idea naturally arises of trying to characterize all the linear mappings of $W_n^1(V)$ into $W_n^1(U)$ that can be obtained in this way. It may be supposed that the mappings having this property should be sought among the mappings that preserve some additional structure in W_n^1. This idea was realized in [107], [88], and [86] by a generalization of the concept of a Royden algebra. Here only continuous functions in the class W_n^1 are considered, so that we are concerned with a certain proper subset of the space W_n^1. In [177] Vodop'yanov and Gol'dshteĭn obtained the required characterization by using the order structure in W_n^1. We present a formulation of their result.

Let U be an arbitrary open domain in $\mathbf{R}^n$. Instead of $W_n^1(U)$ it is convenient to work with the space $L_n^1(U)$; that is, with the set $W_n^1(U)$ of functions endowed with the seminorm

$$\|u\|_{L_n^1(U)} = \left(\int_U |u'(x)|^n \, dx \right)^{1/n}.$$

Let U and V be open domains in $\mathbf{R}^n$. A linear operator $A: L_n^1(V) \to L_n^1(U)$ is called a *structural isomorphism* of the spaces $L_n^1(V)$ and $L_n^1(U)$ if A is a bijective mapping, and the following conditions hold:

(α) $A^*(v) > 0$ if and only if $v > 0$.

(β) If $v \equiv 1$, then $A(v)$ is also the function identically equal to 1.

(γ) If $v_m \to 0$ almost everywhere as $m \to \infty$, then $A(v_m) \to 0$ almost everywhere.

(δ) If $(w_m \in L_n^1(U))$, $m = 1, 2, \ldots$, tends to zero almost everywhere in U, then $A^{-1}(w_m)$ tends to zero almost everywhere in V.

Two domains U_1 and U_2 are said to be $(1, n)$-*equivalent* if the mappings $\theta_i: W_n^1(U_1 \cup U_2) \to W_n^1(U_i)$, where $\theta_i(v) = v|_{U_i}$, $i = 1, 2$, are isometric isomorphisms.

THEOREM 12.39. *For every structural isomorphism* $A: L_n^1(V) \to L_n^1(U)$ *there exists a unique quasiconformal homeomorphism of* U *into* $\mathbf{R}^n$ *that satisfies the following conditions*:

1) *The domain* $\varphi(U)$ *is* $(1, n)$-*equivalent to* V.

2) $(Af)(x) = v[\varphi(x)]$ *almost everywhere in* U *for every function* $v \in L_n^1(V)$.

This theorem is only one of a cycle of investigations carried out by Gol'dshteĭn and Vodop'yanov. The reader can obtain more detailed information about this work in [48], where, in particular, there is a detailed bibliography.

Some Results in the Theory of Functions of a Real Variable and the Theory of Partial Differential Equations

§1. Functions with bounded mean oscillation

Let a number $p \geq 1$ be given. A function $f: U \to \mathbf{R}^d$, where U is an open set in $\mathbf{R}^n$, is called a *function with bounded mean oscillation in the L_p-sense* if there exists a number $\omega > 0$ such that for every cube $Q \subset U$ (only cubes with faces parallel to the coordinate planes are considered) there is a vector $c_Q \in \mathbf{R}^d$ such that

$$\frac{1}{|Q|} \int_Q |f(x) - c_Q|^p \, dx \leq \omega^p.$$

The smallest number ω satisfying the above condition is called the *mean oscillation* of f in L_p. Intuitively, the condition that a function f has bounded mean oscillation means that it can be approximated by a constant on every cube Q in a certain sense, uniformly with respect to the choice of cube.

The concept of a function with bounded mean oscillation was introduced by John and Nirenberg [63]. Our immediate problem is to derive the estimates they obtained for the measures of Lebesgue sets of an arbitrary function with bounded mean oscillation. These estimates will be used below in §2 in proving the Serrin-Moser theorem in connection with the Harnack inequality for elliptic equations.

THEOREM 1.1. *Let $U \subset \mathbf{R}^n$ be an open set, and $f: U \to \mathbf{R}^d$ a function with bounded mean oscillation equal to ω. Let Q be an arbitrary cube contained in U. Denote by $E_Q(t)$ the set of points $x \in Q$ such that $|f(x) - c_Q| > t$. Then for every $t > \omega$ the measure of $E_Q(t)$ admits the estimate*

$$|E_Q(t)| \leq \left[(1/\omega^p) \int_Q |f(x) - c_Q|^p \, dx \right] \exp[\gamma(1 - t/\omega)], \qquad (1.1)$$

where $\gamma > 0$ is the constant

$$\gamma = 1/(1 + 2^{n/p}e). \tag{1.2}$$

PROOF. By a condition of the theorem, for every cube $Q \subset U$

$$\omega^p|Q| \geq \int_Q |f(x) - c_Q|^p \, dx \geq \int_{E_Q(t)} |f(x) - c_Q|^p \, dx,$$

which implies that

$$t^p|E_Q(t)| \leq \int_Q |f(x) - c_Q|^p \, dx. \tag{1.3}$$

The smallest number L such that

$$|E_Q(t)| \leq (L/\omega^p) \cdot \int_Q |f(x) - c_Q|^p \, dx$$

for every cube $Q \subset U$ is denoted by $F(t)$. It is clear from (1.3) that $F(t) \leq (\omega/t)^p$ for all $t > 0$.

Take an arbitrary cube $Q_0 \subset U$, and a number $k > 1$. (The specific value of k will be chosen later.) For brevity we write $c_{Q_0} = c_0$. By assumption,

$$\int_Q |f(x) - c_0|^p \, dx \leq \omega^p|Q_0| < k\omega^p|Q_0|.$$

Let us partition Q_0 into 2^n equal cubes by planes parallel to the coordinate planes, and single out those of the cubes for which the inequality

$$\int_Q |f(x) - c_0|^p \, dx < k\omega^p|Q| \tag{1.4}$$

fails to hold. Each of the remaining cubes is again divided into 2^n equal parts, and again those for which (1.4) fails to hold are singled out, and so on. Continuing this process, we get a finite or infinite sequence of cubes Q_m for which

$$\int_{Q_M} |f(x) - c_0|^p \, dx \geq k\omega^p|Q_m|. \tag{1.5}$$

The cube Q_m is obtained by subdivision of some cube $Q = Q_m^*$ satisfying (1.4). We have that

$$\int_{Q_m} |f(x) - c_0|^p \, dx \leq \int_{Q_m^*} |f(x) - c_0|^p \, dx < k\omega^p|Q_m^*|.$$

Moreover, $|Q_m^*| = 2^n|Q_m|$ and, consequently,

$$\int_{Q_m} |f(x) - c_0|^p \, dx < 2^n k\omega^p|Q_m| \tag{1.6}$$

for each m. The sequence of cubes Q_m can turn out to be empty, of course. Let $H = \bigcup_m Q_m$. For every point $x \in Q_0 \backslash H$ there exists a sequence (Q_ν) of cubes containing x such that the diameter of Q_ν tends to zero as $\nu \to \infty$, and (1.4) holds for each ν. As $\nu \to \infty$,

$$\frac{1}{|Q_\nu|} \int_{Q_\nu} |f(x) - c_0|^p \, dx \to |f(x) - c_0|^p$$

for almost all $x \in Q_0 \backslash H$, and so for almost all $x \in Q_0 \backslash H$

$$|f(x) - c_0|^p \le k^{1/p} \omega. \tag{1.7}$$

Choose a number $t > k^{1/p} \omega$ arbitrarily, and consider the set $E_{Q_0}(t)$. Since (1.7) holds almost everywhere on $Q_0 \backslash H$, $E_Q(t)$ is contained in H to within points forming a set of measure zero. If $|H| = 0$, then $|E_{Q_0}(t)| = 0$, and (1.1) holds for the given t. Assume that $|H| > 0$. We have

$$|E_{Q_0}(t)| = \sum_m |E_{Q_0}(t) \cap Q_m|.$$

Consider the set $E_{Q_0}(t) \cap Q_m$. According to the definition of a function with bounded mean oscillation, to the cube Q_m there corresponds a vector (denote it by c_m) such that

$$\int_{Q_m} |f(x) - c_m|^p \, dx \le \omega^p |Q_m|.$$

For every point $x \in E_{Q_0}(t) \cap Q_m$ we have that $|f(x) - c_0| > t > k^{1/p} \omega$. Also,

$$|f(x) - c_0| \le |f(x) - c_m| + |c_m - c_0|. \tag{1.8}$$

Let us estimate $|c_m - c_0|$. Considering (1.6), we get

$$|c_m - c_0| = \left(\frac{1}{|Q_m|} \int_{Q_m} |c_m - c_0|^p \, dx \right)^{1/p}$$

$$\le \left(\frac{1}{|Q_m|} \int_{Q_m} |c_m - f(x)|^p \, dx \right)^{1/p}$$

$$+ \left(\frac{1}{|Q_m|} \int_{Q_m} |f(x) - c_0|^p \, dx \right)^{1/p} < \omega + 2^{n/p} k^{1/p} \omega,$$

and, thus,

$$|c_m - c_0| \le (2^{n/p} k^{1/p} + 1) \omega.$$

By (1.8), this entails that for all $x \in E_{Q_0}(t) \cap Q_m$

$$|f(x) - c_m| \ge |f(x) - c_0| - (2^{n/p} k^{1/p} + 1) \omega > t - h, \tag{1.9}$$

where

$$h = (2^{n/p}k^{1/p} + 1)\omega. \tag{1.10}$$

This means that if $x \in E_{Q_0}(t) \cap Q_m$, then $x \in E_{Q_m}(t - h)$ for $t > h$. Consequently,

$$\begin{aligned}
|E_{Q_0}(t) \cap Q_m| &\le |E_{Q_m}(t - h)| \\
&\le \frac{F(t - h)}{\omega^p} \int_{Q_m} |f(x) - c_m|^p \, dx \le |Q_m| F(t - h).
\end{aligned}$$

Summing over m, we get

$$|E_{Q_0}(t)| \le |H| F(t - h). \tag{1.11}$$

Inequality (1.5) implies the following estimate for the measure of H:

$$\begin{aligned}
|H| &\le (1/k\omega^p) \sum_m \int_{Q_m} |f(x) - c_0|^p \, dx \\
&\le (1/k\omega^p) \int_{Q_0} |f(x) - c_{Q_0}|^p \, dx,
\end{aligned}$$

which, by (1.11), gives us

$$|E_{Q_0}(t)| \le [F(t - h)/k\omega^p] \int_{Q_0} |f(x) - c_0|^p \, dx. \tag{1.12}$$

The cube $Q_0 \subset U$ was taken arbitrarily, and it thus follows from (1.12) that

$$F(t) \le (1/k)F(t - h). \tag{1.13}$$

The functional equation $\varphi(t) = (1/k)\varphi(t - h)$ has a special solution $\varphi(t) = Ce^{-\lambda t}$, where $\lambda = (\ln k)/h$. If the inequality $F(t) \le \varphi(t)$ holds on $[\omega, \omega + h]$, then it is not hard to see that $F(t) \le \varphi(t)$ for all $t > \omega$. This reduces the matter to choosing C such that $F(t) \le Ce^{-\lambda t}$ for all $t \in [\omega, \omega + h]$. We have that $F(t) \le (\omega/t)^p$ for all t. The function $g(t) = \omega^p e^{\lambda t}/t^p$ has the unique extremum point $t = p/\lambda$ on $(0, \infty)$, and it is a minimum point. This implies that $g(t)$ takes its largest value in each interval $[a, b] \subset \mathbf{R}$ at an endpoint. Comparing $g(\omega)$ and $g(\omega + h)$, we get that $g(\omega) > g(\omega + h)$. Thus, $\omega^p e^{\lambda t}/t^p \le g(\omega) = e^{\lambda \omega}$ and $F(t) \le e^{\lambda(\omega - t)}$ for all $t \ge \omega$.

The estimate obtained is better, the larger λ is. We have that $\lambda = (\ln k)/\omega[2^{n/p}k^{1/p} + 1]$. The function of k on the right-hand side attains a maximum on $[1, \infty)$ when k is the root of the equation $1 + 1/2^{n/p}k = \ln k$. Approximately, $k = e + 2^{-n/p} + \dots$. The estimate in the theorem is obtained if we take $k = e$.

COROLLARY. *Let $f: U \to \mathbf{R}^d$ ($U \subset \mathbf{R}^n$ an open set) be a function with bounded mean oscillation in L_p equal to ω. Then for any $\alpha < \gamma/\omega$ the function $e^{\alpha|f(x)|}$ is integrable on every cube $Q \subset U$, and for each such cube*

$$\int_Q \exp[\alpha|f(x) - c_Q|]\, dx$$

$$\leq |Q| + (\alpha e^{\gamma}/[\gamma - \alpha\omega]\omega^{p-1}) \int_Q |f(x) - c_Q|^p\, dx. \quad (1.14)$$

The function f is integrable to any power $q > p$. For any q_1 and q_2 with $q_1 > q_2 \geq p$,

$$\int_Q |f(x) - c_Q|^{q_1}\, dx \leq C\omega^{q_1 - q_2} \int_Q |f(x) - c_Q|^{q_2}\, dx, \quad (1.15)$$

where $C > 0$ is a constant depending only on q_1, q_2, and p.

PROOF. We first make some simple remarks about the Lebesgue integral. Let $u(x)$ be a nonnegative function integrable in the cube Q, and let $A(t)$ ($t \geq 0$) be the Lebesgue set $\{x \in Q | u(x) \geq t\}$. Then

$$\int_Q u(x)\, dx = \int_0^{\infty} |A(t)|\, dt.$$

The condition $e^{\alpha u(x)} \geq t$, where $t \geq 0$, holds for all $x \in Q$ when $0 \leq t \leq 1$, and is equivalent to $u(x) > (\ln t)/\alpha$ when $t > 1$. This allows us to conclude that the Lebesgue set $\{x | e^{\alpha u(x)} \geq t\}$ of the function $e^{\alpha u}$ coincides with Q for $0 < t \leq 1$ and with $A[(\ln t)/\alpha]$ for $t > 1$. From this we get

$$\int_Q e^{\alpha u(x)}\, dx = |Q| + \int_1^{\infty} |A[(1/\alpha) \ln t]|\, dt$$

$$= |Q| + \alpha \int_0^{\infty} |A(t)| e^{\alpha t}\, dt. \quad (1.16)$$

Further, the Lebesgue set $\{x | [u(x)]^p \geq t\}$ coincides with $A(t^{1/p})$, which leads to the conclusion that

$$\int_Q [u(x)]^p\, dx = \int_0^{\infty} |A(t^{1/p})|\, dt = p \int_0^{\infty} |A(u)| u^{p-1}\, du. \quad (1.17)$$

Let $u(x) = |f(x) - c_Q|$ and $I_p = \int_Q [u(x)]^p\, dx$. By (1.1),

$$|A(t)| \leq (I_p/\omega^p) \exp[\gamma(1 - t/\omega)].$$

Applying this estimate to the right-hand side of (1.16), we get that $\exp \alpha[u(x)]$ is integrable in Q if $\alpha < \gamma/\omega$. Further ,

$$\int_Q \exp \alpha|f(x) - c_Q|\, dx \leq |Q| + \frac{\alpha e \gamma}{(\gamma - \alpha\omega)\omega^{p-1}} I_p,$$

and this establishes (1.14). Let $q > p$ be given. Using (1.17), we find that

$$\int_Q |f(x) - c_Q|^q \, dx = q \int_0^\infty |E_Q(t)| t^{q-1} dt$$

$$= q \int_0^\omega |E_Q(t)| t^{q-1} dt + q \int_\omega^\infty |E_Q(t)| t^{q-1} dt.$$

The first integral on the right-hand side can be estimated as follows. For $0 < t \leq \omega$ we have that $t^{q-1} \leq \omega^{q-p} t^{p-1}$, which implies that

$$q \int_0^\omega |E_Q(t)| t^{q-1} dt \leq q\omega^{q-p} \int_0^\omega |E_Q(t)| t^{p-1} dt$$

$$\leq q\omega^{q-p} \int_0^\infty |E_Q(t)| t^{p-1} dt = (q/p)\omega^{q-p} I_p.$$

By the estimate in the theorem,

$$q \int_\omega^\infty |E_Q(t)| t^{q-1} dt \leq (qI_p/\omega^p) \int_\omega^\infty t^{q-1} \exp[\gamma(1 - t/\omega)] \, dt.$$

The change $t/\omega = u$ of the variable of integration gives us that

$$q \int_\omega^\infty |E_Q(t)| t^{q-1} dt \leq q\omega^{q-p} I_p e^\gamma \int_1^\infty t^{q-1} e^{-\gamma t} dt.$$

From this,

$$\int_Q |f(x) - c_Q|^q \, dx \leq C(q)\omega^{q-p} I_p. \tag{1.18}$$

This is the second inequality of the corollary for the case when $q_2 = p$.

Since f is a function with bounded mean oscillation, $I_p \leq \omega^p |Q|$. We get from this that

$$\int_Q |f(x) - c_Q|^q \, dx \leq C^q \omega^q |Q|$$

for every $q > p$ for any cube $Q \subset U$. This means that f is a function with bounded mean oscillation in the L_q-sense for any $q > p$. Further, the mean oscillation of f in L_q is equal to $\omega_1 = C\omega$, where $C = $ const. Setting $p = q_2$ and $q = q_1$ in the above arguments, we get the desired inequality

$$\int_Q |f(x) - c_Q|^{q_1} \, dx \leq C\omega^{q_1 - q_2} \int_Q |f(x) - c_Q|^{q_2} \, dx.$$

The corollary is proved.

§2. Harnack's inequality for quasilinear elliptic equations

2.1. Preliminary remarks. Let U be an open set in $\mathbf{R}^n$, and $A(x, q)$ an $\mathbf{R}^n$-valued function such that for almost all $x \in U$ the vector $A(x, q)$ is defined for any vector $q \in \mathbf{R}^n$, and the following conditions hold:

a) For every measurable function $q: U \to \mathbf{R}^n$ the function $x \to A[x, q(x)]$ is measurable.

b) There exist constants $a_1, a_2 > 0$ and a number $p > 1$ such that for almost all $x \in U$ and any $q \in \mathbf{R}^n$

$$|A(x, q)| \le a_1 |q|^{p-1}, \tag{2.1}$$

$$\langle q, A(x, q) \rangle \ge a_2 |q|^p. \tag{2.2}$$

Let u be an arbitrary vector-valued function of class $W^1_{p,\text{loc}}(U)$. By condition a), the function $x \to A[x, u'(x)]$ is measurable. In view of (2.1) it admits the estimate

$$|A[x, u'(x)]| \le a_1 |u'(x)|^{p-1},$$

and, consequently, the function $A[x, u'(x)]$ is locally integrable to the power $p' = p/(p - 1)$ in U. The function u is called a *generalized solution* of the equation

$$\operatorname{div} A[x, u'(x)] = 0 \tag{2.3}$$

if for every function $\varphi \in C_0^\infty(U)$

$$\int_U \langle \varphi'(x), A[x, u'(x)] \rangle dx = 0. \tag{2.4}$$

Note that if u is a function of class $W^1_{p,\text{loc}}(U)$, then, since $x \to A[x, u'(x)]$ is locally integrable to the power p', it is easy to get from (2.4) by passing to the limit that (2.4) holds also for every function $\varphi \in W^1_p(U)$ with compact support in U.

Let $F(x, q)$ be a normal kernel in the sense of §5 in Chapter II, i.e., a function satisfying all the conditions A–E in §5.1. Then the vector-valued function

$$A(x, q) = F_q(x, q) = \left(\frac{\partial F}{\partial q_1}(x, q), \frac{\partial F}{\partial q_2}(x, q), \dots, \frac{\partial F}{\partial q_n}(x, q) \right)$$

is defined. It follows from A–E in §5.1 of Chapter II that the function satisfies conditions a) and b) just introduced. A function $u \in W^1_{p,\text{loc}}$ is *stationary* for the functional I_F if and only if u is a generalized solution of (2.3), where $A = F_q$.

Our purpose is to prove Harnack's inequality (Theorem 5.7 in §5 of Chapter II). It gives us as a corollary that generalized solutions of (2.3) have the Hölder property. The following remark is useful below. Let $f: \mathbf{R}^n \to \mathbf{R}^n$ be a given similarity transformation. Then $l = f'(x)$ is a general orthogonal transformation. Let $k = |l|$. Assume that $A(x, q)$ is a given vector-valued function defined for almost all $x \in U$ and any q in $\mathbf{R}^n$ and satisfying conditions a) and b). Let $V = f^{-1}(U)$. For $x \in V$ and $q \in \mathbf{R}^n$ let

$$\tilde{A}(x, q) = (1/k^{p-1})A[f(x), l(q)].$$

It is easy to verify that $\tilde{A}$ also satisfies a) and b), and with the same values of the constants a_1 and a_2 as A. If $u(x)$ is a generalized solution of the equation $\operatorname{div} A[x, u'(x)] = 0$ in U, then the function $v(x) = u[f(x)]$ is a generalized solution of the equation $\operatorname{div} \tilde{A}[x, v'(x)] = 0$.

There are two competing methods for investigating the question of an estimate for the moduli of continuity of solutions of an equation with the form (2.3) and of certain more general equations. DeGiorgi's well-known article [31] is fundamental for one of them. The second method is based on an idea of Moser [103], [104]. A far-reaching development of DeGiorgi's method has been given by Ladyzhenskaya and her students (see, for example, the monographs [78] and [77]). Moser's method in its original form was applied to the investigation of an equation of the form

$$\sum_{i=1}^{n} \frac{\partial}{\partial x_i} \left(\sum_{j=1}^{n} a_{ij}(x) \frac{\partial u}{\partial x_j}(x) \right) = 0. \tag{2.5}$$

An extension of Moser's method to the case of quasilinear equations of the form $\operatorname{div} A(x, u, u') = 0$ was given by Serrin in [155] and [156]. The arguments presented below mainly follow Moser's method. The results obtained here make up a special case of results of Serrin. We do not touch on the general case, since it is not needed for the purposes of this book. On the other hand, the arguments simplify considerably in the special case we need. We remark that, in contrast to Moser and Serrin, we do not use the concept of a subsolution of equation (2.3).

2.2. Main inequalities. The proof that generalized solutions of (2.3) are continuous and the proof of Harnack's inequality for these equations are based on certain integral inequalities, the derivation of which is our immediate task. Recall that a function $F: \mathbf{R} \to \mathbf{R}$ is said to be *piecewise smooth* if there exists a finite set $E \subset \mathbf{R}$ such that F is continuously differentiable at each point not in E, and for all $u \in E$ the finite limits $F'(u-0)$ and $F'(u+0)$ exist. If F is a piecewise smooth function, then at each point $u \in \mathbf{R}$ the derivatives $F'_L(u)$ and $F'_R(u)$ exist and are finite.

If $u(x)$ is a function of class $W^1_{p,\text{loc}}(U)$ and F is a piecewise smooth function such that the derivative $F'_L(u)$ is bounded in $\mathbf{R}$, then, by Lemma

5.8 in Chapter II, the function $v = F \circ u$ also belongs to $W^1_{p,\text{loc}}$. Further, $v'(x) = F'_L[u(x)]u'(x)$ for almost all $x \in U$. The right-hand derivative can be taken here instead of the left-hand derivative.

LEMMA 2.1. *Suppose that F is a piecewise smooth nondecreasing function on $\mathbf{R}$ such that if $F(u) \neq 0$, then $F'_L(u) \neq 0$ and $F'_R(u) \neq 0$. Assume that $F'_L(u)$ is a bounded function on $\mathbf{R}$. Define the auxiliary function Φ by setting $\Phi(u) = 0$ if $F(u) = 0$, and $\Phi(u) = [F(u)]^p/[F'_L(u)]^{p-1}$ if $F(u) \neq 0$. Let M be an arbitrary compact set in $\mathbf{R}^n$, let $G \supset M$ be an open set lying strictly inside U, and let $\delta = \delta(M, \partial G)$ be the infimum of the distances of the points in M from the boundary of G. Then every generalized solution $u(x)$ of (2.3) on U satisfies*

$$\int_M F'[u(x)]|u'(x)|^p \, dx \leq \frac{C_1}{\delta p} \int_G \Phi[u(x)] \, dx, \qquad (2.6)$$

where the constant C_1 depends only on n, p, and the constants a_1 and a_2 in conditions a) and b) of §2.1.

PROOF. Let $u(x)$ be an arbitrary generalized solution of (2.3) on U, and let the function $F: \mathbf{R} \to \mathbf{R}$ and the sets M and G satisfy all the conditions of the lemma. The function $v = F \circ u$ belongs to the class $W^1_{p,\text{loc}}(U)$, and $v'(x) = F'_L[u(x)]u'(x)$ for almost all $x \in U$. We construct an auxiliary function τ. Let $\eta > 0$ be any number such that $3\eta < \delta$. Let $A_\eta = U_\eta(A)$ be the η-neighborhood of a set A. Define

$$\tilde{\tau}(x) = [1 - \rho(x, M_\eta)/(\delta - 2\eta)]^+.$$

Obviously, $\tilde{\tau}(x) = 1$ for $x \in M_\eta$, and $\tilde{\tau}(x) = 0$ if $\rho(x, M_\eta) > \delta - 2\eta$. For any $x_1, x_2 \in \mathbf{R}^n$ we have that

$$|\rho(x_1, M_\eta) - \rho(x_2, M_\eta)| \leq |x_1 - x_2|,$$

which leads us to conclude that

$$|\tilde{\tau}(x_1) - \tilde{\tau}(x_2)| \leq |x_1 - x_2|/(\delta - 2\eta)$$

for any $x_1, x_2 \in \mathbf{R}^n$. Applying to the function $\tilde{\tau}$ the operation of averaging by means of some nonnegative kernel with averaging parameter $h < \eta$, we get a function τ such that $\tau(x) = 1$ for $x \in M$, $\tau(x) = 0$ for $x \notin G$, and $|\tau'(x)| \leq 1/(\delta - 2\eta)$ for all x.

For every function φ of class $W^1_p(U)$ with compact support in U,

$$\int_U \langle \varphi'(x), A[x, u'(x)]\rangle dx = 0. \qquad (2.7)$$

Here let $\varphi(x) = [\tau(x)]^p F[u(x)]$. The given function φ is in $W_p^1(U)$, and its support is in the open set G. Further,

$$\varphi'(x) = p[\tau(x)]^{p-1}\tau'(x)F[u(x)] + [\tau(x)]^p F'[u(x)]u'(x).$$

(Here and below, we write F' instead of F_L'.) For $A = A(x, u')$

$$\langle \varphi', A \rangle = \tau^p F'(u)\langle u', A \rangle + p\tau^{p-1}F(u)\langle \tau', A \rangle.$$

By (2.2) and (2.1),

$$\langle u', A(x, u') \rangle \geq a_2 |u'|^p,$$
$$|\langle \tau', A(x, u') \rangle| \leq a_1 |\tau'| |u'|^{p-1},$$

which leads us to conclude that

$$\langle \varphi', A \rangle \geq a_2 \tau^p F'(u)|u'|^p - a_1 p\tau^{p-1}|F(u)| |u'|^{p-1}|\tau'|. \tag{2.8}$$

Let x be such that $F[u(x)] \neq 0$. Then $F'[u(x)] \neq 0$. We have Young's inequality $XY \leq X^p/p + Y^{(p-1)/p}(p-1)/p$, where $X, Y \geq 0$. Let x be such that $F'[u(x)] \neq 0$. Setting

$$X = t|\tau'| |F(u)|/[F'(u)]^{(p-1)/p},$$
$$Y = \frac{1}{t}[F'(u)]^{(p-1)/p}|u'|^{p-1}\tau^{p-1},$$

where $t > 0$ will be chosen later, we get that

$$a_1 p\tau^{p-1}|F(u)| |\nabla u|^{p-1}|\nabla \tau| \leq a_1 t^p |\tau'|^p |F(u)|^p/(F'(u))^{p-1}$$
$$+ a_1(p-1)[F'(u)]|u|^p \tau^p t^{(p-1)/p}.$$

Substituting this into (2.8), we arrive at

$$\langle \varphi', A \rangle \geq [a_2 - a_1(p-1)/t^{(p-1)/p}]F'(u)|u'|^p\tau^p - a_1 t^p|\tau'|^p\Phi[u(x)]. \tag{2.9}$$

The last inequality was derived under the assumption that $F[u(x)] \neq 0$. It is not hard to see that it remains true also in the case when $F[u(x)] = 0$.

The quantity t is determined from the condition

$$a_2 - a_1(p-1)/\tau^{(p-1)/p} = a_2/2.$$

Integrating (2.9) termwise, with (2.7) in view, we get that

$$0 \geq (a_2/2) \int_U F'(u)|u'|^p\tau^p \, dx - a_1 t^p \int_U |\tau'|^p\Phi[u(x)] \, dx.$$

Since $\tau(x) = 1$ for $x \in M$ and $|\tau'(x)| \leq 1/(\delta - 2\eta)$, this gives us that

$$\int_M [F'(u)]|u'|^p \, dx \leq \frac{C_1}{(\delta - 2\eta)^p} \int_G \Phi[u(x)] \, dx,$$

where C_1 depends only on a_1, a_2, and p. The number $\eta > 0$ is arbitrary. Passing to the limit as $\eta \to 0$ in the displayed inequality, we arrive at the desired inequality. The lemma is proved.

COROLLARY 1. *Suppose that $u(x)$ is a generalized solution of (2.3) on U, M is a compact set, and $G \supset M$ is an open set lying strictly inside U, and let $\delta = \delta(M, \partial G)$. Then for every $m \geq p$*

$$\int_M |u(x)|^{m-p} |u'(x)|^p \, dx \leq \frac{C_1}{[(m-p+1)\delta]^p} \int_G |u(x)|^m \, dx, \qquad (2.10)$$

PROOF. Let $\nu > 0$ be an arbitrary integer, and define $F_\nu(u) = u^{m-p+1}$ for $0 \leq u \leq \nu$, $F_\nu(u) = (m - p + 1)\nu^{m-p}u - (m - p)\nu^{m-p+1}$ for $u > \nu$ and $F_\nu(-u) = -F_\nu(u)$. The function F_ν is piecewise smooth, and its derivative is bounded. Let $\Phi_\nu(u) = |F_\nu(u)|^p / [F_\nu'(u)]^{p-1}$. Then $\Phi_\nu(u) = u^m/(m - p + 1)^{p-1}$ for $0 < u \leq \nu$, $\Phi_\nu(u) = \nu^{m-p}(m - p + 1)(u - \gamma\nu)^p$, where $\gamma = (m - p)/(m - p + 1)$, and $\Phi_\nu(-u) = \Phi_\nu(u)$. Setting $F = F_\nu$ and $\Phi = \Phi_\nu$ in (2.6), we get that

$$\int_M F_\nu'[u(x)]|u'(x)|^p \, dx \leq \frac{C_1}{\delta^p} \int_G \Phi_\nu[u(x)] \, dx. \qquad (2.11)$$

As $\nu \to \infty$ we have that

$$F_\nu'[u'(x)]|u'(x)| \to (m - p + 1)|u(x)|^{m-p}|u'(x)|^p$$

for all x. From this,

$$(m - p + 1) \int_M |u(x)|^{m-p} |u'(x)|^p \, dx$$

$$\leq \lim_{\nu \to \infty} \int_M F_\nu'[u(x)]|u'(x)|^p \, dx. \qquad (2.12)$$

Inequality (2.10) is true if its right-hand side is equal to ∞. Assume that the integral on the right-hand side of (2.10) is finite. As $\nu \to \infty$ we have that

$$\Phi_\nu[u(x)] \to |u(x)|^m/(m - p + 1)^{p-1}$$

and

$$0 \leq \Phi_\nu[u(x)] \leq |u(x)|^m(m - p + 1)$$

for each ν for all $x \in U$. In this case the functions $\Phi_\nu \circ u$ are majorized by an integrable function, and thus, by Lebesgue's theorem,

$$\int_G \Phi_\nu[u(x)] \, dx \to \frac{1}{(m - p + 1)^{p-1}} \int_G |u'(x)|^m \, dx. \qquad (2.13)$$

Passing to the limit in (2.11) and considering (2.12) and (2.13), we get (2.10).

COROLLARY 2. *Let $u(x)$ be a generalized solution of* (2.3) *on an open set U. Assume that $u(x)$ is nonnegative. Then, for any sets M and G such that $M \subset G$, M is compact, G is open, and G lies strictly inside U, and for any numbers $m \neq p - 1$ and $h > 0$ the inequality*

$$\int_M [u(x) + h]^{m-p} |u'(x)|^p \, dx \leq \frac{C}{\delta |m - p + 1|^p} \int_G [u(x) + h]^m \, dx \quad (2.14)$$

holds, where G_1 and δ have the same meaning as in the lemma.

PROOF. Three cases are possible: 1) $m < p - 1$; 2) $p - 1 < m \leq p$; 3) $m > p$. In case 1) inequality (2.14) is obtained at once if we substitute in (2.6) the function F defined as follows: $F(u) = u - h^{m-p+1}$ for $u < 0$, and $F(u) = -(u + h)^{m-p+1}$ for $u \geq 0$. This function F clearly satisfies all the conditions of the lemma. In the case $p - 1 < m \leq p$ we define $F(u)$ as follows: $F(u) = u + h^{m-p+1}$ for $u < 0$, and $F(u) = (u + h)^{m-p+1}$ for $u \geq 0$. Clearly F satisfies all the conditions of the lemma in this case.

In the case $m > p$ inequality (2.14) is a consequence of (2.10), since $u(x)$ is a generalized solution of (2.3), then $u(x) + h$ is also a generalized solution of this equation for any $h \in \mathbf{R}$.

We remark that DeGiorgi's method for investigating generalized solutions of (2.3) is based on the use of an inequality which can be obtained from (2.6) by setting $F(u) = (u - k)^+$ or $F(u) = -(u - k)^-$, where k is an arbitrary real number.

2.3. Consequences of the integral inequalities in §2.2.   Let $a \in \mathbf{R}^n$ be an arbitrary point. For $r > 0$ denote the open cube $Q(a, r)$ by Q_r. For an arbitrary measurable set A with $0 < |A| < \infty$ and any function f integrable on A let

$$\overline{\int_A} f(x) \, dx = (1/|A|) \int_A f(x) \, dx.$$

We mention another corollary to the Sobolev imbedding theorem. Let f be a function of class W_p^1, where $p \leq n$, in the cube $Q_r = Q(a, r) \subset \mathbf{R}^n$. Let s be such that $1 < s < n/(n - p)$. Then, by Lemma 2.1 in Chapter I,

$$\|f - \overline{f}\|_{ps, Q_r} \leq Cr \|\, |f'| \,\|_{p, Q_r}. \quad (2.15)$$

Here

$$\|f\|_{t, A} = \left(\frac{1}{|A|} \cdot \int_A |f(x)|^t \, dx \right)^{1/t},$$

$t \geq 1$. We have that

$$\|f\|_{ps, Q_r} \leq \|f - \overline{f}\|_{ps, Q_r} + |\overline{f}| \leq \|f - \overline{f}\|_{ps, Q_r} + \|f\|_{p, Q_r}.$$

From this, by (2.15),

$$\|f\|_{ps,Q_r} \le Cr\|f'\|_{p,Q_r} + \|f\|_{p,Q_r}.$$

Finally, using the inequality $(a+b)^p \le 2^{p-1}(a^p + b^p)$ for $a, b \ge 0$ and $p \ge 1$, we get that

$$(\|f\|_{ps,Q_r})^p \le C(r^p \|f'\|_{p,Q_r}^p + \|f\|_{p,Q_r}^p),$$

i.e.,

$$\left(\fint_{Q_r} |f(x)|^{ps} dx \right)^{1/s} \le C \left(r^p \fint_{Q_r} |f'(x)|^p \, dx + \fint_{Q_r} |f(x)|^p \, dx \right). \quad (2.16)$$

LEMMA 2.2. *Let $v \ge 0$ be a function of class W_p^1 on the cube $Q_{r_0} = Q(a, r_0)$ in $\mathbf{R}^n$. Assume that there exist constants $C_1 > 0$, $C_1 < \infty$, and $l \in \mathbf{R}$, $l \ne p$, such that for any r_1 and r_2 with $0 < r_2 < r_1 \le r_0$*

$$\int_{Q_{r_2}} [v(x)]^{m-p} |v'(x)|^p \, dx \le \left(\frac{C_1}{|m-l|(r_1 - r_2)} \right)^p \int_{Q_{r_1}} [v(x)]^m \, dx. \quad (2.17)$$

Then v is bounded in any cube Q_r, where $0 < r < r_0$. Further,

$$\operatorname*{ess\,sup}_{x \in Q_r} v(x) \le C_2 \psi \left(\frac{r_0}{r} \right) \left(\fint_{Q_{r_0}} [v(x)]^p \, dx \right)^{1/p}, \quad (2.18)$$

where C_2 depends only on the constants C_1 and l in (2.17), and $\psi(u) = u^\gamma / (\ln u)^{s/(s-1)}$, where $\gamma = (n+p)/p$.

PROOF. Let r_1 and r_2 be arbitrary numbers with $0 < r_1 < r_2 \le r_0$. By the condition of the lemma, (2.17) holds for every $m > p$.

Let $m \ge p$ be such that for any m the right-hand side of (2.17) is finite. Let $w(x) = [v(x)]^{m/p}$. We show that $w \in W_p^1(U)$, and

$$w'(x) = (m/p)[v(x)]^{m/p-1} v'(x).$$

To do this we introduce the auxiliary function $w_k(x) = F_k[v(x)]$, where $F_k(v) = 0$ for $v < 0$, $F_k(v) = v^{m/p}$ for $0 \le v \le k$, and

$$F_k(v) = \frac{m}{p} k^{m/p-1}(v-k) + k^{m/p}$$

for $v > k$. The function F_k is piecewise smooth, and its derivative is bounded. By Lemma 5.8 in Chapter II, this implies that $w_k \in W_p^1(U)$. Further, $w_k'(x) = F_k'[v(x)]v'(x)$. We have that $F_k'(v) = 0$ for $v < 0$, $F_k'(v) = (m/p)v^{(m/p)-1}$ for $0 \le v \le k$, and $F_k'(v) = (m/p)k^{(m/p)-1}$ for $v \ge k$. The inequalities $0 \le w_k(x) \le w(x)$ and $|w_k'(x)| \le (m/p)[v(x)]^{(m/p)-1}|v'(x)|$ hold for each k. Since the right-hand side of (2.17) is assumed to be finite, this implies that the sequence (w_k) is bounded in $W_p^1(Q_{r_2})$. As $k \to \infty$ we have that $w_k(x) \to w(x)$ for all x. From this, $w \in W_p^1(Q_r)$. Further,

$$w'(x) = \lim_{k \to \infty} w_k'(x) = (m/p)[v(x)]^{(m/p)-1} v'(x).$$

In view of the foregoing, (2.17) gives us after obvious transformations that

$$\int_{Q_{r_2}} |w'(x)|^p \, dx \le \left(\frac{C_1 m}{p|m-l|\delta} \right)^p \int_{Q_{r_1}} |w(x)|^p \, dx,$$

where $\delta = r_1 - r_2$. For $s > 1$ let $\gamma(s)$ be the smallest value of $|ps^k - l|$, where k runs through all the nonnegative integers. Let γ_0 be the supremum of the numbers $\gamma(s)$ for s in the interval $(1, s_0)$, where $s_0 = n/(n-p)$. Obviously, $\gamma_0 > 0$. We choose an arbitrary value of s such that $1 < s < s_0$ and $\gamma(s) \ge \gamma_0/2$. Then $|ps^k - l| \ge \gamma_0/2$ for all integers $k \ge 0$. Denote by H the set of all numbers m of the form $m = ps^k$, $k \ge 0$, k an integer. As $m \to \infty$ we have that $m^p/|m-l|^p \to 1$, and $|m-l| \ge \gamma_0/2$ for all $m \in H$. This implies that the ratio $m^p/|m-l|^p$ is bounded on the set H. Consequently,

$$\int_{Q_{r_2}} |w'(x)|^p \, dx \le \frac{C}{\delta^p} \int_{Q_{r_1}} |w(x)|^p \, dx$$

for every $m \in H$ for $w = v^{m/p}$. Using inequality (2.16) for the chosen value of $s \in (1, s_0)$, we get that

$$\left(\overline{\int}_{Q_{r_2}} |w(x)|^{ps} \, dx \right)^{1/s} \le C'' \left(r_2^p \overline{\int}_{Q_{r_2}} [|w'(x)|^p + |w(x)|^p] \, dx \right)$$

$$\le \frac{C'''}{|Q_{r_2}|} \left\{ \frac{r_2^p}{(r_1-r_2)^p} \int_{Q_{r_1}} [w(x)]^p \, dx + \int_{Q_{r_2}} [w(x)]^p \, dx \right\}.$$

In the last term on the right-hand side we replace the domain of integration by Q_{r_1} (this does not decrease the integral), and get as a result that

$$\left(\overline{\int}_{Q_{r_2}} [w(x)]^{ps} \, dx \right)^{1/s} \le \varphi(r_1/r_2) \overline{\int}_{Q_{r_1}} [w(x)]^p \, dx,$$

where $C\rho^n[1 + 1/(\rho-1)^p] = \varphi(\rho)$. Considering that $w(x) = [v(x)]^{m/p}$, we get the inequality

$$\left(\overline{\int}_{Q_{r_2}} [v(x)]^{ms} \, dx \right)^{1/s} \le \varphi(r_1/r_2) \overline{\int}_{Q_{r_1}} [v(x)]^m \, dx. \qquad (2.19)$$

Here it is assumed that $m \in H$, i.e., $m = ps^k$, where $k \ge 0$ is an integer.

Fix an arbitrary value r, $0 \le r < r_0$. For $1 \le \rho \le r_0/r$ we have that

$$\varphi(\rho)(\rho-1)^p = C\rho^n[(\rho-1)^p + 1] \le C(r_0/r)^{n+p},$$

where $C =$ const. Let $h = \ln(r_0/r)$, and define $q_\nu = ps^\nu$ and $\rho_\nu = r\exp(h/2^\nu), \nu = 0, 1, 2, \ldots$. Then $\rho_0 = r_0$, $\rho_\nu > \rho_{\nu+1}$ for each ν, and $\rho_\nu \to r$ as $\nu \to \infty$. Let us write out (3.5), setting $r_1 = \rho_\nu$, $r_2 = \rho_{\nu+1}$, and $m = q_\nu$ in it. We have that

$$\frac{\rho_\nu}{\rho_{\nu+1}} = \exp\left(\frac{h}{2^\nu} - \frac{h}{2^{\nu+1}}\right) = \exp\frac{h}{2^{\nu+1}},$$

$$\varphi\left(\frac{\rho_\nu}{\rho_{\nu+1}}\right) \leq \frac{C(r_0/r)^{n+p}}{[(\rho_\nu/\rho_{\nu+1}) - 1]^p} = \frac{C(r_0/r)^{n+p}}{[\exp(h2^{-\nu-1}) - 1]^p}$$

$$\leq \frac{2^{(\nu+1)p}C}{h^p}\left(\frac{r_0}{r}\right)^{n+p}.$$

This leads to the conclusion that for each ν

$$\left(\overline{\int}_{Q_{\rho_{\nu+1}}} [v(x)]^{q_{\nu+1}}\, dx\right)^{1/s} \leq \frac{C2^{(\nu+1)p}}{h^p}\left(\frac{r_0}{r}\right)^{n+p}\overline{\int}_{Q_{\rho_\nu}} [v(x)]^{q_\nu}\, dx. \quad (2.20)$$

Let

$$\left(\overline{\int}_{Q_\nu} [v(x)]^{q_\nu}\, dx\right)^{1/q_\nu} = P_\nu.$$

Then it follows from (2.20) that for each $\nu = 0, 1, 2, \ldots$

$$P_{\nu+1} \leq \left[C\left(\frac{r_0}{r}\right)^{n+p} h^p\right]^{1/q_\nu} 2^{(\nu+1)p/q_\nu} P_\nu.$$

From this we conclude by induction that for all ν

$$P_\nu \leq \left[C\left(\frac{r_0}{r}\right)^{n+p} h^p\right]^{\sigma_\nu} 2^{\theta_\nu} P_0,$$

where

$$\sigma_\nu = \frac{1}{q_0} + \frac{1}{q_1} + \cdots + \frac{1}{q_{\nu-1}} \leq \sum_{\nu=0}^{\infty}\frac{1}{ps^\nu} = \frac{s}{p(s-1)},$$

$$\theta_\nu = \frac{p}{q_0} + \frac{2p}{q_1} + \cdots + \frac{\nu p}{q_{\nu-1}} \leq \sum_{\nu=0}^{\infty}\frac{\nu+1}{s^\nu} = \theta < \infty.$$

From this, for all ν,

$$P_\nu \leq C_2\left(\frac{r_0}{r}\right)^\gamma h^{-s/(s-1)} P_0, \quad (2.21)$$

where $C_2 =$ const and $\gamma = (n/p) + 1$. We now observe that

$$\lim_{\nu\to\infty} P_\nu = \lim_{\nu\to\infty}\left(\frac{1}{|Q_{\rho_\nu}|}\int_{Q_{\rho_\nu}} [v(x)]^{q_\nu}\, dx\right)^{1/q_\nu} = \operatorname*{ess\,sup}_{x\in Q_r} v(x).$$

Passing in (2.21) to the limit as $\nu \to \infty$, we thus get that

$$\operatorname*{ess\,sup}_{x \in Q_r} v(x) \le C_2 \psi(r_0/r) \left(\overline{\int}_{Q_{r_0}} [v(x)]^p \, dx \right)^{1/p},$$

where $\psi(u) = u^\gamma/(\ln u)^{s/(s-1)}$ and $\gamma = (n+p)/p$. This proves the lemma.

2.4. Boundedness of generalized solutions of equation (2.3). Harnack's inequality. It follows immediately from results in §§2.2 and 2.3 that every generalized solution of (2.3) is bounded. Namely, we have

LEMMA 2.3. *Let U be an open set in $\mathbf{R}^n$, and $u(x)$ a generalized solution of the equation* $\operatorname{div} A[x, u'(x)] = 0$ *in U, where the function A satisfies conditions* a) *and* b) *in §2.1. Then $u(x)$ is equivalent in the Lebesgue sense to a function $\tilde{u}(x)$ bounded on every set M lying strictly inside U. Further, for any cubes $Q(a, r)$ and $Q(a, r_0)$ with $0 < r < r_0$ and $Q(a, r_0)$ strictly inside U the estimate*

$$\operatorname*{ess\,sup}_{x \in Q(a,r)} |u(x)| \le C\psi\left(\frac{r_0}{r}\right) \left(\overline{\int}_{Q(a,r_0)} |u(x)|^p \, dx \right)^{1/p} \tag{2.22}$$

is valid, where C depends on n, p, and the constants a_1 and a_2 in conditions a) *and* b)*, and $\psi(u)$ depends only on n and p.*

PROOF. Let $u(x)$ be a generalized solution of (2.3) on U. Then, by Corollary 1 to Lemma 2.1, $u(x)$ satisfies the inequality

$$\int_M |u(x)|^{m-p} |u'(x)|^p \, dx \le \frac{M}{[(m-p+1)\delta]^p} \int_G [u(x)]^m \, dx,$$

where $m > p$ is arbitrary, $M \subset G$, M is closed, G is open, and

$$\delta = \inf_{x \in A} p(x, \mathbf{R}^n \backslash G).$$

Setting $M = \overline{Q}(a, r_2)$ and $G = Q(a, r_1)$, where $0 < r_2 < r_1 \le r_0$, we get that the function $v = |u|$ satisfies the conditions of Lemma 2.2, which imply the boundedness of $u(x)$ in $Q(a, r)$ and (2.22).

The set U can be represented as a union of countably many cubes lying strictly inside U. By modifying the values of u in each of these cubes on a set of measure zero, we can clearly get a function bounded on every compact subset $M \subset U$, and the lemma is proved.

LEMMA 2.4. *There exists a real function $C(\lambda)$ that is bounded in $(0, 1)$ and has the following property. Let $u(x)$ be an arbitrary nonnegative solution of (2.3) in the cube $Q(a, r_0) \subset \mathbf{R}^n$, where the vector-valued function A*

satisfies conditions a) *and* b) *of* §2.1. *For* $\lambda \in (0, 1)$ *let*

$$m(\lambda) = \operatorname*{ess\,inf}_{x \in Q(a, \lambda, r_0)} u(x),$$

$$M(\lambda) = \operatorname*{ess\,sup}_{x \in Q(a, \lambda, r_0)} u(x)$$

Then $M(\lambda) \le C(\lambda) m(\lambda)$. *For fixed* $\lambda \in (0, 1)$ *the quantity* $C(\lambda)$ *depends only on* n, p, *and the constants* a_1 *and* a_2 *in conditions* a) *and* b).

PROOF. Take a cube $Q(a, r_0) \subset \mathbf{R}^n$ and a generalized solution $u(x)$ of (2.3) on this cube, with $u(x) \ge 0$ almost everywhere on $Q(a, r_0)$. We introduce the following notation. If $\lambda \in (0, 1)$, then let $H_\lambda = Q(a, \lambda r_0)$. Take an arbitrary number λ with $0 < \lambda < 1$, and let $\lambda_1 = (1 + \lambda)/2$. Then $\lambda < \lambda_1 < 1$. For any cube $Q = Q(x, r)$ denote by $\hat{Q}$ the cube $Q(x, r/\lambda_1)$ concentric with Q. Note that if $Q \subset H_{\lambda_1}$, then $\hat{Q} \subset H_1 = Q(a, r_0)$. Take an arbitrary cube $Q = Q(x, r) \subset H_{\lambda_1}$. We use the inequality (2.8) with $M = \overline{Q}$ and $G = \hat{Q}$. For this pair of sets M and G we have that $\delta = (r/\lambda_1) - r = r(1 - \lambda)(1 + \lambda) = r\delta_1$, and we get that for any $m \le p$ with $m \ne p - 1$ and for any $h > 0$

$$\int_Q \frac{|u'(x)|^p}{[u(x) + h]^{p-m}} \, dx \le \frac{C}{[(p - m - 1)\delta_1 r]^p} \int_{\hat{Q}} [u(x) + h]^m \, dx. \tag{2.23}$$

Setting $m = 0$ here, we arrive at the inequality

$$\int_Q \left(\frac{|u'(x)|}{u(x) + h} \right)^p dx \le \frac{C}{[(p - 1)\delta_1]^p} \frac{|\hat{Q}|}{r^p} \frac{C_1 |Q|}{r^p}, \tag{2.24}$$

where $C_1 = C/\lambda_1^n (p - 1)^p \delta_1^p$.

Lemma 5.8 in Chapter II allows us to conclude that the function $w(x) = \ln[u(x) + h]$ is in $W_p^1(Q)$, and

$$w'(x) = u'(x)/[u(x) + h].$$

In view of this, (2.24) can be rewritten as follows:

$$\int_Q |w'(x)|^p \, dx \le C_1 |Q|/r^p.$$

Let $w_Q = \overline{\int}_Q w(x) \, dx$. Then, by Lemma 2.1 in Chapter I,

$$\int_Q |w(x) - w_Q|^p \, dx \le Cr^p \int_Q |w'(x)|^p \, dx \le C_2 |Q|,$$

i.e.,

$$\overline{\int}_Q |w(x) - w_Q|^p \, dx \le C_2, \tag{2.25}$$

where the constant C_2 depends only on p, n, a_1, a_2, and λ, and C_2 admits the estimate $C_2 \leq C/(1 - \lambda)^p$, where C now depends only on p, n, a_1, and a_2. Inequality (2.25) means that w is a function with bounded mean oscillation in the cube H_{λ_1}. By the corollary to Theorem 1.1 in §1 of this chapter, there exists a constant $\gamma > 0$ such that for every $\alpha < \gamma/M_2$ the function $v(x) = \exp \alpha |w(x) - w_Q|$ is integrable for any cube $Q \subset H_{\lambda_1}$, and

$$\overline{\int}_Q v(x)\, dx \leq 1 + \frac{e^\gamma}{\gamma - \alpha C_2} \overline{\int}_Q |w(x) - w_{H_{\lambda_1}}|^p\, dx.$$

Fix the value $\alpha = \min\{\gamma/2C_2, (p-1)/2\}$ here. Then, considering (2.25), we get

$$\overline{\int}_Q v(x)\, dx \leq C_3, \tag{2.26}$$

where C_3 depends only on n, p, λ, and the constants a_1 and a_2 in conditions a) and b) of §2.1. In particular, (2.26) gives us

$$\overline{\int}_Q \exp[\alpha w(x) - \alpha w_Q]\, dx \leq C_3,$$

$$\overline{\int}_Q \exp[-\alpha w(x) + \alpha w_Q]\, dx \leq C_3.$$

Multiplying these inequalities termwise, we get that

$$\overline{\int}_Q \exp[-\alpha w(x)]\, dx \overline{\int}_Q \exp[-\alpha w(x)]\, dx \leq C_3^2,$$

i.e.,

$$\overline{\int}_Q [u(x) + h]^\alpha\, dx \overline{\int}_Q [u(x) + h]^{-\alpha}\, dx \leq C_3^2. \tag{2.27}$$

This inequality holds for any $h > 0$ for any cube $Q \subset H_{\lambda_1}$.

Now let r_1 and r_2 be arbitrary numbers with $0 < r_2 < r_1 \leq r\lambda_1$, and use (2.8) once again, with $M = \overline{Q}_{r_2} = \overline{Q}(a, r_2)$ and $G = Q_{r_1} = Q(a, r_1)$. Then for every $m \neq p - 1$

$$\int_{Q_{r_2}} [u(x) + h]^{m-p} |u'(x)|^p\, dx \leq \frac{C}{[(r_1 - r_2)(p - 1 - m)]^p}$$

$$\times \int_{Q_{r_1}} [u(x) + h]^m\, dx. \tag{2.28}$$

We introduce an auxiliary function z defined by $z(x) = [u(x) + h]^{-\alpha/p}$. Since $u(x)$ is nonnegative, while $F(u) = (u + h)^{-\alpha/p}$ has a bounded continuous derivative on $[0, \infty)$, it follows that $z \in W_p^1$ and

$$z'(x) = -(\alpha/p)[u(x) + h]^{-(\alpha/p)-1}.$$

Substituting in (4.7) the expression for $u(x) + h$ in terms of $z(x)$, after setting $-\alpha m / p = q$ and making obvious transformations we get

$$\int_{Q_{r_2}} [z(x)]^{q-p} |z'(x)|^p \, dx \le \frac{C}{[(r_1 - r_2)(q + l)]^p} \int_{Q_{r_1}} |z(x)|^q \, dx, \qquad (2.29)$$

where $l = \alpha(p - 1)/p$. By (2.27), the function z is integrable to the power p on Q_r. On the basis of Lemma 2.3 this implies that z is bounded on every cube Q_r with $0 < r < r_0 \lambda_1$. Further,

$$\operatorname*{ess\,sup}_{x \in Q_r} z(x) \le C \psi \left(\frac{r_0 \lambda_1}{r} \right) \left(\overline{\int_{H_{\lambda_1}}} [z(x)]^p \, dx \right)^{1/p}, \qquad (2.30)$$

where $\psi(u) = O[1/u - 1)^{s/(s-1)}]$ as $u \to 1 + 0$, and C depends only on the constants C and l in (2.29). Let $r = r_0 \lambda$ in (2.30). Note further that

$$\operatorname*{ess\,sup}_{x \in Q_r} z(x) = [\operatorname*{ess\,inf}_{x \in Q_r} u(x) + h]^{-\alpha/p},$$

which, by (2.30), gives us that

$$\operatorname*{ess\,inf}_{x \in Q_r} u(x) + h \ge C(\lambda) \Big/ \left(\overline{\int_{H_{\lambda_1}}} [u(x) + h]^{-\alpha} \, dx \right)^{1/\alpha}$$

$$\ge C_2(\lambda) \left(\overline{\int_{H_{\lambda_1}}} [u(x) + h]^{\alpha} \, dx \right)^{1/\alpha}. \qquad (2.31)$$

We again use Corollary 2 to Lemma 2.1. For every $m \ne p - 1$ and any concentric cubes Q_{r_1} and Q_{r_2}, where $0 < r_2 < r_1$, we have that

$$\int_{Q_{r_1}} [u(x) + h]^{m-p} |u'(x)|^p \, dx \le \frac{C}{(\delta |m - p + 1|)^p}$$

$$\times \int_{Q_{r_2}} [u(x) + h]^m \, dx. \qquad (2.32)$$

Let $(u(x) + h)^{\alpha/p} = w(x)$. Since $\alpha < p, \alpha/p < 1$. The function $F(u) = (u + h)^{\alpha/p}$ is continuously differentiable in $[0, \infty)$, and its derivative is bounded. This implies that $w \in W_p^1(Q_{r_1})$, and

$$w'(x) = (\alpha/p)[u(x) + h]^{(\alpha/p)-1} u'(x).$$

Substituting in (2.32) the expression for $u(x) + h$ in terms of $w(x)$, we get after obvious transformations that

$$\int_{Q_{r_1}} [w(x)]^{h-p} |w'(x)|^p \, dx \le \frac{C}{(\delta |k - (p - 1)p/\alpha|)^p}$$

$$\times \int_{Q_{r_2}} [w(x)]^k \, dx,$$

where $k = pm/\alpha$. The last inequality holds for every $k \geq p$. On the basis of Lemma 2.2, this implies that w is bounded on every cube Q_r with $0 < r < r_0\lambda_1$. By (2.18),

$$\operatorname*{ess\,sup}_{x\in H_\lambda} w(x) \leq C\psi\left(\frac{\lambda_2}{\lambda}\right)\left(\overline{\int}_{H_{\lambda_1}} [w(x)]^p\,dx\right)^{1/p}$$

$$= C\left(\overline{\int}_{H_{\lambda_1}} [u(x)+h]^\alpha\,dx\right)^{1/p}. \tag{2.33}$$

We now observe that

$$\operatorname*{ess\,sup}_{x\in H_\lambda} w(x) = \left(\operatorname*{ess\,sup}_{x\in H_\lambda}[u(x)+h]\right)^{\alpha/p}.$$

Substituting the estimate in (2.33) into (2.31) and setting $r = \lambda r$ in (2.31), we get that

$$h + \operatorname*{ess\,sup}_{x\in H_\lambda} u(x) \leq C(\lambda)\left[\operatorname*{ess\,sup}_{x\in H_\lambda} u(x) + h\right].$$

The number $h > 0$ is arbitrary. Letting h go to zero, we arrive at the inequality of the lemma.

COROLLARY 1. *Every generalized solution u of (2.3) is continuous and satisfies a Hölder condition on every set V lying strictly inside its domain U.*

PROOF. Let u be a generalized solution of (2.3) in an open set $U \subset \mathbf{R}^n$. Let $a \in U$ be arbitrary, and let $\delta > 0$ be such that $Q(a,r) \subset U$ for $0 < r < \delta$. By Lemma 2.3, $u(x)$ is bounded on every such cube. Let

$$M(r) = \operatorname*{ess\,sup}_{x\in Q(a,r)} u(x), \quad m(r) = \operatorname*{ess\,sup}_{x\in Q(a,r)} u(x).$$

The functions $u(x) - m(r)$ and $M(r) - u(x)$ are generalized solutions of (2.3). In view of the lemma, this implies that

$$M(r/2) - m(r) \leq C[m(r/2) - m(r)],$$
$$M(r) - m(r/2) \leq C[M(r) - M(r/2)],$$

where $C > 1$. Multiplying these inequalities termwise, we get that

$$(C + 1)[M(r/2) - m(r/2)] \leq (C - 1)[M(r) - m(r)],$$

which implies that

$$M\left(\frac{r}{2}\right) - m\left(\frac{r}{2}\right) \leq [(C-1)/(C+1)][M(r) - m(r)].$$

Let $\omega(r)$ denote the oscillation $M(r) - m(r)$ of $u(x)$ in the cube $Q(a, r)$. We have that $\omega(r/2) \leq \theta\omega(r)$, where $\theta = (C - 1)/(C + 1) < 1$. This gives us by induction that

$$\omega(r2^{-\nu}) \leq \theta^{\nu}\omega(r), \qquad \nu = 1, 2, \ldots.$$

The last inequality allows us to conclude that

$$\omega(\rho) \leq \theta^{-1}(\rho/r)^{\alpha}\omega(r),$$

where $\alpha = (\ln 1/\theta)/\ln 2$, $0 < \rho < r$, and it is thereby proved that every generalized solution of (2.3) satisfies a Hölder condition.

COROLLARY 2. *Let* $U \subset \mathbf{R}^n$ *be an open set. Then for every compact set* $M \subset U$ *there exists a constant* $C > 1$ *such that for any nonnegative solution* $u(x)$ *of* (2.3) *on* U

$$\max_{x \in M} u(x) \leq C \min_{x \in M} u(x). \tag{2.34}$$

The proof of the corollary is based on the Borel theorem, and is left to the reader because it is clear.

The *Harnack constant* of the pair (M, U) of sets is defined to be the smallest number $C \geq 1$ such that (2.34) holds for any equation of the form (2.3) with A satisfying conditions a) and b) of §2.1 and any nonnegative generalized solution of (2.3) on U. By definition, this constant C depends only on the constants $p > 1$ and $a_1, a_2 > 0$ in conditions a) and b) of §2.1 and on the choice of the pair (M, U). The next statement is used in the study of mappings with bounded distortion.

LEMMA 2.5. *Let the pairs* (M_1, U_1) *and* (M_2, U_2) *be given, where* U_1 *and* U_2 *are open,* M_1 *and* M_2 *are compact, and* $U_i \supset M_i$ *for* $i = 1, 2$. *If there exists a similarity* f *of* $\mathbf{R}^n$ *such that* $f(U_1) = U_2$ *and* $f(M_1) = M_2$, *then the Harnack constants of the pairs* (M_1, U_1) *and* (M_2, U_2) *coincide.*

PROOF. Let $f'(x) = l$ and $k = |l|$. Assume that the equation $\operatorname{div} A[x, u'(x)] = 0$ is given in U_2, where the vector-valued function A satisfies conditions a) and b) of §2.1. Let

$$\tilde{A}(x, q) = 1/k^{p-1} \cdot A[x, l(q)].$$

The vector-valued function $\tilde{A}(x, q)$ is defined for almost all $x \in U_1$ and any $q \in \mathbf{R}^n$, and satisfies conditions a) and b) of §2.1 with the same values of the constants a_1 and a_2. If $u(x)$ is a solution of the equation $\operatorname{div} A[x, u'(x)] = 0$, then $\tilde{u}(x) = u[f(x)]$ is a solution of $\operatorname{div} \tilde{A}[x, u'(x)] = 0$. The conclusion of the lemma follows in an obvious way from these remarks.

§3. Theorems on semicontinuity and convergence with a functional for functionals of the calculus of variations

3.1. Weak convergence of sequences of functions in measure spaces. Let us consider functions defined on a locally compact metric space $\mathfrak{R}$ and taking values in $\mathbf{R}^l$. It is assumed that a measure μ is given on the σ-algebra of Borel subsets of $\mathfrak{R}$, with $\mu(\mathfrak{R}) < \infty$. A function $u: \mathfrak{R} \to \mathbf{R}^l$ is said to belong to the class $L_p(\mathfrak{R}, \mu)$ $(p \geq 1)$ if

$$\|u\|_{p,\mathfrak{R}} = \left(\int_{\mathfrak{R}} |u(x)|^p \mu(dx) \right)^{1/p} < \infty.$$

We introduce some more notation. Let $C_0(\mathfrak{R})$ denote the collection of all compactly supported real functions on $\mathfrak{R}$, and $S(\mathfrak{R})$ the set of indicator functions of Borel subsets of $\mathfrak{R}$ with compact closures.

Let $p > 1$ and let $(u_m: \mathfrak{R} \to \mathbf{R}^l)$, $m = 1, 2, \ldots$, be an arbitrary sequence of functions in $L_p(\mathfrak{R}, \mu)$. The sequence (u_m) is said to converge weakly in $L_p(\mathfrak{R})$ to a function $u_0 \in L_p(\mathfrak{R})$ if

$$\int_{\mathfrak{R}} \varphi(x) u_m(x) \mu(dx) \to \int_{\mathfrak{R}} \varphi(x) u_0(x) \mu(dx)$$

as $m \to \infty$ for every $\varphi: \mathfrak{R} \to R$ of class $L_q(\mathfrak{R})$, where $q = p/(p-1)$. Assume that $(u_m: \mathfrak{R} \to \mathbf{R}^l)$, $m = 1, 2, \ldots$, is a sequence of functions in $L_1(\mathfrak{R})$. Then the sequence (u_ν), $\nu = 1, 2, \ldots$, is said to converge weakly to u_0 in $L_1(\mathfrak{R})$ if

$$\int_{\mathfrak{R}} u_\nu(x) \varphi(x) \mu(dx) \to \int_{\mathfrak{R}} u_0(x) \varphi(x) \mu(dx)$$

as $\nu \to \infty$ for every function $\varphi \in C_0(\mathfrak{R})$.

We describe a general situation that encompasses various types of weak convergence. Let K be a class of B-measurable bounded real functions on $\mathfrak{R}$. For $u \in L_1(\mathfrak{R})$, $u: \mathfrak{R} \to \mathbf{R}^l$, and $\varphi \in K$ we set

$$\langle u, \varphi \rangle = \int_{\mathfrak{R}} \varphi(x) u(x) \mu(dx).$$

Let $(u_\nu: \mathfrak{R} \to \mathbf{R}^l)$, $\nu = 1, 2, \ldots$, be a sequence of functions of class $L_1(\mathfrak{R}, \mu)$. The sequence (u_ν), $\nu = 1, 2, \ldots$, is said to *converge in the w_K-sense* to a function $u_0 \in L_1(\mathfrak{R}, \mu)$ if:

1) the norm sequence $(\|u_\nu\|_{p,\mathfrak{R}})$, $\nu = 1, 2, \ldots$, is bounded, and
2) $\langle u_m, \varphi \rangle \to \langle u_0, \varphi \rangle$ as $m \to \infty$ for every $\varphi \in K$.

We mention some simple properties of convergence in the w_K-sense. Denote by K^* the collection of all linear combinations of functions of the class K. Let $Z_p(K)$ $(1 \leq p < \infty)$ be the closure of the set K^* in the space $L_p(\mathfrak{R}, \mu)$. Denote by $Z_\infty(K)$ the collection of all functions that are limits of uniformly convergent sequences of functions in K^*.

LEMMA 3.1. *Let* $(u_\nu : \mathfrak{R} \to \mathbf{R}^l)$, $\nu = 1, 2, \ldots$, *be the sequence of* $L_1(\mathfrak{R}, \mu)$-*functions converging in the* w_K-*sense to a function* $u_0 \in L_1(\mathfrak{R}, \mu)$. *Then* $u_\nu \to u_0$ *also in the* $w_{Z_\infty(K)}$-*sense. Suppose that* $p > 1$ *and* $q = p/(p-1) > 1$. *If the functions* u_ν *are all in* $L_p(\mathfrak{R}, \mu)$ *and the sequence of norms* $\|u_\nu\|_{p,\mathfrak{R}}$ *is bounded, then* $u_\nu \to u_0$ *in the* $w_{Z_q(K)}$-*sense.*

PROOF. Suppose that $u_\nu \to u_0$ in the w_K-sense, i.e., $\langle u_\nu, \varphi \rangle \to \langle u_0, \varphi \rangle$ for every $\varphi \in K$. Then it is obvious that $\langle u_\nu, \varphi \rangle \to \langle u_0, \varphi \rangle$ also for every function that is a linear combination of functions in K, i.e., $u_\nu \to u_0$ in the w_{K^*}-sense. Let $\varphi \in Z_\infty(K)$, and let $M < \infty$ be such that $\|u_\nu\|_{1,\mathfrak{R}} \le M$ for all ν. Take an arbitrary $\varepsilon > 0$ and choose $\varphi_1 \in K^*$ such that $|\varphi_1(x) - \varphi(x)| < \varepsilon/3M$ for all x. Then

$$|\langle u_\nu, \varphi \rangle - \langle u_0, \varphi \rangle| \le |\langle u_\nu, \varphi \rangle - \langle u_\nu, \varphi_1 \rangle|$$
$$+ |\langle u_\nu, \varphi_1 \rangle - \langle u_0, \varphi_1 \rangle| + |\langle u_0, \varphi_1 \rangle - \langle u_0, \varphi \rangle|.$$

It remains to observe that

$$|\langle u_\nu, \varphi \rangle - \langle u_\nu, \varphi_1 \rangle| \le M\varepsilon/3M = \varepsilon/3, |\langle u_0, \varphi \rangle - \langle u_0, \varphi_1 \rangle| \le M\varepsilon/3M,$$

and $\langle u_\nu, \varphi_1 \rangle \to \langle u_0, \varphi_1 \rangle$ as $\nu \to \infty$, since $\varphi_1 \in K^*$. This leads to the conclusion that $|\langle u_\nu, \varphi \rangle - \langle u_0, \varphi \rangle| < \varepsilon$ for sufficiently large ν, i.e., $\langle u_\nu, \varphi \rangle \to \langle u_0, \varphi \rangle$ as $\nu \to \infty$.

Assume that the norm sequence $(\|u_\nu\|_{p,\mathfrak{R}})$ is bounded, where $p > 1$, and let $q = p/(p-1)$. Suppose that $\|u_\nu\|_{p,\mathfrak{R}} \le M$ for all ν. Let $\varphi \in Z_q(K)$ be arbitrary, and for a given $\varepsilon > 0$ let $\varphi_1 \in K^*$ be such that $\|\varphi_1 - \varphi\|_{q,\mathfrak{R}} < \varepsilon/3M$. Arguing as in the preceding case, we then get that

$$|\langle u_\nu, \varphi \rangle - \langle u_0, \varphi \rangle| \le |\langle u_\nu, \varphi \rangle - \langle u_\nu, \varphi_1 \rangle|$$
$$+ |\langle u_\nu, \varphi_1 \rangle - \langle u_0, \varphi_1 \rangle| + |\langle u_0, \varphi_1 \rangle = \langle u_0, \varphi \rangle|$$
$$\le \|u_\nu\|_{p,\mathfrak{R}} \|\varphi - \varphi_1\|_{q,\mathfrak{R}} + |\langle u_\nu, \varphi_1 \rangle| - \langle u_0, \varphi_1 \rangle|$$
$$+ \|u_0\|_{p,\mathfrak{R}} \|\varphi - \varphi_1\|_{q,\mathfrak{R}} < (2/3)\varepsilon + |\langle u_\nu, \varphi_1 \rangle - \langle u_0, \varphi_1 \rangle|.$$

We have that $\langle u_\nu, \varphi_1 \rangle - \langle u_0, \varphi_1 \rangle \to 0$ as $\nu \to \infty$, and this implies that there exists a ν_0 such that $|\langle u_\nu, \varphi \rangle - \langle u_0, \varphi \rangle| < \varepsilon$ for $\nu > \nu_0$. This establishes that $\langle u_\nu, \varphi \rangle \to \langle u_0, \varphi \rangle$ as $\nu \to \infty$, and the lemma is proved.

LEMMA 3.2. *Let* (u_ν), $\nu = 1, 2, \ldots$, *be a sequence of* $L_1(\mathfrak{R}, \mu)$-*functions converging in the* w_K-*sense to a function* u_0, *and let* (φ_ν), $\nu = 1, 2, \ldots$, *be a uniformly convergent sequence of functions in* $Z_\infty(K)$, *with* $\varphi_0 = \lim_{\nu \to \infty} \varphi_\nu$. *Then* $\langle u_\nu, \varphi_\nu \rangle \to \langle u_0, \varphi_0 \rangle$ *as* $\nu \to \infty$.

PROOF. Let

$$\delta_\nu = \sup_{x \in \mathfrak{R}} |\varphi_\nu(x) - \varphi_0(x)|.$$

Then $\delta_\nu \to 0$ as $\nu \to \infty$. For each ν,

$$|\langle u_\nu, \varphi_\nu \rangle - \langle u_0, \varphi_0 \rangle| \leq |\langle u_\nu, \varphi_\nu - \varphi_0 \rangle|$$
$$+ |\langle u_\nu, \varphi_0 \rangle - \langle u_0, \varphi_0 \rangle|$$
$$\leq \delta_\nu \|u_\nu\|_{1,\mathfrak{R}} + |\langle u_\nu, \varphi_0 \rangle - \langle u_0, \varphi_0 \rangle|.$$

Then $\delta_\nu \to 0$ and $\langle u_\nu, \varphi_0 \rangle \to \langle u_0, \varphi_0 \rangle$ as $\nu \to \infty$, because $\varphi_0 \in Z_\infty(K)$. This leads to the conclusion that $\langle u_\nu, \varphi_\nu \rangle \to \langle u_0, \varphi_0 \rangle$ as $\nu \to \infty$, which is what was required to prove.

We remark that if $K = C_0(\mathfrak{R})$, and also if $K = S(\mathfrak{R})$, then $Z_q(K) = L_q(\mathfrak{R}, \mu)$ for $q > 1$. This yields the following assertion.

COROLLARY. *Let* $(u_\nu : \mathfrak{R} \to \mathbf{R}^l)$, $\nu = 1, 2, \ldots$, *be a sequence of* $L_p(\mathfrak{R}, \mu)$-*functions such that the norm sequence* $(\|u_\nu\|_{p,\mathfrak{R}})$ *is bounded. If* (u_ν) *converges weakly in* L_1 *to a function* $u_0 \in L_p(\mathfrak{R}, \mu)$, *then* $u_\nu \to u_0$ *weakly in* L_p. *If the sequence* (u_ν), $\nu = 1, 2, \ldots$, *converges to* u_0 *in the* $w_{S(\mathfrak{R})}$-*sense, then* (u_ν) *converges weakly to* u_0 *in* $L_p(\mathfrak{R}, \mu)$.

3.2. Some lemmas about convex functions

LEMMA 3.3. *Let* U *be a convex open subset of* $\mathbf{R}^l$, *and let* (F_m), $m = 1, 2, \ldots$, *be a sequence of convex functions on* U *such that there is a finite limit* $\lim_{m \to \infty} F_m(x) = F_0(x)$ *for all* $x \in U$. *The function* F_0 *defined in this way is convex, and* $F_m \to F_0$ *locally uniformly in* U.

PROOF. Obviously F_0 is convex. Let $Q = Q(a, t)$ be an arbitrary closed cube contained in U, with vertices $u_1, \ldots, u_r$ $(r = 2^l)$. Every point $x \in Q$ can clearly be represented in the form $x = \lambda_1 u_1 + \cdots + \lambda_r u_r$, where λ_i is nonnegative and $\lambda_1 + \cdots + \lambda_r = 1$. Using Jensen's inequality, we get that

$$F_m(x) \leq \sum_{i=1}^{r} \lambda_i F_m(u_i).$$

The sequences $(F_m(u_i))$ are bounded. Let $F_m(u_i) \leq M$ for all m and any $i = 1, \ldots, r$. It is then obvious that $F_m(x) \leq M$ for all $x \in Q$.

We show that the functions F_m are also bounded below on Q. The sequence $(F_m(a))$ (a is the center of Q) is clearly bounded. Suppose that $|F_m(a)| \leq M_1$ for all m. Let L_m be an affine function such that $L_m(a) = F_m(a)$ and $L_m(x) \leq F_m(x) \leq M$ for all $x \in Q$. Then

$$L_m(x) = \langle p_m, x - a \rangle + F_m(a).$$

We prove that $|p_m| \leq 2(M + M_1)/t$. Indeed, if $|p_m| = 0$, then this is obvious. Let $|p_m| \neq 0$. Setting $x = a + te/2$, where $e = p_m/|p_m|$, we get that

$$L_m(x) = (t/2)\langle p_m, e \rangle \leq M - F_m(a) \leq M + M_1,$$

which implies that $t|p_m|/2 \leq M + M_1$, as required. It is obvious that

$$L_m(x) \geq -M_1 - |p_m||x - a|,$$

for all $x \in Q$, which leads to the conclusion that $L_m(x) \geq M_0$ for all $x \in Q(a, r)$, where $M_0 = $ const, and hence $M_0 \leq F_m(x) \leq M$ for all $m = 1, 2, \ldots$ and all $x \in Q$.

We now consider the cube $Q_1 = Q(a, t/2)$ concentric with Q and having sides half as long as those of Q. Let $x_1, x_2 \in Q_1$ be two arbitrary points in Q_1, and let K_m be an affine function such that $K_m(x_1) = F_m(x_1)$, and $K_m(x) \leq F_m(x)$ for all $x \in U$. We have that

$$K_m(x) = \langle q_m, x - x_1 \rangle + F_m(x_1), \qquad F_m(x_1) \geq M_0.$$

Suppose that $q_m \neq 0$. Let $x = x_1 + te/4$, where $e = (1/|q_m|)q_m$. Obviously, $x \in Q$, and this gives us that

$$t|q_m|/4 = \langle q_m, x - x_1 \rangle \leq M - F_m(x_0),$$

which implies that $|q_m| \leq 4(M - M_0)/t$. This inequality is true also when $q_m = 0$. We have that

$$F_m(x_2) - F_m(x_1) \geq K_m(x_2) - K_m(x_1) = \langle q_m, x_2 - x_1 \rangle \geq -L|x_2 - x_1|,$$

where $L = 4(M - M_0)/t$. Interchanging the roles of x_1 and x_2, we get in exactly the same way that

$$F_m(x_1) - F_m(x_2) \geq -L|x_1 - x_2|,$$

which leads to the conclusion that

$$|F_m(x_1) - F_m(x_2)| \leq L|x_1 - x_2|$$

for any points $x_1, x_2 \in Q_1$.

It follows from what has been proved that the sequence (F_m) of functions is uniformly equicontinuous in Q_1, and hence $F_m \to F_0$ uniformly in Q_1.

Thus, every point $a \in U$ has a neighborhood in which $F_m \to F_0$ uniformly. This obviously implies that $F_m \to F_0$ uniformly on every compact set $A \subset U$. The lemma is proved.

We now describe a construction for approximating convex functions on $\mathbf{R}^l$ by affine functions.

Let F be an arbitrary function on a set $E \subset \mathbf{R}^l$. Denote by A_F the set of all points $(x, z) \in \mathbf{R}^{l+1}$ such that $x \in E$ and $z \geq F(x)$. The set A_F is called the *supergraph* of F.

Let $E = \mathbf{R}^l$. If the function $F: \mathbf{R}^l \to \mathbf{R}$ is continuous, then A_F is closed. The function $F: \mathbf{R}^l \to \mathbf{R}$ is convex if and only if A_F is a convex set.

Let H be an arbitrary closed convex set in $\mathbf{R}^m$, and let a be a point not in H. Denote by b the point in H closest to a. Such a point exists because H is closed. The point b closest to a is unique in view of the convexity of H. Let $p = b - a$ and $c = \frac{1}{2}(a + b)$, and let $L(x) = \langle x - c, p \rangle$. The equation $L(x) = 0$ determines a plane P in $\mathbf{R}^{l+1}$. This plane passes through the midpoint c of $[a, b]$ and is perpendicular to $[a, b]$. For all $x \in H$ we have that $L(x) > 0$, while $L(a) < 0$. Intuitively, this means that P separates a from the set H, with a on one side of this plane and H on the other.

Let $F: \mathbf{R}^l \to \mathbf{R}$ be a convex function. Take an arbitrary point $x_0 \in \mathbf{R}^l$ and a number $\varepsilon > 0$. Consider the points $b_1 = (x_0, F(x_0))$ and $a_1 = (x_0, F(x_0) - \varepsilon)$ in $\mathbf{R}^{l+1}$. Obviously, $a_1 \notin A_F$. Let b_2 be the point of A_F closest to a_1. Let P be the plane through the midpoint of $[a_1, b_2]$ and perpendicular to this segment. The plane P is defined in $\mathbf{R}^{l+1}$ by an equation $L(x, z) = \langle h, x \rangle + kz + B = 0$ with $h \in \mathbf{R}^l$ and $B, k \in \mathbf{R}$, and $L(x, z) > 0$ for all $(x, z) \in A_F$, while $L(x_0, z_0 - \varepsilon) < 0$, where $z_0 = F(x_0)$. This implies that $k < 0$, and the equation of P can be rewritten in the form $z = \langle p, x \rangle + c$. The condition that $L(x, z) > 0$ for all $(x, z) \in A_F$ and the condition that $L(x, z) < 0$ for $(x, z) = a_1$ give us that $F(x) > \langle p, x \rangle + c$ for all $x \in \mathbf{R}^l$, and $F(x_0) - \varepsilon < \langle p, x_0 \rangle + c$. The vector p and the number c are uniquely determined by F, the point x_0, and the number $\varepsilon > 0$. The notation $p = p_F(x_0, \varepsilon)$ and $c = c_F(x_0, \varepsilon)$ will be used in this connection.

LEMMA 3.4. *Let* $(F_m), m = 1, 2, \ldots$, *be a sequence of convex functions on* $\mathbf{R}^l$ *that converges to a convex function* F_0 *on* $\mathbf{R}^l$ *as* $m \to \infty$, *and let* $(x_m), m = 1, 2, \ldots$, *be a sequence of points in* $\mathbf{R}^l$ *converging to* x_0 *as* $m \to \infty$. *Then*

$$p_{F_m}(x_m, \varepsilon) \to p_{F_0}(x_0, \varepsilon),$$
$$c_{F_m}(x_m, \varepsilon) \to c_{F_0}(x_0, \varepsilon)$$

as $m \to \infty$.

PROOF. Let $a_m = (x_m, F_m(x_m) - \varepsilon)$, and let b_m be the point of A_{F_m} closest to a_m. It obviously suffices to establish that $a_m \to a = (x_0, F_0(x_0) - \varepsilon)$ and $b_m \to b$ as $m \to \infty$, where b is the point of A_{F_0} closest to a. In view of Lemma 2.3, F_m converges to F_0 uniformly on every compact subset $E \subset \mathbf{R}^l$. This implies that $a_m \to a$ as $m \to \infty$. Let $b' = (x_0, F(x_0)), b'_m(x_0, F_m(x_0))$. This it is clear that $b'_m \to b$ as $m \to \infty$.

For each m we have that $|b'_m - a_m| \geq |b_m - a_m|$, because b_m is the point of A_{F_m} closest to a_m; $|b'_m - a_m| \to |b - a|$ as $m \to \infty$, and hence the sequence $(|b'_m - a_m|), m = 1, 2, \ldots$, of distances is bounded. This implies that the sequence $(|b_m - a_m|)$, $m = 1, 2, \ldots$, is also bounded, and hence (b_m), $m = 1, 2, \ldots$, is a bounded sequence. Let (b_{m_k}), $m_1 < m_2 < \ldots$, be an arbitrary convergent subsequence of this sequence, and let $b^* = \lim_{k \to \infty} b_{m_k}$. Since $F_m \to F$ uniformly on every bounded subset of $\mathbf{R}^l$, it follows that $b^* = (y, F(y))$ for some $y \in \mathbf{R}^l$. This implies that $|b^* - a| \geq |b - a|$. On the other hand,

$$|b_{m_k} - a| \leq |b'_{m_k} - a_{m_k}|,$$

for each k, and

$$|b'_{m_k} - a_{m_k}| \to |b - a|$$

as $k \to \infty$; by passage to the limit this implies that $|b^* - a| \leq |b - a|$. Hence, $|b^* - a| = |b - a|$, and since the point of A_F closest to a is unique, it follows that $b^* = b$. We thus get that any convergent subsequence of (b_m) converges to b. Hence, $b_m \to b$ as $m \to \infty$. This obviously proves the lemma.

3.3. Theorems about semicontinuity of functionals of the calculus of variations. Below, $\mathfrak{R}$ denotes a locally compact metric space, and μ a measure on the σ-ring of all Borel subsets of $\mathfrak{R}$ such that $\mu(\mathfrak{R}) < \infty$.

THEOREM 3.1 [27]. *Let $(F_m \colon \mathfrak{R} \times \mathbf{R}^l \to \mathbf{R})$, $m = 0, 1, 2, \ldots$, be a sequence of nonnegative continuous functions such that $F_m \to F_0$ locally uniformly in $\mathfrak{R} \times \mathbf{R}^l$ as $m \to \infty$. If for each m the function $q \to F_m(t, q)$ is convex for every $t \in \mathfrak{R}$, then every sequence $(u_m \colon \mathfrak{R} \to \mathbf{R}^l)$, $m = 1, 2, \ldots$, of functions of the class $L_1(\mathfrak{R}, \mu)$ that converges in the w_{C_0}-sense to a function $u_0 \colon \mathfrak{R} \to \mathbf{R}^l$, $u_0 \in L_1(\mathfrak{R}, \mu)$, satisfies the inequality*

$$\int_{\mathfrak{R}} F_0[x, u_0(x)]\mu(dx) \leq \overline{\lim_{m \to \infty}} \int_{\mathfrak{R}} F_m[x, u_m(x)]\mu(dx).$$

PROOF. For any set $A \subset \mathfrak{R}$ let

$$\int_A F_m[x, u_m(x)]\mu(dx) = (u_m, F_m, A), \qquad m = 0, 1, 2, \ldots.$$

Let $(u_m, F_m, \mathfrak{R}) = I_m$, $m = 0, 1, 2, \ldots$.

Given an arbitrary $\varepsilon > 0$, we construct a compact set $K \subset \mathfrak{R}$ such that u_0 is continuous on K and $(u_0, F_0, K) > I_0 - \varepsilon$ in the case when $I_0 < \infty$, and $(u, F_0, K) > 1/\varepsilon$ when $I_0 = \infty$.

We construct a decreasing sequence of open sets $U_1 \supset U_2 \supset \ldots$ such that the closure of each is compact, and their intersection coincides with K. Let $\varphi_\nu \colon \mathfrak{R} \to \mathbf{R}$ be a continuous function such that $\varphi_\nu(x) = 0$ for $x \notin U_\nu$, $\varphi_\nu(x) = 1$ for $x \in K$, and $0 \leq \varphi_\nu(x) \leq 1$ for all x. Let $u_0 = (u_{01}, \ldots, u_{0l})$, where the u_{0i} are real functions. For each i we construct according to the

Urysohn extension theorem a continuous function $v_i: \mathfrak{R} \to \mathbf{R}$ such that $v_i(x) = u_{0i}(x)$ for $x \in K$ and

$$|v_i(x)| \le \max_{x \in K} |u_{0i}(x)|,$$

and let $v(x) = (v_1(x), \dots, v_l(x))$. Then

$$|v(x)| \le l \max_{x \in K} |u(x)|$$

for all $x \in \mathfrak{R}$.

For each $m = 0, 1, \dots$ and each $x \in \mathfrak{R}$ we have a convex function $F_{m,x}: q \to F_m(x, q)$ of the variable $q \in \mathbf{R}^l$ and a point $v = v(x)$. Given an arbitrary $\eta > 0$, we construct a vector $p = p_{F_{m,x}}$ and a number $c = c_{F_{m,x}}$ corresponding to the function $F_{m,x}$, the point $v = v(x)$, and the number η in accord with the constructions of §2.1. Let $p_m(x) = p_{F_{m,x}}$ and $c_m(x) = c_{F_{m,x}}$. Then $F_m(x, q) \ge \langle p_m(x), q \rangle + c_m(x)$ for all m, x, and q, and

$$F_0[x, v(x)] - \eta < \langle p_0(x), v(x) \rangle + c_0(x). \tag{3.1}$$

Each of the functions p_m and c_m is continuous. Indeed, let (x_k), $k = 1, 2, \dots$, be an arbitrary convergent sequence of points of $\mathfrak{R}$, with limit x_0. We have that

$$v(x_k) \to v(x_0), \qquad F_m(x_k, q) \to F_m(x_0, q)$$

as $k \to \infty$, and this implies in view of Lemma 3.4 that $p_m(x_k) \to p_m(x_0)$ and $c_m(x_k) \to c_m(x_0)$ as $k \to \infty$. This proves that p_m and c_m are continuous.

By arguing similarly it can be proved that

$$\lim_{m \to \infty} p_m(x_m) = p_0(x_0), \qquad \lim_{m \to \infty} c_m(k_m) = c_0(x_0)$$

for every convergent sequence (x_m), $m = 1, 2, \dots$, with $x_m \to x_0$ as $m \to \infty$. This implies that p_m and c_m converge as $m \to \infty$ to functions p_0 and c_0 uniformly on every compact subset of $\mathfrak{R}$.

Let $q_{m,\nu}(x) = \varphi_\nu(x) p_m(x)$ and $d_{m,\nu}(x) = \varphi_\nu(x) c_m(x)$. By the foregoing, the functions $q_{m,\nu}$ and $d_{m,\nu}$ are continuous and compactly supported, and $q_{m,\nu} \to q_{0,\nu}$ and $d_{m,\nu} \to d_{0,\nu}$ uniformly on $\mathfrak{R}$ as $m \to \infty$. It is obvious that for each m

$$F_m(x, u) \ge \langle q_{m,\nu}(x), u \rangle + d_{m,\nu}(x).$$

Note that $q_{0,\nu}(x) = p_0(x)$ for $x \in K$. Integrating (3.1) termwise, we get that

$$(u_0, F_0, K) \le \int_K \{\langle p_0(x), u_0(x) \rangle + c_0(x)\} \, \mu(dx) + \eta \mu(K).$$

The number $\eta > 0$ is arbitrary. It will be assumed that η is chosen so that $(u_0, F_0, K) - \eta\mu(K) > I_0 - \varepsilon$ when $I_0 < \infty$, and $(u_0, F_0, K) - \eta\mu(K) > 1/\varepsilon$ when $I_0 = \infty$. (The quantity (u_0, F_0, K) is clearly finite, because K is compact, F is continuous, and u_0 is continuous on K.)

As $\nu \to \infty$ the integral

$$\int_{\mathfrak{R}} \{\langle q_{0,\nu}(x), u_0(x)\rangle + d_{0,\nu}(x)\}\mu(dx)$$
$$= \int_{\mathfrak{R}} \varphi_\nu(x)\{\langle p_0(x), u_0(x)\rangle + c_0(x)\}\mu(dx)$$

tends to a limit equal to

$$\int_K \{\langle p_0(x), u_0(x)\rangle + c_0(x)\}\mu(dx),$$

since $\varphi_\nu(x) \to 0$ for $x \notin K$, $\varphi_\nu(x) \to 1$ for $x \in K$, $0 \leq \varphi_\nu(x) \leq 1$ for all x, the functions $p_0(x)$ and $c_0(x)$ are continuous, hence bounded on U_1 (because the closure of U_1 is compact), and u_0 is integrable. We fix a ν_0 such that

$$\int_{\mathfrak{R}} \{\langle q_{0,\nu_0}(k), u_0(x)\rangle + d_{0,\nu_0}(x)\}\mu(dx)$$

is larger than $I_0 - \varepsilon$ in the case when $I_0 < \infty$, and larger than $1/\varepsilon$ when $I_0 = \infty$. Let $q_m = q_{m,\nu_0}$ and $d_m = d_{m,\nu_0}$, $m = 0, 1, 2, \ldots$. For each m

$$I_m \geq \int_{\mathfrak{R}} \{\langle q_m(x), u_m(x)\rangle + d_m(x)\}\mu(dx). \tag{3.2}$$

We have that $q_m \to q_0$ and $d_m \to d_0$ uniformly as $m \to \infty$. The functions q_m and d_m are all compactly supported and vanish outside the compact set $\overline{U}_1$. The functions u_m converge to u_0 in the w_C-sense as $m \to \infty$. In view of Lemma 3.2, this implies that the right-hand side of (3.2) tends to the limit

$$\int_{\mathfrak{R}} \{\langle q_0(x), u_0(x)\rangle + d_0(x)\}\mu(dx)$$

as $m \to \infty$. Passing in (3.2) to the limit as $m \to \infty$, we get from this that

$$\varliminf_{m\to\infty} I_m \geq I_0 - \varepsilon \quad \text{if } I_0 < \infty,$$

and

$$\varliminf_{m\to\infty} I_m \geq 1/\varepsilon \quad \text{if } I_0 = \infty.$$

Since $\varepsilon > 0$ is arbitrary, this establishes that

$$\varliminf_{m\to\infty} I_m \geq I_0,$$

which is what was required.

The continuity conditions imposed on F in Theorem 3.1 can be essentially weakened. Here it is obvious that convergence in the w_{C_0}-sense must be replaced by convergence of a stronger type.

We present the corresponding formulations. Let $F: \mathfrak{R} \times \mathbf{R}^l \to \mathbf{R}$ be given. We say that F is a function of type Σ, and write $F \in \Sigma$, if for every $\varepsilon > 0$ there is a compact set $A \subset \mathfrak{R}$ such that $\mu(\mathfrak{R} \backslash A) < \varepsilon$ and the restriction of F to $A \times \mathbf{R}^l$ is continuous on the metric space $A \times \mathbf{R}^l$.

Let $F \in \Sigma$. Then we say that $F \in \Sigma V$ if for almost all (in the sense of the measure μ) points $x \in \mathfrak{R}$ the function $q \to F(x, q)$ is convex on $\mathbf{R}^l$. By using the known C-property of measurable functions it is not hard to show that if F is a function of type Σ, then for every μ-measurable function $u: \mathfrak{R} \to \mathbf{R}^l$ the function $x \to F[x, u(x)]$ is measurable.

Let $F_m: \mathfrak{R} \times \mathbf{R}^l \to \mathbf{R}$, $m = 1, 2, \ldots$, be an arbitrary sequence of functions of type Σ. We say that this sequence Σ-*converges* to a function $F_0 \in \Sigma$ as $m \to \infty$ (notation: $F_m \to F_0 | \Sigma$ as $m \to \infty$) if every subsequence of it has a subsequence (F_{m_k}), $m_1 < m_2 < \ldots$, such that for any $\varepsilon > 0$ there is a compact set $A \subset \mathfrak{R}$ such that $\mu(\mathfrak{R} \backslash A) < \varepsilon$, each of the functions F_{m_k} and F_0 is continuous on $A \times \mathbf{R}^l$, and $F_{m_k} \to F_0$ locally uniformly on $A \times \mathbf{R}^l$ as $k \to \infty$.

THEOREM 3.2 [127]. *Let* $F_m: \mathfrak{R} \times \mathbf{R}^l \to \mathbf{R}$, *be a sequence of functions of class* ΣV *that* Σ-*converges to a function* $F_0: \mathfrak{R} \times \mathbf{R}^l \to \mathbf{R}$ *as* $m \to \infty$, *and let all the functions* F_m *be nonnegative. Then every sequence* $(u_m: \mathfrak{R} \to \mathbf{R}^l)$ *of* $L_1(\mathfrak{R}, \mu)$-*functions convergent in the* w_S-*sense to a function* $u_0: \mathfrak{R} \to \mathbf{R}^l$ *as* $m \to \infty$ *satisfies the inequality*

$$\int_{\mathfrak{R}} F_0[x, u_0(x)] \mu(dx) \leq \lim_{m \to \infty} \int_{\mathfrak{R}} F_m[x, u_m(x)] \mu(dx).$$

PROOF. For an arbitrary set $A \subset \mathfrak{R}$ let

$$(u_m, F_m, A) = \int_A F_m[x, u_m(x)] \mu(dx).$$

Further, let $(u_m, F_m, \mathfrak{R}) = I_m$, $m = 0, 1, 2, \ldots$. Take a subsequence of (F_m) such that the values I_m tend to $\underline{\lim}_{m \to \infty} I_m$. Given an arbitrary positive integer ν, we find a compact set $K_\nu \subset \mathfrak{R}$ with $\mu(\mathfrak{R} \backslash K_\nu) < 1/\nu$ and a subsequence (F_{m_k}), $m_1 < m_2 < \ldots$, of the constructed subsequence such that $F_{m_k} \to F_0$ locally uniformly on $K_\nu \times \mathbf{R}^l$. For every continuous vector-valued function φ defined on K_ν we have that

$$\int_{K_\nu} \langle \varphi(x), u_m(x) \rangle \mu(dx) \to \int_{K_\nu} \langle \varphi(x), u_0(x) \rangle \mu(dx),$$

as follows from the remarks in §3.1. This means that $u_m(x) \to u_0(x)$ on the subspace K_ν in the $w_{C_0(K_\nu)}$-sense. Hence, we can apply Theorem 3.1

to the space K_ν and the sequences (F_{m_k}) and (u_{m_k}), $k = 1, 2, \ldots$. As a result we get that

$$\lim_{m \to \infty} \int_{\mathfrak{R}} F_m[x, u_m(x)]\mu(dx) = \lim_{k \to \infty} \int_{\mathfrak{R}} F_{m_k}[x, u_{m_k}(x)]\mu(dx)$$

$$\geq \lim_{h \to \infty} \int_{K_\nu} F_{m_k}[x, u_{m_k}(x)]\mu(dx) \geq \int_{K_\nu} F_0[x, u_0(x)]\mu(dx).$$

It remains to see that the integral on the right-hand side of the last inequality tends to

$$\int_{\mathfrak{R}} F_0[x, u_0(x)]\mu(dx)$$

as $\nu \to \infty$, and the theorem is proved.

3.4. Corollaries to Theorems 3.1 and 3.2. Let $U \subset \mathbf{R}^n$ be an arbitrary bounded open set, and let $u: U \to \mathbf{R}^m$ be a mapping of class $W_1^1(U)$. The matrix $(\partial u_i/\partial x_j)$, $i = 1, \ldots, m$, $j = 1, \ldots, n$, is denoted here by $u'(x)$. The vector space of $m \times n$ matrices is identified with the space $\mathbf{R}^{mn}$.

Let $F(x, u, w)$ be a function such that for almost all $x \in U$ the expression $F(x, u, w)$ is defined for every $u \in \mathbf{R}^m$ and $w \in \mathbf{R}^{mn}$. We say that F belongs to the class $\Sigma V(U)$ if for almost all $x \in U$ and any $u \in \mathbf{R}^m$ the function $w \to F(x, u, w)$ is convex on $\mathbf{R}^{mn}$ and for any $\varepsilon > 0$ there is a compact set $A \subset U$ such that $|U \backslash A| < \varepsilon$ and $F(x, u, w)$ is continuous on $A \times \mathbf{R}^m \times \mathbf{R}^{mn}$.

LEMMA 3.5. Let $(f_\nu: U \to \mathbf{R}^m)$ be a sequence of functions of class $W_p^1(U)$, where U is an open set in $\mathbf{R}^n$, and $p \geq 1$. Assume that the norm sequence $(\|f_\nu\|_{1,p,U})$, $\nu = 1, 2, \ldots$, is bounded and the functions f_ν converge in L_1 to a function $f_0: U \to \mathbf{R}^m$ as $\nu \to \infty$. Then $f_0 \in W_p^1(U)$, and the functions f_ν' converge weakly in $L_p(U)$ to f_0'.

PROOF. The fact that the limit function f_0 belongs to $W_p^1(U)$ follows from Theorem 1.1 in Chapter I.

Let K be the collection of all continuously differentiable functions $\varphi: U \to \mathbf{R}$ with compact support in U. Then the class $Z_\infty(K)$ contains all the continuous functions with compact support in U, and the class $Z_q(K)$ coincides with $L_q(U)$ when $q > 1$.

Suppose that the vector-valued function f_ν has components $f_{\nu 1}, \ldots, f_{\nu m}$. Then for any $\varphi \in K$

$$\int_U \frac{\partial f_{\nu, j}}{\partial x_i}(x)\varphi(x)\, dx = -\int_U f_{\nu, j}(x)\frac{\partial \varphi}{\partial x_i}(x)\, dx.$$

Since $f_{\nu,j} \to f_{0,j}$ in $L_1(U)$, it follows that

$$-\int_U f_{\nu,j}(x)\frac{\partial\varphi}{\partial x_i}(x)\,dx \to -\int_U f_{0,j}(x)\frac{\partial\varphi}{\partial x_i}(x)\,dx$$

$$= \int_U \frac{\partial f_{0,j}}{\partial x_i}(x)\varphi(x)\,dx.$$

Thus,

$$\int_U \frac{\partial f_{\nu,j}}{\partial x_i}(x)\varphi(x)\,dx \to \int_U \frac{\partial f_{0,j}}{\partial x_i}(x)\varphi(x)\,dx$$

as $\nu \to \infty$. This means that

$$\frac{\partial f_{\nu,j}}{\partial x_i} \to \frac{\partial f_{0,j}}{\partial x_i}$$

in the w_K-sense. By Lemma 3.1, this implies the statement of the lemma.

THEOREM 3.3 [127]. *Let* $F\colon (x,u,w) \to F(x,u,w)$ $(x \in U, v \in \mathbf{R}^m, w \in \mathbf{R}^{mn})$ *be a function that is defined and continuous on the set* $U \times \mathbf{R}^n \times \mathbf{R}^{mn}$, *and convex in* w *for any* $x \in U$ *and* $u \in \mathbf{R}^m$. *Let* (f_ν), $\nu = 1, 2, \ldots$, *be a sequence of vector-valued functions of class* W_1^1 *such that the functions* f_ν *are all continuous,* $f_\nu \to f_0$ *locally uniformly on* U *as* $\nu \to \infty$, *and the norm sequence* $(\|f_\nu\|_{1,1,U})$, $\nu = 1, 2, \ldots$, *is bounded. Then*

$$\int_U F[x, f_0(x), f_0'(x)]\,dx \le \lim_{\nu\to\infty} \int_U F[x, f_\nu(x), f_\nu'(x)]\,dx.$$

PROOF. The theorem is a corollary to Theorem 3.1. Let

$$f_\nu'(x) = w_\nu(x), \qquad F_\nu(x,w) = F[x, f_\nu(x), w].$$

Then $F_\nu(x,w) \to F_0 - (x,w)$ for all $x \in U$ and $w \in \mathbf{R}^{mn}$ as $\nu \to \infty$. Since F is convex and continuous and since $f_\nu \to f_0$ locally uniformly as $\nu \to \infty$, we have that $F_\nu(x,w) \to F_0(x,w)$ locally uniformly on $U \times \mathbf{R}^{mn}$. By Lemma 3.5, the functions f_ν' converge to f_0' in the w_{C_0}-sense. Thus, we have shown that all the conditions of Theorem 3.1 hold here, and the theorem is thus proved.

THEOREM 3.4 [127]. *Let* $(x,u,w) \to F(x,u,w)$ *be a nonnegative function of class* $\Sigma V(U)$ *defined for all* $x \in U$, $u \in \mathbf{R}^m$ *and* $w \in \mathbf{R}^{mn}$, *where* U *is an open set in* $\mathbf{R}^n$, *and let* $(f_\nu\colon U \to \mathbf{R}^m)$, $\nu = 1, 2, \ldots$, *be a sequence of vector-valued functions of class* $W_p^1(U)$ $(p > 1)$ *that is bounded in* $W_p^1(U)$. *Assume that the functions* f_ν *converge in* $L_1(U)$ *to a function* $f_0\colon U \to \mathbf{R}^m$ *as* $\nu \to \infty$. *Then*

$$\int_U F[x, f_0(x), f_0'(x)]\,dx \le \lim_{\nu\to\infty} \int_U F[x, f_\nu(x), f_\nu'(x)]\,dx.$$

PROOF. Take a subsequence of (f_ν) for which the integrals

$$\int_U F[x, f_\nu(x), f_\nu'(x)]\,dx$$

converge to their limit inferior, and then take a subsequence of that sequence such that $f_\nu \to f_0$ almost everywhere in U. In order not to complicate the notation it will be assumed that (f_ν), $\nu = 1, 2, \ldots$, is this second subsequence. Let $F_\nu(x, w) = F[x, f_\nu(x), w]$. Take an arbitrary $\varepsilon > 0$, and for it take a compact set $K \subset U$ such that F is continuous on $K \times \mathbf{R}^m \times \mathbf{R}^{mn}$, $|U \backslash K| < \varepsilon$, the functions f_ν are all continuous on K, and $f_\nu \to f_0$ uniformly on K. The existence of such a compact set K follows from Lusin's theorem on the C-property of a measurable function, Egorov's theorem on sequences of measurable functions converging almost everywhere, and the fact that F is a function of class ΣV. Then it is clear that $F_\nu(x, w) \to F_0(x, w)$ locally uniformly on $K \times \mathbf{R}^{mn}$. Since $\varepsilon > 0$ is arbitrary, this proves that the sequence (F_ν) of functions Σ-converges to $F_0(x, w)$ as $\nu \to \infty$.

In view of Lemma 3.5, the functions $w_\nu = f_\nu'$ converge to $w_0 = f_0'$ weakly in $L_p(U)$ as $\nu \to \infty$, and hence $w_\nu \to w_0$ in the $w_{S(U)}$-sense. Now Theorem 3.4 is an immediate consequence of Theorem 3.2.

3.5. The convex envelope of a function. Let $\Phi = \{F\}$ be an arbitrary nonempty set of functions on $\mathbf{R}^l$. For $x \in \mathbf{R}^l$ let $\overline{F}_\Phi(x) = \sup_{F \in \Phi} F(x)$, a real function defined on $\mathbf{R}^l$ with values in the extended real line $\overline{\mathbf{R}} = \mathbf{R} \cup \{-\infty, \infty\}$. This function is called the *upper envelope* of the set Φ of functions. If all the functions $F \in \Phi$ are convex and $\overline{F}_\Phi(x) < \infty$ for all $x \in \mathbf{R}^l$, then the function $\overline{F}_\Phi$ is easily shown also to be convex.

Let $F: \mathbf{R}^l \to \mathbf{R}$ be an arbitrary function. Denote by V_F the collection of all convex functions G such that $G(x) < F(x)$ for all x. Assume that V_F is nonempty. Then $\sup_{G \in V_F} G(x) < \infty$ for any $x \in \mathbf{R}^l$, and $\overline{F}_{V_F}$ is called the *convex envelope* of F in this case. A function $F: \mathbf{R}^l \to \mathbf{R}$ is said to be *essentially convex* if F is convex and

$$F[\lambda x_0 + (1 - \lambda)x_1] < \lambda F(x_0) + (1 - \lambda)F(x_1)$$

(strict inequality!) for any two distinct points x_0 and x_1 in $\mathbf{R}^l$ and for any $0 < \lambda < 1$.

LEMMA 3.6. *Let F be a positive essentially convex function on $\mathbf{R}^l$ and let $G: \mathbf{R}^l \to \mathbf{R}$ be a continuous function such that $|G(u)| \leq kF(u) + M$, where $k > 0$ and M are constants. Let F_λ be the convex envelope of the function $\lambda F(u) + G(u)$, and let $H_\lambda(u) = \lambda F(u) + G(u) - F_\lambda(u)$. The function F_λ is* *defined for any $\lambda > k$, $H_\lambda(u) \to 0$ as $\lambda \to \infty$ for all $u \in \mathbf{R}^l$, and*

$$0 \le H_\lambda(u) \le 2kF(u) + 2M. \tag{3.3}$$

PROOF. For all $u \in \mathbf{R}^l$ we have

$$\lambda F(u) + G(u) \ge (\lambda \to k)F(u) - M,$$

by a condition of the lemma. The right-hand side of this inequality is a convex function for $\lambda > k$; hence the convex envelope of $\lambda F + G$ exists for every $\lambda > k$. Further, $F_\lambda(u) \ge (\lambda - k)F(u) - M$ for all u. This implies that

$$\begin{aligned}
H_\lambda(u) &= \lambda F(u) + G(u) - F_\lambda(u) \\
&\le (\lambda + k)F(u) + M - (\lambda - k)F(u) + M \\
&= 2kF(u) + 2M
\end{aligned}$$

for $\lambda > k$, and thus it is established that $H_\lambda(u) \le 2kF(u) + 2M$. It is obvious by definition that H_λ is nonnegative, and (3.3) is proved.

It remains to prove that $H_\lambda(u) \to 0$ as $\lambda \to \infty$ for all $u \in \mathbf{R}^l$. Let $u_0 \in \mathbf{R}^l$ be an arbitrary point. Since F is a convex function, there is an affine function L such that $F(u_0) = L(u_0)$ and $F(u) \ge L(u)$ for all u.

Take an arbitrary $\varepsilon > 0$. Since G is continuous at u_0, for the given ε there exists a $\delta > 0$ such that if $|u - u_0| < \delta$, then $|G(u) - G(u_0)| < \varepsilon$. Let $M_1 = M + |G(u_0) - \varepsilon|$. Then for all $u \in \mathbf{R}^l$

$$kF(u) + M_1 + G(u) \ge G(u_0) - \varepsilon.$$

We now prove that there is an $\alpha > 0$ such that for every $u \in \mathbf{R}^l$ with $|u - u_0| \ge \delta$

$$F(u) - L(u) \ge \alpha(kF(u) + M_1). \tag{3.4}$$

Assume on the contrary that there is no such α. Then for every $m = 1, 2, \ldots$ there is a point $u_m \in \mathbf{R}^l$ for which $|u_m - u_0| \ge \delta$, while

$$F(u_m) - L(u_m) \le (1/m)kF(u_m) + (1/m)M_1. \tag{3.5}$$

The ray $[u_0, u_m]$ intersects the sphere $S(u_0, \delta)$ at some point v_m. We have that

$$v_m = u_0 + \lambda_m(u_m - u_0), \quad \text{where } \lambda_m = \delta/|u_m - u_0|.$$

By (3.5),

$$(1 - (k/m))[F(u_m) - L(u_m)] \le (k/m)L(u_m) + (M_1/m). \tag{3.6}$$

Since L is an affine function, $|L(u)| \le A|u - u_0| + B$, where A and B are constants. By (3.6), this tells us that

$$0 \le F(u_m) - L(u_m) \le \frac{Ak}{m-k}|u_m - u_0| + \frac{Bk}{m-k} + \frac{M_1}{m-k}. \tag{3.7}$$

The convexity of $F - L$ implies that

$$F(v_m) - L(v_m) \le \lambda_m[F(u_m) - L(u_m)],$$

which, by (3.7), gives us that

$$F(v_m) - L(v_m) \le \frac{Ak\delta}{m - k} + \frac{Bk + M_1}{m - k}. \qquad (3.8)$$

Choose a subsequence of (v_m) that converges to some point $v_0 \in S(u_0, \delta)$. Then, since $F - L$ is nonnegative, (3.8) implies that $F(v_0) - L(v_0) = 0$. Thus, the nonnegative convex function $F - L$ vanishes at two different points u_0 and v_0. This implies that $F - L$ vanishes on the whole segment $[u_0, v_0]$, obviously contradicting the essential convexity of F. Thus, a contradiction has resulted from the assumption that there is no $\alpha > 0$ such that (3.4) holds when $|u - u_0| \ge \delta$.

Accordingly, let $\alpha > 0$ be such that $F(u) - L(v) \ge \alpha(kF(u) + M_1]$ for all u with $|u - u_0| \ge \delta$. Let $\lambda > 1/\alpha$. Then for $|u - u_0| \ge \delta$

$$\begin{aligned} \lambda F(u) + G(u) &= \lambda[F(u) - L(u)] + G(u) + \lambda L(u) \\ &\ge \lambda\alpha[kF(u) + M_1] + G(u) + \lambda L(u) \\ &\ge kF(u) + M_1 + G(u) + \lambda L(u) \ge \lambda L(u) + G(u_0) - \varepsilon. \end{aligned}$$

For $|u - u_0| < \delta$

$$\begin{aligned} \lambda F(u) + G(u) &= \lambda[F(u) - L(u)] + G(u) + \lambda L(u) \\ &\ge G(u) + \lambda L(u) \ge \lambda L(u) + G(u_0) - \varepsilon. \end{aligned}$$

Thus, $\lambda F(u) + G(u) \ge \lambda L(u) + G(u_0) - \varepsilon$ for all $u \in \mathbf{R}^l$. The function $u \to \lambda L(u) + G(u_0) - \varepsilon$ is convex. Hence, $F_\lambda(u) \ge \lambda L(u) + G(u_0) - \varepsilon$ for all u. In particular, setting $u = u_0$ and noting that $L(u_0) = F(u_0)$, we get that $0 \le H_\lambda(u_0) \le \varepsilon$. Therefore, $F_\lambda(u_0) \ge \lambda F(u_0) + G(u_0) - \varepsilon$. Since $\varepsilon > 0$ is arbitrary and the last inequality requires only that λ be greater than $1/\alpha$, this establishes that $H_\lambda(u_0) \to 0$ as $\lambda \to \infty$. The lemma is proved.

LEMMA 3.7. *Let* (F_m), $m = 1, 2, \ldots$, *be a sequence of functions on* $\mathbf{R}^l$ *such that the following conditions hold:*

a) $\underline{\lim}_{m \to \infty} F_m(u) = F_0(u)$ *is finite for any* $u \in \mathbf{R}^l$, *and* $F_m \to F_0$ *in* $\mathbf{R}^l$ *locally uniformly.*

b) *There exists a function* $\theta: \mathbf{R}^l \to \mathbf{R}$ *that is bounded below and such that* $F_m(u) \ge \theta(u)$ *for all* $u \in \mathbf{R}^l$, $m = 1, 2, \ldots$, *and* $\theta(u)/(|u| + 1) \to \infty$ *as* $|u| \to \infty$.

Let G_m *and* G_0 *be the convex envelopes of the functions* F_m *and* F_0. *Then* $G_m \to G_0$ *as* $m \to \infty$ *locally uniformly in* $\mathbf{R}^l$.

PROOF. All the functions G_m and G_0 are clearly convex, and to prove that $G_m \to G_0$ locally uniformly it suffices by Lemma 3.3 to establish that $G_m(u) \to G_0(u)$ for all $u \in \mathbf{R}^l$.

Take an arbitrary point $u_0 \in \mathbf{R}^l$, and let L be an affine function such that $G(u_0) = L(u_0)$ and $G(u) \geq L(u)$ for all $u \in \mathbf{R}^l$; $L(u) = \langle p, u \rangle + A$, where $p \in \mathbf{R}^l$ and A is a real number. Let $\varepsilon > 0$ be arbitrary, and let K be the collection of all $u \in \mathbf{R}^l$ such that $\theta(u) \leq L(u) - \varepsilon$. The set K is bounded, because $\theta(u)/(|u| + 1) \to \infty$ as $|u| \to \infty$. For all $u \in K$ we have that $F_0(u) \geq L(u)$, and since F_m converges to F locally uniformly, there is an index m_0 such that $F_m(u) > L(u) - \varepsilon$ for $m \geq m_0$ and for all $u \in K$. Since $F_m(u) \geq \theta(u)$ for all u, the inequality $F_m(u) > L(u) - \varepsilon$ holds also for $u \notin K$ when $m \geq m_0$. From this, $G_m(u) \geq L(u) - \varepsilon$ for all u when $m \geq m_0$, and, in particular, $G_m(u_0) \geq L(u_0) - \varepsilon = G(u_0) - \varepsilon$ for $m \geq m_0$. Since $\varepsilon > 0$ is arbitrary, what has been proved implies that

$$\lim_{m \to \infty} G_m(u_0) \geq G(u_0).$$

We now prove that

$$\varlimsup_{m \to \infty} G_m(u_0) \leq G(u_0). \tag{3.9}$$

For each m we construct an affine function L_m such that $G_m(u_0) = L_m(u_0)$ and $G_m(u_0) \geq L_m(u)$ for all u. Then

$$L_m(u) = \langle p_m, u - u_0 \rangle + G_m(u_0).$$

For each m we have that $G_m(u_0) \leq F_m(u_0)$, which implies that the sequence $(G_m(u_0))$, $m = 1, 2, \ldots$, is bounded above. Let $\theta(u) \geq M$ for all u, where $M \in \mathbf{R}$. Then $G_m(u_0) \geq M$ for all m, and hence the sequence $(G_m(u_0)$, $m = 1, 2, \ldots$, is also bounded below. Let $p_m = (p_{m1}, \ldots, p_{mn})$ and $p_{mi} = \langle p_m, e_i \rangle$. We have that

$$L_m(u_0 + e_i) = p_{mi} + G_m(u_0) \leq F_m(u_0 + e_i),$$
$$L_m(u_0 - e_i) = -p_{mi} + G_m(u_0) \leq F_m(u_0 - e_i).$$

This leads to the conclusion that each of the sequences (p_{mi}), $m = 1, 2, \ldots$, is bounded. Let us first construct a strictly increasing sequence of numbers m such that $G_m(u_0)$ tends to the limit superior $\varlimsup_{m \to \infty} G_m(u_0)$, and then choose a subsequence (denoted by (m_k), $m_1 < m_2 < \ldots$) of it such that $G_{m_k}(u_0)$ and $p_{m_k i}$ tend to finite limits as $k \to \infty$. Then $F_{m_k}(u) \geq L_{m_k}(u)$ for each k and each u, and the functions L_{m_k} converge to some affine function L_0 as $k \to \infty$. Passing to the limit, we get that $F_0(u) \geq L_0(u)$ for all u, and hence $G_0(u_0) \geq L_0(u_0)$. But

$$L_0(u_0) = \lim_{k \to \infty} G_{m_k}(u_0) = \varlimsup_{m \to \infty} G_m(u_0)$$

by construction, and this establishes (3.9). The lemma is proved.

3.6. A theorem on convergence with a functional. Corresponding to the two variants of the semicontinuity theorem we have proved, there are two possible variants of a theorem on convergence with a functional; to avoid repeating awkward formulations we introduce concepts allowing us to unify the different formulations. We distinguish two cases, Situation I and Situation II.

As above, let $\mathfrak{R}$ be a locally compact metric space, and μ a measure defined on the σ-algebra of Borel subsets of $\mathfrak{R}$ such that $\mu(\mathfrak{R}) < \infty$. Consider a sequence $(F_m: \mathfrak{R} \times \mathbf{R}^l \to \mathbf{R})$, $m = 1, 2, \ldots$, of functions. In Situation I it is assumed that all the functions F_m are continuous on $\mathfrak{R} \times \mathbf{R}^l$, nonnegative for all (x, u), and convex in U for each $x \in \mathfrak{R}$, and that they converge locally uniformly as $m \to \infty$ to a function $F_0(x, u)$ that is essentially convex in u for almost all values of $x \in \mathbf{R}$ in the sense of μ. In Situation II it is assumed that $F_m \in \Sigma V(\mathfrak{R})$, the functions F_m are nonnegative, $F_m \to F_0$ in the sense of Σ, and for each $\varepsilon > 0$ there exists a compact set $K \subset \mathfrak{R}$ such that $\mu(\mathfrak{R} \backslash K) < \varepsilon$ and F_m is continuous on $K \times \mathfrak{R}^l$ and essentially convex in u for $x \in K$.

In both situations it is assumed that there exists a continuous function $\theta: \mathbf{R}^l \to \mathbf{R}$ such that $\theta(u)/(|u| + 1) \to \infty$ as $|u| \to \infty$, $\theta(u) > 0$ for all u, and

$$F_m(x, u) \geq \theta(u) \qquad (3.10)$$

for any x and u.

Finally, let $(K_m: \mathfrak{R} \times \mathbf{R}^l \to \mathbf{R})$ be another sequence of functions such that for $m = 1, 2, \ldots$

$$|K_m(x, u)| \leq kF_m(x, u) + M_m(x), \qquad (3.11)$$

where $k > 0$ is a constant, and $M_m \to M_0$ as $m \to \infty$. In Situation I it will be assumed that the functions K_m are all continuous on $\mathfrak{R} \times \mathbf{R}^l$ and converge to K_0 locally uniformly on $\mathfrak{R} \times \mathfrak{R}^l$, $M_1 = M_2 = \ldots$, and all the functions M_m are constant. In Situation II it is assumed that all the K_m belong to the class Σ and Σ-converge to K_0, while the functions M_m are μ-integrable and converge in $L_1(\mathfrak{R}, \mu)$ to the function M_0.

Finally, let (u_m), $m = 1, 2, \ldots$, be a sequence of $L_1(\mathfrak{R}, \mu)$-functions with values in $\mathbf{R}^l$ such that $u_m \to u_0$ in the $w_{C_0(\mathfrak{R})}$-sense in Situation I as $m \to \infty$, and $u_m \to u_0$ in the $w_{S(\mathfrak{R})}$-sense in Situation II.

THEOREM 3.5 [127]. *Let (F_m), (K_m), and (u_m) be sequences of functions satisfying all the above conditions. If*

$$\int_{\mathfrak{R}} F_m[x, u_m(x)]\mu(dx) \to \int_{\mathfrak{R}} F_0[x, u_0(x)]\mu(dx)$$

as $m \to \infty$, and the integrals here are all finite, then

$$\int_{\Re} K_m[x, u_m(x)]\mu(dx) \to \int_{\Re} K_0[x, u_0(x)]\mu(dx)$$

as $m \to \infty$.

PROOF. We assume that $M_m(x) \to M_0(x)$ almost everywhere in Situation II. Let $F_m(x, u, \lambda)$, where $\lambda > k$, be the convex envelope of the function $\lambda F_m(x, u) + K_m(x, u)$. We prove that in Situation I the function $(x, u) \to F_m(x, u, \lambda)$ is continuous on $\Re \times \mathbf{R}^l$ and that $F_m(x, u, \lambda) \to F_0(x, u, \lambda)$ locally uniformly on $\Re \times \mathbf{R}^l$ as $m \to \infty$. In Situation II, $F_m(x, u, \lambda)$ is a $\Sigma(\Re)$-function that Σ-converges to $F_0(x, u, \lambda)$ as $m \to \infty$. Take an arbitrary $\varepsilon > 0$. Let $A \subset \Re$ be a set such that the functions F_m and K_m, $m = 0, 1, \ldots$, are continuous on $A \times \mathbf{R}^l$ and locally uniformly convergent to F_0 and K_0, respectively, and the M_m, $m = 0, 1, \ldots$, are continuous on A and converge to M_0 uniformly on A as $m \to \infty$. In Situation I we let $A = \Re$, while in Situation II the set A is compact and $\mu(\Re \backslash A) < \varepsilon$. For $\lambda > k$ we have for $x \in A$ that

$$\lambda F_m(x, u) + K_m(x, u) \geq (\lambda - k)\theta_0 - q_m,$$

where $\theta_0 = \inf \theta(u)$ and $q_m = \sup_{x \in A} M_m(x)$. The functions $\lambda F_m + K_m$ are nonnegative on A for $\lambda > (q_m/\theta_0) + k$.

For every convergent sequence (x_ν) of points in A the functions $F_\nu(u) = \lambda F_m(x_\nu, u) + K_m(x_\nu, u)$ converge to $F_0(u) = \lambda F_m(x_0, u) + K_m(x, u)$, where $x_0 = \lim_{\nu \to \infty} x_\nu$. By (3.10), the sequence F_ν satisfies all the conditions of Lemma 3.7, and this implies that the convex envelopes of the F_ν converge to the convex envelope of F_0. This proves that the function $(x, u) \to F_m(x, u, \lambda)$ is continuous on $A \times \Re^l$.

The sequence (q_m), $m = 1, 2, \ldots$, is bounded. Let $q_0 = \sup q_m$. Using Lemma 3.7 once more for an arbitrary convergent sequence (x_m), $m = 1, 2, \ldots$, of points in A, we get that $F_m(x_m, u, \lambda) \to F_0(x_0, u, \lambda)$ as $m \to \infty$, where $x_0 = \lim_{m \to \infty} x_m$. This proves that the functions $(x, u) \to F_m(x, u, \lambda)$ converge locally uniformly to $(x, u) \to F_0(x, u, \lambda)$ on $A \times \mathbf{R}^l$ as $m \to \infty$.

Let

$$P_m = \int_{\Re/A} F_m[x, u_m(x)]\mu(dx), \qquad Q_m = \int_A F_m[x, u_m(x)]\mu(dx).$$

Then $P_m + Q_m \to P_0 + Q_0$ as $m \to \infty$, by a condition of the theorem. Let (m_k), $m_1 < m_2 < \ldots$, be a sequence of positive integers such that

$$\lim_{k \to \infty} P_{m_k} = \overline{\lim_{m \to \infty}} P_m.$$

Since P_m and Q_m are nonnegative and $P_0 + Q_0 < \infty$, this implies that $\lim_{k\to\infty} P_{m_k} < \infty$, and the sequence (Q_{m_k}) has a limit. We get that

$$\overline{\lim_{m\to\infty}}\, P_m = P_0 + Q_0 - \lim_{k\to\infty} Q_{m_k}.$$

However, Theorems 3.1 and 3.2 give us that $\lim_{k\to\infty} Q_{m_k} \geq Q_0$, which leads us to conclude that $\overline{\lim}_{m\to\infty} P_m \leq P_0$. On the other hand, since $\lim_{m\to\infty} P_m \geq P_0$, it thus follows that $P_m \to P_0$ as $m \to \infty$. This implies that $Q_m \to Q_0$ as $m \to \infty$. For each m

$$\int_A \{\lambda F_m[x, u_m(x)] + K_m[x, u_m(x)]\}\mu(dx)$$

$$\geq \int_A F_m[x, u_m(x), \lambda]\mu(dx). \tag{3.12}$$

For $\lambda > k + q_0/\theta_0$ all the functions $x \to F_m[x, u_m(x), \lambda]$ are nonnegative. Passing to the limit in (3.12), we get that

$$\lim_{m\to\infty} \int_A [\lambda F_m(x, u_m) + K_m(x, u_m)]\mu(dx)$$

$$\geq \overline{\lim_{m\to\infty}} \int_A F_m(x, u_m, \lambda)\mu(dx).$$

In view of Theorem 3.1,

$$\overline{\lim_{m\to\infty}} \int_A F_m(x, u_m, \lambda)\mu(dx) \geq \int_A F_0(x, u_0, \lambda)\mu(dx). \tag{3.13}$$

Further, by what has been proved,

$$\lambda \int_A F_m(x, u_m)\mu(dx) \to \lambda \int_A F_0(x, u_0)\mu(dx) \tag{3.14}$$

as $m \to \infty$. Comparing (3.12)–(3.14), we conclude that

$$\int_A \{\lambda F_0(x, u_0) + K(x, u_0) - F_0(x, u_0, \lambda)\}\mu(dx)$$

$$\geq \int_A K_0(x, u_0)\mu(dx) - \lim_{m\to\infty} \int_A K_m(x, u_m)\mu(dx). \tag{3.15}$$

By Lemma 3.6, the absolute value of the integrand on the left-hand side here does not exceed the function $2kF_0[x, u_0(x)] + 2M_0(x)$, which is integrable over A, and it tends to zero as $\lambda \to \infty$. This implies that the integral on the left-hand side of (3.15) tends to zero as $\lambda \to \infty$. Passing to the limit in (3.15) as $\lambda \to \infty$, we find that

$$\lim_{m\to\infty} \int_A K_m[x, u_m(x)]\mu(dx) \geq \int_A K_0[x, u_0(x)]\mu(dx). \tag{3.16}$$

Replacing K_m by $-K_m$ $(m = 1, 2, \dots)$ and K_0 by $-K_0$ in these arguments, we get

$$\varlimsup_{m \to \infty} \int_A K_m[x, u_m(x)]\mu(dx) \leq \int_A K_0[x, u_0(x)]\mu(dx). \qquad (3.17)$$

It obviously follows from (3.16) and (3.17) that

$$\int_A K_m[x, u_m(x)]\mu(dx) \to \int_A K_0[x, u_0(x)]\mu(dx)$$

as $m \to \infty$. The proof of the theorem is complete in Situation I, since $A = \mathfrak{R}$ in this case.

Consider Situation II. Let $\eta > 0$ be arbitrary, and choose A such that

$$\int_{\mathfrak{R}\backslash A} F_0[x, u_0(x)]\mu(dx) < \eta, \qquad \int_{\mathfrak{R}\backslash A} M_0(x)\mu(dx) < \eta.$$

As $m \to \infty$,

$$\int_{\mathfrak{R}\backslash A} F_m[x, u_m(x)]\mu(dx) = Q_m \to \int_{\mathfrak{R}\backslash A} F_0[x, u_0(x)]\mu(dx) = Q_0,$$

$$\int_{\mathfrak{R}\backslash A} M_m(x)\mu(dx) \to \int_{\mathfrak{R}\backslash A} M_0(x)\mu(dx).$$

From this, there is an m_1 such that for $m \geq m_1$

$$\int_{\mathfrak{R}\backslash A} F_m[x, u_m(x)]\mu(dx) < \eta, \qquad \int_{\mathfrak{R}\backslash A} M_m(x)\mu(dx) < \eta. \qquad (3.18)$$

It is obvious that for $m \geq m_1$

$$\left| \int_{\mathfrak{R}\backslash A} K_m[x, u_m(x)]\mu(dx) \right| < (k + 1)\eta. \qquad (3.19)$$

It is also clear that

$$\left| \int_{\mathfrak{R}\backslash A} K_0[x, u_0(x)]\mu(dx) \right| < (k + 1)\eta. \qquad (3.20)$$

Let $m_2 \geq m_1$ be such that for $m > m_2$

$$\left| \int_A K_m[x, u_m(x)]\mu(dx) - \int_A K_0[x, u_0(x)]\mu(dx) \right| < \eta.$$

By (3.18)–(3.20),

$$\left| \int_{\mathfrak{R}} K_m[x, u_m(x)]\mu(dx) - \int_{\mathfrak{R}} K_0[x, u_0(x)]\mu(dx) \right|$$
$$< \eta + 2(k + 1)\eta$$

for $m \geq m_2$. Since $\eta > 0$ is arbitrary, this proves that

$$\int_{\mathfrak{R}} K_0[x, u_0(x)]\mu(dx) = \lim_{m \to \infty} \int_{\mathfrak{R}} K_m[x, u_m(x)]\mu(dx). \qquad (3.21)$$

Thus, the theorem is proved in Situation II under the temporary assumption that $M_m(x) \to M_0(x)$ almost everywhere.

We take an arbitrary sequence of positive integers and select from it a subsequence (m_k), $m_1 < m_2 < \ldots$, such that $M_{m_k}(x) \to M_0(x)$ almost everywhere. In view of what has been proved, (3.21) holds for this sequence. Since the original sequence of positive integers was arbitrary, this establishes that

$$\lim_{m \to \infty} \int_{\mathfrak{R}} K_m[x, u_m(x)]\mu(dx) = \int_{\mathfrak{R}} K_0[x, u_0(x)]\mu(dx).$$

The proof of the theorem is complete.

3.7. Corollaries to the theorem on convergence with a functional. Below, U denotes a bounded open set in $\mathbf{R}^n$. We consider functions of the class W_p^1 ($p > 1$) defined on U and with values in $\mathbf{R}^m$. For every such function f let f' denote the Jacobi matrix of the mapping f. The space of $m \times n$ matrices is denoted by $\mathbf{R}^{mn}$. Let $F: \mathbf{R}^{mn} \to \mathbf{R}$ be an arbitrary function satisfying the following conditions:

a) F is an essentially convex function on $\mathbf{R}^{mn}$.

b) There exist constants $k_1, k_2 > 0$ such that $k_1|v|^p \leq F(v) \leq k_2|v|^p$ for every $v \in \mathbf{R}^{mn}$.

THEOREM 3.6. *Let $(f_\nu: U \to \mathbf{R}^m)$, $\nu = 1, 2, \ldots$, be a sequence of $W_p^1(U)$-functions converging in L_1 to some function $f_0 \in W_p^1$. Assume that*

$$\int_U F[f_\nu'(x)]dx \to \int_U F[f_0'(x)]dx$$

as $\nu \to \infty$, where F satisfies conditions a) *and* b). *Then $f_\nu \to f_0$ in $W_p^1(U)$.*

PROOF. Assume that the sequence (f_ν) of functions satisfies the conditions in the theorem. By Lemma 3.5, the functions f_ν' converge weakly to f_0' in $L_p(U)$ as $\nu \to \infty$. For $v \in \mathbf{R}^{mn}$ let $K(x, v) = |v - f_0'(x)|^p$. For any x and u,

$$|v - f_0'(x)|^p \leq (|v| + |f_0'(x)|)^p \leq 2^{p-1}(|v|^p + |f_0'(x)|^p)$$
$$\leq (2^{p-1}/k_1)F(v) + 2^{p-1}|f_0'(x)|^p.$$

Let $F_m(x, v) = F(v)$ and $K_m(x, v) = K(x, v)$. The functions K_m and F_m satisfy all the conditions in Theorem 3.5 relating to Situation II; hence

$$\int_U K[x, f_\nu'(x)]dx \to \int_U K[x, f_0'(x)]dx = 0,$$

as $\nu \to \infty$, i.e., $\|f_\nu' - f_0'\|_{p,U}^p \to 0$ as $\nu \to \infty$. This proves the theorem.

§4. Some properties of functions with generalized derivatives

4.1. A theorem on differentiability almost everywhere. The concept of a function differentiable in the sense of convergence in R was defined in Chapter II, §1.3, where $\mathfrak{R}$ is a topological vector space. By analogy it is possible to introduce the concept of the limit of a function in the sense of convergence in $\mathfrak{R}$ and the concept of continuity in the sense of convergence in $\mathfrak{R}$. These concepts are needed below in studying the question of differentiability almost everywhere. We present the definitions.

Let $U \subset \mathbf{R}^n$ be an arbitrary open set whose elements are real functions defined in the unit ball $B(0, 1)$ of $\mathbf{R}^n$.

A number $k \in \mathbf{R}$ is called the limit of a function $f: U \to \mathbf{R}$ at a point a in the sense of convergence in $\mathfrak{R}$, or, briefly, the $\mathfrak{R}$-*limit* of f as $x \to a$, if there exists a number $r_0 > 0$ such that for $0 < r < r_0$ the function

$$F_r: X \in B(0, 1) \to f(a + rX) - k$$

belongs to the class $\mathfrak{R}$, and F_r tends to the function identically equal to zero as $r \to 0$, in the sense of the topology in $\mathfrak{R}$. (We assume that the zero function belongs to $\mathfrak{R}$.) If, in particular, the value $f(a)$ of f at a is the limit of f in the sense of convergence in $\mathfrak{R}$, then we say that f is continuous at a in the sense of convergence in $\mathfrak{R}$.

By choosing $\mathfrak{R}$ in different ways we get different variants of the concept of limit and continuity at a point for a function. Suppose, for example, that $\mathfrak{R}$ is the space M of bounded functions $\varphi: B(0, 1) \to \mathbf{R}$ and the topology in M is defined by the norm

$$\|\varphi\|_M = \sup_{x \in B(0,1)} |\varphi(x)|.$$

It is not hard to show that the limit in the sense of convergence in M coincides with the limit in the usual sense. The concept of a function continuous at a point $a \in U$ in the sense of convergence in M is equivalent to the usual concept of a function continuous at a. A function $f: U \to \mathbf{R}$ is said to be L_p-*continuous* at a point $a \in U$ if

$$\lim_{h \to 0} \int_{B(0,1)} |f(a + hX) - f(a)|^p \, dx = 0. \tag{4.1}$$

If f is L_p-continuous at a, then we say also that a is a *Lebesgue L_p-point* of f.

Performing a change of the variable of integration in the integral in (4.1) according to the formula $a + rX = t$, we get that (4.1) is equivalent to the following condition:

$$\lim_{r \to 0} \frac{1}{|B(a,r)|} \int_{B(a,r)} |f(t) - f(a)|^p \, dt = 0.$$

Let $U \subset \mathbf{R}^n$ be an open set. Denote by $\mathfrak{B}_0(U)$ the collection of all Borel sets lying strictly inside U. A *measure* on U is defined to be any countably additive function $\mu: \mathfrak{B}_0(U) \to \mathbf{R}$. Let f be a function of class $L_{1,\text{loc}}(U)$. Then a certain measure μ_f is defined by

$$\mu_f(A) = \int_A f(x)dx, \qquad A \in \mathfrak{R}_0(U).$$

The measure μ_f is called the *indefinite integral* of f.

Let $x \in U$, and let μ be a measure on U. The limit

$$\lim_{r \to 0} \frac{\mu[B(x,r)]}{|B(x,r)|} = (D\mu)(x),$$

if it exists, is called the *density* of μ at the point x. The measure μ is said to be *differentiable* at a point a if its density $(D\mu)(x)$ at this point is defined and finite.

THEOREM 4.1. *Every measure μ in an open set $U \subset \mathbf{R}^n$ is differentiable almost everywhere in U. If μ is the indefinite integral of $f \in L_{1,\text{loc}}(U)$, then $D\mu(x) = f(x)$ for almost all $x \in U$.*

(See, for example, [149] for a proof of the theorem.)

Let $E \subset U$ be a measurable set, and let $\mu_E(A) = |A \cap E|$ for $A \in \mathfrak{B}_0(U)$. The function μ_E is a measure. Obviously, μ_E is the indefinite integral of the indicator function χ_E of E. It follows from Theorem 4.1 that for almost all $x \in E$ the density of the measure μ_E is equal to 1, and hence the set of $x \in E$ such that $D\mu_E(x) < 1$ is a set of measure zero. In particular, if $D\mu_E(x) < 1$ for all $x \in E$, then E itself has measure zero; this fact was used in proving Theorem 11.1 of Chapter II.

We have the following theorem, also due to Lebesgue.

THEOREM 4.2. *Suppose that U is an open subset of $\mathbf{R}^n$, and $f: U \to \mathbf{R}^n$ is a function of class $L_{p,\text{loc}}(U)$. Then f is continuous in the sense of convergence in L_p almost everywhere in U. In other words, if $f \in L_{p,\text{loc}}(U)$, then for almost all $x \in U$*

$$\int_{B(0,1)} |f(x + hX) - f(x)|^p \, dX \to 0$$

as $h \to 0$.

Note that Theorem 4.1 implies only that for almost all $x \in U$

$$f(x) = \lim_{h \to 0} \frac{1}{\sigma_n} \int_{B(0,1)} f(x + hX)dX.$$

For the proof of Theorem 4.2 see, for example, [183].

LEMMA 4.1. *Let $u: U \to \mathbf{R}^n$ (U an open subset of $\mathbf{R}^n$) be a function in $l_{p,\text{loc}}(U)$. If $x \in U$ is a Lebesgue L_p-point of u, then*

$$\frac{1}{h^n} \int_U u(z) \psi \left(\frac{z - x}{h} \right) dz \to u(x) \int_{B(0,1)} \psi(y) dy$$

for every function $\psi \in C^\infty$ compactly supported in $\mathbf{R}^n$ inside the unit ball $B(0, 1)$ as $h \to 0$.

PROOF. Performing a change of variable in the integral by setting $(z - x)/h = y$, we get that

$$\frac{1}{h^n} \int_U u(z) \psi \left(\frac{z - x}{h} \right) dz = \int_{B(0,1)} u(x + hy) \psi(y) dy.$$

Assume that x is a Lebesgue L_p-point of f. Then

$$\int_{B(0,1)} |u(x + hy) - u(x)|^p \, dy \to 0$$

as $h \to 0$; this clearly gives us the statement of the lemma.

LEMMA 4.2. *Suppose that u is a function of class $W^1_{1,\text{loc}}(U)$, where U is an open subset of $\mathbf{R}^n$, and $\varphi \geq 0$ is a function of class C^∞ with compact support in $B(0, 1)$. Then for almost all $x \in U$*

$$\int_{B(0,1)} \frac{u(x + hy) - u(x)}{h} \varphi(y) dy$$

$$= \frac{1}{h} \int_0^h \left\{ \sum_{i=1}^n \frac{1}{t^n} \int_U \frac{\partial u}{\partial x_i}(z) \frac{z_i - x_i}{l} \varphi \left(\frac{z - x}{t} \right) dz \right\} dt \qquad (4.2)$$

when $0 < h < \rho(x, \partial U)$.

PROOF. Let $u \in W^1_{p,\text{loc}}(U)$, and let

$$\theta(h) = \int_{B(0,1)} u(x + hy) \varphi(y) dy = \frac{1}{h^n} \int_{B(x,y)} u(z) \varphi \left(\frac{z - x}{h} \right) dz.$$

For $h < \rho(x, \partial U)$ the closed ball $\overline{B}(x, h)$ is entirely contained in U, and the function $z \to \varphi((z - x)/h)$ vanishes outside this ball; hence

$$\theta(h) = \frac{1}{h^n} \int_U u(z) \varphi \left(\frac{z - x}{h} \right) dz.$$

From this it is clear, in particular, that θ is a C^∞-function in the interval $(0, \rho(x, \partial U))$. Differentiating with respect to h, we get that

$$h^n \theta'(h) + nh^{n-1}\theta(h) = -\frac{1}{h^2}\int_U \sum_{i=1}^n \frac{\partial\varphi}{\partial y_i}\left(\frac{z-x}{h}\right)(z_i - x_i)$$
$$\times u(z)dz. \tag{4.3}$$

Let $\lambda(z) = \varphi((z-x)/h)$. Then

$$\frac{\partial\lambda}{\partial z_i} = \frac{1}{h}\frac{\partial\varphi}{\partial y_j}\left(\frac{z-x}{h}\right),$$

and hence

$$\frac{\partial\varphi}{\partial y_i}\left(\frac{z-x}{h}\right)(z_i - x_i) = h\frac{\partial}{\partial z_i}\left[(z_i - x_i)\varphi\left(\frac{z-x}{h}\right)\right] - h\varphi\left(\frac{z-x}{h}\right).$$

We substitute this into (4.3). The function

$$z \to \varphi\left(\frac{z-x}{h}\right)(z_i - x_i)$$

belongs to C^∞, and is compactly supported in U for $h < \rho(x, \partial U)$. Transferring the differentiation operator $\partial/\partial z_i$ in the well-known way to the function f, we get from this that

$$h^n \theta'(h) + nh^{n-1}\theta(h) = \int_U \sum_{i=1}^n \frac{z_i - x_i}{h}\frac{\partial u}{\partial z_i}(z)\varphi\left(\frac{z-x}{h}\right)dz$$
$$+ \frac{n}{h}\int_U u(z)\varphi\left(\frac{z-x}{h}\right)dz.$$

The last term on the right-hand side is equal to $nh^{n-1}\theta(h)$; thus,

$$\theta'(h) = \frac{1}{h^n}\sum_{i=1}^n \int_U \frac{\partial u}{\partial z_i}(z)\frac{z_i - x_i}{h}\varphi\left(\frac{z-x}{h}\right)dz.$$

From this we find by integration that

$$\theta(h) - \theta(\varepsilon) = \int_\varepsilon^h \sum_{i=1}^n \frac{1}{t^n}\int_U \frac{\partial f}{\partial x_i}(z)\frac{z_i - x_i}{t}\varphi\left(\frac{z-x}{t}\right)dz\,dt. \tag{4.4}$$

Further,

$$\theta(\varepsilon) = \int_{B(0,1)} u(x + \varepsilon y)\varphi(y)dy.$$

By Lemma 4.1, at each Lebesgue point x of u (almost all $x \in U$ are such points)

$$\int_{B(0,1)} u(x + \varepsilon y)\varphi(y)dy \to u(x)\int_{B(0,1)} \varphi(y)dy.$$

Passing to the limit in (4.4), we get that if x is a Lebesgue point of u, then (4.2) holds for this x. The lemma is proved.

THEOREM 4.3. *Let $f: U \to \mathbf{R}$ be a function of class $W_{p,\mathrm{loc}}^1(U)$. Then the linear mapping $f'(x)$ is the differential of f at the point $x \in U$ in the sense of convergence in W_p^1 for almost all $x \in U$.*

This is a special case of a certain general theorem established by the author in [131].

PROOF. For $x \in U$ and $h > 0$ let $R_{h,x}$ be the function

$$Y \in B(0,1) \to \frac{f(x+hY) - f(x)}{h} - \sum_{i=1}^{n} \frac{\partial f}{\partial x_i}(x)Y_i.$$

The theorem asserts that for almost all $x \in U$ the quantity $\|R_{h,x}\|_{W_p^1[B(0,1)]}$ tends to zero as $h \to 0$. Let φ be an arbitrary C_0^∞-function with compact support in $B(0,1)$ and such that $\int_{\mathbf{R}^n} \varphi(y)dy = 1$. For $u \in W_p^1[B(0,1)]$ let

$$\|u\|_{W_p^1} = \left| \int_{B(0,1)} \varphi(z)u(x)dx \right| + \sum_{i=1}^{n} \left\| \frac{\partial u}{\partial x_i} \right\|_{L_p[B(0,1)]} \tag{4.5}$$

This defines a norm which, in view of Sobolev's classical results, is equivalent to $\|u\|_{1,p,B(0,1)}$. To prove the theorem it thus suffices to establish the following statements:

A. $\|\frac{\partial}{\partial y_i}\mathbf{R}_{h,x}\|_{L_p[B(0,1)]} \to 0$ as $h \to 0$ for almost all $x \in U$ for each $i = 1,\dots,n$.

B. For almost all $x \in U$

$$\int_{B(0,1)} R_{h,x}(y)\varphi(y)dy \to 0$$

as $h \to 0$.

Let us prove A. For each $i = 1,\dots,n$

$$\frac{\partial}{\partial y_i} R_{h,x}(y) = \frac{\partial f}{\partial x_i}(x+hy) - \frac{\partial f}{\partial x_i}(x),$$

and, hence,

$$\left\| \frac{\partial}{\partial y_i} R_{h,x} \right\|_{L_p} = \left(\int_{B(0,1)} \left| \frac{\partial f}{\partial x_i}(x+hy) - \frac{\partial f}{\partial x_i}(x) \right|^p dy \right)^{1/p}.$$

In view of Theorem 4.2 the integral on the right-hand side tends to zero as $h \to 0$ for almost all $x \in U$, and statement A is proved.

We now prove B. The proof is based on the equality in Lemma 4.2. Let x be a Lebesgue L_p-point of f and each of its derivatives. Let $0 < h < p(x, \partial U)$. Then, by Lemma 4.2,

$$\int_{B(0,1)} (1/h)[f(x+hy) - f(x)]\varphi(y)dy$$

$$= (1/h) \int_0^h \left\{ \sum_{i=1}^n \frac{1}{t^n} \int_U \frac{\partial f}{\partial x_i}(z) \frac{z_i - x_i}{t} \varphi\left(\frac{z-x}{t}\right) dz \right\} dt. \quad (4.6)$$

Let us now use Lemma 4.1, with

$$u = \frac{\partial f}{\partial x_i} \psi(y) = y_i \varphi(y).$$

We get that as $t \to 0$

$$\frac{1}{t^n} \int_U \frac{\partial f}{\partial x_i}(z) \frac{z_i - x}{t} \varphi\left(\frac{z-x}{t}\right) dz \to \int_{B(0,1)} \frac{\partial f}{\partial x_i}(x) y_i \varphi(y) dy.$$

This implies that the right-hand side of (4.6) tends to the limit

$$\int_{B(0,1)} \sum_{i=1}^n \frac{\partial f}{\partial x_i}(x) y_i \varphi(y) dy$$

as $h \to 0$. Thus, for the given $x \in U$

$$\lim_{h \to 0} \int_{B(0,1)} \frac{f(x+hy) - f(x)}{h} \varphi(y) dy = \int_{B(0,1)} \sum_{i=1}^n \frac{\partial f}{\partial x_i}(x) y_i \varphi(y) dy.$$

This establishes B, and the theorem is proved.

4.2. Proof of Lemma 1.1 in Chapter II. The lemma can be proved by means of a certain integral representation due to Sobolev. We present the corresponding formula here. A derivation of it can be found, for example, in [155] or [156]. Let us first consider the case of functions on the unit ball.

Take an arbitrary function $\varphi \geq 0$ of class $C^\infty(\mathbf{R}^n)$ such that $S(\varphi)$ is contained in the ball $B(0, 1/3)$ and $\int_{\mathbf{R}^n} \varphi(y) dy = 1$. For every vector-valued function $f: B(0,1) \to \mathbf{R}^k$ of class C^1 we have

$$f(x) = \int_{B(0,1)} \varphi(y) f(y) dy + \sum_{i=1}^n \int_{B(0,1)} K_i(x,y) \frac{\partial f}{\partial x_i}(y) dy, \quad (4.7)$$

where the functions K_i are such that $|K_i(x,y)| \leq L/|x-y|^{n-1}$ for any $x, y \in B(0,1)$, where $L < \infty$ is a constant, and K_i belongs to C^∞ in the subdomain of $\mathbf{R}^{2n}$ consisting of all pairs (x,y) with $x, y \in B(0,1)$ and $x \neq y$. (See, for example, [138], [161], or [162] for a proof of (4.7).) It is usual to consider real functions in deriving (4.7). The result can be obtained for vector-valued functions by applying (4.7) to the individual components of the function.

Let f be an arbitrary function of class C^1 on the ball $B(a, r)$. Applying (4.7) to the function $\tilde{f}(X) = f(a + rX)$, we get after obvious transformations that

$$f(x) = (1/r^n) \int_{B(a,r)} \varphi[(y - a)/r] f(y) dy$$

$$+ \sum_{i=1}^{n} \int_{B(a,r)} H_i(x, y) \frac{\partial f}{\partial x_i}(y) dy, \qquad (4.8)$$

where

$$H_i(x, y) = \frac{1}{r^{n-1}} K_i \left(\frac{x - a}{r}, \frac{y - a}{r} \right).$$

The function H_i is a function of class C^∞ on the set of all pairs $(x, y) \in \mathbf{R}^{2n}$ such that $x, y \in B(a, r)$ and $x \neq y$, and it admits the estimate $|H_i(x, y)| \leq L/|x - y|^{n-1}$.

Let U be an open domain in $\mathbf{R}^n$ and $f: U \to \mathbf{R}^k$ a function of class $W_m^1(U)$, where $1 \leq m \leq n$. Assume that there exist numbers $\alpha > 0$, $\delta > 0$, and $M < \infty$ such that for every ball $B(a, r) \subset U$ with radius less than δ

$$\int_{B(a,r)} \|f'(x)\|^m \, dx \leq M r^{n-m+m\alpha}. \qquad (4.9)$$

Assume first that f is a C^∞-function. Let $a \in U$ be an arbitrary point, and let $r < \rho(a)/3$, where $\rho(a) = \rho(a, \partial U)$. We use the integral representation (4.8). Define

$$(1/r^n) \int_{B(a,r)} \varphi[(y - a)/r] f(y) dy = k.$$

Then

$$|f(x) - k| \leq \sum_{i=1}^{n} \int_{B(a,r)} |H_i(x, y)| \left| \frac{\partial f}{\partial x_i}(y) \right| dy.$$

We have that

$$|H_i(x, y)| \leq L|x - y|^{n-1}, \qquad \left| \frac{\partial f}{\partial x_i}(y) \right| \leq \|f'(y)\|.$$

From this,

$$|f(x) - k| \leq \int_{B(a,r)} (dL \|f'(y)\| dy / |x - y|^{n-1}).$$

For $x \in B(a, r)$ the inclusion $B(x, 2r) \supset B(a, r)$ holds, and at the same time $B(x, 2r) \subset B(a, 3r) \subset U$. Hence,

$$|f(x) - k| \leq nL \int_{B(x, 2r)} \|f'(y)\|(dy/|x - y|^{n-1})$$

$$= nL \int_{B(0, 2r)} \|f'(x + z)\| dz/|z|^{n-1}. \tag{4.10}$$

Let

$$v(t) = \int_{B(0, t)} \|f'(x + z)\| dz, \qquad \lambda(t) = \int_{S(0, t)} \|f'(x + z)\| d\sigma_z,$$

where $d\sigma_z$ is the area element of the sphere. We have that $v(t) = \int_0^t \lambda(t) dt$. In our notation, (4.10) gives us that

$$|f(x) - k| \leq nL \int_0^{2r} (\lambda(t)/t^{n-1}) dt = nL \int_0^{2r} v'(t) dt/t^{n-1}. \tag{4.11}$$

Let us estimate $v(t)$. By the Hölder inequality,

$$v(t) = \int_{B(0, t)} \|f'(x + z)\| dz$$

$$\leq \left(\int_{B(0, t)} \|f'(x + z)\|^m dz \right)^{1/m} \sigma_n^{1/m} t^{n - n/m}$$

$$\leq (\sigma_n M)^{1/m} t^{n/m - 1 + \alpha} t^{n - n/m} = (\sigma_n M)^{1/m} t^{n - 1 + \alpha}. \tag{4.12}$$

We transform the integral on the right-hand side of (4.11) by integration by parts. The use of (4.12) gives us

$$\int_0^{2r} \frac{v'(t)}{t^{n-1}} dt = \frac{v(t)}{t^{n-1}} \Big|_0^{2r} + (n - 1) \int_0^{2r} \frac{v(t) dt}{t^n} \leq C_1 M^{1/m} r^\alpha.$$

From this,

$$|f(x) - k| \leq C_2 M^{1/m} r_\alpha, \tag{4.13}$$

where $C_2 = 2^\alpha nL\sigma_n^{1/m}(1 + (n - 1)/\alpha)$ is a constant. Since the point $x \in B(a, r)$ was arbitrary, it follows from (4.13) that the oscillation of f on the set $B(a, r) \backslash E$ does not exceed $CM^{1/m} r^\alpha$.

We now get rid of the condition that $f \in C^\infty$. Let $f: U \to \mathbf{R}^k$ be an arbitrary function satisfying the conditions of Lemma 4.1. Take an arbitrary averaging kernel $\alpha \geq 0$, and let f_h $(h > 0)$ be the mean function for f constructed from this kernel. For every $x \in \hat{U}_h$

$$\|f_h'(x)\|^m = \left\| \int_{B(0, 1)} f'(x + hz)\alpha(z) dz \right\|^m$$

$$\leq \int_{(0, 1)} \|f'(x + hz)\|^n \alpha(z) dz.$$

Integrating both sides over the ball $B(a, r) \subset \hat{U}_h$, we get that

$$\int_{B(a,r)} \|f'_n(x)\|^m \, dx \leq \int_{B(0,1)} \left(\int_{B(a,r)} \|f'(x + hz)\|^m \, dx \right) \alpha(z) dz.$$

If $x \in B(a, r)$, then $x + hz \in U$ for $|z| \leq 1$. Estimating the inside integral, we find from (4.9) that

$$\int_{B(a,r)} \|f'_h(x)\|^m \, dx \leq Mr^{n-m+m\alpha}. \tag{4.14}$$

Thus, f_h satisfies the same conditions as f. By what has been proved, this gives us that the oscillation of f on every ball $B(a, r)$ with $a \in U_h$ and $r < \rho(a, \partial U)/3$ does not exceed Cr^α, where C is a constant. This enables us to conclude that the family (f_h) of functions is uniformly equicontinuous on every set lying strictly inside U. As $h \to 0$ we have that $\|f_h - f\|_{L_1(A)} \to 0$ for every set A lying strictly inside U. This implies that the family of functions f_h converges locally uniformly to some function $\tilde{f}$ as $h \to 0$. The function $\tilde{f}$ is continuous in U, and $\tilde{f}(x) = f(x)$ almost everywhere in U. Passing to the limit, we get that the oscillation of $\tilde{f}$ on $B(a, r)$ does not exceed $CM^{1/m}r^\alpha$ for $r < \rho(a)/3$. The proof of the lemma is complete.

4.3. An estimate of the modulus of continuity of a monotone function of class W_n^1. Let U be an open subset of R^n, and $f: U \to \mathbf{R}^m$ a continuous mapping. We say that f *satisfies condition* $E(T)$, where $T \geq 1$ is a constant, if

$$d[f(\overline{B}(x, r)] \leq Tdf[S(x, r)]$$

for every closed ball $\overline{B}(x, r) \subset U$. In particular, if f is a homeomorphism, then f satisfies condition $E(T)$ with the constant $T \equiv 1$.

LEMMA 4.3. *Let* $f: S(x_0, r) \to \mathbf{R}^m$ *be a continuous function of class* W_n^1 *on the sphere* $S(x_0, r)$ *in* $\mathbf{R}^n$. *Then*

$$d(f[S(x_0, r)]) \leq C_n r^{1/n} \left(\int_{S(x_0,r)} \|f'(x)\|^n \, d\sigma \right)^{1/n},$$

where C_n *is a constant,* $0 < C_n < \infty$.

PROOF. Since $n > \dim S(x_0, r) = n - 1$, the Sobolev imbedding theorem gives us the estimate

$$d(f[S(x_0, r)]) \leq C(r) \left(\int_{S(x_0,r)} \|f'(x)\|^n \, d\sigma \right)^{1/n}.$$

It is not hard to conclude from similarity considerations that $C(r) = C_n r^{1/n}$, which is what was required to prove (see also [48]).

THEOREM 4.4. *Let U be an open subset of $\mathbf{R}^n$, $K \subset U$ a compact set, and $V \supset K$ an open set strictly interior to U. Then there is a constant $\delta > 0$ such that*

$$|f(x_1) - f(x_2)| \leq C_n(M)^{1/n} T\theta(|x_1 - x_2|)$$

for every mapping $f: U \to \mathbf{R}^m$ of class $W^1_{n,\mathrm{loc}}(U)$ satisfying condition $E(T)$ and for any two points x_1 and x_2 in K with $|x_1 - x_2| < \delta$, where C_n is the constant in Lemma 4.3,

$$M = \int_V \|f'(x)\|^n \, dx,$$

and

$$\theta(t) = (2/\ln(2/t))^{1/n};$$

$\theta(t) \to 0$ *as $t \to 0$.*

PROOF. Let $\rho = \mathrm{dist}(K, \mathbf{R}^n \backslash V)$. Denote by δ the largest number t such that $t \leq 2$ and $(t/2) + \sqrt{t/2} \leq \rho$.

Let x_1 and x_2 be two arbitrary points in K such that $|x_1 - x_2| = 2h < \delta$. Then $h < \delta/2 \leq 1$. Let $x_0 = \frac{1}{2}(x_1 + x_2)$. We construct the balls $B(x_0, h)$ and $B(x_0, \sqrt{h})$. For every $x \in B(x_0, \sqrt{h})$ we have that

$$|x - x_1| \leq |x - x_0| + |x_0 - x_1| < \sqrt{h} + h < \delta/2 + \sqrt{\delta/2} < \rho,$$

which implies that $x \in V$, and hence $B(x_0, \sqrt{h}) \subset V$. Obviously, $B(x_0, h) \subset B(x_0, \sqrt{h})$.

Let (f_ν), $\nu = 1, 2, \ldots$, be a sequence of C^∞-functions such that $f_\nu(x) \to f(x)$ uniformly on V as $\nu \to \infty$ and

$$\int_V |f'_\nu(x) - f'(x)|^n \, dx < 2^{-\nu}$$

for every ν. We have that

$$\int_{B(x_0, \sqrt{h})} |f'_\nu(x) - f'(x)|^n \, dx = \int_0^{\sqrt{h}} \left(\int_{S(x_0, r)} |f'_\nu(x) - f'(x)|^n \, d\sigma \right) dr.$$

Let

$$\int_{S(x_0, r)} |f'_\nu(x) - f'(x)|^n \, dx = F_\nu(r).$$

For each $\nu = 1, 2, \ldots$

$$\int_0^{\sqrt{h}} F_\nu(r) dr < \frac{1}{2^\nu}$$

and hence

$$\sum_{\nu=1}^{\infty} \int_0^{\sqrt{h}} F_\nu(r) dr < \infty.$$

This implies that $F_\nu(r) \to 0$ as $\nu \to \infty$ for almost all $r \in (0, \sqrt{h})$. For each ν such that this holds, the restriction of f to $S(x_0, r)$ belongs to the class $W_n^1(S(x_0, r))$. For $r \in (0, \sqrt{h})$ let

$$\gamma(r) = d(f[S(x_0, r)]).$$

By the estimate of Lemma 4.3,

$$\frac{[\gamma(r)]^n}{r} \le C_n^n \int_{S(x_0, r)} |f'(x)|^n \, d\sigma$$

for almost all $r \in (0, \sqrt{h})$. Integrating with respect to r, we get that

$$\int_h^{\sqrt{h}} \frac{[\gamma(r)]^n}{r} \, dr \le C_n^n \int_h^{\sqrt{h}} \left(\int_{S(x_0, r)} |f'(x)|^n \, d\sigma \right) dr$$

$$\le C_n^n \int_{B(x_0, \sqrt{h})} |f'(x)|^n \, dx \le C_n^n M.$$

Let A be the set of $t \in [h, \sqrt{h}]$ such that

$$[\gamma(t)] \int_h^{\sqrt{h}} \frac{dr}{r} \le \int_h^{\sqrt{h}} \frac{[\gamma(r)]^n}{r} \, dr. \tag{4.15}$$

The measure of A is positive. Indeed, assume not. Then for almost all t

$$\frac{\gamma(t)}{t} \int_h^{\sqrt{h}} \frac{dr}{r} > \frac{1}{t} \int_h^{\sqrt{h}} \frac{[\gamma(r)]^n}{r} \, dr.$$

Integrating termwise, we get a contradiction. Fix an arbitrary $t \in [h, \sqrt{h}]$ for which (4.15) holds. For this t

$$[\gamma(t)]^n \ln \frac{\sqrt{h}}{h} \le C_n^n M,$$

which implies that

$$\gamma(t) \le \frac{C_n (2M)^{1/n}}{[\ln(1/h)]^{1/n}} = C_n M^{1/n} \theta(|x_1 - x_2|).$$

We have that $f(x_1) \in f[B(x_0, t)]$ and $f(x_2) \in f(B(x_0, t))$; hence $|f(x_1) - f(x_2)|$ does not exceed the diameter of $f[B(x_0, t)]$. Since f satisfies condition $E(T)$, the diameter of $f[B(x_0, t)]$ does not exceed $Td(f[S(x_0, t)]) = T\gamma(t)$. We get finally that

$$|f(x_1) - f(x_2)| \le T\gamma(t) \le C_n M^{1/n} T\theta(|x_1 - x_2|),$$

which is what was required to prove.

COROLLARY. *Let U be an open set in $\mathbf{R}^n$. Assume that the mapping $f: U \to \mathbf{R}^m$ has property $E(T)$. If $f \in W^1_{n,\text{loc}}(U)$, then f is differentiable almost everywhere in U.*

PROOF. For $x \in U$, $h > 0$ and $Y \in \mathbf{R}^n$ let

$$F_h(Y) = \frac{f(x + hY) - f(x)}{h}.$$

On the basis of Theorem 4.3 the functions F_h converge in $W^1_n[B(0, 1)]$ to the function

$$f'(x)(Y) = \sum_{i=1}^{n} \frac{\partial f}{\partial x_i}(x) Y_i$$

for almost all $x \in U$ as $h \to 0$. A simple change of variables enables us to conclude that the functions $F_h(Y)$ converge to $f'(x)(Y)$ in $W^1_n[B(0, r)]$ for any $r > 0$; in particular, $F_h(Y) \to f'(x)(Y)$ in $W^1_n[B(0, 3)]$ as $h \to 0$. The function F_h has property $E(T)$ for each h, and there are constants $h_0 > 0$ and $M < \infty$ such that for $0 < h < h_0$

$$\int_{B(0,3)} |F'_h(Y)| dY \leq M.$$

We use the theorem, setting $K = \overline{B}(0, 1)$, $V = B(0, 2)$, and $U = B(0, 3)$ in it; we get that the family (F_h), $0 < h < h_0$, of functions is uniformly equicontinuous on $\overline{B}(0, 1)$. Consequently, $F_h(Y) \to f'(x)(Y)$ uniformly in $B(0, 1)$ as $h \to 0$. However, this means that $f'(x)$ is the usual differential of f at x. The corollary is proved.

§5. On the degree of a mapping

We present proofs of Lemmas 2.1 and 2.2 in §2 of Chapter II. All references below relate to that section.

Let G be a compact domain in $\mathbf{R}^n$, and let $f_0: G \to \mathbf{R}^n$ and $f_1: G \to \mathbf{R}^n$ be mappings of class C^k, $k \geq 1$. The mappings f_0 and f_1 are said to be *smoothly homotopic* as mappings of the pair $(G, \partial G)$ to the pair $(\mathbf{R}^n, \mathbf{R}^n \backslash \{y\})$ if there exists a homotopy φ of f_0 and f_1 which is a C^k-mapping on $G \times [0, 1]$.

LEMMA 5.1. *If the C^k-mappings $(k \geq 1) f_0: G \to \mathbf{R}^n$ and $g_1: G \to \mathbf{R}^n$ are homotopic as mappings of the pair $(G, \partial G)$ to the pair $(\mathbf{R}^n, \mathbf{R}^n \backslash \{y\})$, where $y \in \mathbf{R}^n$, then they are also smoothly homotopic.*

PROOF. Let $f: G \times [0, 1] \to \mathbf{R}^n$ be an arbitrary homotopy of f_0 and f_1 as mappings of the pair $(G, \partial G)$ to the pair $(\mathbf{R}^n, \mathbf{R}^n \backslash \{y\})$. For each $t \in [0, 1]$, $f_t: x \mapsto f(x, t)$ is a mapping of $(G, \partial G)$ to $(\mathbf{R}^n, \mathbf{R}^n \backslash \{y\})$. Let $H = f(\partial G \times [0, 1])$. The set H is compact, and $y \notin H$. This implies that there exists a $\delta > 0$ such that $|z - y| < \delta$ for all $z \in H$. Let $\varphi: G \times [0, 1] \to \mathbf{R}^n$ be a C^∞-mapping such that $|\psi(x, t) - f(x, t)| < \delta/2$ for $(x, t) \in G \times [0, 1]$. A mapping ψ satisfying this condition can be constructed by extending f to $\mathbf{R}^{n+1}$ with preservation of continuity and applying the averaging operation to the resulting extension. Let

$$\varphi(x, t) = \psi(x, t) + (1 - t)[f_0(x) - \psi(x, 0)] + t[f_1(x) - \psi(x, 1)].$$

Obviously, φ is a mapping of the same smoothness class as f_0 and f_1, and also $\varphi(x, 0) = f_0(x)$ and $\varphi(x, 1) = f_1(x)$. We let $\varphi_t(x) = \varphi(x, t)$ and show that φ_t is a mapping of $(G, \partial G)$ to $(\mathbf{R}^n, \mathbf{R}^n \backslash \{y\})$. Take an arbitrary point $x_0 \in \partial G$. Let $z_0 = f_t(x)$ and $z_1 = \varphi_t(x_0)$. Take an arbitrary point $x_0 \in \partial G$. Let $z_0 = f_t(x)$ and $z_1 = \varphi_t(x_0)$. Then

$$|z_1 - z_0| \le |\psi(x_0, t) - f(x_0, t)| + (1 - t)|f_0(x_0) - \psi(x_0, 0)|$$
$$+ t|f_1(x_0) - \psi(x_0, t)| < (\delta/2) + (1 - t)\delta/2 + t\delta/2 = \delta.$$

The point z_0 belongs to H, and hence $|z_0 - y| > \delta$. Since $|z_1 - z_0| < \delta$, this implies that $\varphi_t(x_0) \ne y$. The point $x_0 \in \partial G$ was chosen arbitrarily, so this proves that φ_t is a mapping of $(G, \partial G)$ to $(\mathbf{R}^n, \mathbf{R}^n \backslash \{y\})$, and thus φ is a homotopy of f_0 and f_1. The lemma is proved.

LEMMA 5.2 (SARD'S THEOREM). *Let $U \subset \mathbf{R}^n$ be an open set, $f: U \to \mathbf{R}^n$ a C^1-mapping, and $E \subset U$ the set of all points $x \in U$ at which the Jacobian of f vanishes. Then $f(E)$ is a set of measure zero in $\mathbf{R}^n$.*

PROOF. Let a be an arbitrary point of E, and let $r > 0$ be such that the closed cube $Q_0 = \overline{Q}(a, r)$ is contained in U. For $x, y \in Q_0$ let $y(t) = (1 - t)x + ty$. Then

$$f(y) - f(x) = f[y(1)] - f[y(0)]$$
$$= \int_0^1 \frac{d}{dt} (f[y(t)])dt = \int_0^1 \sum_{i=1}^n \frac{\partial f}{\partial x_i} [y(t)](y_i - x_i)dt. \quad (5.1)$$

Further,

$$f(y) - f(x) - f'(r)(y - x)$$
$$= \int_0^1 \sum_{i=1}^n \left(\frac{\partial f}{\partial x_i} [y(t)] - \frac{\partial f}{\partial x_i} (x) \right) (y_i - x_i)dt. \quad (5.2)$$

The functions $\partial f/\partial x_i$ are continuous. Since Q_0 is compact, this implies that they are bounded and uniformly continuous on Q_0. Let $|(\partial f/\partial x_i)(x)| \le M_i$ for all $i = 1, \ldots, n$ and $x \in Q_0$, and let ε_i be the modulus of continuity of $\partial f/\partial x_i$. Then (5.1) gives us that

$$|f(y) - f(x)| \le \sum_{i=1}^{n} M_i|y_i - x_i| \le M|y - x|, \qquad (5.3)$$

where $M = \sqrt{\sum_1^n M_i^2}$. We conclude similarly from (5.2) that

$$|f(y) - f(x) - f'(x)(y - x)| \le \varepsilon(|y - x|)|y - x|, \qquad (5.4)$$

where $\varepsilon = \sqrt{\sum_1^n \varepsilon_i^2}$.

Let ν be an arbitrary positive integer, and divide the cube Q_0 into ν^n equal closed cubes by planes parallel to the coordinate planes. The diameter of each cube in the subdivision is equal to $r\sqrt{n}/\nu$, and its volume is $(r/\nu)^n$.

Let Q be a cube in this subdivision such that $Q \cap E$ is nonempty. We estimate the measure of the set $f(Q)$. Take an arbitrary point $y \in Q \cap E$ and let $v = f(y)$. Then $v \in f(Q)$, and

$$|f(x) - v| \le M|x - y| \le M\sqrt{u}r/\nu$$

for every $x \in Q$.

According to the definition of E, $J(y, f) = 0$. Let Y be the set of all points $u \in \mathbf{R}^n$ of the form $u = f(y) + f'(y)(h)$, where $h \in \mathbf{R}^n$. Since $J(y, f) = 0$, Y is a k-dimensional plane passing through the point $v = f(y)$, where $0 \le k \le n - 1$. We construct an arbitrary $(n - 1)$-dimensional plane X containing Y. For $x \in Q$ we have that

$$|f(x) - f(y) - f'(y)(x - y)| \le \varepsilon(|x - y|)|x - y|$$
$$\ge \varepsilon(r\sqrt{n}/\nu)(r\sqrt{n})/\nu.$$

The point $f(y) + f'(y)(x - y)$ belongs to X; hence the distance from $f(x)$ to this plane is at most $\varepsilon[r\sqrt{n}/\nu](r\sqrt{n}/\nu)$. Let $K = B(v, Mr\sqrt{n}/\nu) \cap X$. It follows from what has been proved that for $x \in Q$ the point $f(x)$ lies in the right circular cylinder of height $2\varepsilon(r\sqrt{n}/\nu)(r\sqrt{n}/\nu)$ whose middle section is the $(n - 1)$-dimensional disk K. This implies that

$$|f(Q)| \le C\varepsilon(r\sqrt{n}/\nu)(r/\nu)^n = C\varepsilon(r\sqrt{n}/\nu)|Q|. \qquad (5.5)$$

Let $Q_1, \ldots, Q_m$ be the cubes in our subdivision of Q_0 which contain points of E. Obviously,

$$f(E \cap Q_0) \subset \bigcup_{i=1}^{m} f(Q_i).$$

Applying (5.5) to each cube Q_i, we get that

$$|f(E \cap Q_0)| \leq C\varepsilon(r\sqrt{n}/\nu)|Q_0|.$$

The right-hand side of this inequality tends to zero as $\nu \to \infty$. This tells us that $|f(E \cap Q_0)| = 0$.

Every point $a \in E$ thus has a neighborhood Q_0 such that $f(Q_0 \cap E)$ is a set of measure zero. Covering E by countably many such neighborhoods, we get that $f(E)$ is a set of measure zero. The lema is proved.

PROOF OF LEMMA 2.1. Let G be a compact domain, and $f: G \to \mathbf{R}^n$ a mapping of class C^k, where $k > 1$ and a is an (f, G)-admissible point. Let $A_1 = f^{-1}(a)$, and let $A_0 = \mathbf{R}^n \backslash G^0$. Take an arbitrary function $\zeta \in \widetilde{W}(A_0, A_1)$. The function ζ is equal to 1 in some neighborhood U_1 of the set A_1. Let $P = f(G \backslash U_1)$. The set P is compact, and $a \notin P$. Hence, there is a $\delta > 0$ such that the ball $B(a, \delta)$ does not contain points of P. According to (2.6), for every $y \in \mathbf{R}^n$ the exterior form θ_y is defined. We construct the exterior form $f^*\theta_y$ from it. For $y \in B(a, \delta)$ the form $d\zeta \wedge f^*\theta_y$ is defined on $G^0 \backslash U_1$. On U_1 we have that $d\zeta \equiv 0$. We extend the definition of the form $d\zeta \wedge f^*\theta_y$ by setting it equal to zero on U_1. As a result we get a form defined and continuous everywhere in G^0. The coefficients of this form are continuous functions of x and y. Let

$$\mu^*(y, f, G) = -\frac{1}{\omega_n} \int_{G^0} d\zeta \wedge f^*\theta_y.$$

The integral on the right-hand side does not depend on the choice of ζ, as can be shown by arguments repeating the beginning of the proof of Lemma 2.4. By the same Lemma 2.4, if f is regular with respect to y, then

$$\mu^*(y, f, G) = \mu(y, f, G).$$

On the basis of classical theorems on integrals depending on a parameter, $\mu^*(y, f, G)$ is continuous. Lemma 5.2 allows us to conclude that the set of $y \in \mathbf{R}^n$ such that $y \notin f(\partial G)$ and f is regular with respect to y is a set of measure zero. This implies, in particular, that $\mu^*(y, f, G)$ takes integer values on a dense subset of $B(a, \delta)$, which, since μ^* is continuous, is possible only if μ^* is constant and equal to an integer for $y \in B(a, \delta)$. Thus, $\mu^*(y, f, G)$ is an integer for $y \in B(a, \delta)$.

Let f_0 and f_1 be two homotopic C^k-mappings $(k > 1)$ of the pair $(G, \partial G)$ to the pair $(\mathbf{R}^n, \mathbf{R}^n \backslash \{a\})$. By Lemma 5.1, there exists a smooth homotopy f of f_0 and f_1 as mappings of pairs. Let $\widetilde{A}_1 = f^{-1}(a)$, and let A_1 be the image of $\widetilde{A}_1$ under the mapping $(x, t) \in \mathbf{R}^{n+1} \mapsto x$. The set A_1 is compact, and $G^0 \supset A_1$. Take an arbitrary $\zeta \in \widetilde{W}(A_0, A_1)$, where $A_0 = \mathbf{R}^n \backslash G^0$. For

$t \in [0, 1]$ let f_t be the mapping $x \mapsto f(x, t)$, and let

$$\mu(t) = -\frac{1}{\omega_n} \int_{G^0 \setminus A_1} d\xi \wedge f_t^* \theta_a.$$

The coefficients of the form $d\xi \wedge f_t^* \theta_a$ depend continuously on t, and this implies that μ is continuous. For each $t \in [0, 1]$ the quantity $\mu(t)$ is an integer, and hence $\mu(t) \equiv \text{const}$ in $[0, 1]$. If f_0 and f_1 are regular with respect to a, then

$$\mu(0) = \mu(a, f_0, G), \qquad \mu(1) = \mu(a, f_1, G),$$

and thus

$$\mu(a, f_0, G) = \mu(a, f_1, G).$$

The lemma is proved.

PROOF OF LEMMA 2.2. Suppose that $f: G \to \mathbf{R}^n$ is continuous, and a is an (f, G)-admissible point. We construct a sequence of C^∞-functions $\varphi_\nu: G \to \mathbf{R}^n$ converging uniformly to f. By Lemma 5.2, for each ν there is a point y_ν such that $|y_\nu - a| < 1/\nu$ and f_ν is regular with respect to y_ν. Let $h_\nu = y_\nu - a$, and let $f_\nu = \varphi_\nu + h_\nu$. As $\nu \to \infty$ the mappings f_ν converge to f uniformly, and each of them is regular, with respect to a. The lemma is proved.

Bibliography

ROBERT A. ADAMS
1. *Sobolev spaces*, Academic Press, 1975.

LARS V. AHLFORS
2. *Quasiconformal reflections*, Acta Math. **109** (1963), 291–301.
3. *Kleinsche Gruppen in der Ebene und im Raum*, Festband 70. Geburtstag R. Nevanlinna, Springer-Verlag, 1966, pp. 7–15.
4. *Conditions for quasiconformal deformations in several variables*, Contributions to Analysis (Collection Dedicated to Lipman Bers), Academic Press, 1974, p. 19–25.
5. *Invariant operators and integral representations in hyperbolic space*, Math. Scand. **36** (1975), 27–43.
6. *Quasiconformal deformations and mappings in* $\mathbf{R}^n$, J. Analyse Math. **30** (1976), 74–97.
7. *A singular integral equation connected with quasiconformal mappings in space*, Enseign. Math. (2) **24** (1978), 225–236.
8. *A singular operator in hyperbolic space*, Complex Analysis and Its Applications (I. N. Vekua Seventieth Birthday Vol.), "Nauka", Moscow, 1978, pp. 40–44.
9. *The Hölder continuity of quasiconformal deformations*, Amer. J. Math. **101** (1979), 1–9.

LARS AHLFORS AND ARNE BEURLING
10. *Conformal invariants and function-theoretic null-sets*, Acta Math. **83** (1950), 101–129.

A. D. ALEKSANDROV
11. *The existence almost everywhere of the second differential of a convex function and some properties of convex surfaces connected with it*, Leningrad Gos. Univ. Uchen. Zap. Ser. Mat. **6** (1939), 3–35. (Russian)

P. P. BELINSKIĬ

12. *On the continuity of spatial quasiconformal mappings and Liouville's theorem*, Dokl. Akad. Nauk SSSR **147** (1962), 1003–1004; English transl. in Soviet Math. Dokl. **3** (1962).

13. *Stability in the Liouville theorem on spatial quasiconformal mappings*, Some Problems in Mathematics and Mechanics (M. A. Lavrent'ev Seventieth Birthday Vol.), "Nauka", Leningrad, 1970, pp. 88–102; English transl. in Amer. Math. Soc. Transl. (2) **104** (1976).

14. *On the order of proximity of a spatial quasiconformal mapping to a conformal mapping*, Sibirsk. Mat. Zh. **14** (1973), 475–483; English transl. in Siberian Math. J. **14** (1973).

P. P. BELINSKIĬ AND M. A. LAVRENTIEV [LAVRENT'EV]

15. *On locally quasiconformal mappings in n-space ($n = 3$)*, Contribution to Analysis (Collection Dedicated to Lipman Bers), Academic Press, 1974, pp. 27–30.

O. V. BESOV

16. *Integral representations of functions in a domain with the flexible horn condition, and imbedding theorems*, Dokl. Akad. Nauk SSSR **273** (1983), 1294–1297; English transl. in Soviet Math. Dokl. **28** (1983).

17. *Estimates of the L_p-moduli of continuity and imbedding theorems for domains with the flexible horn condition*, Dokl. Akad. Nauk SSSR **275** (1984), 1036–1041; English transl. in Soviet Math. Dokl. **29** (1984).

O. V. BESOV, V. P. IL'IN, AND S. M. NIKOL'SKIĬ

18. *Integral representations of functions and imbedding theorems*, "Nauka", Moscow, 1975; English transl., Vols. I, II, Wiley, 1979.

B. V. BOYARSKIĬ [BOGDAN BOJARSKI]

19. *Homeomorphic solutions of Beltrami systems*, Dokl. Akad. Nauk SSSR. **102** (1955), 661–664. (Russian)

20. *Generalized solutions of a system of first-order differential equations of elliptic type with discontinuous coefficients*, Mat. Sb. **43** (85) (1957), 451–503. (Russian)

EUGENIO CALABI AND PHILIP HARTMAN

21. *On the smoothness of isometries*, Duke Math. J. **37** (1970), 741–750.

ALBERTO P. CALDERÓN

22. *On the differentiability of absolutely continuous functions*, Riv. Mat. Univ. Parma **1951**, 203–213.

A. P. CALDERÓN AND A. ZYGMUND

23. *Singular operators and differential equations*, Amer. J. Math. **79** (1957), 901–921.

E. DAVID CALLENDER

24. *Hölder continuity of n-dimensional quasi-conformal mappings*, Pacific J. Math. **10** (1960), 499–515.

PETRU CARAMAN

25. *n-dimensional quasiconformal mappings*, Ed. Acad. Rep. Soc. România, Bucharest, 1968; rev. English transl., Ed. Acad. Române, Bucharest, Abacus Press, Tunbridge Wells, and Haessner, Newfoundland, N. J., 1974.

A. V. CHERNAVSKIĬ

26. *Finite-to-one open mappings of manifolds*, Mat. Sb. **65** (**107**) (1964), 357–369; English transl. in Amer. Math. Soc. Transl. (2) **100** (1972).

27. Addendum to [26], Mat. Sb. **66** (**108**) (1964), 471–472; English transl. in Amer. Math. Soc. Transl. (2) **100** (1972).

N. S. DAIRBEKOV

28. *On the stability of classes of conformal mappings on the plane and in space*, Sibirsk. Mat. Zh. **27** (1986), no. 5, 188–191. (Russian)

N. S. DAIRBEKOV AND A. P. KOPYLOV

29. *On stability in the C-norm of classes of solutions of systems of linear partial differential equations*, Ninth School Theory of Operators in Function Spaces (September 1984), Abstracts of Reports, Ternopol', 1984, p. 137. (Russian)

30. *ξ-stability of classes of mappings, and systems of linear partial differential equations*, Sibirsk. Mat. Zh. **26** (1985), no. 2, 73–90; English transl. in Siberian Math. J. **26** (1985).

ENNIO DE GIORGI

31. *Sulla differenziabilità e l'analiticità delle estremali degli integrali multipli regolari*, Mem. Accad. Sci. Torino Cl. Sci. Fis. Mat. Nat. (3) **3** (1957), 25–43.

K. O. FRIEDRICHS

32. *On the boundary-value problems of the theory of elasticity and Korn's inequality*, Ann. of Math. (2) **48** (1947), 441–471.

F. W. GEHRING

33. *The definitions and exceptional sets for quasiconformal mappings*, Ann. Acad. Sci. Fenn. Ser. A I No. 281 (1960).

34. *Rings and quasiconformal mappings in space*, Proc. Nat. Acad. Sci. U.S.A. **47** (1961), 98–105.

35. *Symmetrization of rings in space*, Trans. Amer. Math. Soc. **101** (1961), 499–519.

36. *A remark on the moduli of rings*, Comment. Math. Helv. **36** (1961), 42–46.

37. *Rings and quasiconformal mappings in space*, Trans. Amer. Math. Soc. **103** (1963), 353–393.

38. *The L^p-integrability of the partial derivatives of a quasiconformal mapping*, Acta Math. **130** (1973), 265–277.

39. *Injectivity of local quasi-isometries*, Comment. Math. Helv. **57** (1982), 202–220.

40. *Characteristic properties of quasidisks*, Sém. Math. Sup., vol. 84, Presses Univ. Montréal, Montréal, 1982.

F. W. Gehring and F. Huckemann

41. *Quasiconformal mappings of a cylinder*, Proc. Romanian-Finnish Sem. Teichmüller Spaces and Quasiconformal Mappings (Braşov, 1969), Publ. House Acad. Socialist Republic Romania, Bucharest, 1971, pp. 171–186.

F. W. Gehring and J. Väisälä

42. *The coefficients of quasiconformality of domains in space*, Acta Math. **114** (1965), 1–70.

V. M. Gol'dshteĭn

43. *A homotopy property of mappings with bounded distortion*, Sibirsk. Mat. Zh. **11** (1970), 999–1008; English transl. in Siberian Math. J. **11** (1970).

44. *On the behavior of mappings with bounded distortion when the distortion coefficient is close to* 1, Sibirsk. Mat. Zh. **12** (1971), 1250–1258; English transl. in Siberian Math. J. **12** (1971).

45. *Extension of functions in the classes $B_{p,q}^l$ across quasiconformal boundaries*, Theory of Cubature Formulas and Applications of Functional Analysis to Problems of Mathematical Physics, (Proc. Sem. S. L. Sobolev, 1979, vol. 1), Inst. Mat. Sibirsk. Otdel. Akad. Nauk SSSR, Novosibirsk, 1979, pp. 12–32.

Vladimir Mikailovitch Gol'dšteĭn [V. M. Gol'dshteĭn] and Sergei Konstantinovitch Vodop'janov [S. K. Vodop'yanov]

46. *Prolongement des fonction de classe $\mathscr{L}_p^1$ et applications quasi conforme*, C. R. Acad. Sci. Paris Sér. A–B **290** (1980), A453–A456.

V. M. Gol′dshteĭn

47. *Extension of functions with first generalized derivatives from plane domains*, Dokl. Akad. Nauk SSSR **257** (1981), 268–271; English transl. in Soviet Math. Dokl. **23** (1981).

V. M. Gol′dshteĭn and Yu. G. Reshetnyak

48. *Introduction to the theory of functions with generalized derivatives and quasiconformal mappings*, "Nauka", Novosibirsk, 1983; English transl., *Foundations of the theory of functions with generalized derivatives and space quasiconformal mappings*, Reidel (to appear).

V. M. Gol′dshteĭn, V. I. Kuz′minov, and I. A. Shvedov

49. *Classes of differential forms analogous to classes of Sobolev functions*, Preprint, Inst. Mat. Sibirsk. Otdel. Akad. Nauk SSSR, Novosibirsk, 1981. (Russian) R. Zh. Mat. **1982**, 2A655.

L. G. Gurov

50. *On the stability of Lorentz mappings*, Dokl. Akad. Nauk SSSR **213** (1973), 267–269; English transl. in Soviet Math. Dokl. **14** (1973).

51. *Stability estimates for Lorentz mappings*, Sibirsk. Mat. Zh. **15** (1974), 498–515; English transl. in Siberian Math. J. **15** (1974).

52. *On the stability of Lorentz transformations. Estimates for the derivatives*, Dokl. Akad. Nauk SSSR **220** (1975), 273–276; English transl. in Soviet Math. Dokl. **16** (1975).

53. *On the stability of Lorentz transformations in space*, Sibirsk. Mat. Zh. **21** (1980), no. 2, 51–60; English transl. in Siberian Math. J. **21** (1980).

L. G. Gurov and Yu. G. Reshetnyak

54. *On an analogue of the concept of a function with bounded mean oscillation*, Sibirsk. Mat. Zh. **17** (1976), 540–546; English transl. in Siberian Math. J. **17** (1976).

Sze-tsen Hu

55. *Homotopy theory*, Academic Press, 1959.

V. P. Il′in

56. *On an imbedding theorem for a limiting exponent*, Dokl. Akad. Nauk SSSR **96** (1954), 905–908. (Russian)

V. V. Ivanov

57. *Smooth isomorphisms not having quasiconformal fractional powers*, Sibirsk. Mat. Zh. **27** (1986), no. 3, 103–111; English transl. in Siberian Math. J. **27** (1986).

TADEUSZ IWANIEC

58. *On L^p-integrability in PDEs and quasiregular mappings for large exponents*, Ann. Acad. Sci. Fenn. Ser. A I Math. **7** (1982), 301–322.
59. *Regularity theorems for solutions of partial differential equations for quasiconformal mappings in several dimensions*, Dissertation, Polish Acad. of Sci., Warsaw, 1982.

FRITZ JOHN

60. *Rotation and strain*, Comm. Pure Appl. Math. **14** (1961), 391–413.
61. *On quasi-isometric mappings*. I, Comm. Pure Appl. Math. **21** (1968), 77–110.
62. *On quasi-isometric mappings*. II, Comm. Pure Appl. Math. **22** (1969), 265–278.

F. JOHN AND L. NIRENBERG

63. *On functions of bounded mean oscillation*, Comm. Pure Appl. Math. **14** (1961), 415–426.

PETER W. JONES

64. *Quasiconformal mappings and extendibility of functions in Sobolev spaces*, Acta Math. **147** (1981), 71–88.

A. P. KOPYLOV

65. *On the behavior of quasiconformal space mappings close to conformal mappings on hyperplanes*, Dokl. Akad. Nauk SSSR **209** (1973), 1278–1280; English transl. in Soviet Math. Dokl. **14** (1973).
66. *On approximation of quasiconformal space mappings close to conformal mappings by smooth quasiconformal mappings*, Sibirsk. Mat. Zh. **13** (1972), 94–106; English transl. in Siberian Math. J. **13** (1972).
67. *Integral averagings and quasiconformal mappings*, Dokl. Akad. Nauk SSSR **231** (1976), 289–291; English transl. in Soviet Math. Dokl. **17** (1976).
68. *On the removability of a ball for space mappings close to conformal mappings*, Dokl. Akad. Nauk SSSR **234** (1977), 525–527; English transl. in Soviet Math. Dokl. **18** (1977).
69. *Stability of classes of multidimensional holomorphic mappings*. I: *The concept of stability. Liouville's theorem*, Sibirsk. Mat. Zh. **23** (1982), no. 2, 83–111; English transl. in Siberian Math. J. **23** (1982).
70. *Stability of classes of multidimensional holomorphic mappings*. II: *Stability of classes of holomorphic mappings*, Sibirsk. Mat. Zh. **23** (1982), no. 4, 65–89; English transl. in Siberian Math. J. **23** (1982).

71. *Stability of classes of multidimensional holomorphic mappings.* III: *Properties of mappings close to holomorphic mappings,* Sibirsk. Mat. Zh. **24** (1983), no. 3, 70–91; English transl. in Siberian Math. J. **24** (1983).

72. *On boundary values of mappings of the half-space that are close to being conformal,* Sibirsk. Mat. Zh. **24** (1983), no. 5, 76–93; English transl. in Siberian Math. J. **24** (1983).

M. KREĬNES

73. *Sur une classe de fonctions de plusieurs variables,* Mat. Sb. **9** (**51**) (1941), 713–720.

S. L. KRUSHKAL'

74. *On the absolute continuity and differentiability of certain classes of mappings of multidimensional domains,* Sibirsk. Mat. Zh. **6** (1965), 692–696. (Russian)

S. L. KRUSHKAL', B. N. APANASOV, AND N. A. GUSEVSKIĬ

75. *Kleinian groups and uniformization in examples and problems,* "Nauka", Novosibirsk, 1981; English transl., Amer. Math. Soc., Providence, R. I., 1986.

K. KURATOWSKI

76. *Topology.* Vol. 2, 4th ed., PWN, Warsaw, and Academic Press, New York, 1968.

O. A. LADYZHENSKAYA AND N. N. URAL'TSEVA

77. *Linear and quasilinear elliptic equations,* 2nd ed., "Nauka", Moscow, 1973; English transl. of 1st ed., Academic Press, 1968.

O. A. LADYZHENSKAYA, V. A. SOLONNIKOV, AND N. N. URAL'TSEVA

78. *Linear and quasilinear equations of parabolic type,* "Nauka", Moscow, 1967; English transl., Amer. Math. Soc., Providence, R. I., 1968.

M. LAVRENTIEFF [M. A. LAVRENT'EV]

79. *Sur une classe de représentations continues,* Mat. Sb. **42** (1935), 407–424.

80. *Sur un critère différentiel des transformations homéomorphes des domaines à trois dimensions,* C. R. (Dokl.) Acad. Sci. URSS **20** (1938), 241–242.

81. *Stability in Liouville's theorem,* Dokl. Akad. Nauk SSSR **95** (1954), 925–926. (Russian)

82. *On the theory of quasi-conformal mappings of three-dimensional domains*, J. Analyse Math. **19** (1967), 217–225.

M. A. LAVRENT'EV AND P. P. BELINSKIĬ
83. *Some problems of geometric function theory*, Trudy Mat. Inst. Steklov. **128** (1972), 34–40; English transl. in Proc. Steklov Inst. Mat. **128** (1972).

OLLI LEHTO
84. *Remarks on the integrability of the derivatives of quasiconformal mappings*, Ann. Acad. Sci. Fenn. Ser. A I No. 371 (1965).
85. *Univalent functions and Teichmüller spaces*, Springer-Verlag, 1987.

JACQUELINE LELONG-FERRAND
86. *Interprétation fonctionnelle des homéomorphismes quasi-conformes*, C. R. Acad. Sci. Paris Sér. A–B **272** (1971), A1390–A1392.
87. *Geometrical interpretations of scalar curvature and regularity of conformal homeomorphisms*, Differential Geometry and Relativity (A. Lichnerowicz Sixtieth Birthday Vol.), Reidel, 1976, pp. 91–105.

LAWRENCE G. LEWIS
88. *Quasiconformal mappings and Royden algebras in space*, Trans. Amer. Math. Soc. **158** (1971), 481–492.

CHARLES LOEWNER
89. *On the conformal capacity in space*, J. Math. and Mech. **8** (1959), 411–414.

A. MARKOUCHEVITCH [A. I. MARKUSHEVICH]
90. *Sur certaines classes de transformations continues*, C. R. (Dokl.) Acad. Sci. URSS **28** (1940), 301–304.

O. MARTIO
91. *A capacity inequality for quasiregular mappings*, Ann. Acad. Sci. Fenn. Ser. A I No. 474 (1970).
92. *Definitions for uniform domains*, Ann. Acad. Sci. Fenn. Ser. A I Math. **5** (1980), 197–205.

O. MARTIO AND J. SARVAS
93. *Injectivity theorems in plane and space*, Ann. Acad. Sci. Fenn. Ser. A I Math. **4** (1979), 383–401.

O. MARTIO, S. RICKMAN, AND J. VÄISÄLÄ
94. *Definitions for quasiregular mappings*, Ann. Acad. Sci. Fenn. Ser. A I No. 448 (1969).

95. *Distortion and singularities for quasiregular mappings*, Ann. Acad. Sci. Fenn. Ser. A I No. 465 (1970).

96. *Topological and metric properties of quasiregular mappings*, Ann. Acad. Sci. Fenn. Ser. A I No. 488 (1971).

D. MENCHOFF [D. E. MEN'SHOV]
97. *Sur une généralisation d'un théorème de M. H. Bohr*, Mat. Sb. 2 (**44**) (1937), 339–356.

NORMAN G. MEYERS AND ALAN ELCRAT
98. *Some results on regularity for solutions of non-linear elliptic systems and quasi-regular functions*, Duke Math. J. **42** (1975), 121–136.

S. G. MIKHLIN
99. *Multidimensional singular integrals and integral equations*, Fizmatgiz, Moscow, 1962; English transl., Pergamon Press, 1965.

V. M. MIKLYUKOV
100. *On removable singularities of quasiconformal mappings in space*, Dokl. Akad. Nauk SSSR **188** (1969), 525–527; English transl. in Soviet Math. Dokl. **10** (1969).

101. *On the asymptotic properties of subsolutions of quasilinear elliptic equations and mappings with bounded distortion*, Mat. Sb. **111** (**153**) (1980), 42–66; English transl. in Math. USSR Sb. **39** (1981).

CHARLES B. MORREY, JR.
102. *Multiple integral problems in the calculus of variations and related topics*, Univ. California Publ. Math. (N.S.) **1** (1943), 1–130.

JÜRGEN MOSER
103. *A new proof of De Giorgi's theorem concerning the regularity problem for elliptic differential equations*, Comm. Pure Appl. Math. **13** (1960), 457–468.

104. *On Harnack's theorem for elliptic differential equations*, Comm. Pure Appl. Math. **14** (1961), 577–591.

P. P. MOSOLOV AND V. P. MYASNIKOV
105. *A proof of Korn's inequality*, Dokl. Akad. Nauk SSSR **201** (1971), 36–39; English transl. in Soviet Math. Dokl. **12** (1971).

G. D. MOSTOW
106. *Quasi-conformal mappings in n-space and the rigidity of hyperbolic space forms*, Inst. Hautes Études Sci. Publ. Math. No. 34 (1968), 53–104.

MITSURU NAKAI

107. *Algebraic criterion on quasiconformal equivalence of Riemann surfaces*, Nagoya Math. J. **16** (1960), 157–184.

JINDŘICH NEČAS

108. *Sur les normes équivalentes dans* $W_p^{(k)}(\Omega)$ *et sur la coercitivité des formes formellement positives*, Équations aux Dérivées Partielles (1965), Sém. Math. Sup., vol. 19, Presses Univ. Montréal, Montréal, 1966, pp. 102–128.

ROLF NEVANLINNA

109. *Eindeutige analytische Funktionen*, Springer-Verlag, 1936.

S. M. NIKOL'SKIĬ

110. *Properties of certain classes of functions of several variables on differentiable manifolds*, Mat. Sb. **33 (75)** (1953), 261–326; English transl. in Amer. Math. Soc. Transl. (2) **80** (1969).

LOUIS NIRENBERG

111. *On nonlinear elliptic partial differential equations and Hölder continuity*, Comm. Pure Appl. **6** (1953), 103–156, 395.

I. S. OVCHINNIKOV AND G. D. SUVOROV

112. *Transformations of the Dirichlet integral and space mappings*, Sibirsk. Mat. Zh. **6** (1965), 1292–1314. (Russian)

E. A. POLETSKIĬ

113. *The modulus method for nonhomeomorphic quasiconformal mappings*, Mat. Sb. **83 (125)** (1970), 261–272; English transl. in Math. USSR Sb. **12** (1970).

114. *On the removal of singularities of quasiconformal mappings*, Mat. Sb. **92 (134)** (1973), 242–256; English transl. in Math. USSR Sb. **21** (1973).

G. PÓLYA AND G. SZEGÖ

115. *Isoperimetric inequalities in mathematical physics*, Princeton Univ. Press, Princeton, N. J., 1951.

T. RADO AND P. V. REICHELDERFER

116. *Continuous transformations in analysis. With an introduction to algebraic topology*, Springer-Verlag, 1955.

P. K. RASHEVSKIĬ

117. *Riemannian geometry and tensor analysis*, 3rd ed., "Nauka", Moscow, 1967; German transl. of 1st ed., VEB Deutscher Verlag Wiss., Berlin, 1959.

H. M. Reimann

118. *Ordinary differential equations and quasiconformal mappings*, Invent. Math. **33** (1976), 247–270.

Yu. G. Reshetnyak

119. *On a sufficient condition for Hölder continuity of a mapping*, Dokl. Akad. Nauk SSSR **130** (1960), 507–509; English transl. in Soviet Math. Dokl. **1** (1960).

120. *On conformal mappings of space*, Dokl. Akad. Nauk SSSR **130** (1960), 1196–1198; English transl. in Soviet Math. Dokl. **1** (1960).

121. *Stability in the Liouville theorem on conformal mappings*, Some Problems in Mathematics and Mechanics (M. A. Lavrent'ev, Sixtieth Birthday Vol.), Izdat. Sibirsk. Otdel. Akad. Nauk SSSR, Novosibirsk, 1961, pp. 219–223. (Russian)

122. *On stability in Liouville's theorem on conformal mappings of space*, Dokl. Akad. Nauk SSSR **152** (1963), 286–287; English transl. in Soviet Math. Dokl. **4** (1963).

123. *Some geometric properties of functions and mappings with generalized derivatives*, Sibirsk. Mat. Zh. **7** (1966), 886–919; English transl. in Siberian Math. J. **7** (1966).

124. *Estimates of the modulus of continuity for certain mappings*, Sibirsk. Mat. Zh. **7** (1966), 1106–1114; English transl. in Siberian Math. J. **7** (1966).

125. *Space mappings with bounded distortion*, Sibirsk. Mat. Zh. **8** (1967), 629–658; English transl. in Siberian Math. J. **8** (1967).

126. *Liouville's conformal mapping theorem under minimal regularity assumptions*, Sibirsk. Mat. Zh. **8** (1967), 835–840; English transl. in Siberian Math. J. **8** (1967).

127. *General theorems on semicontinuity and convergence with functionals*, Sibirsk. Mat. Zh. **8** (1967), 1051–1069; English transl. in Siberian Math. J. **8** (1967).

128. *On the set of singular points of solutions of certain nonlinear elliptic equations*, Sibirsk. Mat. Zh. **9** (1968), 354–367; English transl. in Siberian Math. J. **9** (1968).

129. *On a condition for boundedness of the index for mappings with bounded distortion*, Sibirsk. Mat. Zh. **9** (1968), 368–374; English transl. in Siberian Math. J. **9** (1968).

130. *Mappings with bounded distortion as extremals of integrals of Dirichlet type*, Sibirsk. Mat. Zh. **9** (1968), 652–666; English transl. in Siberian Math. J. **9** (1968).

131. *Generalized derivatives and differentiability almost everywhere*, Mat. Sb. **75** (**117**) (1968), 323–334; English transl. in Math. USSR Sb. **4** (1968).

132. *On the concept of capacity in the theory of functions with generalized derivatives*, Sibirsk. Mat. Zh. **10** (1969), 1109–1138; English transl. in Siberian Math. J. **10** (1969).

133. *On extremal properties of mappings with bounded distortion*, Sibirsk. Mat. Zh. **10** (1969), 1300–1310; English transl. in Siberian Math. J. **10** (1969).

134. *The local structure of mappings with bounded distortion*, Sibirsk. Mat. Zh. **10** (1969), 1311–1333; English transl. in Siberian Math. J. **10** (1969).

135. *Estimates for certain differential operators with finite-dimensional kernels*, Sibirsk. Mat. Zh. **11** (1970), 414–428; English transl. in Siberian Math. J. **11** (1970).

136. *On a stability estimate in Liouville's theorem on conformal mappings of multidimensional spaces*, Sibirsk. Mat. Zh. **11** (1970), 1112–1139; English transl. in Siberian Math. J. **11** (1970).

137. *On the set of branch points of mappings with bounded distortion*, Sibirsk. Mat. Zh. **11** (1970), 1333–1339; English transl. in Siberian Math. J. **11** (1970).

138. *Certain integral representations of differentiable functions*, Sibirsk. Mat. Zh. **12** (1971), 420–432; English transl. in Siberian Math. J. **12** (1971).

139. *Lectures on mathematical analysis. Mathematical analysis on manifolds*, Novosibirsk. Gos. Univ., Novosibirsk, 1972. (Russian)

140. *Stability in Liouville's theorem on conformal mappings of a space for domains with nonsmooth boundary*, Sibirsk. Mat. Zh. **17** (1976), 361–369; English transl. in Siberian Math. J. **17** (1976).

141. *Stability estimates in Liouville's theorem and L_p-integrability of the derivatives of quasiconformal mappings*, Sibirsk. Mat. Zh. **17** (1976), 868–896; English transl. in Siberian Math. J. **17** (1976).

142. *Stability estimates in the class W_p^1 for Liouville's theorem on conformal mappings for a closed domain*, Sibirsk. Mat. Zh. **17** (1976), 1382–1394; English transl. in Siberian Math. J. **17** (1976).

143. *Differentiability properties of quasiconformal mappings and conformal mappings of Riemannian spaces*, Sibirsk. Mat. Zh. **19** (1978), 1166–1183; English transl. in Siberian Math. J. **19** (1978).

144. *Stability theorems in geometry and analysis*, "Nauka", Novosibirsk, 1982. (Russian)

Yu. G. Reshetnyak and B. V. Shabat

145. *On quasiconformal mappings in space*, Proc. Fourth All-Union Math. Congr. (Leningrad, 1961), Vol. II, "Nauka", Leningrad, 1964, pp. 672–680. (Russian)

Georges de Rham

146. *Variétés différentiables. Formes, courants, formes harmoniques*, Actualités Sci. Indust., no. 1222, Hermann, Paris, 1955.

Seppo Rickman

147. *On the number of omitted values of entire quasiregular mappings*, J. Analyse Math. **37** (1980), 100–117.

Frédéric Riesz and Béla Sz. Nagy

148. *Leçons d'analyse fonctionnelle*, 2nd ed., Akad. Kiadó, Budapest, 1953; English transl., Ungar, New York, 1955.

Stanisław Saks

149. *Theory of the integral*, 2nd rev. ed., PWN, Warsaw, 1937; reprint, Dover, New York, 1964.

V. I. Semenov

150. *On one-parameter groups of quasiconformal homeomorphisms in a Euclidean space*, Sibirsk. Mat. Zh. **17** (1976), 177–193; English transl. in Siberian Math. J. **17** (1976).

151. *Semigroups of quasi-Lorentz mappings*, Sibirsk. Mat. Zh. **18** (1977), 713; English transl. in Siberian Math. J. **18** (1977).

152. *Uniform estimates of stable isometries*, Sibirsk. Mat. Zh. **27** (1986), no. 3, 193–199; English transl. in Siberian Math. J. **27** (1986).

153. *On uniform stability estimates for quasiconformal and quasi-isometric mappings in space*, Dokl. Akad. Nauk SSSR **286** (1986), 295–297; English transl. in Soviet Math. Dokl. **33** (1986).

154. *Stability estimates for quasiconformal space mappings of a star-shaped domain*, Sibirsk. Mat. Zh. **28** (1987), no. 6, 102–118; English transl. in Siberian Math. J. **28** (1987).

James Serrin

155. *A Harnack inequality for nonlinear equations*, Bull. Amer. Math. Soc. **69** (1963), 481–486.

156. *Local behavior of solutions of quasi-linear equations*, Acta. Math. **111** (1964), 247–302.

B. V. Shabat

157. *The modulus method in space*, Dokl. Akad. Nauk SSSR **130** (1960), 1210–1213; English transl. in Soviet Math. Dokl. **1** (1960).

158. *On the theory of quasiconformal mappings in space*, Dokl. Akad. Nauk SSSR **132** (1960), 1045–1048; English transl. in Soviet Math. Dokl. **1** (1960).

S. Z. SHEFEL'

159. *Conformal correspondence of metrics, and the smoothness of isometric immersions*, Sibirsk. Mat. Zh. **20** (1979), 397–401; English transl. in Siberian Math. J. **20** (1979).

160. *Smoothness of a conformal mapping of Riemannian spaces*, Sibirsk. Mat. Zh. **23** (1982), no. 1, 153–159; English transl. in Siberian Math. J. **23** (1982).

V. I. SMIRNOV

161. *A course in higher mathematics*, Vol. 5, rev. ed., Fizmatgiz, Moscow, 1959; English transl., Pergamon Press, Oxford, and Addison-Wesley, Reading, Mass., 1964.

S. L. SOBOLEV

162. *Applications of functional analysis in mathematical physics*, Izdat. Leningrad. Gos. Univ., Leningrad, 1950; English transl., Amer. Math. Soc., Providence, R. I., 1963.

W. STEPANOFF [V. V. STEPANOV]

163. *Sur les conditions de l'existence de la différentielle totale*, Mat. Sb. **32** (1924/25), 511–527.

SHLOMO STERNBERG

164. *Lectures on differential geometry*, Prentice-Hall, Englewood Cliffs, N. J., 1964.

G. D. SUVOROV

165. *Extensions of topological structures and metric properties of mappings*, Contributions to Extension Theory of Topological Structures (Proc. Sympos., Berlin, 1967; J. Flachsmeyer et al., editors), VEB Deutscher Verlag Wiss., Berlin, 1969, pp. 257–273; English transl. in Amer. Math. Soc. Transl. (2) **134** (1987).

A. V. SYCHEV

166. *Moduli and n-dimensional quasiconformal mappings*, "Nauka", Novosibirsk, 1983. (Russian)

HANS TRIEBEL

167. *Interpolation theory, function spaces, differential operators*, VEB Deutscher Verlag Wiss., Berlin, 1977, and North-Holland, Amsterdam, 1978.

D. A. TROTSENKO

168. *Properties of domains with nonsmooth boundary*, Sibirsk. Mat. Zh. **22** (1981), no. 4, 221–224. (Russian)

169. *Extension from a domain and approximation of spatial quasiconformal mappings with small distortion coefficient*, Dokl. Akad. Nauk SSSR **270** (1983), 1331–1333; English transl. in Soviet Math. Dokl. **27** (1983).

170. *Extension of spatial quasiconformal mappings that are close to conformal mappings*, Sibirsk. Mat. Zh. **28** (1987), no. 6, 126–133; English transl. in Siberian Math. J. **28** (1987).

JUSSI VÄISÄLÄ

171. *On quasiconformal mappings in space*, Ann. Acad. Sci. Fenn. Ser. A I No. 298 (1961).

172. *Discrete open mappings on manifolds*, Ann. Acad. Sci. Fenn. Ser. A I No. 392 (1966).

173. *Removable sets for quasiconformal mappings*, J. Math. and Mech. **19** (1969/70), 49–51.

174. *Modulus and capacity inequalities for quasiregular mappings*, Ann. Acad. Sci. Fenn. Ser. A I No. 509 (1972).

175. *A survey of quasiregular maps in R^n*, Proc. Internat. Congr. Math. (Helsinki 1978), vol. 2, Acad. Sci. Fenn., Helsinki, 1980, pp. 685–691.

M. YU. VASIL'CHIK

176. *On a lower estimate of the distortion coefficient for infinitely close domains*, Sibirsk. Mat. Zh. **19** (1978), 547–554; English transl. in Siberian Math. J. **19** (1978).

S. K. VODOP'YANOV AND V. M. GOL'DSHTEĬN

177. *Lattice isomorphisms of the spaces W_n^1 and quasiconformal mappings*, Sibirsk. Mat. Zh. **16** (1975), 224–246; English transl. in Siberian Math. J. **16** (1975).

178. *Quasiconformal mappings and spaces of functions with first generalized derivatives*, Sibirsk. Mat. Zh. **17** (1976), 515–531; English transl. in Siberian Math. J. **17** (1976).

179. *Sobolev spaces and special classes of mappings*, Novosibirsk. Gos. Univ., Novosibirsk, 1981. (Russian)

S. K. VODOP'YANOV, V. M. GOL'DSHTEĬN, AND T. G. LATFULLIN

180. *A criterion for extendibility of L_2^1-functions from unbounded plane domains*, Sibirsk. Mat. Zh. **20** (1979) 416–420; English transl. in Siberian Math. J. **20** (1979).

WILLIAM P. ZIEMER

181. *Extremal length and p-capacity*, Michigan Math. J. **16** (1969), 43–51.

V. A. ZORICH

182. *A theorem of M. A. Lavrent'ev on quasiconformal space maps*, Mat. Sb. **74 (116)** (1967), 417–433; English transl. in Math. USSR Sb. **3** (1967).

A. ZYGMUND

183. *Trigonometric series*, 2nd rev. ed., Vol. I, Cambridge Univ. Press, 1959.

COPYING AND REPRINTING. Individual readers of this publication, and nonprofit libraries acting for them, are permitted to make fair use of the material, such as to copy an article for use in teaching or research. Permission is granted to quote brief passages from this publication in reviews, provided the customary acknowledgment of the source is given.

Republication, systematic copying, or multiple reproduction of any material in this publication (including abstracts) is permitted only under license from the American Mathematical Society. Requests for such permission should be addressed to the Executive Director, American Mathematical Society, P.O. Box 6248, Providence, Rhode Island 02940.

The owner consents to copying beyond that permitted by Sections 107 or 108 of the U.S. Copyright Law, provided that a fee of $1.00 plus $.25 per page for each copy be paid directly to the Copyright Clearance Center, Inc., 21 Congress Street, Salem, Massachusetts 01970. When paying this fee please use the code 0065-9282/89 to refer to this publication. This consent does not extend to other kinds of copying, such as copying for general distribution, for advertising or promotion purposes, for creating new collective works, or for resale.

ABCDEFGHIJ – 89